About IFPRI

The International Food Policy Research Institute (IFPRI), a research center of CGIAR, provides research-based policy solutions to sustainably reduce poverty and end hunger and malnutrition in low- and middle-income countries. IFPRI was established in 1975 to identify and analyze alternative national and international strategies and policies for meeting the food needs of the developing world, with particular emphasis on low-income countries and on the poorer groups in those countries. Partnerships, communications, capacity strengthening, and data and knowledge management are essential components for translating IFPRI's research to action and impact. The Institute's regional and country programs play a critical role in responding to demand for food policy research and in delivering holistic support to country-led development. IFPRI collaborates with partners around the world.

www.ifpri.org

About CGIAR

CGIAR is a global research partnership for a food-secure future, dedicated to transforming food, land, and water systems in a climate crisis. CGIAR science aims to reduce poverty, enhance food and nutrition security, and improve natural resources and ecosystem services. As the world's largest agricultural innovation network, its research is carried out by 15 CGIAR Centers working around the world in close collaboration with hundreds of partners, including national and regional research institutes, civil society organizations, academia, development organizations, and the private sector.

www.cgiar.org

Food Policy

Lessons and Priorities for a Changing World

Edited by Johan Swinnen and Christopher B. Barrett

A Peer-Reviewed Publication

International Food Policy Research Institute
Washington, DC

ISBN: 978-0-89629-473-8
ISSN: 2329-2873
Handle: https://hdl.handle.net/10568/174108
Library of Congress Control Number: 2025940425

Recommended citation: Swinnen, Johan, and Barrett, Christopher B., editors. *Food Policy: Lessons and Priorities for a Changing World. 2025 Global Food Policy Report.* Washington, DC: International Food Policy Research Institute.

International Food Policy Research Institute
1201 Eye Street, NW, 12th floor
Washington, DC 20005 USA
Telephone: +1-202-862-5600
www.ifpri.org

Cover photo and art credits: G.M.B. Akash/Panos Pictures; SALMONNEGRO-STOCK/Shutterstock; AU USAnakul/Shutterstock; Chris de Bode/CGIAR; AJP/Shutterstock; Tavarius/Shutterstock; Freepik.
Cover Design and Book Layout: Jason Chow
Project Manager: Pamela Stedman-Edwards

Contents

Tables and Figures

Tables

Figures

Foreword

As IFPRI marks its 50th anniversary, we congratulate the Institute on its tremendous impact on food policy knowledge, capacity, and action. Since the Institute's beginnings, its work has emphasized the importance of an enabling policy environment for agricultural growth, better livelihoods, and improved food security, nutrition, and well-being for all. The central role of policy in shaping our agrifood systems is now widely recognized, with IFPRI's role serving as a continued testament to the importance of policy research.

IFPRI has long been a leader in supporting agrifood systems-led development, identifying investment priorities and projecting policy impacts, informing international processes, evaluating social protection programs, testing nutrition policies, highlighting community-based resource governance, bringing gender differences and women's empowerment into mainstream policies and evaluations, identifying innovations that support small farmers, and refocusing the fight against hunger from providing enough calories to supporting equitable, sustainable, healthy food systems. This year's Global Food Policy Report chronicles this impressive history of identifying, evaluating, and promoting understanding of food policies that improve lives in low- and middle-income countries (LMICs).

Since its creation, IFPRI has built strong relations with a wide range of partners and stakeholders, including local and national governments, regional bodies and international organizations, CGIAR Centers, universities, and donors, both meeting demand for policy research and contributing to a substantial increase in research capacity in countries working to end poverty and malnutrition. To better meet national and subnational needs, IFPRI continues to pursue decentralization, with 14 country and regional offices and half of its staff now based in LMICs.

IFPRI's dedication to research quality is reflected in its wide range of highly rated and cited research, which constitutes a major public good for policy- and decision-makers and researchers. This work takes a cross-sectoral approach to global and context-specific research and impact that aims to clarify the situation and outlook, test and scale solutions, and shape enabling environments

Over this half-century, the world has undergone major economic, demographic, environmental, and social changes that have shaped our food systems. As an independent organization and CGIAR Center, IFPRI's work has often led the way in addressing changing contexts, emerging trends, and new crises, and its interdisciplinary work has contributed to growing recognition of the complexity and interconnected nature of food, health, social, and environmental systems.

In this time of rapid and sometimes dramatic global change—from the rapid advent of artificial intelligence to climate change to increasing conflict and a weakened international system—this report also looks to the future, considering which trends and policy needs will be most pressing and how food policy research can best contribute to a better world.

IFPRI is a small organization with an outsized impact, thanks to the talent, dedication, and drive of its researchers and staff to meet the Institute's vision of a world free of hunger and malnutrition. We expect that it will continue to serve as a leader in food policy research in the years to come. On this milestone anniversary, we invite you to read this informative report, both to understand how food policy has advanced over the last half-century and to consider how research can contribute to a better future for people everywhere.

Pascal Lamy
Chair, IFPRI Board
of Trustees

Lindiwe Sibanda
Chair, CGIAR Integrated
Partnership Board

Acknowledgments

The *2025 Global Food Policy Report* was prepared under the overall leadership of Johan Swinnen, Christopher Barrett, and a core team comprising Charlotte Hebebrand, Pamela Stedman-Edwards, Claire Davis, Sivan Yosef, Jamed Falik, and Jason Chow.

Contributions were made by Kibrom A. Abay, Kamiljon Akramov, Kate Ambler, Joaquín Arias Segura, Suresh Babu, Ousmane Badiane, Samuel Benin, Jeffrey R. Bloem, Antoine Bouët, Elizabeth Bryan, Francisco Ceballos, Kevin Z. Chen, Uchendu Eugene Chigbu, Kristin Davis, S. Mahendra Dev, Xinshen Diao, Eugenio Díaz-Bonilla, Maria DiGiovanni, Shenggen Fan, Rikin Gandi, Madhur Gautam, Stuart Gillespie, Joseph Glauber, Ganna Hassan, Peter Hazell, Jessica Heckert, Ruth Hill, P.K. Joshi, Nina Jovanovic, Jawoo Koo, Katrina Kosec, Ke Kou, Berber Kramer, Sikandra Kurdi, Jordan Kyle, David Laborde, Isabel Lambrecht, Jef L. Leroy, John Lynam, Margaret McMillan, Brian McNamara, Hazel Malapit, Will Martin, Ruth Meinzen-Dick, Purnima Menon, Alesha Miller, Jonathan Mockshell, Iliana Monterroso, Ephraim Nkonya, Deanna Olney, David Orden, Steven Were Omamo, Karl Pauw, Valeria Piñeiro, Frank Place, Agnes Quisumbing, Catherine Ragasa, Sharanya Ragiv, Sarobidy Rakotonarivo, Shahidur Rashid, Jona Repishti, Danielle Resnick, Claudia Ringler, Sherman Robinson, Mark Rosegrant, Marie T. Ruel, Khalid Siddig, Anusara Singhkumarwong, David J. Spielman, Linda Steinhuebel-Rasheed, Diana Suhardiman, Rashid Sulaiman V., Timothy Thomas, James Thurlow, Fousseini Traoré, John Ulimwengu, Bjorn Van Campenhout, Rob Vos, Keith Wiebe, Kris Wyckhuys, Sivan Yosef, Tamsin Zandstra, Wei Zhang, and Yunyi Zhou.

Production of the report was led by Pamela Stedman-Edwards. Jason Chow was responsible for design and layout. Editorial assistance was provided by Claire Davis, Gillian Hollerich, and Cara Aldridge Young.

We would like to thank IFPRI's donors, including those who supported IFPRI's research through their contributions to the CGIAR Fund: https://cgiar.org/funders/

SECTION I

Pathways to Progress

Chapter 1

FOOD POLICY RESEARCH IN LOW- AND MIDDLE-INCOME ECONOMIES

Past, Present, and Future

Christopher B. Barrett, Maria DiGiovanni, and Johan Swinnen[1]

Christopher B. Barrett is a professor of Applied Economics and Management, Cornell University, and chair of IFPRI's Strategy and Program Council; **Maria DiGiovanni** is a research analyst, Director General's Office, IFPRI; and **Johan Swinnen** is director general, IFPRI.

As an integral part of daily life around the world, food systems affect people and societies in many ways. Everyone participates in food systems through making choices about what to purchase, cook, eat, sell, store, or grow, as well as through political and social actions and the management of natural resources. Food systems are critical to economies as a major source of both private and public revenue and to societies as a basis for political power. Almost half of the world's population earns a livelihood from farming, fishing, or processing and distributing food (Davis et al. 2023). Throughout history, governments have developed and enacted policies to regulate, shape, and support the production, consumption, and trade of food, with crises often precipitating significant reforms. In both the distant past and today, these policies aim to address a range of objectives, such as raising tax revenue, protecting consumers, supporting farmers' incomes, protecting powerful landowners and agribusinesses, and addressing climate, environmental, or health externalities, among others.

IFPRI was founded in 1975 at a time when food systems had been disrupted by the global food crisis of 1972–1975. The Institute's mission to end

1 The authors thank Kamiljon Akramov, Claire Davis, Kristin Davis, Xinshen Diao, Eugenio Díaz-Bonilla, Dan Gilligan, Stuart Gillespie, Charlotte Hebebrand, Ruth Meinzen-Dick, Purnima Menon, Valeria Piñeiro, Frank Place, Agnes Quisumbing, Tom Reardon, Claudia Ringler, Pamela Stedman-Edwards, John Ulimwengu, and Rob Vos for comments and suggestions on earlier drafts of this chapter.

hunger and poverty[2] remains the same today, but in a much different scientific, political, and global context. In 1975, 35 percent of people in low- and middle-income countries (LMICs) were experiencing hunger (Ritchie et al. 2023). Geopolitics were driven by the Cold War and the energy crisis, China was poor and still largely closed to the world, many African countries were newly independent, and the digital revolution had barely begun. The Club of Rome's newly published report, *Mankind at the Turning Point*, called for an interdependent world system to confront modern energy and food challenges (Mesarovic and Pestel 1974). Although much has changed since then, the world faces yet another global food crisis. Once again, rapid growth in food demand—due to population and income growth, as well as urbanization—is outpacing expansion of the food supply, generating price shocks that imperil the health and nutrition of billions, while production processes generate unsustainable environmental impacts and increase the risk of novel infectious diseases. Food policy research helped overcome the challenges of half a century ago. It can likewise help empower consumers, producers, and policymakers to make better decisions that will facilitate healthy, equitable, resilient, and sustainable food systems transformation in the coming decades.

Looking back over the past half-century can help identify key lessons that will better prepare us to move forward to 2050. History shows that agricultural development evolves continuously, driven by producers, consumers, workers, policymakers, and entrepreneurs working to improve their own and their communities' living conditions. Achieving sustainable and equitable solutions to hunger, malnutrition, and poverty demands a shift in the way we study, implement, and evaluate innovations in technologies, programming, governance, investments, and markets: the food systems transformation we need will require a continued evolution in our approach to food policy research.

This Global Food Policy Report reviews and reflects on food policy research over the past 50 years and presents recommendations for a forward-looking research agenda. While many of today's challenges existed in 1975, they are contextually different and, in some cases, carry greater importance today—and moving forward—than in the past. In this chapter, we first review the evolution of food policy as well as the events and theories that drove that

2 It is important to note that our understanding and approach to hunger and malnutrition has been greatly refined since 1975, shifting from a singular focus on increasing caloric intake to a focus on promoting healthy diets to address all forms of malnutrition, including micronutrient deficiencies and overweight and obesity, based substantially on evidence generated by IFPRI, as described later in this chapter and in Chapter 12.

evolution, many of which are explored in more detail in the topical chapters. We then look specifically at IFPRI's contribution and evolution across these years. The final section of the chapter briefly outlines key directions for future research and IFPRI's strategy for tackling this work, before concluding with a summary of the report's topical and regional chapters.

Food policy: 1975–2025

Starting in the mid-1960s, the widespread release of high-yielding varieties of rice, wheat, and maize in Asia and Latin America marked the advent of the Green Revolution. Significant research and development (R&D) investments by philanthropies and bilateral and multilateral donor agencies spurred the diffusion of science and technology, inducing unprecedented growth in agricultural productivity. Through the technological innovations of the Green Revolution, policymakers and scientists aimed to meet basic caloric needs by boosting yields of staple crops. Indeed, growth in the world food supply outpaced demand growth, leading to lower real food prices. However, the 1972–1975 world food crisis showed that the prevailing supply-side focus was insufficient, though necessary, to guarantee food security.

At the time, the primary role of food policy research was to complement and facilitate the beneficial diffusion of new agricultural technologies. It was widely understood that agricultural development was closely linked to broader economic growth, especially in the heavily agrarian economies of LMICs, where most people lived in rural areas and worked on farms (Johnston and Mellor 1961). Agricultural productivity growth was considered the key catalyst for development.[3]

However, the success of the Green Revolution relied on far more than improved crop varieties, requiring complementary advances in rural infrastructure, inputs and irrigation, services, natural resource management, and sectoral and macroeconomic policies (Barrett 2021). By the early 1970s, attention was turning to the social and environmental spillovers resulting from rapid agricultural transformation. No purely technical fix could sustainably resolve the food crisis that emerged in the early 1970s—the moment called for appropriate policy designs and reforms.

---

3 For more on this theory of development, see Christiaensen et al. (2011), Christiaensen and Martin (2018), de Janvry and Sadoulet (2010), Dorosh and Thurlow (2018), Johnston and Mellor (1964), and Timmer (1988).

The food crisis also accelerated understanding of the causes of hunger. The 1974 World Food Conference, convened by the Food and Agriculture Organization of the United Nations (FAO), was the first high-level global event to recognize food security as a pressing policy issue. The work of Amartya Sen (1981) showed that adequate food intake depends not only on food availability in the broader economy but also on individuals' access to food. Poverty was—and remains—the primary driver of food insecurity, necessitating policies that targeted it for the joint goals of hunger reduction and economic development (World Bank 1986). Ultimately, it was recognized that food systems outcomes concern people's agency in their daily lives—eating, working, raising families, and exercising voice in their communities (Sen 2000).

This expanded understanding of food security and the necessary policy environment was transformational. Raising incomes, especially among the poor, became the lead objective in global development strategies, and technological change was appreciated as a means to achieve that goal, not a goal unto itself. Moreover, policy responses—including those meant to foster agricultural technology development and diffusion, as well as extending far beyond—were recognized as a primary vehicle to advance that objective.

A major rethinking was also underway regarding the state's role in economic development and poverty reduction. In the 1960s and 1970s, state control of agricultural markets was widespread, especially in LMICs. This was most extreme in the communist countries in Asia and Eastern Europe, but many African, Caribbean, and Latin American states also controlled marketing and trade in agricultural products and food. Farmers were generally taxed, either explicitly or implicitly through export controls and exchange rate manipulation, to keep food prices low for urban consumers or to raise government revenue (Anderson et al. 2013). Governments in higher-income countries increased farm subsidies, which distorted domestic and global markets (Johnson 1991). These interventionist economic policies were called into question by the global macroeconomic crises of the 1970s and the resulting LMIC debt crisis of the early 1980s.

In response to these crises and the waning of the Cold War, the 1980s and 1990s saw a wave of market-oriented liberalization. In almost every country, the state's role in agricultural input and output markets shrank dramatically. Governments were expected to end market-distorting policies to unleash the economic growth potential of the private sector. Large-scale, market-oriented economic reforms in China starting in the late 1970s and in Viet Nam in the mid-1980s vividly demonstrated the potential for policy reform to support agricultural transformation, economic growth, poverty reduction, and rapid improvements in food security (Liu et al. 2020; Rozelle and Swinnen 2004).

In the 1990s, pressure from global agricultural exporters also led to reforms in high-income countries following years of multilateral trade negotiations that resulted in the Uruguay Round Agreement on Agriculture and the World Trade Organization. These trade agreements further intensified globalization that had been stimulated by advances in transportation and information and communication technologies, as well as the liberalization of domestic markets.

On the supply side, globalization was reflected in growing trade in agricultural commodities and in cross-border investments in the food and retail sector by private companies (Barrett, Reardon et al. 2022; Reardon et al. 2003). On the consumer side, by the turn of the millennium, governments shifted away from large-scale food subsidies toward social protection programs, often built around cash transfer programs, recognizing the need to reduce food insecurity by directly addressing poverty. International humanitarian food assistance transitioned sharply from in-kind shipments originating in high-income agricultural exporting countries to cash-based models that leveraged commercial markets.

Liberalization of the agriculture sector necessitated a broader understanding of agricultural transformation. The focus of technology innovation, policies, and investments had to expand from on-farm agricultural production to the whole agrifood system, including the rapidly growing nonfarm economy (Haggblade et al. 2007, 2010; Hazell et al. 2024). The off-farm economy served as a source of employment and income for rural households, and the post-farmgate segments of agrifood value chains increasingly connected consumers and producers. Policy attention to nonfarm components was expected to strengthen on- and off-farm linkages to accelerate diffusion of improved agricultural practices and technologies, as well as to add value to postharvest farm outputs. As agrifood value chains expanded, innovations in processing, transport, food environments, and other services were required to improve value chain productivity, satisfy changing consumer preferences, and improve household welfare.

Despite earlier concerns about the impact of rapid population growth, the world saw a remarkable reduction in hunger and poverty between the 1970s and 2000s. Falling food prices—partly thanks to gains in agricultural productivity—led to a global decline in the prevalence of undernourishment, even as the world population rose to 6 billion people by the mid-2000s. Global poverty rates declined rapidly, with the share of those in extreme poverty[4] falling from 44 percent to 21.5 percent between 1981 and 2005 (World Bank n.d.-d). The most significant progress occurred in Asia, where concerns that population growth would outstrip the food supply had been most pronounced (Ehrlich 1968). In China, the number of hungry people fell from around 300 million in 1980 to 120 million in 2005, even as the population grew from 981 million to 1.3 billion (Spielman and Pandya-Lorch 2009; World Bank n.d.-a). In India, the number of people facing chronic hunger declined from 210 million in 1990 to 194 million in 2015, while the population climbed from 865 million to 1.3 billion (FAO et al. 2015; World Bank n.d.-c). No country's path to agricultural development was the same, and progress stemmed from many different factors, including reforms to property rights, improved productivity in staple crops and livestock, better management of natural resources, expansion of markets, diversification of agricultural production, reformed trade and fiscal policies, and improved nutrition.

Unfortunately, the astounding achievements of these decades led to a sense of complacency in policymaking. Agriculture sectors had contracted sharply as a share of macroeconomic output, populations were urbanizing, and workforces were shifting off-farm.[5] With the problems of agricultural development and hunger seemingly well on the way to resolution, policymakers turned their attention and resources elsewhere. Starting in the 1990s, governments in high-income countries reduced public investments in agricultural R&D (Fuglie 2016), and foreign aid for agricultural R&D became scarcer and less flexible, including CGIAR funding (Barrett et al. 2009). Some emerging economies, such as China and Brazil, increased their own investments in agricultural R&D after 2000, but for most African countries, public investments remained well below their agreed-upon official target of 1 percent of agricultural value added.

Despite considerable progress, however, poverty and food insecurity were far from eradicated, and concerns about the sustainability of the food system were growing rapidly. Food security had improved in many LMICs, but in the mid-2000s, hunger persisted among 45 million people in Latin America, and the population facing hunger in sub-Saharan Africa had almost doubled—from 125 to 212 million people—since 1979. The impact of

4 Defined as living on less than $2.15 a day per person in 2017 PPP terms.

5 The global labor force in agriculture peaked in 2003 at just over 1 billion workers and has been declining steadily ever since (Fuglie et al. 2024), although the workforce in downstream agrifood industries grows at the same pace as economies overall, on average (Yi et al. 2025; Corong et al. 2024).

market-based reforms on agricultural development in Africa was mixed and, in some cases, produced worse outcomes for the poor (Kherallah et al. 2002; Swinnen et al. 2010).

Amid these persistent challenges, market volatility has also become a key characteristic of the world food system over the past 20 years. A surge in international grain prices caused two additional global food crises, the first in 2007/08 and then again in 2011/12. The disruption of livelihoods and supply chains by the COVID-19 pandemic reversed global progress in reducing poverty and malnutrition, and many low- and lower-middle-income countries still face a double burden of high food price inflation and large external debts resulting from the pandemic. The Russia–Ukraine War, begun in 2022, induced another spike in food, fuel, and fertilizer prices that further complicated pandemic recovery and impeded progress on food security goals. While global agrifood trade has tripled since 1975 and can compensate for domestic supply disruptions, increased trade can also expand the impact of national and regional shocks, especially in today's more volatile trade environment. Agrifood value chains have expanded rapidly in LMICs with scant oversight from the public sector, and concerns about concentrated corporate power—from agricultural input markets to consumer retail—are becoming more prominent in policy debates (Clapp 2025; Crespi and MacDonald 2022).

Overall, progress in reducing food insecurity and poverty has stagnated since the 2010s. The striking, steady progress of the 1960s to1990s has slowed and even reversed course. Income inequality is on the rise, including in Asia, where the benefits of the Green Revolution and liberalization were strongest (Fosu 2017). Globally, average agricultural productivity growth fell by nearly half between the 2000s and 2010s (Fuglie et al. 2024). The fragility of our food systems is also increasingly apparent. The past decade has been marked by a rise in both extreme weather events and environmental degradation and biodiversity loss related to foods systems and other human activities. In addition, the global increase in conflicts and displaced populations, including a threefold increase in the number of people living in conflict-affected areas since 1990 (World Bank n.d.-b), has greatly increased the prevalence of poverty and acute malnutrition in these places. The agricultural technology and market-oriented policy research of the prior decades now seems insufficient to manage the myriad shocks and stressors suddenly confronting the world's poor, record numbers of whom have migrated in search of better lives.

The nature of malnutrition challenges has also evolved over the past half-century. At the time of IFPRI's creation, concerns revolved around undernourishment from inadequate supplies of and access to dietary energy and protein and around famine and hunger-related morbidity and mortality. While the share of undernourished people in the global population has declined rapidly, both undernourishment and poverty have become increasingly concentrated in regions facing complex emergencies and conflict (FAO et al. 2024). At the same time, nutrition and public health research has called attention to the drastic rise in obesity rates, especially in middle-income countries (Shekar and Popkin 2020). Research has also brought growing awareness to the formidable challenges of micronutrient deficiencies (vitamins and minerals), especially among women and children, and highlighted the presence of multiple coinciding nutrition burdens—undernourishment, micronutrient deficiencies, and/or overweight/obesity—within individuals, households, and communities. In response, in addressing food policy questions, researchers and policymakers have increasingly shifted their attention to the relationship between farm and post-farmgate production and marketing systems, including to commercial drivers of dietary choices, and chronic disease and to gender and youth issues.

IFPRI's role, achievements, and challenges: 1975–2025

In the years leading up to IFPRI's creation, international food policy research was a relatively novel concept. The primary justification for an institution devoted to this research centered on the pressing need to expand the adoption of agricultural technologies, which policy was expected to facilitate. Experts including John Mellor, who was Chief Economist at the U.S. Agency for International Development when IFPRI was launched and, a few years later, became IFPRI's Director General, promoted agricultural development as an engine of economic growth and structural transformation, especially among smallholders in LMICs.[6] High-level officials in national governments and international organizations also sought better understanding of the world food situation and the international management of agricultural inputs and food in the wake of the 1972–1975 food crisis. With key support from the Ford Foundation, the Rockefeller Foundation, and the International Development Research Centre of Canada (IDRC), IFPRI was established in 1975 to address these needs (Farrar 2000). In 1979, the Institute joined CGIAR to facilitate policy research support to the existing Research Centers (see Box 1.3).

6 For more on this theory of development, see Barrett et al. (2010), Christiaensen et al. (2011), Christiaensen and Martin (2018), de Janvry and Sadoulet (2010), Dorosh and Thurlow (2018), Johnston and Mellor (1961), Mellor and Johnston (1984), and Timmer (1988, 1992).

In its early years, IFPRI's research focused on four global issues: world food trends, agricultural production policies, food subsidies, and agricultural trade policies.[7] It emphasized the creation of international public goods, using case studies to generate knowledge that could be applied to many, if not all, low-income countries. In a role that continues today, IFPRI made major contributions to developing theory, methods, strategies, tools, and especially empirical evidence to support sound policymaking for food systems, although the concept of "systems" was relatively uncommon 50 years ago. A key contribution during this period involved addressing questions related to world food production and consumption data and projections, collecting household survey data, and establishing more reliable data and forecasting methods (Pinstrup-Andersen 2000). In the 1980s and 1990s, IFPRI researchers studied the nuances of food access through research on intrahousehold resource allocation, with attention to gender as a cross-cutting issue (Alderman et al. 1995; Haddad et al. 1997). The Institute's research in the 1990s examined the sector's role in providing nutritious and diversified diets, not solely sufficient calories (Kennedy et al. 1992; von Braun and Kennedy 1994). This work helped influence the evaluation and design of conditional cash transfer programs and gender-sensitive targeting around the world. IFPRI research also focused increasingly on the environmental footprint of agriculture, including land and water resources, and later the relationship between agriculture and climate change. By 1993, IFPRI took on a more visible and direct role in shaping global food policy agendas. The *2020 Vision Initiative*, for which Director General Per Pinstrup-Andersen was awarded the 2001 World Food Prize, called for multistakeholder action on the intersecting issues of agriculture, nutrition, poverty, and the environment (Paarlberg 1999).

This evolution was reflected in and enhanced by the increasingly multidisciplinary range of IFPRI's research expertise and efforts. In the early years, most staff were economists, but the subsequent recruitment of experts from other disciplines contributed to a stronger focus on inter- or transdisciplinary research appropriate to policy issues that rarely adhere to disciplinary boundaries. Today, nutrition and health specialists make up 12 percent of IFPRI's research staff, and social and political scientists make up 5 percent. While economists remain the largest group, at roughly 70 percent (Hazell and Place 2025), the nature of IFPRI's economic analyses has evolved over the past 50 years with the growing influence of institutional economics, behavioral economics, and the use of randomized controlled trials in impact evaluation.

7 See Hazell and Place (2025) for a detailed analysis of how research programs at IFPRI have evolved over the past 50 years.

IFPRI's research aims to provide evidence and data that inform policy efforts to reduce malnutrition and poverty. While measuring the impact of this work is complex, the Institute's work has been shown to improve millions of lives (see Chapter 2). We highlight here IFPRI's work on liberalization, market and trade analysis, collective action, nutrition, and gender, and its support for capacity building.

IFPRI played a key role in informing the transition to agricultural trade liberalization in several countries. For instance, its analysis of Viet Nam's rice policies between 1996 and 1998 contributed to the government's decision to relax rice export quotas and internal restrictions on rice trade. The market-oriented policy changes generated substantial economic benefits for Viet Nam, including greater opportunities, incomes, and food security for 24 million people dependent on rice farming (Ryan 2000). When a severe flood in Bangladesh in 1998 caused a drastic reduction in rice production, IFPRI researchers helped the government decide to allow the private sector to import rice, thus avoiding a price spike (del Ninno et al. 2001). More stable prices for this staple crop likely benefited the 65 million people living below the national poverty line at the time.

Market and trade analysis continues to be a hallmark of IFPRI research, providing country-level, regional, and global insights. For instance, IFPRI and the African Growth and Development Policy (AGRODEP) Modeling Consortium developed the general equilibrium model MIRAGRODEP, which allows researchers to simulate the effect of trade policy on poverty, hunger, diets, and global greenhouse gas emissions. IFPRI also coordinates the AgIncentives Consortium and Database, a collaborative effort among predominant international organizations to monitor agricultural incentives. IFPRI's investments in cutting-edge modeling assets also include the International Model for Policy Analysis of Agricultural Commodities and Trade (IMPACT), launched in the 1990s. More recently, IFPRI has complemented its global trade and market models with domestic economy models (with disaggregated food systems and value chain segments) for a large group of LMICs, particularly the Rural Investment and Policy Analysis (RIAPA) model designed in 2016, which can be used to assess the interaction of global shocks, domestic policy reforms, and structural changes.

Although market-based reforms have changed the role of the state in agriculture, public investment remains fundamental to the transformation of agriculture and broader rural development. Starting in the mid-1990s, IFPRI began influential work using evidence to inform the design of cost-effective public policies that focused on smallholders, poor populations, and other marginalized groups. Significantly, IFPRI conducted research for India and China in the early 2000s that informed public investments in rural development, enabling millions of people in these countries to exit poverty within the decade (Fan and Gulati 2008; Renkow 2010). During the same period, IFPRI started to evaluate social protection programs, conducting landmark evaluations such as those in Bangladesh, Ethiopia, and Mexico that inspired a culture of accountability in social policy (Nelson et al. 2015). Evaluations of the benefits and costs of different food assistance modalities (cash versus food versus vouchers) affected the strategies of countries and international organizations, such as the World Food Programme. IFPRI also led important work on biosafety and intellectual property policies related to biotechnology in food and agriculture (El-Chichakli et al. 2016; Pardey 2001; Falck-Zepeda and Zambrano 2011; Pinstrup-Andersen 1999; Spielman 2007).

IFPRI spearheaded research on community-led collective action related to natural resource management. The Institute provided evidence to inform policymakers on issues of property rights, especially around management of common property resources such as land and water. These efforts also helped civil society organizations and marginalized subpopulations resist elite capture of valuable resources (Markelova et al. 2009; Meinzen-Dick 2007; Meinzen-Dick et al. 1997).

Research on nutrition policy has long been central to IFPRI's work on food security. In the 1980s, IFPRI showed that increased caloric intake as a result of poverty reduction did not automatically lead to improved nutrition. Work considering the role of gender found that putting additional income in the hands of women was associated with higher per capita calorie and protein intake, as well as faster increases in children's growth. IFPRI also provided evidence on the importance of early nutrition for development and well-being later in life. For example, in the 2000s, IFPRI's nutrition research in Latin America and the Caribbean and in other countries provided high-quality evidence that improving nutrition in children's first 1,000 days is critical for their development, findings that have contributed to widespread adoption of preventative nutrition approaches (Grajeda et al. 2005; Hoddinott et al. 2008). At the turn of the century, IFPRI researchers were among the first to document and draw wide attention to the "double burden" of malnutrition, referring to undernutrition and obesity. In 2003, IFPRI together with other CGIAR Centers established HarvestPlus, a multicountry program focused on facilitating policy for the development and scaling of improved and biofortified crop varieties. Since the 2010s, IFPRI's research linking nutrition with agriculture, social protection, and extension programming in Bangladesh and

BOX 1.1 IFPRI and the World Food Prize

IFPRI's research and thought leadership has been recognized three times by the World Food Prize. In 2001, Dr. Per Pinstrup-Andersen (then Director General of IFPRI) received the award in recognition of his role as "the catalyst behind the groundbreaking 2020 Vision Initiative, and for his contribution to agricultural research, food policy, and uplifting the status of the poor and starving citizens of the world." In 2016, four scientists, including Dr. Howarth Bouis from IFPRI, received the award for their "development and implementation of biofortification, breeding critical vitamins and nutrients into staple crops, thereby dramatically reducing 'hidden hunger.'" In 2018, Dr. Lawrence Haddad (then with IFPRI) received the award with Dr. David Nabarro for their "individual and complementary global leadership in elevating maternal and child undernutrition to a central issue within the food security and development dialogue at national and international levels."

Source: World Food Prize Foundation, accessed May 2025. www.worldfoodprize.org/en/laureates/20002009_laureates/

other parts of South Asia has demonstrated the potential for multisectoral approaches to improve diets and health (Ruel and Alderman 2013). These initiatives reflect not only the widening scope of nutrition and food security in research and practice but also IFPRI's ability to lead and collaborate on generating such change. Several of these programs have contributed to major nutrition and health policy changes in countries where IFPRI was active, and two IFPRI researchers were awarded the World Food Prize for their work in these areas (see Box 1.1).

For decades, IFPRI has also emphasized the role of gender both in its research strategy and analyses. IFPRI's transformative work on gender issues, both conceptually and empirically, has been widely recognized. Early work helped to push economists to recognize how neglect of intrahousehold decision-making dynamics could lead to poorly designed policies and project interventions (Haddad et al. 1997; Quisumbing 2003; and Udry et al. 1995). The IFPRI-developed Women in Agricultural Empowerment Index (WEAI) and related tools are now widely used in surveys, program evaluations, and analyses throughout LMICs (Alkire et al. 2013; Malapit et al. 2019; Meinzen-Dick et al. 2019; Moore et al. 2023; Sraboni et al. 2014).

In the early 21st century, the concept of food systems began to take hold as scholars and policymakers recognized the need to grapple with the close interconnections between natural systems and human activities and across all actors along food value chains. IFPRI's solution-focused, interdisciplinary research teams were well-positioned to undertake the integrative work required to tackle complex food systems challenges.

IFPRI's broader research focus and the shift to more integrated and interdisciplinary research coincided with another important transition in the Institute. IFPRI began to supplement its generalizable knowledge products with demand-driven, country-specific capacity strengthening work and more contextualized policy research. First started on an ad hoc basis, this expansion evolved into a more deliberate decentralization strategy with the establishment of country strategy support programs in 2003. Although IFPRI began as a small institution with most staff based in Washington, DC, more than half of its multinational staff now work in LMIC offices, where they engage with civil society, government, and private sector partners. The network of regional and country offices provides IFPRI with unsurpassed capacity to integrate deep local knowledge and policy influence with evidence from global research across diverse contexts to inform policies (Benin et al. 2018). IFPRI researchers routinely co-design and co-develop research projects with partners, elevating their ideas and voices and ensuring they are recognized for their intellectual leadership and innovative contributions.

Throughout its 50-year history, IFPRI has been in the vanguard of changes in research objectives and strategies, as well as addressing unexpected crises. When the 2007/08 food crisis struck, IFPRI launched a unique communications campaign that engaged high-level policymakers and the media with findings generated by its models, evaluations, and open-access data. As part of its mandate, IFPRI has remained active in providing rapid analyses on food crises, their drivers, and effective responses, as reflected by past and ongoing work related to the COVID-19 pandemic, the Russia–Ukraine war, and a host of other conflicts, using its unique combination of global and local models, accumulated knowledge, and analytical capacity (Glauber and Laborde 2023; Swinnen and McDermott 2020).

Alongside its modeling and quantitative analysis, IFPRI has made significant contributions to research and innovation in governance and the political economy of food systems, which are crucial factors in policymaking. Its capacity strengthening work has enhanced analytical capacity among institutions and individuals in numerous countries (Kuyvenhoven 2014). Together, these have positioned IFPRI as a recognized thought leader and convener around policy design, formulation, and implementation related to food systems, directly and indirectly shaping both policies and lives.

Food policy for the future

Fifty years after IFPRI's founding, the world again faces a global food crisis. But today, unlike 50 years ago, climate change, including more frequent extreme weather events, is slowing productivity growth (Ortiz-Bobea et al. 2021) and increasing the volatility of agricultural production and prices. Food systems are major contributors of greenhouse gas emissions, but they also face harm from the changing climate (IPCC 2023; Rohr et al. 2019; Urban 2024). Climate change has a disproportionate impact on tropical regions where productivity growth and poverty reduction have been slowest, creating unique risks for the world's poor and vulnerable (Crippa 2021; Lipper et al. 2014; Rezaei 2023; Willett et al. 2019). The global population has more than doubled since IFPRI's creation, straining the limits of land and water resources. Meeting rising food demand will require expanding supply while also identifying and scaling up novel means to reduce the water, land, and chemical footprints of agrifood systems. But just as policy research helped overcome major challenges half a century ago, it can also address these overlapping crises in the coming years. While the chapters of this report point to various specific areas for research, many highlight the imperative for policy research to address several broad challenges, as described here.

Improve diets, nutrition, and food environments

The challenge of meeting minimal caloric needs in the 1960s and 1970s is no longer salient, outside of complex emergencies arising from climate and conflict. In these contexts, the primary research questions must focus on improving the speed and cost-effectiveness of humanitarian response. For most of the world, the main nutritional challenges now involve micronutrient deficiencies and overweight and obesity, even in LMICs, not a shortage of calories. Food policy research must identify and address both structural (commercial and political) drivers of unhealthy food environments and ways to most effectively shift the behaviors of all food systems actors toward healthier dietary patterns. R&D investments and policy need to move production and consumer behavior toward a healthy balance of staple cereals, roots, and tubers and relatively more expensive, more nutrient-dense foods—such as fruits, nuts, vegetables, and animal-source foods—and make healthy diets affordable for the roughly 3 billion people for whom they are unaffordable today (FAO et al. 2024). Increased investment is needed in R&D throughout agrifood value chains—both in primary production and in processing methods that affect diets and health (Schneider et al. 2023)—and in food systems more broadly to boost the availability and accessibility of more sustainable, healthier foods.

Focus on resilience and inclusion for vulnerable people and places

Food policy research must continue to shift its geographic focus to sub-Saharan Africa and the conflict- and disaster-affected areas of Asia and Latin America and the Caribbean. Poverty and malnutrition are concentrated in these places, where growth in food demand over the coming decades will be fastest but productivity is lowest and slowest growing. Renewed attention to resilience has emerged amid increased disorder and volatility in food systems, not only from weather-related shocks but also those related to conflict, disease, and macroeconomic policy shocks. The recognition of differences in vulnerability to malnutrition and various other risks has also elevated attention to inclusion, particularly of women, leading to a call for contextualized solutions that are sensitive to the often-intersecting needs of the poor, women, youth, and other marginalized and vulnerable populations.

The rise of conflict and fragile environments over the past 15 years affects both the focus of research and the logistics of its local engagement. IFPRI is currently active in several conflict-affected areas in Asia and Africa, where its work is especially relevant, as it is sometimes the sole on-the-ground source of high-quality, objective information on the food and economics situation. Working in these areas obviously implies major challenges and risks for staff and for research design and implementation. IFPRI's research on fragility and conflict-affected areas and societies is likely to grow in importance in the coming years, as multiple crises exacerbate fragility in weak institutional environments, and as the world increasingly recognizes that most people at risk of famine are concentrated in conflict-affected areas.

Take on the challenge of new technologies

The rapid development of new technologies, combined with the dramatic growth of the private sector, has profound implications for the distribution and ownership of innovation in agricultural technologies and innovations. The rapid diffusion of information and communications technologies and the digitization of technologies have extended the private sector's reach into even remote, low-income rural areas, raising a host of new privacy concerns and widening the gap between those who are digitally connected and those who are not. The dawning era of artificial intelligence (AI) within food systems magnifies these concerns, while also multiplying opportunities for accelerated innovation.

Engage with the private sector and value chains to meet food systems goals

The private sector has become an essential partner in ending poverty and malnutrition. The financing needs for R&D and innovations now far exceed what governments and philanthropies can and will provide, while pre- and post-farmgate value chains have grown immensely. Moreover, the private sector plays a key role in developing technologies that could radically reshape food systems, from AI to alternative proteins and vertical farming (Swinnen et al. 2024). As a result, off-farm companies have great potential to increase access to technologies and finance across the whole food system (Yi et al. 2021). In fact, most agricultural R&D now occurs in the private sector (Fuglie 2016), and research organizations such as IFPRI can contribute to better outcomes through collaboration. Nevertheless, policy researchers must be mindful of the potential mismatch between the commercial interests of private sector firms and the broader societal goals espoused by not-for-profit organizations.

Leverage public investments

With important gaps in private sector investment in R&D (UN et al. 2021), key public investments will remain important. Novel public–private partnerships and new forms of financing are needed to mobilize and leverage these resources. Significant room remains for the food policy research community to engage public sector and civil society institutions, and to use available sources of public funding more effectively. One high priority is reform of existing public support for agriculture, which remains highly distortionary and often promotes unsustainable practices (Gautam et al. 2022). While distortionary subsidies generate short-run gains for politicians, this money could be better used for future-oriented R&D. Research is greatly needed to help design these policy reforms to address political constraints while providing more efficient and sustainable outcomes.

Partner for interdisciplinary research and policymaking

Achieving policy reforms is complicated by the legacy institutional structures of governments and donors that make it difficult to identify institutional entry points and feasible political strategies. All countries have ministries of agriculture, environment, health, and trade, among others, but none have ministries or agencies with authority that spans food systems. Likewise, donor agencies rarely organize internally to cover the broad policy spaces relevant to food systems issues, although some local and civil society organizations undertake this work. These broad issues can best be addressed by developing a range of partnerships and maintaining a local presence in key countries, while remaining connected globally.

Build research capacity and decolonize research

As improved technologies and market access become ever more knowledge intensive, building human capital and research capacity remains critical for progress. This includes the technical capacity of experts in regions where performance improvements to agrifood systems are most urgently needed. As lessons from the past 50 years show, technologies, policies, and institutions developed in and for high-income countries rarely transfer easily, or with similar effectiveness, to food systems in Africa, Asia, or Latin America and the Caribbean. The future agenda for food policy research must therefore be closely linked to capacity sharing and capacity maintenance activities that enhance partners' ability to adapt and scale research findings, as well as to identify follow-on research needs.

The imperative for capacity sharing reflects the shifting landscape of global governance and finance. Growing skepticism about both globalization and multilateralism, as well as increased political polarization and rejection of modern science, increasingly complicates coordinated international responses to various crises. Yet growth in South–South cooperation and the economic rise of Brazil, China, India, South Africa, Viet Nam, and other countries suggest that new geopolitical developments also provide opportunities.

Development research and practice is also increasingly sensitive to the need for decolonization and local engagement. IFPRI was founded at the end of the 20-year period in which former colonies won their independence from European rule. Today, the persistent, pernicious legacy of colonization remains all too real, manifested in both economic performance (Acemoglu et al. 2001) and the composition of the scientific workforce (Beintema 2020). IFPRI is thus committed to research decentralization, which entails working with local researchers in country-based offices, and to capacity sharing with partners throughout the Global South. For an international research organization such as IFPRI, success should involve becoming steadily less central or assuming a different role as local partners take on more of the work. While local partners still vary considerably in their human capital, analytical skills, and policymaker outreach capacity, these capabilities have improved tremendously. Continued advances by local partners will require that IFPRI's role evolves over time.

IFPRI's strategy

Looking ahead, IFPRI's primary task is still to marshal the best possible research—but now, that research must help chart a transformation from current crises toward healthier, more equitable, resilient, and sustainable agrifood systems throughout the world (Barrett, Benton et al. 2022). Just as the initiatives begun 50 years ago confronted the challenges of that moment and generated unprecedented advances in agricultural productivity, food security, and poverty reduction, today's research can help us navigate the challenges and opportunities of the decades ahead.

At the end of 2025, IFPRI will launch a new institutional Strategy, organizing its future work around four research components (Box 1.2) that target the five impact areas defined in the CGIAR Research and Investment Strategy (Box 1.3). The four components are (1) to develop a clear understanding of the context and outlook of a particular situation; (2) to test, adapt, and scale potential solutions; (3) to build an enabling environment to facilitate change and reforms; and (4) to build and strengthen local capacities.

For IFPRI, crucial conditions for success will include continued attention to research quality, a well-organized communications strategy, and a sound financial foundation. To secure the quality of its research and disseminate findings to broader academic audiences, the Institute maintains a strong commitment to publishing its research in top peer-reviewed journals. IFPRI consistently ranks among the world's top producers of peer-reviewed research related to food systems. Through its open-access policy, IFPRI makes its research and datasets freely available online to enable more stakeholders to employ its evidence-based insights and recommendations. The Institute also engages with policymakers and external stakeholders through multiple channels to ensure its research fills critical evidence gaps and aligns with local assets and needs.

Today's challenging and rapidly changing global funding landscape not only affects food systems transformation itself but also the funding of organizations such as IFPRI. Changes include a radical shift from core funding to project-based funding over the past 15 years and recent sharp reductions in overseas development assistance from traditional donors, which have accelerated a long downward trend in public investment in agrifood R&D by the high-income world. The limited capacity of other funding sources, for example, from middle-income countries and private finance, to replace bilateral donors poses a major challenge for the future. While private finance has the potential to provide ample resources for some much-needed investments,

BOX 1.2 IFPRI's approach to research

IFPRI works with its partners in four ways: to develop a clear understanding of the context and outlook of a particular situation; to test, adapt, and scale potential technological and policy solutions; to build an enabling environment to facilitate change and reforms; and, finally, to strengthen research methods and capacity.

CLARIFYING THE SITUATION AND OUTLOOK. To inform relevant and effective programs and policies, IFPRI works to understand the immediate regional, country, or local context and formulate a vision of its future. This helps illuminate immediate challenges, especially among the most vulnerable populations, and provides a baseline for long-term planning. IFPRI's foresight analytics and policy models also enable decision-makers to imagine how potential future scenarios such as shifts in public expenditures, trade shocks, or climate uncertainties will affect hunger, poverty, and development in the long term—and provide tools and frameworks to plan and prioritize policies accordingly.

TESTING, ADAPTING, AND SCALING DIVERSE SOLUTIONS. After identifying promising pathways for food systems transformation, IFPRI develops corresponding institutional, technical, governance, policy, and behavior change interventions and innovations. By implementing solutions with local and global partners, the Institute can evaluate their effectiveness, their adaptability to varied contexts, and the feasibility of scaling them to reach a broader population. IFPRI's implementation and evaluation work is often accompanied by concurrent efforts to identify and advance policy and incentives that will facilitate scaling. Once tested, evaluated, improved, and scaled, these interventions may go on to improve livelihoods, well-being, equity, and inclusion in and beyond the targeted communities.

SHAPING ENABLING ENVIRONMENTS. To achieve large-scale and lasting impact, innovations and interventions require enabling environments, which include governance, trade and markets, and investments. IFPRI studies the institutional, governance, and regulatory frameworks, as well as the cultural and financial contexts that can put into place sustainable solutions to malnutrition and poverty. It engages policymakers, donors, the private sector, and civil society to ensure that research supports decision-making and that food security, healthy diets, livelihoods, and gender equality are at the forefront of global agendas.

STRENGTHENING RESEARCH METHODS AND BUILDING CAPACITY. IFPRI supports researchers and organizations by developing research methods and tools and providing training and opportunities for collaborative work. The Institute also collects and disseminates unique databases and analyzes organizational performance. These efforts aim to build the capability of researchers and organizations to provide evidence that supports agenda setting, tests innovations, and changes policies.

BOX 1.3 IFPRI and CGIAR

IFPRI officially became part of CGIAR in 1979. CGIAR, then known as the Consultative Group on International Agricultural Research, was founded in 1971 to coordinate international agricultural research in low- and middle-income countries. CGIAR evolved from efforts that date back to the 1940s, when breeding work by predecessor programs of the International Maize and Wheat Improvement Center (CIMMYT) in Mexico and the International Rice Research Institute (IRRI) in the Philippines set the stage for the Green Revolution. These first CGIAR Centers engaged with social scientists, though only for specific purposes. For instance, CIMMYT and IRRI undertook a limited range of studies on the social and economic obstacles to farmers' adoption of more productive agricultural technologies. As an institution devoted to policy research, IFPRI could address questions related to policies and technology, as well as broader ones related to prices, trade, taxes, and public investment. IFPRI therefore filled a gap in CGIAR's overarching strategy to achieve global food security.

Today, IFPRI and its food policy research remain important within CGIAR. It has aligned its work to the five CGIAR Impact Areas, as documented in the 2030 CGIAR Research and Innovation Strategy:

NUTRITION, HEALTH, AND FOOD SECURITY. IFPRI advances research and recommendations to diversify diets, enhance agrifood value chains, boost productivity of nutrient-dense foods, and advocate for biofortification.

POVERTY REDUCTION, LIVELIHOODS, AND JOBS. IFPRI's policy research helps to strengthen markets and institutions, test and scale up social protection programs, and facilitate the development and adoption of improved crop varieties.

CLIMATE ADAPTATION AND MITIGATION. IFPRI integrates sustainability into food systems transformation planning and policies, explores innovative financing to fund climate research and investments, and disseminates climate-resilient production methods and varieties to help producers, communities, and countries mitigate and adapt to the disruptions that come with a changing climate.

GENDER EQUALITY, YOUTH, AND SOCIAL INCLUSION. IFPRI works to develop and evaluate gender-transformative interventions, improve data on inclusion and empowerment, and enhance opportunities for youth employment and entrepreneurship.

ENVIRONMENTAL HEALTH AND BIODIVERSITY. As part of its aim to strengthen ecosystem health, IFPRI generates models to estimate the impact of food systems on the environment and devise alternative pathways, explores improved resource governance mechanisms, and creates enabling environments for the adoption of sustainable agricultural technologies.

especially in infrastructure and R&D, there is a pressing need for policies and instruments that can combine private sector incentives with societal returns, while avoiding or carefully and transparently managing conflicts of interest. Policy research on creating an enabling environment to mobilize private finance for sustainable development initiatives, especially in sub-Saharan Africa, is just beginning (Barrett 2023; Kremer et al. 2020). This evolving funding landscape will require IFPRI to continuously evaluate and hone its comparative advantage to efficiently provide the greatest positive impact in a changing world.

IFPRI's relevance to first-order global challenges has never been greater. Food systems transformation is inevitable, and its pace in the years ahead will be brisk. An ambitious research agenda will be essential to inform actions by governments, donors, civil society organizations, and firms, as well as by individual consumers, producers, resource managers, and entrepreneurs. People throughout food systems seek to learn what works, why, how, and for whom, including not only government ministries and donor agencies but also farmers' and women's groups, those responding to disasters, and those making daily choices about how to manage limited budgets and the planet's finite resources so that their families and communities flourish. IFPRI's task in the years ahead, as it has been over the past five decades, is to provide the evidence that supports effective policies to sustainably reduce poverty and end hunger and malnutrition.

Organization of the book

In the chapters that follow, this Global Food Policy Report analyzes topic areas crucial to agrifood systems development over the past 50 years. Each chapter provides both a historical review and a current assessment of the challenges and opportunities anticipated up to 2050, featuring IFPRI's past and present contributions to research, policy, and practice.

Chapters 2 and 3 examine the role of food policy research and agrifood system development in creating pathways for reducing poverty and malnutrition. **Chapter 2** examines IFPRI's impacts over the last 50 years, synthesizing case studies of impact and exploring how policy research organizations can measure their impact and overcome impact assessment challenges for the future. Through an assessment of Asia's Green Revolution and its lessons for other LMICs, **Chapter 3** examines how agriculture's role in structural change and economic development is evolving in ways that continue to redefine discourse, research, and action.

Chapters 4 to 7 focus on sustainability and natural resources along with the tenure rules and markets that shape farming decisions and practices. **Chapter 4** reviews the evolution of food policy research on climate change, food security, and food systems, reflecting on IFPRI's major contributions to understanding climate change impacts and identifying promising policies and investments for mitigation and adaptation. **Chapter 5** considers the broader intersection of agrifood systems and ecosystem health, including land, water, and energy systems. The chapter highlights how taking a systems view can best provide evidence and policy approaches to balance agricultural production with environmental sustainability.

Pivoting to the secure tenure of land and natural resources, **Chapter 6** examines foundational concepts and key lessons from the research, reviewing the variety of tenure arrangements, relationships between different forms of tenure and farming practices, and investments in resource management, and the effects of reforms, especially for women and youth. **Chapter 7** focuses on food value chains, with a discussion of the major drivers and revolutions that have shaped the growth, structure, and importance of these chains to economies, employment, and diets, as well as policy research contributions.

Chapters 8 to 10 explore key instruments used to support livelihoods at the farm level. **Chapter 8** applies IFPRI's "best fit" conceptual framework to examine the global evolution of agricultural extension and rural advisory services over the past 50 years, as well as the shift from a "transfer of technology" approach to a more sophisticated "facilitation for innovation" paradigm. **Chapter 9** looks at how evidence on crop genetic improvement has contributed to productivity, nutrition, environment, and poverty outcomes, as well as the hurdles to farmer uptake of new varieties and ways to overcome this challenge. Turning to agricultural insurance, **Chapter 10** examines how policy-oriented research has assessed and refined insurance options to facilitate farmer use and uptake, and explores how new technologies and approaches are creating opportunities for increasing coverage.

Chapters 11 to 14 analyze how polices can reduce vulnerability and inequities, including the role of social protection, ways to address different forms of malnutrition, the impacts of conflict, and the importance of gender. **Chapter 11** examines the development of increasingly sophisticated social protection programs and discusses the contribution of research to changing program approaches and policies, highlighting IFPRI's role in providing evaluations and recommendations. **Chapter 12** reviews the evolution of nutrition in both policy and programming, with particular attention to the intersection of nutrition with agriculture, food systems, gender, and multisectoral approaches. **Chapter 13** focuses on hunger and undernourishment in fragile and conflict-affected areas, reflecting on changes in food policy research in these areas and highlighting how evidence has helped to clarify the linkages between conflict and hunger. **Chapter 14** examines the evolution of gender research in the context of the development discourse on gender, with a focus on agrifood systems, and highlights the need for intentional gender programming and transformative approaches to improve women's empowerment and achieve gender equality.

Chapters 15 to 18 examine macro-level factors related to governance, trade, public investment, and finance that determine whether and how policy change occurs. With a focus on political economy and governance, **Chapter 15** examines the key areas of decentralization, agriculture and food policy reform processes, political economy of distribution, and state capacity, as well as the need to build effective and legitimate global institutions for food systems governance. **Chapter 16** reviews the evolution of research on agrifood trade, looking closely at the contributions made by IFPRI and others, and considers how trade policy analysis can contribute to more stable and sustainable food systems in the future. **Chapter 17** addresses innovation in agriculture, exploring the evolution of research on technical change and public policy, including how research has supported priority-setting and policymaking, multistakeholder innovation platforms and partnerships, and scaling up of promising innovations. Looking at the financing of agricultural and food production, **Chapter 18** examines the various actors and channels in LMIC food systems funding flows, describes related policy research, and highlights possible policy options to mobilize future financing for food systems transformation.

Chapters 19 to 24 examine food policy and food policy research in six major world regions over the past 50 years and up to 2050. **Chapter 19** explores how Africa's agrifood policy landscape has evolved in response to complex challenges, including food insecurity, climate change, and socioeconomic disparities, and the research-based solutions offered to address these challenges. Looking to Central Asia, **Chapter 20** examines how food policy research informed market-oriented reforms and agricultural transformation, and assesses ongoing challenges, including climate change, land use, and markets and incentives. **Chapter 21** reviews the evolution of South Asia's food systems, covering its agricultural growth linkages, policymaking and investments in agriculture, and institutional innovations, and highlights how policy research has played a critical role in shaping national policies on food security, rural development, and nutrition.

Turning to East and Southeast Asia, **Chapter 22** analyzes how the alignment of research outputs, financial programs, and regional cooperation initiatives with national policy frameworks in the region has led to improvements in food security, nutrition, and livelihoods, though important challenges remain. **Chapter 23** draws on research by IFPRI and partners to outline the evolving food systems landscape in Latin America and the Caribbean, with attention to the region's role in supplying food exports and environmental services. With a focus on the Middle East and North Africa, **Chapter 24** synthesizes historical trends in policies and outcomes related to agricultural production, consumers, and food and nutrition programs, and assesses emerging policy issues and research priorities.

References

Acemoglu, D., S. Johnson, and J.A. Robinson. 2001. "The Colonial Origins of Comparative Development: An Empirical Investigation." *American Economic Review* 91 (5): 1369–1401.

Alderman, H., P.A. Chiappori, L. Haddad, J. Hoddinott, and R. Kanbur. 1995. "Unitary versus Collective Models of the Household: Is It Time to Shift the Burden of Proof?" *World Bank Research Observer* 10 (1): 1–19.

Alkire, S., R. Meinzen-Dick, A. Peterman, A. Quisumbing, G. Seymour, and A. Vaz. 2013. "The Women's Empowerment in Agriculture Index." *World Development* 52: 71–91.

Anderson, K., G. Rausser, and J. Swinnen. 2013. "Political Economy of Public Policies: Insights from Distortions to Agricultural and Food Markets." *Journal of Economic Literature* 51 (2): 423–477.

Barrett, C.B. 2021. "Overcoming Global Food Security Challenges through Science and Solidarity." *American Journal of Agricultural Economics* 103 (2): 422–447.

Barrett, C.B. 2023. "'Benevolent' Patent Extensions Could Raise Billions for R&D in Poorer Countries." *Nature* 621: 687–690.

Barrett, C.B., A. Agrawal, O.T. Coomes, and J.-P. Platteau. 2009. "Stripe Review of Social Sciences in the CGIAR." *SSRN Electronic Journal* (August 1).

Barrett, C.B., T. Benton, J. Fanzo, et al. 2022. *Socio-Technical Innovation Bundles for Agri-Food Systems Transformation*. Sustainable Development Goals Series. Cham, Switzerland: Palgrave Macmillan.

Barrett, C.B., M.R. Carter, and P.C. Timmer. 2010. "A Century-Long Perspective on Agricultural Development." *American Journal of Agricultural Economics* 92 (2): 447–468.

Barrett, C.B., T. Reardon, J. Swinnen, and D. Zilberman. 2022. "Agri-Food Value Chain Revolutions in Low- and Middle-Income Countries." *Journal of Economic Literature* 60 (4): 1316–1377.

Benin, S., F. Place, and P. Hazell. 2018. *Has IFPRI's Decentralization Strategy Made a Difference? An Econometric Study of African and Asian Countries, 1981–2014.* Independent Impact Assessment Report 44. Washington, DC: IFPRI.

Christiaensen, L., L. Demery, and J. Kuhl. 2011. "The (Evolving) Role of Agriculture in Poverty Reduction: An Empirical Perspective." *Journal of Development Economics* 96 (2): 239–254.

Christiaensen, L., and W. Martin. 2018. "Agriculture, Structural Transformation and Poverty Reduction: Eight New Insights." *World Development* 109: 413–416.

Clapp, J. 2025. *Titans of Industrial Agriculture.* Cambridge, MA: MIT Press.

Corong, E., M. Gautam, W. Martin, and R. Vos. 2024. *Measuring Employment and Job Quality in Agrifood Systems: A Comprehensive Approach.* CGIAR Initiative on Rethinking Food Markets. Washington, DC: IFPRI.

Crespi, J.M., and J.M. MacDonald. 2022. "Concentration in Food and Agricultural Markets." *Handbook of Agricultural Economics* 6: 4781–4843.

Crippa, M., E. Solazzo, D. Guizzardi, F. Monforti-Ferrario, F.N. Tubiello, and A. Leip. 2021. "Food Systems Are Responsible for a Third of Global Anthropogenic GHG Emissions." *Nature Food* 2: 198–209.

Curtis, F. 2000. *IFPRI's First 10 Years.* Washington, DC: IFPRI.

Davis, B., E. Mane, L.Y. Gurbuzer, et al. 2023. "Estimating Global and Country-Level Employment in Agrifood Systems." FAO Statistics Working Paper Series Issue 23/24. Food and Agriculture Organization of the United Nations, Rome.

de Janvry, A., and E. Sadoulet. 2009. "Agricultural Growth and Poverty Reduction: Additional Evidence." *World Bank Research Observer* 25 (1): 1–20.

Dorosh, P., and J. Thurlow. 2018. "Beyond Agriculture versus Non-Agriculture: Decomposing Sectoral Growth–Poverty Linkages in Five African Countries." *World Development* 109: 440–451.

Ehrlich, P.R. 1968. *The Population Bomb.* New York: Ballantine Books.

El-Chichakli, B., J. von Braun, C. Lang, D. Barben, and J. Philp. 2016. "Policy: Five Cornerstones of a Global Bioeconomy." *Nature* 535 (7611): 221–223.

Falck-Zepeda, J.B., and P. Zambrano. 2011. "Socio-Economic Considerations in Biosafety and Biotechnology Decision Making: The Cartagena Protocol and National Biosafety Frameworks." *Review of Policy Research* 28 (2): 171–195.

Fan, S., and A. Gulati. 2008. "The Dragon and the Elephant: Learning from Agricultural and Rural Reforms in China and India." *Economic and Political Weekly* 43 (26/27): 137–144.

FAO (Food and Agriculture Organization of the United Nations), IFAD (International Fund for Agricultural Development), UNICEF, WFP (World Food Programme), and WHO (World Health Organization). 2024. *The State of Food Security and Nutrition in the World 2024.* Rome.

FAO, IFAD, and WFP. 2015. *The State of Food Security in the World 2015*. Rome.

Fosu, A.K. 2017. "Growth, Inequality, and Poverty Reduction in Developing Countries: Recent Global Evidence." *Research in Economics* 71 (2): 306–336.

Farrar, C. 2000. *IFPRI's First 10 Years*. Washington, DC: IFPRI.

Fuglie, K., S. Morgan, and J. Jelliffe. 2024. *World Agricultural Production, Resource Use, and Productivity, 1961–2020*. Report No. EIB-268. Washington, DC: U.S. Department of Agriculture, Economic Research Service.

Fuglie, K. 2016. "The Growing Role of the Private Sector in Agricultural Research and Development World-Wide." *Global Food Security* 10: 29–38.

Gautam, M., D. Laborde, A. Mamun, W. Martin, V. Pineiro, and R. Vos. 2022. *Repurposing Agricultural Policies and Support: Options to Transform Agriculture and Food Systems to Better Serve the Health of People, Economies, and the Planet*. Washington, DC: World Bank and IFPRI.

Glauber, J., and D. Laborde, eds. 2023. *The Russia-Ukraine Conflict and Global Food Security*. Washington, DC: IFPRI.

Grajeda, R., J.R. Behrman, R. Flores, J.A. Maluccio, R. Martorell, and A.D. Stein. 2005. "The Human Capital Study 2002-04: Tracking, Data Collection, Coverage, and Attrition." *Food and Nutrition Bulletin* 26 (2 Suppl. 1): S15–S24.

Haddad, L., J. Hoddinott, and H. Alderman. 1997. *Intrahousehold Resource Allocation in Developing Countries: Models, Methods, and Policies*. Baltimore: Johns Hopkins University Press.

Haggblade, S., P. Hazell, and T. Reardon. 2007. *Transforming the Rural Nonfarm Economy: Opportunities and Threats in the Developing World*. Baltimore: Johns Hopkins University Press; Oxford, UK: Oxford University Press.

Haggblade, S., P. Hazell, and T. Reardon. 2010. "The Rural Non-Farm Economy: Prospects for Growth and Poverty Reduction." *World Development* 38 (10): 1429–1441.

Hazell, P., S. Haggblade, and T. Reardon. 2024. "Transformation of the Rural Nonfarm Economy During Rapid Urbanization and Structural Transformation in Developing Regions." *Annual Review of Resource Economics* 16: 277–299.

Hazell, P., and F. Place. 2025. *Taking Stock: Impacts of 50 Years of Policy Research at IFPRI*. Independent Impact Assessment Report No. 42. Washington, DC: IFPRI. Forthcoming.

Hoddinott, J., J.A. Maluccio, J.R. Behrman, R. Flores, and R. Martorell. 2008. "Effect of a Nutrition Intervention during Early Childhood on Economic Productivity in Guatemalan Adults." *Lancet* 371 (9610): 411–416.

IPCC (Intergovernmental Panel on Climate Change). 2023. "Summary for Policymakers." In *Climate Change 2023: Synthesis Report. Contribution of Working Groups I, II and III to the Sixth Assessment Report of the Intergovernmental Panel on Climate Change*, eds. H. Lee and J. Romero. Geneva: IPCC.

Johnson, D.G. 1991. *World Agriculture in Disarray*, 2nd ed. London: Palgrave Macmillan.

Johnston, B.F., and J.W. Mellor. 1960. "The Nature of Agriculture's Contributions to Economic Development." *Food Research Institute Studies* 1 (3): 335–356.

Johnston, B.F., and J.W. Mellor. 1961. "The Role of Agriculture in Economic Development." *The American Economic Review* 51 (4): 566 –593.

Kennedy, E., H. Bouis, and J. von Braun. 1992. "Health and Nutrition Effects of Cash Crop Production in Developing Countries: A Comparative Analysis." *Social Science and Medicine* 35 (5): 689–697.

Kherallah, M., C.L. Delgado, E. Zaude Gabre-Madhin, N. Minot, and M.E. Johnson. 2002. *Reforming Agricultural Markets in Africa: Achievements and Challenges*. Baltimore and London: Johns Hopkins University Press.

Kremer, M., J. Levin, and C.M. Snyder. 2020. "Advance Market Commitments: Insights from Theory and Experience." *AEA Papers and Proceedings* 110: 269–273.

Lipper, L., P. Thornton, B.M. Campbell, et al. 2014. "Climate-Smart Agriculture for Food Security." *Nature Climate Change* 4: 1068–1072.

Liu, Y., C.B. Barrett, T. Pham, and W. Violette. 2020. "The Intertemporal Evolution of Agriculture and Labor over a Rapid Structural Transformation: Lessons from Vietnam." *Food Policy* 94: 101913.

Malapit, H., A. Quisumbing, R. Meinzen-Dick, G. Seymour, E.M. Martinez, J. Heckert, D. Rubin, A. Vaz, and K.M. Yount. 2019. "Development of the Project-Level Women's Empowerment in Agriculture Index (pro-WEAI)." *World Development* 122: 675–692.

Markelova, H., R. Meinzen-Dick, J. Hellin, and S. Dohrn. 2009. "Collective Action for Smallholder Market Access." *Food Policy* 34 (1): 1–7.

Meinzen-Dick, R. 2007. "Beyond Panaceas in Water Institutions." *Proceedings of the National Academy of Sciences of the United States of America* 104 (39): 15200–15205.

Meinzen-Dick, R.S., L.R. Brown, H.S. Feldstein, and A.R. Quisumbing. 1997. "Gender, Property Rights, and Natural Resources." *World Development* 25 (8): 1303–1315.

Meinzen-Dick, R.S., D. Rubin, M. Elias, A.A. Mulema, and E. Myers. 2019. "Women's Empowerment in Agriculture: Lessons from Qualitative Research." IFPRI Discussion Paper 1797. IFPRI, Washington, DC.

Mellor, J.W. 1984. "The World Food Equation: Interrelations among Development, Employment, and Food Consumption." *Journal of Economic Literature* 22 (2): 531–574.

Mesarovic, M., and E. Pestel. 1974. *Mankind at the Turning Point: The Second Report to the Club of Rome*. New York: New American Library.

Moore, L., M. Dissanayake, H.J. Malapit, and A. Go. 2023. *Uncovering More Than a Decade of WEAI Use in USAID Projects.* WEAI Applications and Insights 2. Washington, DC: USAID and IFPRI.

Nelson, S., T. Frakenberger, V. Brown, C. Presnall, and J. Downen. 2015. *Ex-Post Impact Assessment Review of IFPRI's Research Program on Social Protection, 2000–2012.* Independent Impact Assessment Report 40. Washington, DC: IFPRI.

Ninno, C.D., P.A. Dorosh, L.C. Smith, and D.K. Roy. 2001. *The 1998 Floods in Bangladesh: Disaster Impacts, Households Coping Strategies, and Response.* Research Report 122. Washington, DC: IFPRI.

Ortiz-Bobea, A., T.R. Ault, C.M. Carrillo, R.G. Chambers, and D.B. Lobell. 2021. "Anthropogenic Climate Change Has Slowed Global Agricultural Productivity Growth." *Nature Climate Change* 11 (4): 306–312.

Paarlberg, R.L. 1999. *External Impact Assessment of IFPRI's 2020 Vision for Food, Agriculture and the Environment Initiative.* Independent Impact Assessment Report 10. Washington, DC: IFPRI.

Pardey, P.G., ed. 2001. *The Future of Food: Biotechnology Markets and Policies in an International Setting.* Washington, DC: IFPRI.

Pinstrup-Andersen, P. 1999. "Agricultural Biotechnology, Trade, and the Developing Countries." *AgBioForum* 2 (3–4): 215–217.

Pinstrup-Andersen, P. 2000. *25 Years of Food Policy Research: Reflections by Per Pinstrup-Anderson.* Washington, DC: IFPRI.

Quisumbing, A.R., ed. 2003. *Household Decisions, Gender, and Development: A Synthesis of Recent Research.* Washington, DC: IFPRI.

Reardon, T., P.C. Timmer, C.B. Barrett, and J. Berdegué. 2003. "The Rise of Supermarkets in Africa, Asia, and Latin America." *American Journal of Agricultural Economics* 85 (5): 1140–1146.

Renkow, M. 2010. "Impacts of IFPRI's 'Priorities for Pro-Poor Public Investment' Global Research Program." Washington, DC: IFPRI.

Rezaei, E.E., H. Webber, S. Asseng, et al. 2023. "Climate Change Impacts on Crop Yields." *Nature Reviews Earth and Environment* 4: 831–846.

Ritchie, H., P. Rosado, and M. Roser. 2023. "Hunger and Undernourishment." Our World in Data. https://ourworldindata.org/hunger-and-undernourishment

Rohr, J.R., C.B. Barrett, D.J. Civitello, et al. 2019. "Emerging Human Infectious Diseases and the Links to Global Food Production." *Nature Sustainability* 2: 445–456.

Rozelle, S., and J.F.M. Swinnen. 2004. "Success and Failure of Reform: Insights from the Transition of Agriculture." *Journal of Economic Literature* 42 (2): 404–456.

Ruel, M.T., and H. Alderman. 2013. "Nutrition-Sensitive Interventions and Programmes: How Can They Help to Accelerate Progress in Improving Maternal and Child Nutrition?" *Lancet* 382 (9891): 536–551.

Ryan, J. 2002. "Assessing the Impact of Food Policy Research: Rice Trade Policies in Viet Nam." *Food Policy* 27 (1): 1–29.

Schneider, K.R., J. Fanzo, L. Haddad, et al. 2023. "The State of Food Systems Worldwide in the Countdown to 2030." *Nature Food* 4: 1090–1110.

Sen, A. 1983. *Poverty and Famines: An Essay on Entitlement and Deprivation*. Oxford, UK: Oxford University Press.

Sen, A. 2000. *Development as Freedom*. New York: Alfred A. Knopf.

Shekar, M., and B. Popkin, eds. 2020. *Obesity Health and Economic Consequences of an Impending Global Challenge*. Washington, DC: World Bank Group.

Spielman, D.J. 2007. "Pro-Poor Agricultural Biotechnology: Can the International Research System Deliver the Goods?" *Food Policy* 32 (2): 189–204.

Spielman, D.J., and R. Pandya-Lorch, eds. 2009. *Millions Fed: Proven Successes in Agricultural Development*. Washington, DC: IFPRI.

Sraboni, E., H.J. Malapit, A.R. Quisumbing, and A.U. Ahmed. 2014. "Women's Empowerment in Agriculture: What Role for Food Security in Bangladesh?" *World Development* 61: 11–52.

Swinnen, J., and J. McDermott. 2020. "Covid-19 and Global Food Security." *EuroChoices* 19 (3): 26–33.

Swinnen, J., L. Ronchi, and T. Reardon. 2024. "Harness Agrifood Value Chains to Help Farmers Be Climate-Smart." *Science* 386 (6725): 974–977.

Swinnen, J., A. Vandeplas, and M. Maertens. 2010. "Liberalization, Endogenous Institutions, and Growth: A Comparative Analysis of Agricultural Reforms in Africa, Asia, and Europe." *World Bank Economic Review* 24 (3): 412–445.

Timmer, P.C. 1988. "Chapter 8: The Agricultural Transformation." In *Handbook of Development Economics* 1: 275–331.

Timmer, C. 1992. "Agriculture and Economic Development Revisited." Agricultural Systems 40 (1–3): 21–58.

Udry, C., J. Hoddinott, H. Alderman, and L. Haddad. 1995. "Gender Differentials in Farm Productivity: Implications for Household Efficiency and Agricultural Policy." *Food Policy* 20 (5): 407–423.

UN (United Nations), World Bank, Food and Land Use Coalition, and IFPRI. 2021. *Food Finance Architecture Financing a Healthy, Equitable, and Sustainable Food System*, Washington, DC: World Bank Group.

Urban, M.C. 2024. "Climate Change Extinctions." *Science* 386 (6726): 1123–1128.

von Braun, J., and E. Kennedy, eds. 1994. *Agricultural Commercialization, Economic Development, and Nutrition*. Baltimore and London: John Hopkins University Press.

Willett, W., J. Rockström, B. Loken, et al. 2019. "Food in the Anthropocene: The EAT–Lancet Commission on Healthy Diets from Sustainable Food Systems." *Lancet* 393 (10170): 447–492.

World Bank. 1986. *Poverty and Hunger: Issues and Options for Food Security in Developing Countries*. Washington, DC.

World Bank. n.d.-a. "Population, Total - China." Accessed April 2025. https://data.worldbank.org/indicator/SP.POP.TOTL?locations=CN

World Bank. n.d.-b. "Population, Total - Fragile and Conflict-Affected Situations." Accessed April 2025. https://data.worldbank.org/indicator/SP.POP.TOTL?locations=F1

World Bank. n.d.-c. "Population, Total - India." Accessed April 2025. https://data.worldbank.org/indicator/SP.POP.TOTL?locations=IN

World Bank. n.d.-d. "Poverty Headcount Ratio at $2.15 a Day (2017 PPP) (% of Population)." Accessed April 2025. https://data.worldbank.org/indicator/SI.POV.DDAY

Yi, J., E.M. Meemken, V. Mazariegos-Anastassiou, et al. 2021. "Post-Farmgate Food Value Chains Make Up Most of Consumer Food Expenditures Globally." *Nature Food* 2: 417–425.

Yi, J., S. Jiang, D. Tran, M. Gómez, P. Canning, J.R. Bloem, and C.B. Barrett. 2025. "How Agri-Food Value Chain Employment and Compensation Evolve with Structural Transformation." *Nature Food*. Forthcoming.

Chapter 2

IMPACT
Assessing the Outcomes of IFPRI's Research[1]

Sivan Yosef, Tamsin Zandstra, Frank Place, and Suresh Babu

Sivan Yosef is manager, Strategic Outreach, Director General's Office (DGO), IFPRI; **Tamsin Zandstra** is a research assistant, DGO, IFPRI; **Frank Place** is a consultant, CGIAR; and **Suresh Babu** is head of Capacity Building and a senior research fellow, Development Strategies and Governance Unit, IFPRI.

Key messages

- IFPRI's approach to research entails working with partners to understand the immediate regional, country, and local contexts and outlooks; testing, adapting, and scaling diverse solutions; shaping enabling environments; and strengthening research methods and sharing capacity.
- IFPRI's assessment of its own work has yielded a range of evidence on its impact and that of its partner organizations. These benefits have included improvements in policies, programs, innovations, and investments that have, in turn, positively impacted food security, healthy and diverse diets, livelihoods, and environmental sustainability. IFPRI's capacity strengthening work has also improved the ability of government ministries and local research organizations in low- and middle-income countries to collect data and analyze policy.
- Obtaining evidence of impact can be difficult, with challenges in attributing influence to specific research outputs or organizations, evaluating the importance of specific research on impact, and understanding gaps in research and methodology.

1 Portions of this chapter draw on the 50th Anniversary Impact Assessment Report, *Taking Stock: Impacts of 50 Years of Policy Research at IFPRI*, by Peter Hazell and Frank Place, publication forthcoming.

To improve impact in the future, research and policy organizations will need to consider:

- There is increased pressure for policy organizations to improve their impact. Policy research organizations can do so by ensuring that the **demand for their research** is driven by governments and other external stakeholders, maintaining **close relationships with policymakers**, and articulating a **deliberate strategy of influence** at the beginning of research projects.
- As development funding becomes scarce, proof of impact will become even more important. Organizations can work with funders to **forge a culture of impact** and plan for impact assessments and evaluation at the project design phase.
- Policy research organizations should continue to **prioritize high-quality science** so that they remain a trusted source of evidence on policies, programs, and investments that can advance development goals.

Throughout 2024, the world continued to face challenges to achieving food security, ensuring nutritious, diverse, and sustainable diets for all, and securing other development goals. Major challenges included conflicts in Burkina Faso, Chad, Gaza, Mali, Myanmar, and Sudan; climate change-induced disasters, which displaced 123 million people by the end of June 2024 alone (USA for UNHCR 2024); and shocks such as high inflation and cross-border migration. Some of these challenges result from significant demographic, economic, and social shifts that have occurred since IFPRI was founded 50 years ago: during this period, the global population has more than doubled, and urbanization and international trade have changed the locus and dynamics of food supply chains and markets, giving rise to a dietary transition in many countries. The world faces increasing pressures on natural resources and the environment, a changing climate, biodiversity loss, and the spread of infectious diseases, all of which are becoming increasingly interconnected. IFPRI's goals have similarly evolved to address these challenges: although the Institute was founded to address famines and lack of calories, it now has a much broader set of priorities related to ending hunger and malnutrition globally, strengthening livelihoods, ensuring environmental sustainability, and transforming entire food systems.

Food policy research plays an essential role in helping food systems meet these various aims. It can provide timely data on rapidly evolving situations, generate and evaluate evidence, and develop metrics and frameworks to analyze and monitor impacts on the poorest and most marginalized populations. But research cannot exist in a vacuum: food policy research must provide independent, high-quality evidence to inform policy options, programs, and investments. Researchers can work with policymakers, both to help them test alternative interventions and weigh the results and to promote adoption of effective policies and help prevent bad ideas from shaping government interventions. Policy research can also complement the scientific and engineering research that underlies technological advances, creating "socio-technical innovation bundles" (Barrett 2022) and identifying options to promote inclusivity in technology use and benefits. The ability to measure a technology's prospective impact, model its scaling, and assess relevant human, institutional, political, and cultural interactions is integral to its success in the real world. Evidence-based policies, interventions, and investments have a track record of helping to achieve food security for all, promote sustainable and healthy diets, build efficient markets, transform economies, and strengthen food systems institutions and governance.

Research with a real-world impact

IFPRI's approach to research (see Chapter 1) entails working with partners to understand the immediate regional, country, or local contexts and outlooks; testing, adapting, and scaling diverse solutions; shaping enabling environments; and strengthening research methods and sharing capacity. Articulating these modes of work helps the Institute assess more clearly whether its food policy research has made a positive difference or affected policies, investments, and programmatic choices that shape food systems. It also allows IFPRI to determine whether this research has led to further downstream impacts on food and nutrition security, poverty, and the environment, even if these outcomes are not directly attributable to the research activities.

Policy research institutes employ several methods to collect data on their influence and undertake robust impact assessments. IFPRI uses a range of methods, including analysis of large-scale datasets and modeling. Available methods, detailed on the next page, examine influence in countries where research activities were conducted, as well as spillover or global effects of the research. Research organizations can also undertake quantitative and qualitative impact assessments to examine downstream impacts of research on development goals.

On-the-ground engagement. Ongoing relationships with stakeholders and major philanthropic or for-profit investors within countries and at a global scale can show direct evidence of the use of research within policies and strategies. One-on-one meetings, interviews, and policy events can uncover these influences.

Academic citation analysis. Undertaking literature reviews, tracking outputs such as publications and citations, and monitoring Altmetric scores (readership metrics) and institutional rankings can assess influence on academic discourse. During its 50 years, IFPRI has published more than 4,000 journal articles; these are generally well-cited and downloaded, and the top ones attract much international attention every year. Its datasets have been downloaded more than 2.2 million times in the past decade.

Capacity building. If a research organization engages in building human and institutional capacity through, for example, trainings or co-creation of tools and governmental policy units, as IFPRI does, these successes can be directly monitored (see capacity building section below).

Online trends. Tracking specialized search terms on internet search engines, such as "agriculture+nutrition+health," can yield insights on whether research has influenced global policy agenda setting.

Policy document scoping. Research organizations can systematically examine policy or institutional documents and government websites for research citations or acknowledgments.

Media scoping. News articles with quotes and research citations from decision-makers can point to research uptake.

Challenges in measuring policy research impact

Despite the methods described above, assessing policy-oriented research poses several major challenges (Place and Hazell 2015).

Assessing policy influence and attributing influence to specific research outputs or organizations

Obtaining evidence to demonstrate policy influence is often hindered by the lack of a paper trail. In some cases, making claims of influence can create frictions in researchers' relationships with policymakers. Impact claims may undercut policymakers' incentives to pursue recommended reforms, as they may not want to be seen as simply championing others' ideas, especially if these ideas come from a foreign or international institution. Policy research may also recommend retaining the status quo, making it impossible to point to any "change." In contrast to reporting on innovation use, policy outcomes achieved by research institutes may go underreported. Given the many actors involved in policymaking processes, measuring the influence of research or evidence is generally difficult; thus, contribution of evidence or of a research organization is a more useful concept than attribution for policy outcomes.

Evaluating the importance of specific research outputs to a policy decision

In addition to documenting contributions, policy researchers and their funders need to assess how their research affects policy decisions. The importance of any individual piece of evidence or research actor is often difficult to gauge, raising the following questions: Would the evidence have been generated by another actor, possibly at a later date? Would the policy decision have occurred without the scientific evidence? Qualitative evaluations that capture the viewpoints of key decision-makers can help shed light on these questions, but given the cost of conducting such exercises across many policy influences, these evaluations need to be selective and strategic.

Addressing methodological and counterfactual challenges to measuring and attributing impacts from policy outcomes

Once a contribution to a policy decision has been established, another area of interest encompasses the extent of the decision's impact on objectives such as poverty reduction, improved nutrition, better environmental outcomes, or more gender equality, among others. For IFPRI, this focus is equivalent to assessing the extent of an agricultural innovation's impact, once its adoption has been established. The first challenge, however, is that it often takes a long time for policy changes to be implemented and effects to materialize so they can be evaluated. Second, if the policy change occurs at the national level, there is no obvious counterfactual to the change, so one must be constructed, often through the use of modeling. In theory, it should be more feasible to identify counterfactuals to policy changes at the subnational level, but data availability and modeling efforts at that level are more difficult. In some cases, the counterfactual is the absence of a policy change, while in other cases it is a different policy change. Third, it is difficult for any one model to capture the full range of a policy's effects; some models are better for assessing economic impacts while others are better for assessing environmental impacts. Therefore, a blending of quantitative and qualitative approaches may be needed.

The road to demand-driven research

IFPRI has been measuring its institutional impact since the mid-1990s and has completed 46 independent impact assessment studies covering many countries for which it carries out research. These studies range in scope from the evaluation of relatively small research projects to evaluations of entire research programs.

IFPRI's approach to setting research priorities has directly influenced the scope of its impact. In the early years, its research agenda was based largely on literature reviews and discussions with donors. Its early emphasis was on generating knowledge products, or international public goods, about regional and global policy problems that could be extrapolated to multiple countries. Many researchers were trained at universities of the Global North, especially US land grant colleges, which have long been among IFPRI's key research partners. Today, IFPRI's research is much more demand-driven and determined at the country and regional levels, in consultation with local policymakers, researchers, and donors, in order to contextualize policy research and develop deep engagement with local partners. For example, there is great demand for ex ante impact assessments to help partners and stakeholders assess options; the high quality and credibility of these analyses in predicting ex post outcomes influences impact. Demand-driven research is facilitated by the Institute's long-term commitment to decentralization and building a strong international presence. In 2003, most of IFPRI's staff were based at its Washington, DC, headquarters. As of 2024, more than half of its staff were based outside of headquarters, and 59 percent hailed from LMICs. Through country programs in Bangladesh, China, Egypt, Ethiopia, Ghana, Kenya, Malawi, Myanmar, Nigeria, Papua New Guinea, Rwanda, Senegal, and Sudan, along with regional offices in Africa and South Asia, IFPRI has a unique opportunity to engage and contribute at local and subregional levels, as well as to facilitate cross-country learning.

IFPRI's growing research on political economy and governance (Chapter 15) has also yielded insights on favorable environments for the uptake of research that can lead to desirable development outcomes. This work has illuminated the complicated process by which evidence is—or is not—considered in decision-making, as a result of power dynamics, interest groups and coalitions, differing timetables, and more. IFPRI has developed practical tools, such as the Kaleidoscope Model, for assessing factors that drive policy change (Haggblade and Resnick 2018).

IFPRI has long aligned its data collection and impact reporting with CGIAR, which it joined in 1979. Between 2010 and 2021, IFPRI reported its own outcomes and impacts to CGIAR, as well as those of the CGIAR

Research Programs it led with other Centers during this period. Since 2021, IFPRI has aligned its work with the five CGIAR impact areas, which aim to achieve a range of interlinked impacts: climate adaptation and mitigation; environmental health and biodiversity; gender equality, youth, and social inclusion; nutrition, health, and food security; and poverty reduction, livelihoods, and jobs. These impact areas consider the interconnectedness between climate change and food-related nutritional and health challenges, as well as the direct linkages between agrifood system performance and broader economic growth and poverty reduction outcomes, as well as the challenges related to a range of natural resource limits to system expansion.

Attempts to quantify impact

Due to the challenges associated with impact assessment, a definitive quantitative summary of IFPRI's total impact will remain elusive. For example, an impact assessment report in 2015 using estimates from IFPRI's work on six distinct projects in Bangladesh, Ethiopia, India, Mexico, Viet Nam, and Kenya and Tanzania found that the economic benefits from these projects alone could exceed US$1 billion. However, the authors acknowledged that the full benefits were likely much larger, as these estimates did not include cross-country spillovers and benefits associated with global public goods. In addition, the evidence-informed policy and program changes continued to generate benefits after the estimates were calculated. This example demonstrates that policy research can have great impact, but the limited quantification of such impacts highlights both the challenges to measuring them and the limitations of using a single figure, or dollar amount, to convey the full range of impacts generated by policy research.

Another quantification exercise summed up the numbers of beneficiaries reported for many of IFPRI's "big successes"—projects or initiatives that produced reasonable evidence of influence on programs, policies, or investments impacting beneficiaries. This exercise concluded that IFPRI's work has indirectly benefited around half a billion people, including individuals benefiting from social protection schemes, biofortification, commodity policy reforms, and agricultural strategies. This figure is likely conservative, as it does not account for the vast amount of research that did not estimate the number of beneficiaries, smallholders affected by IFPRI's more general policy support in specific countries, or people who benefited when countries refrained from enacting harmful policies, such as export bans, due to IFPRI's research. Regardless, these types of estimates reflect long-term relationships with hundreds of partners, including governments, national agricultural research systems (NARS), universities of the Global North and Global South, development agencies, funders, multilateral organizations, the private sector, and civil society organizations. Many of these partners directly serve beneficiaries, and without them, development impact would be impossible.

Despite the challenges, nuanced and thoughtful communication of research results, impacts, and outcomes is key to highlighting research findings as well as demonstrating to partners the value of investing in impact assessment. IFPRI publishes a wide range of impact-focused blogs, brochures, and briefs that synthesize lengthy technical reports for decision-makers. These communication efforts have won several awards. Most recently, IFPRI researchers were twice awarded the Quality of Communication Award from the Agricultural and Applied Economics Association (2021 and 2023) for coverage and analysis of how the COVID-19 pandemic the and Russia–Ukraine war negatively impacted global food security.

Case studies of impact

A forthcoming assessment of IFPRI's 50 years of impact (Hazell and Place 2025) concludes that IFPRI has been successful as a research institution in terms of its publications and international recognition. It then uses case studies to illustrate IFPRI's deeper policy impact. This chapter adopts the same approach, summarizing case studies of impact covering different themes, impact pathways, and geographies, including portfolios of work across Africa, Latin America, and South Asia (see also Chapters 19–24 on regional impacts). These cases illustrate how policy research can shape both development and research methodologies.

Addressing sustainability and climate change

IFPRI's work has provided knowledge, modeling assessments and tools, and policy guidance that have influenced sustainability and climate change efforts globally (see Section 2 of this volume). This body of work has helped countries commit to reducing emissions, trained communities to manage scarce natural resources, and led to the design of holistic approaches that can be integrated into policies.

IMPACT model for climate change mitigation in Colombia. In the early 1990s, IFPRI developed the International Model for Policy Analysis of Agricultural Commodities and Trade (IMPACT) to help policymakers explore sustainable ways to address long-term challenges in reducing hunger and poverty under changing demographic and climate conditions. The IMPACT model has since been used many times to guide transformations of agrifood systems under climate change (see Chapter 4). In Colombia, for example, IFPRI researchers used the model to identify agricultural policy options that supported climate change mitigation and resilience while maintaining compatibility with national development goals. The Colombian government used the results to develop and submit its Intended Nationally Determined Contribution in 2015, which committed the country to slashing emissions by 20 percent by 2030 (UNFCCC 2015). Using IMPACT, IFPRI and the International Fund for Agricultural Development (IFAD) found that of all Viet Nam's crops, rice held the greatest potential for climate change mitigation, accounting for 57 percent of the country's agricultural greenhouse gas emissions (Yu et al. 2013). IFPRI presented a Low Emissions Development Strategy (Salas et al. 2012), which took Viet Nam's national growth plans for rice production into account. The strategy was endorsed by the government.

Water-energy-food-ecosystem nexus in the Niger River Basin. The Niger Water Basin Authority (NBA) coordinates development projects across the nine West African countries that share the river basin, which faces challenges from extreme water insecurity and civil strife. To support the NBA in better understanding possible trade-offs and synergies among its 350 projects (Autorité du Bassin du Niger 2017), which include a range of investments and funding sources, IFPRI collaborated with GIZ (Deutsche Gesellschaft für Internationale Zusammenarbeit, the German development agency) and Canada's University of Ottawa to create tools and guidelines for a systematic water-energy-food-ecosystem nexus analysis in 2018 (see Chapter 5). After several workshops and further capacity building on nexus analysis methodologies, in 2022 the NBA's ministers formally adopted a nexus guidance policy, which is anticipated to benefit 160 million people, across 2.23 million square km, who rely on the basin (Mounkaila et al. 2023).

Experimental games for sustainable water management in India. In 2012, IFPRI partnered with the Foundation for Ecological Security and the International Crops Research Institute for the Semi-Arid Tropics in India to enhance sustainable water management and collective action for dwindling groundwater supplies (see Chapter 6) by using experimental games and learning tools. Participating communities were more likely to alter water-use decisions and water infrastructure maintenance than nonparticipating communities (Meinzen-Dick et al. 2018). As a result of these outcomes, the approach was scaled up to 6,500 rural communities across six Indian states (Falk et al. 2023). In addition, Atal Bhujal Yojana, the government-run groundwater management program, subsequently adopted the games approach across seven Indian states.

Supporting smallholder farmers

The Institute has long supported smallholder farmers by identifying ways to improve extension services; collaborating with breeding partners within CGIAR, government agencies, and NARS to improve uptake of crop varieties; and partnering with the private sector on innovative insurance and credit products (see Section 3).

Agricultural extension in Africa and Asia. From 2016 to 2021, IFPRI worked with the U.S. Agency for International Development (USAID), the public and private sector, and civil society to draft recommendations to improve service delivery and extension program operations (see Chapter 8). The Developing Local Extension Capacity (DLEC) project was implemented by Digital Green, an NGO, in partnership with IFPRI and several other organizations, reaching 1.3 million farming households and influencing more than 77 other partners across 17 countries, including Bangladesh, Guinea, Honduras, Liberia, Malawi, Mali, Mozambique, Nigeria, Rwanda, and Senegal. In Bangladesh, Digital Green worked to bridge gaps in neglected services and provide farmers with financial, market, and extension support, boosting incomes by 15 percent (Digital Green Trust 2019). In Nigeria, a video-enabled approach doubled the number of processors being reached with information, which helped improve the quality and quantity of milk and reduced milk spoilage by 40 percent (USAID and IFPRI n.d.).

Nutrition and public health through biofortified crops. HarvestPlus has rapidly scaled up the production and consumption of biofortified staple crops (iron pearl millet and bean; vitamin A cassava, orange sweet potato, and maize; and zinc maize, rice, and wheat) while also working to improve farmer uptake (see Chapter 9) and consumer acceptance. By 2023, about 3,000 scaling partners had helped release more than 443 varieties of biofortified crops in more than 40 countries (HarvestPlus 2023). HarvestPlus served as the technical partner for Enhancing Nutrition Services to Improve Maternal and Child Health in Africa (ENRICH), a five-year program that aimed to reduce maternal and child mortality by addressing malnutrition in the first 1,000 days of life. From 2016 to 2021, ENRICH delivered a series of gender-responsive interventions to train smallholder farmers on the cultivation of biofortified staple crops. ENRICH is estimated to have directly improved the diets of more than 1 million people in farming families in Bangladesh, Kenya, and Tanzania, and yielded nutritional benefits for another 1 million people through communitywide sharing of seeds, vines, and agronomic benefits.

Improving the lives of vulnerable people

IFPRI has a long history of working to transform the lives and livelihoods of vulnerable people around the world, with these efforts represented by a wide range of causal impact evaluations (Section 4).

Productive Safety Net Program (PSNP) in Ethiopia. In 2005, the government of Ethiopia launched its PSNP (see Chapter 11) to support populations facing chronic food insecurity, poverty, and shocks. IFPRI played the primary role in monitoring and evaluating the program's effectiveness, as well as supporting continued engagement, financial support, and coordination among donors (EDRI and IFPRI 2014). IFPRI's targeted recommendations included encouraging the government to strengthen the program's nutrition objectives, such as through temporary direct support payments to pregnant women. As of 2024, the PSNP had reached 8 million beneficiaries, making it one of the largest safety net programs in sub-Saharan Africa (Hirvonen 2018).

Nutrition Embedded Evaluation Program in Malawi. The Nutrition Embedded Evaluation Program was established to use community-based childcare centers in Malawi as a platform to promote child health and nutrition. The University of Malawi and Save the Children provided training on nutrition and agricultural production and distributed seeds of nutritious crops to households. IFPRI worked with Save the Children to undertake an impact evaluation of the program, which showed that preschool children receiving the integrated nutrition and agriculture intervention experienced greater increases in nutrient intake and dietary diversity. Younger siblings of participating children experienced better growth (height-for-age z-scores), resulting in a 17 percentage point difference in the prevalence of stunting between the participating group and the control group. As a result of this evaluation, Malawi's government collaborated with the World Bank in 2018 to scale up the intervention with a $60 million investment, to reach over 700,000 beneficiaries.

Women's Empowerment in Agriculture Index (WEAI). WEAI is a standardized metric to directly measure women's empowerment and inclusion in the agriculture sector (see Chapter 14). As of March 2025, WEAI tools have been used by 279 organizations across 69 countries (IFPRI WEAI n.d.). In Bangladesh, USAID used the WEAI as part of a baseline survey that found 75 percent of rural women were disempowered in the Feed the Future Zone of Influence (the geographic area targeted by Feed the Future interventions). As a result of these findings, the government worked with Feed the Future to refocus its efforts and create more opportunities for women to control productive assets and hold leadership positions in their communities. A follow-up study conducted in 2015 showed that these changes led to significant improvements in women's empowerment, with the share of empowered women rising from 27.4 percent to 41.2 percent (USAID 2016).

Focusing on enabling environments

IFPRI's work around the world allows it to focus on the enabling environments needed to transform food systems, including factors related to political economy, governance, finance, and innovation (see Section 5).

Local government accountability in Uganda. In Uganda, the government initiated a program in 2009 to improve the quality of public service delivery by providing beneficiaries with locally led monitoring and accountability forums known as *barazas* to share information (see Chapter 15). At the government's request, IFPRI conducted an evaluation of the program in 2015, finding barazas to be a valuable investment that directly contributed to finishing delayed projects, correcting substandard work, and realigning priorities to meet the needs of local citizens and communities. The final report was used as a key input for the Government Evaluation Facility Roadmap, and the prime minister's office adopted IFPRI's recommendations to prioritize sectors such as health and agriculture in subcounty baraza programs, build citizens' capacity in planning and budgeting, and produce budgetary information in local languages (CGIAR 2020).

Modeling work in response to crisis. IFPRI has used its modeling work to provide guidance to governments all over the world, as well as multilateral and civil society organizations, in response to global crises such as the COVID-19 pandemic and Russia's invasion of Ukraine. Data from the IFPRI-facilitated Food Security Portal's Food Crisis Risk Monitoring and Vulnerability analysis led the Group of Twenty (G20) and World Trade Organization to call for swiftly phasing out food export restrictions imposed in response to the pandemic and the war in Ukraine. In addition, the Myanmar Agriculture Development Bank used IFPRI's modeling results to inform its decision to allocate US$430 million in loans to farmers to offset the economic impact of COVID-19 (CGIAR 2021).

Building individual and institutional capacity

IFPRI's work goes beyond producing knowledge and testing innovations: through programs around the world, the Institute has been significantly involved in building individual and institutional capacity for 50 years.

In the first decade of IFPRI's existence, it built its knowledge base through engagement with visiting researchers who offered extensive policy expertise on their home countries. Capacity strengthening focused on individuals, including, for example, through training on modeling, data analysis, and research methods. Visiting researchers learned about the latest research methods, while IFPRI's staff gained an understanding of national policy challenges. Partnerships with universities in the Global North also provided a source of human capital and opportunities for the exchange of research methods and tools. Soon after, IFPRI researchers began to collect original data from household surveys, with local researchers and analysts collaborating in these efforts. In the 1980s, the Institute launched extensive research and policy communication activities in select countries. Country programs were established in Bangladesh, Ghana, Malawi, Mozambique, and Pakistan, with a focus on the capacity of local researchers (Babu 2000; Ryan 1999). In Pakistan, for example, datasets collected by IFPRI researchers were used by the Pakistan Institute of Development Economics and resulted in more than 100 master's and doctoral theses.

Establishment of regional programs. The 1990s saw the establishment of regional programs in East Africa, Central Asia, South Asia, and West Africa. In the early 2000s, more country programs emerged, all systematically assessing and serving unique capacity needs. Individual trainings have remained a hallmark throughout the Institute's history: in 2022, IFPRI held 350 events that trained more than 10,600 individuals. In the early 2000s, special thematic programs such as the Regional Strategic Analysis and Knowledge Support System (ReSAKSS) in Africa, Policy Analysis and Advisory Network for South Asia, and African Growth and Development Policy Modeling Consortium (AGRODEP) in West Africa became flagship networks for capacity strengthening to support policy analysis and showcased an approach to building institutional capacity for sustainable development. When IFPRI merged with the International Service for National Agricultural Research (ISNAR) in 2005, the Institute added another core approach—strengthening national systems of research and extension—enabling it to reach out to national food, agricultural, health, and nutrition systems. It also introduced a strong decentralization strategy that greatly increased the number of outposted staff worldwide: this move transformed IFPRI's demand-led research goal into a reality (IFPRI WEAI n.d.). Today, IFPRI's capacity development activities are often amplified by global capacity development activities carried out by the World Bank, Food and Agriculture Organization of the United Nations, the International Fund for Agricultural Development, and World Food Programme, universities, and regional programs such as the Association of Southeast Asian Nations (ASEAN), Bay of Bengal Initiative for Multi-Sectoral Technical and Economic Cooperation (BIMSTEC), and the African Union.

BOX 2.1 Achievements in capacity strengthening in China and India

IFPRI's policy research collaboration with institutions in China began when it welcomed visiting researchers from the Chinese Academy of Agricultural Sciences (CAAS) in the 1980s and 1990s. This engagement led to the establishment of the Center for Chinese Agricultural Policy (CCAP) at the Institute of Agricultural Economics and Development at CAAS, modeled after IFPRI. IFPRI collaborators who were focused on reforms in Chinese agriculture subsequently moved CCAP to the Chinese Academy of Sciences (CAS) and Beijing University, where it remains a key policy think tank advising the Chinese government on agriculture. Collaboration with Chinese institutions also involved training of PhD students, often by IFPRI researchers visiting China (Jikun Huang, Peking University, personal communication, August 2024). Researchers from CAAS, CAS, and various universities in China have also collaborated with IFPRI researchers at the Institute's headquarters, which eventually led to the establishment of IFPRI's country office on the CAAS premises. These collaborations continue to provide timely policy advice for national agricultural consultations and local capacity building.

In the 1980s, IFPRI also received visiting researchers from India, who were interested in gathering empirical evidence on the aftermath of the Green Revolution. Researchers engaged in one-on-one collaboration, joint report writing, and more (Hazell and Ramasamy 1991), and IFPRI's model of policy research and outreach subsequently inspired creation of India's National Institute for Agricultural Policy (Dr. Dayanand Jha, National Institute for Agricultural Policy, personal communication, June 1995). Following regular training and capacity strengthening activities with state agricultural universities, the Indian Council for Agricultural Research Institutions, and other policy think tanks, India's then-prime minister invited IFPRI to establish a regional office in New Delhi to serve the South Asia region. Over the last 40 years, policy researchers visiting IFPRI from India have made key contributions to India's food and agriculture policy and strategy development.

Agricultural Science and Technology Indicators. Agricultural Science and Technology Indicators (ASTI), a program of IFPRI between 2010 and 2020, represents another success in building system-level capacity. ASTI played a key role in providing indicators that were not previously available on agricultural research and development in the Global South. Its data encouraged Ethiopia to double the salary of its agricultural researchers and Kenya to form the Kenya Agricultural and Livestock Research Organization (KALRO). ASTI also supported research that backed the

BOX 2.2 Achievements in capacity strengthening in Africa

IFPRI has been undertaking capacity-strengthening activities in Africa since its founding, in collaboration with partners at the country, regional, and continental levels. This work has helped build robust policy frameworks and address food security and poverty reduction. IFPRI played a key role in the creation of the African Association of Agricultural Economics (AAAE), founded in 2004, providing technical assistance and financial support to AAEA's initial meetings and workshops, and working alongside the Food and Agriculture Organization, World Food Programme, and other partners to connect African economists with global thought leaders. The Association now works to foster knowledge exchange, policy research, and professional development among agricultural economics and development experts in Africa and beyond. IFPRI continues to actively participate and contribute to the Association's biennial conferences by presenting research, participating in discussions, and engaging with stakeholders. IFPRI researchers have also served on the board of the Association's flagship *African Journal of Agricultural and Resource Economics.*

Between 2006 and 2020, IFPRI also facilitated the Regional Strategic Analysis and Knowledge Support System (founded in 2006), the African Growth and Development Policy (AGRODEP) Modeling Consortium (2010), and the Malabo Montpellier Panel (2017). In 2020, these programs transitioned to AKADEMIYA2063, a pan-African research organization established by African researchers who previously led the programs at IFPRI. AKADEMIYA2063 now houses more than 50 staff from 16 countries deployed across eight strategic departments. A fourth program, Africa Agriculture Watch, was also launched by the organization, and uses machine learning and remote-sensing data to predict production for 10 key crops as well as tracking greenhouse gas emissions across Africa. Having successfully supported the transition of the program to African leadership, IFPRI has continued to be a key partner of AKADEMIYA2063 since 2020, representing a case of successful localization for sustainable impact for African countries.

creation of the Agricultural Market Information System (AMIS), influenced Egypt's food security and water strategy, and prompted global donors to invest US$7.5 billion in Yemen. This work helped to underpin other commitments to boosting agricultural research and development such as the Comprehensive Africa Agriculture Development Program (CAADP), an Agenda 2063 initiative that helps African countries eliminate hunger and reduce poverty by increasing economic growth through agriculture-led development.

Recommendations for improving research policy impact

In a world in turmoil and a development landscape marked by declining funding for foreign aid, policy research organizations are under increased pressure to show that they are fulfilling their stated missions. So how can they improve their impact?

IFPRI's investments in assessing its own impact have yielded valuable insights on this question (Hazell and Place 2025). Several independent impact assessment reports have pointed to the importance of aligning policy research with demand and maintaining relationships with policymakers in influencing policies (for example, English and Renkow 2007; Bennett 2013; Lynam 2016). IFPRI had already recognized these concerns and embarked on a process of decentralization; the assessment of the effectiveness of IFPRI's decentralization policy (Benin et al. 2018) found that this has arguably been the most important action taken by the Institute to increase its impact. Close partnerships on the ground can also enable clear pathways for building the capacity of institutional partners to undertake research that has a better chance of being utilized for policymaking.

Another important lesson that emerges for enhancing policy influence is the need to articulate a deliberate strategy of influence from the beginning of a research project. From the outset, it is essential to ensure that there is demand within the country for the research to be undertaken and that a sense of research project "ownership" exists within the policy circles the research is designed to influence (Bennett 2013). Being part of an impact pathway or decision-making framework that is at least partly driven by an external actor such as a government or funding agency provides the clearest avenue for IFPRI research to influence decisions. In these cases, IFPRI uses ex ante models to assess the likely impacts of policy or investment choices or rigorously test an intervention or program (Lowder 2025; Lowder and Regmi 2019; Somwaru 2021). Several external evaluations also point to the importance of proactively communicating evidence to different audiences (for example, Kydd 2015). Therefore, IFPRI outreach and communication strategies are also critical for success and need to be articulated early in a research project.

A more foundational factor in generating impact is conducting and maintaining a strong reputation for high-quality science. As noted earlier, IFPRI has been able to consistently rank highly among peers in its thematic areas. Areas where IFPRI has earned such a reputation include its modeling tools (for example, the IMPACT global model and RIAPA country models), causal impact evaluation work, collection and analyses of household datasets, and policy and political analysis. The Institute has also developed a solid reputation in specific thematic areas, as noted in the other chapters in this volume. Potential new areas for IFPRI to influence policy are discussed in Chapter 1. The ever-improving capability of NARS and universities in LMICs will enable IFPRI to enter into a wider array of collaborations and potentially generate more international public goods. Artificial intelligence could be viewed as a source of potential competition for policy research, especially for certain types of analyses based on existing evidence. However, it can also be harnessed to add value to IFPRI's existing assets—for example, to its suite of modeling tools—as long as proper processes are instituted to ensure quality and academic integrity.

When an impact has been achieved, it also needs to be assessed and documented. IFPRI's experience highlights how policy research organizations can measure their impact and overcome impact assessment challenges for the future. Strong long-term partnerships with funders can help address the costs of high-quality impact assessment within an increasingly fraught financial landscape. Together, organizations and funders can forge a culture of impact, in which the project design phase includes joint planning for impact evaluation at various levels of investment. Similarly, funders can be asked to accommodate flexible funding cycles that include impact assessments to be undertaken—either by the research organization or a third party—after the project and its traditional core funding have ended. Organizations such as IDInsight have created standardized metrics for evaluating the social contribution of programs, based on breadth (the number of beneficiaries), depth (impact on beneficiaries' quality of life), and contribution (the extent to which the evidence shaped a program). These metrics allow for cross comparison and identification of cumulative institution-level impact results. The issue of limited funding can be addressed by systematically measuring only the impact of projects an organization considers to be large.

Impact assessments should be undertaken with the intention of generating scientific nuance and with a sense of humility. Impact assessments sometimes identify failures, which can provide valuable insights on how programs and investments can change course to improve outcomes. When they do identify successes, policy research organizations can use this opportunity to celebrate the achievements of their partners who work on the ground.

IFPRI's five decades of impact underline the importance of policy research in shaping global research and development agendas, testing whether innovations and interventions can and should be scaled up, and advancing knowledge through partnerships and capacity strengthening. As the world grapples with climate change, conflict, and geopolitical tensions, it is only high-quality science that can help decision-makers overcome misinformation and arrive at policies, programs, and investments that safeguard the health of our planet and improve the lives of vulnerable people everywhere.

References

Autorité du Bassin du Niger. 2017. "Plan Opérationnel 2016–2024." www.abn.ne

Babu, S.C. 2000. *Impact of IFPRI's Policy Research on Resource Allocation and Food Security in Bangladesh.* Independent Impact Assessment Report 13. Washington, DC: IFPRI.

Barrett, C.B. 2022. *Socio-Technical Innovation Bundles for Agri-Food Systems Transformation.* London: Palgrave Macmillan.

Benin, S., F.M. Place, and P.B.R. Hazell. 2018. *Has IFPRI's Research Decentralization Strategy Made a Difference? An Econometric Study of African and Asian Countries, 1981–2014.* Independent Impact Assessment Report 44. Washington, DC: IFPRI.

Bennet, J.W. 2013. "An Ex-Post Impact Assessment of IFPRI's GRP22 Program, Water Resource Allocation: Productivity and Environmental Impacts." Impact Assessment Discussion Paper 35. IFPRI, Washington, DC.

CGIAR. 2020. *The Government of Uganda Adopts PIM Recommendations to Improve a Nationwide Program Aimed at Involving Citizens towards Effective Public Service Delivery.* CGIAR Research Program on Policies, Institutions, and Markets. Montpellier, France.

CGIAR. 2021. *Reporting 2021 Evidences.* Study 4077. CGIAR Research Program on Policies, Institutions, and Markets. Montpellier, France.

Digital Green Trust. 2019. *Digital Green: Annual Report 2018–19.* New Delhi.

EDRI (Ethiopian Development Research Institute) and IFPRI. 2014. *The Impact of Research on the Productive Safety Net Programme.* Washington, DC: IFPRI.

English, J., and M. Renkow. 2007. "The Impacts of IFPRI's Global Research Program on the Sustainable Development of Less-Favored Areas." Impact Assessment Discussion Paper 26. IFPRI, Washington, DC.

Falk, T., R. Sanil, R. Meinzen-Dick, and P. Priyadarshini 2023. "Changing the Game: Experiential Learning for Triggering Large Scale Change towards Sustainable Water Management in India." *IFPRI Blog,* November 10.

Haggblade, S., and D. Resnick. 2018. *Conceptualizing Key Drivers of Policy Change. An Introduction to the Kaleidoscope Model.* Feed the Future Innovation Lab for Food Security Policy Research Brief 61. East Lansing, MI: Feed the Future Innovation Lab for Food Security, Michigan State University.

HarvestPlus. 2023. *20 Years of Impact for a Food Systems Transformation. HarvestPlus 2023 Annual Report.* Washington, DC: IFPRI.

Hazell, P.B.R., and F. Place. 2025. *Taking Stock: Impacts of 50 Years of Policy Research at IFPRI.* Washington, DC: International Food Policy Research Institute. Forthcoming.

Hazell, P.B.R., and C. Ramasamy. 1991. *The Green Revolution Reconsidered.* Food Policy Statement 14. Washington, DC: International Food Policy Research Institute.

Hirvonen, K. 2018. "Improving the Resilience of the Poor in Ethiopia through Community-Based Health Insurance." *IFPRI Blog,* June 5.

IFPRI WEAI (Women's Empowerment in Agriculture Index). n.d. "Countries Using WEAI." Accessed April 1, 2025. https://weai.ifpri.info/weai-resource-center/countries-using-weai/

Kydd, J. 2015. *Ex-Post Evaluation Study of IFPRI's Research on High-Value Agriculture, 1994–2010.* Independent Impact Assessment Report 39. Washington, DC: IFPRI.

Lowder, S. 2025. *External Assessment of the Impact of IFPRI's Causal Impact Evaluation Research 2012–2022.* Washington, DC: IFPRI.

Lowder, S.K., and A. Regmi. 2019. *Assessment of Outcomes Based on the Use of PIM-Supported Foresight Modeling Work, 2012–2018.* Washington, DC: IFPRI.

Lynam, J. 2016. *Balancing International Public Goods and Accountability: Exploring the Impact of IFPRI's Policy Research on Science, Technology and Innovation.* Independent Impact Assessment Report 43. Washington, DC: IFPRI.

Meinzen-Dick, R.S., M.A. Janssen, S. Kandikuppa, R. Chaturvedi, K. Rao, and S. Theis. 2018. "Playing Games to Save Water: Collective Action Games for Groundwater Management in Andhra Pradesh, India." *World Development* 107: 40–53.

Mounkaila, N., O. Seidou, C. Ringler, and R. Kranefeld. 2023. "World Water Day 2023: Cooperation for Accelerating Change in the Niger River Basin." *IFPRI Blog,* March 22.

Place, F., and P.B.R. Hazell. 2015. *Workshop on Best Practice Methods for Assessing the Impact of Policy-Oriented Research: Summary and Recommendations for the CGIAR.* Independent Impact Assessment Report 41. Washington, DC: IFPRI.

Ryan, J. G. 1999. *Assessing the Impact of Policy Research and Capacity Building by IFPRI in Malawi.* Independent Impact Assessment Report 11. Washington, DC: IFPRI.

Salas, W., C. Li, P. Ingraham, M. Van Trinh, D.T. Anh, N.N. Mai, and C. Ringler. 2012. *National-Level Crop Mitigation Potential for Key Food Crops in Vietnam.* Washington, DC: IFPRI and International Fund for Agricultural Development.

Somwaru, A. 2021. *Assessment of the Use of Outputs from PIM Supported Work on National SAMs and CGE Models.* Washington, DC: IFPRI.

UNFCCC (United Nations Framework Convention on Climate Change). 2015. "Colombia iNDC." July 9.

USA for UNHCR (The UN Refugee Agency). 2024. "How Climate Change Impacts Refugees and Displaced Communities." *News*, November 12.

USAID (U.S. Agency for International Development). 2016. 2016 *Progress Report: Growing Prosperity for a Food-Secure Future.* Washington, DC.

USAID and IFPRI. n.d. *Feed the Future: Developing Local Extension Capacity (DLEC).* Accessed January 8, 2025.

Yu, B., T. Zhu, C. Breisinger, and N. Manh Hai. 2013. "How Are Farmers Adapting to Climate Change in Vietnam?: Endogeneity and Sample Selection in a Rice Yield Model." IFPRI Discussion Paper 1248. IFPRI, Washington, DC.

Chapter 3

AGRIFOOD SYSTEMS
Transformation, Structural Change, and Development

Xinshen Diao, Margaret McMillan, Karl Pauw, and James Thurlow

Xinshen Diao is a senior research fellow, Foresight and Policy Modeling Unit (FPM), IFPRI; **Margaret McMillan** is a professor of economics, Tufts University; **Karl Pauw** is a senior research fellow, FPM, IFPRI; and **James Thurlow** is director of FPM, IFPRI.

Key messages

- Historically, agricultural growth was considered necessary for development in low-income countries, a perspective that contributed to widespread use of supply-side policies to raise agricultural productivity and output.
- Asia's Green Revolution is considered a model of successful transformation that led to increased rural incomes, greater availability and access to food, and unprecedented economic growth.
- Research on Asia's experience offers lessons for other low- and middle-income countries, but the experience of sub-Saharan Africa shows the challenges of replicating the Green Revolution's successes. Evidence from the region highlights how the role of agriculture in structural change and economic development is evolving as low-income countries become more integrated into global markets, and their consumers become more urbanized and consume more diverse diets.
- Research also underscores the need for policymakers to adopt a broader agrifood systems approach as the agriculture sector becomes integrated with the broader agrifood system and the development mandate of agriculture expands.
- This approach will require innovative tools and indicators to build understanding, such as IFPRI's Agricultural Transformation Index, which comparatively assesses a country's level of agricultural transformation and tracks progress over time.

Moving forward, research on development economics should focus on:

- **Informing strategies** for the design and implementation of multifaceted and tailored interventions that address constraints to agricultural and agrifood systems transformation in different environments and contexts.
- **Identifying labor-saving technologies** that can help achieve agricultural productivity and environmental sustainability, as well as ensuring these technologies are suitable and adaptable to local environments and contexts.
- **Supporting a systems approach** to transformation that "looks beyond the farm," including by evaluating how different actors respond to investment incentives and how policies can ensure positive outcomes for both producers and consumers.
- **Building understanding of how to promote better dietary outcomes**, both through economic policies that change the behavior of producers, processors, and consumers and behavioral change policies that address cultural and social drivers of dietary choices.

Historically, agricultural transformation has been critical to improving access to food, reducing poverty, and stimulating economic growth. Over the last 50 years, these contributions helped lead to the widespread use of supply-side agricultural policies to raise agricultural productivity and output. In the modern context, however, the role of agriculture in structural change and economic development is evolving as low-income countries experience socio-demographic changes, characterized by population growth in urban areas and consumption of more diverse diets, as well as economic integration with global markets.

This chapter examines agrifood systems transformation within the broader process of structural change and development. While *agricultural* transformation is often the initial catalyst for economic development, *agrifood systems* transformation is a broader concept that extends beyond farm-level improvements to encompass (1) better integration of the on-farm (primary agriculture) and off-farm (processing and distribution) components of agrifood value chains, (2) market development and expansion of agriproducts (see Barrett et al. 2022), and (3) policy interventions that shape agrifood systems outcomes, such as food security, nutrition, equity, and environmental sustainability (see Ingram and Thornton 2022).

Agricultural productivity growth, rural labor shifts, and evolving food value chains play a major role in economic transitions. This is exemplified by Asia's Green Revolution, which serves as a model of a successful agricultural transformation. This chapter contextualizes these dynamics and explores how lessons from Asia's experience can inform strategies in other regions, particularly sub-Saharan Africa.

Evolutions in development economics research

Understanding the drivers, trends, and structural patterns of growth has long been a cornerstone of development economics research, and theories of growth and development have evolved substantially as a result of this work. In the 1950s, growth was understood as a process of labor moving from a "traditional" low-productivity agriculture sector to a "modern" high-productivity manufacturing sector (Lewis 1954), known as structural change. By contrast, the Solow model (Solow 1956) emphasizes the accumulation of human and physical capital and ultimately offers exogenous technological progress as an explanation for long-run growth. Most researchers now tend to view growth as an outcome of both, that is, structural transformation or the movement of labor from low- to high-productivity sectors, as well as productivity growth within sectors (McMillan and Rodrik 2011).

Regardless of the theoretical framework, productivity growth in the agriculture sector is widely acknowledged as a catalyst for broader economic development and a shift toward industrialization (De Janvry and Sadoulet 2019). Such productivity growth is usually associated with the transition that takes place within low-income countries' agriculture sectors as they develop from low-productivity, subsistence-oriented agriculture toward modern, commercially oriented agriculture, a process also called agricultural transformation.

Figure 3.1 depicts the typical phases of this transformation process as it is often described in the literature. The adoption of modern production technologies—such as high-yielding varieties of crops and livestock products, other inputs, and agricultural machinery—leads to growth in agricultural productivity and usually provides the initial impetus for agricultural transformation. Further development and expansion of the sector is fostered by market development and rural infrastructure investments, which facilitate increased diversification and commercialization, both on the farm and, eventually, in the off-farm components of agrifood value chains.

FIGURE 3.1 Four components of agricultural transformation

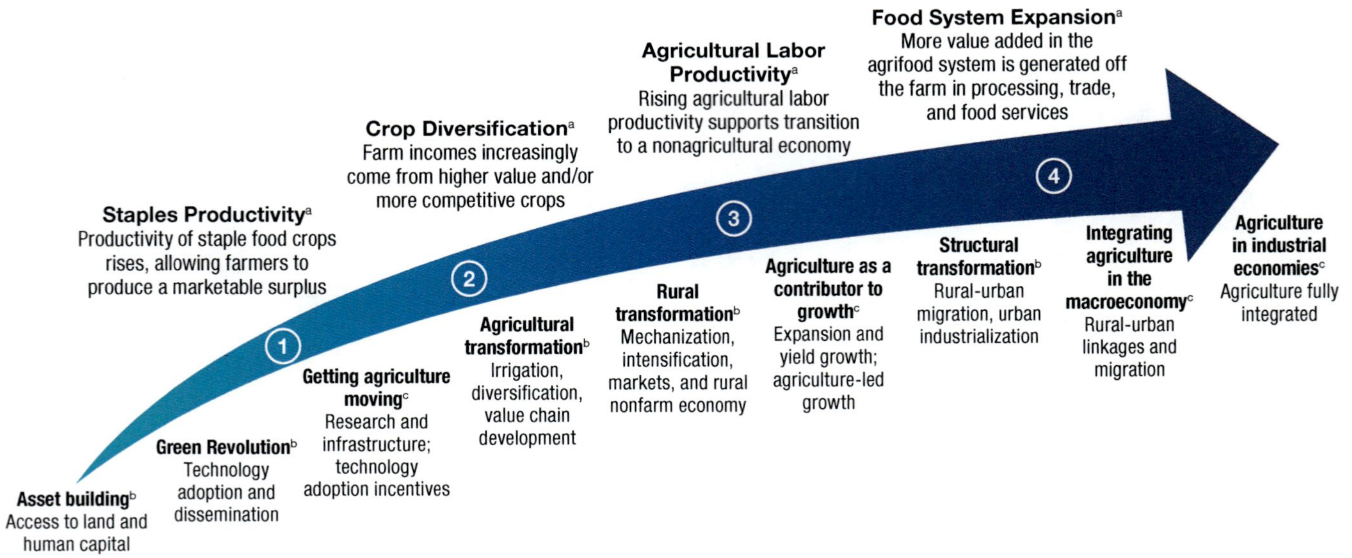

Source: Adapted from Diao et al. (2024).

Note: (a) The top part of the figure shows four stylized components of agricultural transformation as proposed by Diao et al. (2024). A discussion of the Agricultural Transformation Index (ATI) developed by Diao et al. (2024) follows later in the chapter. The ATI is constructed from indicators representing each of the four components of agricultural transformation and can be used to assess the advancement of countries in agricultural transformation. (b) The bottom part of the figure shows how the framework compares to De Janvry and Sadoulet's (2019) five stages of agriculture modernization and (c) Timmer's (1988) four stages of agricultural transformation.

Agricultural transformation in Asia

Perhaps the most-cited example of successful agricultural transformation is the Green Revolution, which spanned from the mid-1960s throughout the 1990s in Asia. At the start of this period, agriculture sectors in most Asian countries were dominated by smallholder farmers. The high population density and relative scarcity of land necessitated or "induced" the development of land-saving technologies—that is, increasing production per unit of land—to meet demand (Hayami and Ruttan 1985). The period saw a rapid rise in productivity on smallholder farms, partly due to the adoption of high-yielding cereal varieties, especially rice and wheat, many of which were developed and disseminated by CGIAR (Hardin 2010). The effectiveness of new varieties was further enhanced by the increased use of modern inputs, including chemical fertilizers that raised soil fertility; pesticides, herbicides, and fungicides that lowered yield losses from weeds, pests, and crop diseases; the increased integration of livestock production for traction, manure, and transport, and as an off-cycle source of cash to pay for other inputs; and irrigation infrastructure that reduced dependence on seasonal rainfall (see Pingali 2012).

Asia's Green Revolution led to growth in rural incomes and increased availability and access to food in both rural and urban areas. As higher incomes drove demand for more diverse products, the region's agriculture sector gradually became more diversified, and rural nonfarm sectors started to develop as rural households diversified their livelihood strategies. This rural nonfarm development further supported agricultural transformation by reducing the size of the labor force engaged in agriculture, thereby raising output per worker in the sector.

As agricultural and rural development progressed, many East Asian countries adopted manufacturing-led, export-oriented growth strategies to build on the success of agricultural transformation. These strategies provided much-needed foreign exchange, promoted learning and innovation, and accelerated structural change by creating employment outside of agriculture, particularly in manufacturing (Birdsall et al. 1993; Stiglitz 2018). The unprecedented economic growth that followed—dubbed the Asian Miracle—helped close the gap between per capita incomes in East Asian countries and those in more advanced economies in the Western world. The success of the region's growth strategies led development economists to promote manufacturing-led, export-oriented growth strategies as a development pathway for other low-income countries to emulate.

Can the Asian experience be replicated in sub-Saharan Africa?

Since the start of the Green Revolution more than 50 years ago, total cereal production in low- and middle-income countries has tripled. Remarkably, this growth in output was supported by an increase in agricultural land of less than one-third. During this same period, population size doubled in low- and middle-income countries. The Green Revolution therefore brought about large increases in both agricultural land productivity and the amount of food produced per capita (Pingali 2012).

For low- and middle-income countries in sub-Saharan Africa, however, the experience was vastly different. Although optimism abounded as many of these countries started gaining independence from colonial rule in the 1950s and 1960s, just as the Green Revolution began in Asia, agricultural growth stagnated for several decades. By the turn of the century, most countries in sub-Saharan Africa produced less food per capita than they did in the 1960s (Frankema 2014).

Many theories attempt to explain why sub-Saharan Africa did not experience a Green Revolution of its own during the early postcolonial era (see discussion in Frankema 2014 and Rosenberg 2014). Certainly, climate conditions in the tropics prevented the rice and wheat technologies that were successful in Asia from being readily transferred to sub-Saharan Africa. Even today, agricultural technologies are often highly specific to the ecological characteristics of the countries that led their development (Moscona and Sastry 2025). The food crops introduced instead, such as maize and cassava, were suited to local conditions and crucial for improving food security in the region, but tended to be of lower value, with limited opportunities for off-farm processing and value addition.

Colonial powers further encouraged reliance on a few traditional export crops such as cocoa, coffee, cotton, or tobacco, which were often highly profitable. Although external capital helped finance transport infrastructure, these developments were concentrated in cash crop production zones and sometimes crowded out economic activity in other regions. Furthermore, restrictive trade policies and vertical integration with overseas processing facilities meant products were mostly exported in unprocessed form. As a result, cash crops were associated with limited diversification and agricultural transformation in sub-Saharan African countries (Roessler et al. 2022). On the institutional front, weak bureaucracies in many of the newly independent states led to poor policy coordination and limited support for—or even bias against—agriculture, which, together with a lack of land tenure security for smallholders, poor infrastructure, and weak markets, resulted in farmers adopting strategies to minimize risk rather than maximize income or productivity.

In the early 2000s, sub-Saharan Africa's fortunes began to change. The region finally experienced its own African Growth Miracle as its economic growth outpaced that of the rest of the world (Diao et al. 2025). During this same period, the structure of sub-Saharan African economies changed as the share of employment in agriculture fell. Although these structural shifts explain most of the region's labor productivity growth in the first two decades of the 21st century, much of the off-farm employment growth was in low-productivity service sectors, while employment growth in the manufacturing sector remained stagnant, reflecting a tendency for relatively productive firms to adopt capital-intensive rather than labor-absorbing manufacturing processes (Diao et al. 2025). This trend is not necessarily unique to sub-Saharan Africa, as evidence shows manufacturing is becoming more skill- and capital-intensive globally (Rodrik 2016). As a result, manufacturing is unlikely to play the same role in income and employment growth in sub-Saharan Africa that it did in South and East Asia following the Green Revolution (AfDB 2024; Rodrik and Stiglitz 2024; Stiglitz 2018).

Within the agriculture sector, output per worker increased due to the rising value of agricultural commodities that accompanied a global surge in commodity prices, along with a decline in the size of the agriculture workforce. However, this recorded growth in labor productivity may simply have been a result of labor shifting from farm to off-farm work due to urbanization. Agricultural land productivity increased modestly in some sub-Saharan African countries, but the pace of growth was slower than the overall increase in agricultural output per worker (see Badiane et al. 2021). Without the land productivity growth seen in Asia during the Green Revolution, agricultural growth in sub-Saharan Africa in the post-2000s was simply not as transformational as it had been in Asia between the 1960s and 1990s.

More recent evidence from farm household surveys conducted in several sub-Saharan African countries paints a more somber picture of declining crop yields and negative total factor productivity growth during the 2008–2019 period (Wollburg et al. 2024), suggesting continued challenges for agricultural transformation in the region. As during the postcolonial era, multiple factors explain these challenges, including credit, insurance, and information constraints; high transaction costs; imperfect land and labor markets; and increased competition in globally integrated product markets. Different combinations of constraints seem to affect different farmers (Suri and Udry 2022), meaning that there is no simple solution to address the myriad challenges—instead, tailored packages of interventions will be required. In practice, achieving this will be difficult, especially without greater clarity on which elements of multifaceted programs are essential in different contexts or environments.

The stark contrast between the Asian and sub-Saharan African experiences, together with the ongoing challenges facing sub-Saharan Africa in transforming agriculture, has led many development economists to question the feasibility of replicating the Green Revolution. As a result of structural change and urbanization, sub-Saharan African countries are generally less dependent on agriculture today than the economies of South and East Asia were in the 1960s. African consumers are also becoming less dependent on local food supplies as economies become more integrated into global food value chains. These trends increase the attractiveness of non-agricultural policies and investments to promote economic development. But does this mean that agriculture should be deprioritized?

A broader agrifood systems perspective is imperative

Despite some pessimism about its prospects, the agriculture sector remains vitally important in sub-Saharan Africa. Compared to the rest of the world, the region has the highest share of employment in agriculture, and the sector represents a large share of gross domestic product (GDP) for many of its countries (Gollin 2023; Suri and Udry 2022). Agriculture should therefore remain a key element of development strategies for sub-Saharan African countries and for low-income countries in general (Stiglitz 2018). However, these development strategies should be multifaceted, and primary agriculture cannot be the only focus. In the past, agricultural development policy largely focused on technology adoption on the farm. Though this focus remains important, we now understand that dramatic transformations of entire agrifood value chains occur alongside both agricultural and industrial revolutions (Barret et al. 2022). This new perspective requires agricultural policymakers to look "beyond the farm" and adopt a more holistic agrifood systems approach to transformation that recognizes the importance of supporting the integration of both on-farm and off-farm components of agrifood value chains. An agrifood systems perspective also considers how transformation shapes not only economic outcomes, such as growth and structural change, but also agrifood systems outcomes, such as food security, nutrition, equity, and environmental sustainability.

To elaborate, a broader agrifood systems perspective is imperative for several reasons: first, systemwide investments in agrifood value chains can strengthen linkages and facilitate the transfer of knowledge and skills between their on-farm and off-farm components. While traditional development strategies often regarded on-farm productivity growth as a necessary precondition for developing off-farm components of the agrifood system, innovations in off-farm components often induce the adoption and diffusion of agricultural technologies on the farm (Barrett et al. 2022). For instance, growing demand for workers in rural off-farm activities creates incentives for the adoption of labor-saving technologies on the farm. Thus, transformation cannot be seen as a linear development pathway along which on-farm productivity growth strictly precedes growth beyond the farm.

Second, urbanization, income growth, and higher participation of women in the labor force are contributing to rapid changes in consumer preferences, particularly the increasing demand for higher quality, more processed, and more varied foods (Barret et al. 2022). The consumption baskets of today's African consumers do not contain any commodities that match the importance of rice and wheat for Asian consumers during the Green Revolution (Gollin 2023). This means on-farm investments in one or two staple crops alone are insufficient and must be complemented by investments in a wider range of agricultural products, as well as in food processing capacity and food distribution services to supply changing diets. Food environments—the places and contexts in which consumers access food—also evolve as more consumers rely on supermarkets, restaurants, or fast-food chains instead of their own food production and home cooking.

Facilitating investments in retail markets and restaurants may help formalize and strengthen agrifood value chains so that they are better equipped to respond to consumer demand. For example, contracting arrangements or the use of mobile phone applications can help link farmers, traders, food service providers, and consumers, thus ensuring more efficient production and delivery of food products demanded by consumers. The trade, transport, and food services sectors are all nontradable sectors with potentially large domestic employment and growth multiplier impacts, which, alongside the expansion of agro-processing sectors, can contribute directly to structural change–led growth and employment. Global evidence confirms that the shares of employment and value added in the off-farm components of agrifood systems increase relative to the share of on-farm components as countries transition from low-income to middle- or high-income status (Thurlow et al. 2025).

Third, although the opportunities for low-income countries to adopt export-oriented manufacturing-led growth strategies are now more limited than in the past, there may be scope to expand agro-processing—a subsector within manufacturing—and successfully compete domestically and regionally with food imports. A recent United Nations report finds that the average food trade deficit among least-developed countries is 50 percent, with large trade deficits for processed foods, particularly in sub-Saharan Africa (UN 2024). However, by the time developing nations move above the least-developed-country threshold, their processed food exports exceed imports by 23 percent. Although these statistics mask substantial heterogeneity across countries, they are broadly indicative of the scope for low-income countries to gradually reduce their reliance on processed food imports through development and transformation of their agrifood systems. A starting point could involve investing in small-scale and traditional food processing sectors or foods tailored to local and regional demand, such as fruits and vegetables, meat and dairy, grain milling, or snack food products. Promoting regional trade to substitute for imports of processed foods from outside sub-Saharan Africa can help smaller countries within the region expand their market for processed food products. An expansion in agro-processing capacity will also likely be accompanied and facilitated by what is described as the "quiet revolution" in food value chains, that is, transformation and growth among midstream intermediaries, such as truckers, warehouse owners, millers, cold storage operators, wholesale traders, and rural brokers (Reardon 2015) (see also Chapter 7).

The fourth point relates to the changing role of the agrifood system in contributing to household welfare outcomes. During the early stages of development, investing in primary agriculture is often an effective strategy for reducing poverty because the livelihoods of many of the poor depend on the sector. However, structural change and rural–urban migration cause those agricultural growth–poverty linkages to weaken (Christiaensen et al. 2011; Dorosh and Thurlow 2018). For many low-income households employed outside of agriculture, particularly the rural poor, agricultural productivity growth can still reduce the cost of primary foods that often make up the bulk of their budgets. However, as lifestyles change and processed or prepared foods become relatively cheaper, these low-income households may also undergo a dietary transition, in which case off-farm agrifood systems investments could gradually become more effective at lowering food costs and increasing household income opportunities from employment in off-farm agrifood sectors. An appreciation for these employment and consumption dynamics, as they pertain to low-income households, can help governments adopt on-farm and off-farm agrifood systems policies that better address development goals, such as poverty reduction.

Last, views on agrifood systems transformation are being shaped by the growing consensus on the importance of promoting sustainability in agrifood systems. These systems should no longer exclusively aim to supply cheap calories to poor consumers but should instead be transformed to promote a wider range of goals, including not only food security, nutrition, human health, and equity but also environmental sustainability (Caron et al. 2018). Put differently, agrifood systems investments must find a balance between making healthy, diverse diets available and affordable to poor consumers and ensuring that foods are produced in an environmentally sustainable way, such that the food security and nutrition of future generations are not compromised (Willett et al. 2019) (see Chapters 5 and 12).

Within this broader context, governments may opt for a wide range of investments across agrifood systems to achieve multiple goals (see Chapter 18). For instance, the underconsumption of plant-based protein sources such as beans and pulses might require traditional on-farm investments in production and marketing channels; the underconsumption of fruits and vegetables in urban markets might best be addressed through off-farm investments in logistics and cold chains; or the underconsumption of milk might require development of dairy processing sectors (see Pauw et al. 2023). On the other hand, the overconsumption of starchy or ultra-refined grains might require reforms to existing staple food subsidy programs to encourage consumption of healthier foods while still ensuring the overall affordability of diets.

Tracking agrifood systems transformation

Agrifood systems are impacted by multiple external drivers, including biophysical, environmental, political, economic, and demographic ones (Fanzo et al. 2020). The challenge faced by policymakers is to design policies that create incentives to produce enough healthy and affordable food for a growing consumer base in a sustainable manner. Because the global agrifood system is already operating outside of several planetary boundaries (Ruggeri Laderechi et al. 2024), urgent transformation of agrifood systems is required.

Agricultural transformation will remain an important element of broader agrifood systems transformation, even though, as argued in this chapter, future contributions of agriculture to broader structural change and development processes will likely differ from the past. While agricultural productivity has always been the gold standard for tracking progress within agriculture, new indicators are required to measure and monitor broader transformation of the agrifood system. One such tool is the Agricultural Transformation Index (ATI) proposed by IFPRI researchers (Diao et al. 2024).

The ATI combines indicators for key *economic outcomes* pertaining to agrifood systems transformation into a composite progress score ranging from 0 to 1, with a higher score indicating a greater degree of transformation. It is designed as a practical indicator for tracking the transformational process of agrifood systems within and across all countries around the world, one which can be easily updated using publicly available longitudinal datasets. The ATI is consistent with the framework presented in Figure 3.1 and includes components that measure progress in (1) staples productivity, (2) crop diversification, and (3) agricultural labor productivity. These three indicators all pertain to the on-farm component of the agrifood system. A fourth indicator, (4) food system expansion, measures the share of total value added in the agrifood system that is generated off the farm (for details, see Diao et al. 2024).

Figure 3.2 presents the default ATI scores and changes over time for different country income groups as an illustration of how the indicator can be used to assess progress. Panel A shows that the ATI increases from low- to high-income countries. Although each component contributes roughly equally to the composite score in high-income countries, the contribution of the crop diversification indicator in low-income countries, for instance, is notably small. Panel B confirms the strong positive correlation between countries' ATI scores and their levels of well-being, as measured by GDP per capita, although there is a significant degree of dispersion of ATI values within country income groupings, and especially among lower-middle-income and upper-middle-income countries.

Panel C shows the point changes in the ATI from 2000 to 2021, the range of datapoints currently available. On average, all country groups experienced increases in their ATI scores, with the largest gains in absolute terms in the lower-middle-income (+0.13) and upper-middle-income (+0.09) groups. Agricultural labor productivity growth was the most important contributor to the change in ATI among low-income countries, while staples productivity contributed most to the change in all other country groups, thus supporting the assertion that agricultural transformation is not strictly a linear modernization sequence (Diao et al. 2024). Whereas increased crop diversification had some impact on ATI scores in middle- and high-income countries, this indicator had a negative impact on ATI scores in low-income countries.

FIGURE 3.2 Agricultural Transformation Index (ATI) composition, correlation, and trends

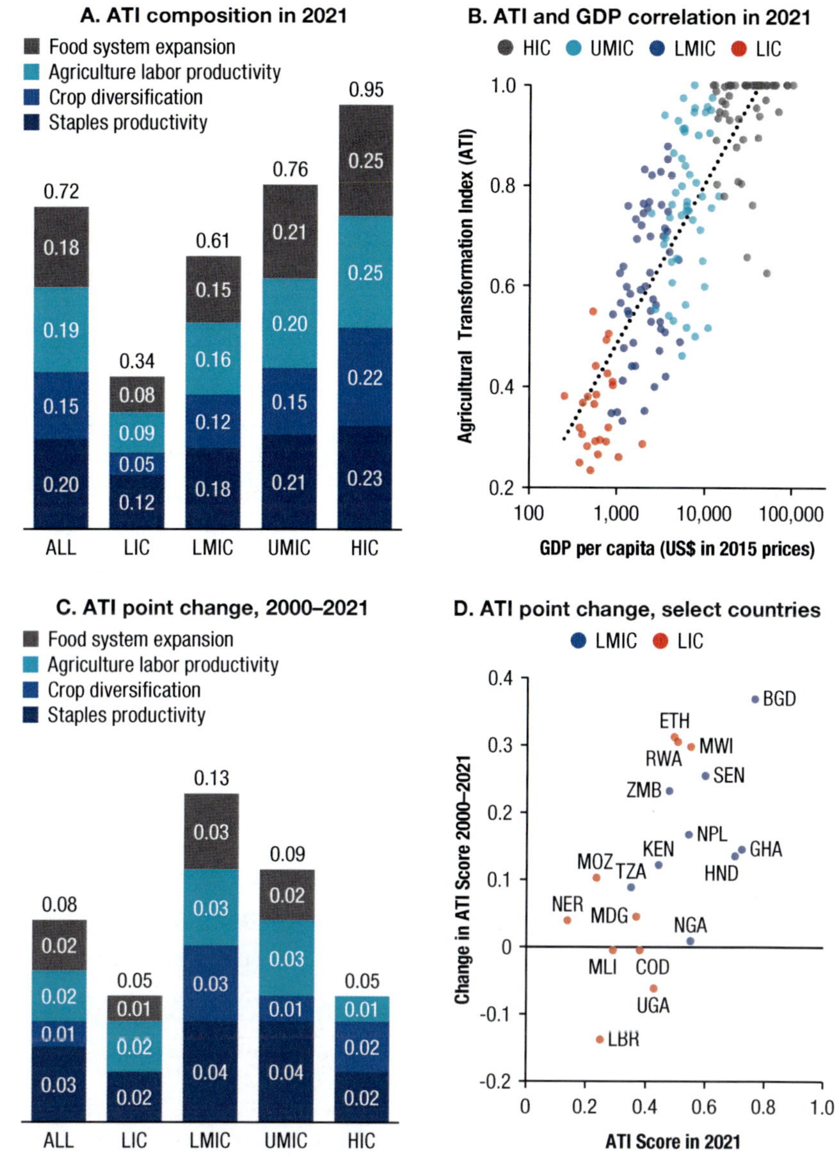

Source: Diao et al. (2024).

Notes: LIC = low-income countries; LMIC = lower-middle-income countries; UMIC = upper-middle-income countries; HIC = high-income countries. The countries included in Panel D are selected developing countries including BGD = Bangladesh; COD = Democratic Republic of Congo (DRC); ETH = Ethiopia; GHA = Ghana; HND = Honduras; KEN = Kenya; LBR = Liberia; MDG = Madagascar; MLI = Mali; MOZ = Mozambique; MWI = Malawi; NER = Niger; NGA = Nigeria; NPL = Nepal; RWA = Rwanda; SEN = Senegal; TZA = Tanzania; UGA = Uganda; ZMB = Zambia.

Panel D presents changes in ATI scores for select countries. These countries display both a wide variation in ATI scores—ranging from 0.14 in Niger to 0.76 in Bangladesh—and in changes in ATI scores from 2000 to 2021. The countries with deteriorating scores, such as the Democratic Republic of Congo, Liberia, Mali, and Uganda, are all low-income countries, although Ethiopia, Malawi, and Rwanda, also low-income countries, were some of the best performers. The average improvement across all countries in the figure was +0.12, with slightly smaller improvements among low-income countries (+0.09) than among lower-middle-income countries (+0.17).

Given its focus on economic outcomes, the ATI does not currently incorporate indicators pertaining to *agrifood systems outcomes* such as food security, nutrition, equity, or sustainability. One of the more prominent efforts to measure, assess, and track agrifood systems outcomes at a global scale is the Food Systems Countdown Initiative (FSCI), which focuses on indicators across five thematic areas, namely (1) diets, nutrition, and health, (2) environment, production, and natural resources, (3) livelihoods, poverty, and equity, (4) governance, and (5) resilience (FSCI 2025). Given the complexity of the indicator architecture, the FSCI makes no attempt at forming a composite index or uniquely ranking countries as is done with the ATI. However, with the availability of this comprehensive database, it is worth exploring opportunities for expanding the scope of the ATI in the future to include selected agrifood systems outcomes, particularly those that are collected in a consistent way across countries and deemed comparable.

Future research

Agriculture's role in structural change and economic development is evolving. Agricultural transformation was historically seen as a necessary precondition for development in low-income countries, a perspective that contributed to widespread use of supply-side agricultural policies to raise agricultural productivity and output. Today, however, low-income countries are more integrated into global markets, and their consumers are more urbanized and consume more diverse diets, an evolution that affects both the supply and demand sides of agrifood systems. Other emerging factors, such as a high burden of noncommunicable disease and global climate and environmental challenges (see Chapters 4, 5, and 12), are also affecting agrifood systems, thereby broadening the development mandate of agriculture (Byerlee et al. 2009). Beyond traditional goals such as improving access to food, reducing poverty, and stimulating growth and structural change, the agriculture sector is now an integrated part of a broader agrifood system that is also expected to contribute to improved diet quality and health outcomes while remaining within planetary boundaries of sustainable production.

This expanded mandate requires agricultural policymakers to adopt a broader "agrifood systems approach" to promoting transformation and development. This approach is characterized by (1) a recognition of the interdependencies across different parts of agrifood value chains; (2) an appreciation for agrifood systems' dynamic nature and role in development as production or processing technologies and patterns of food demand evolve; and (3) an understanding that the way in which agrifood systems transform has important social and environmental implications, both today and for future generations. Policies designed with this broader agrifood systems perspective can help strengthen integration within agrifood value chains and improve the transfer of knowledge and skills among different components of agrifood systems; enable agrifood systems to better meet changing consumer preferences for higher-quality and more diverse foods; compete with and substitute for food imports; and simultaneously address multiple development challenges, including reducing poverty and food insecurity, improving diet quality and health outcomes, and facilitating the adoption of environmentally sustainable production systems.

Research has an important role to play, first and foremost, in tracking agrifood systems transformation and continuously expanding knowledge and understanding of its drivers. In this regard, indicators such as IFPRI's ATI can provide useful information about a country's level of agricultural transformation in relation to other countries and its progress over time. The ATI can also highlight specific components of transformation that may be lagging, allowing policymakers to prioritize certain interventions over others, especially in contexts where resources are limited. While the ATI focuses on economic outcomes, global databases such as that of the FSCI (2025) create opportunities for expanding the ATI to incorporate agrifood systems outcomes such as food security, nutrition, equity, or sustainability.

Drawing on the wide-ranging discussions in this chapter, we conclude by highlighting select areas for future research where knowledge gaps are most pressing. First, as we have discussed, many constraints to agricultural and agrifood systems transformation have been identified, but since different agrifood systems actors are affected by different combinations of these constraints, multifaceted and tailored programs are required to address them in different environments and contexts. These packages of interventions are complex and require implementation by capable practitioners, though research can inform their design and implementation strategies.

Second, agricultural land productivity remains low in sub-Saharan Africa, which has encouraged the adoption of expansive agricultural production systems to increase output. However, with high population growth and growing land and water constraints, this aim is becoming harder to achieve. Rapid rural–urban migration further contributes to rural labor shortages, which means the labor-intensive agricultural technologies promoted during Asia's Green Revolution are less attractive and profitable to sub-Saharan African smallholders today. Labor-saving technologies that simultaneously raise agricultural land productivity can offer a win-win strategy by increasing agricultural productivity and environmental sustainability (see Chapter 17). Research can help identify what those technologies look like and ensure their suitability and adaptability to local environments and contexts.

Third, we have argued that an agrifood systems perspective on transformation requires policymakers to look beyond the farm and adopt policies that also support the development of the off-farm components of agrifood value chains (see Chapter 7). Market incentives are important drivers for private sector actors to invest in these off-farm components. These actors are also often more diverse than their counterparts in primary agriculture. For instance, they may range from small-scale processors or traders to large-scale investors or multinational food processing companies, which differ not only in size or economic influence, but also in political influence. Research can offer insight into how different actors respond to investment incentives and how policies can ensure outcomes that are equitable for all producers—and, ultimately, that support consumers' health and well-being.

Last, the need to shift consumer preferences toward healthy and sustainably produced foods may be the most challenging aspect of agrifood systems transformation. Poor diet quality is a major contributor to noncommunicable diseases and related deaths, yet the consumption of ultra-processed foods high in sugar and fat is becoming more pervasive. This reflects, to a large extent, the high cost of healthy diets compared to energy-sufficient diets and convenience foods (see Chapter 12) (Herforth et al. 2022). More work is needed to identify which strategies are most effective at lowering the overall cost of food, but, beyond that, research can help us understand how policies can change the behavior of producers, processors, or consumers to ultimately promote better dietary outcomes. For instance, price subsidies or production-side incentives could reduce the relative cost and increase the availability of healthy foods. However, the extent to which economic policies can change consumer behavior is limited. Therefore, understanding the cultural and social drivers of human diets will also be important to inform the behavioral change policies that are required along with economic policies.

References

AfDB (African Development Bank Group). 2024. "Taking Stock of Africa's Economic Transformation Progress." In *African Economic Outlook 2024: Driving Africa's Transformation—The Reform of the Global Financial Architecture*, 87–139. Abidjan.

Badiane, O., X. Diao, and J. Jayne. 2021. "Africa's Unfolding Agricultural Transformation." In *Agricultural Development: New Perspectives in a Changing World*, eds. K. Otsuka and S. Fan, 153–192. Washington, DC: IFPRI.

Barrett, C.B., T. Reardon, J. Swinnen, and D. Zilberman. 2022. "Agri-Food Value Chain Revolutions in Low- and Middle-Income Countries." *Journal of Economic Literature* 60 (4): 1316–1377.

Birdsall, N., J.E.L. Campos, C. Kim, et al. 1993. *The East Asian Miracle: Economic Growth and Public Policy*. World Bank Policy Research Report. Washington, DC: World Bank Group.

Byerlee, D., A. de Janvry, and E. Sadoulet. 2009. "Agriculture for Development: Toward a New Paradigm." *Annual Review of Resource Economics* 1: 15–31.

Caron, P., G. Ferrero y de Loma-Osorio, D. Nabarro, et al. 2018. "Food Systems for Sustainable Development: Proposals for a Profound Four-Part Transformation." *Agronomy for Sustainable Development* 38 (41): 1–12.

Christiaensen, L., L. Demery, and J. Kuhl. 2011. "The (Evolving) Role of Agriculture in Poverty Reduction—An Empirical Perspective." *Journal of Development Economics* 96 (2): 239–254.

De Janvry, A., and E. Sadoulet. 2019. "Transforming Developing Country Agriculture: Removing Adoption Constraints and Promoting Inclusive Value Chain Development." Working Paper 253. Foundation for Studies and Research on International Development, Clermont-Ferrand, France.

Diao, X., M. Ellis, M. McMillan, and D. Rodrik. 2025. "Africa's Manufacturing Puzzle: Evidence from Tanzanian and Ethiopian Firms." *World Bank Economic Review* 39 (2): 308–340.

Diao, X., E. Jones, K. Pauw, J. Thurlow, and W. Xu. 2024. "The Agricultural Transformation Index." IFPRI Discussion Paper 2275. IFPRI, Washington, DC.

Dorosh, P., and J. Thurlow. 2018. "Beyond Agriculture versus Non-Agriculture: Decomposing Sectoral Growth–Poverty Linkages in Five African Countries." *World Development* 109: 440–451.

Fanzo, J., L. Haddad, R. McLaren, Q. Marshall, C. Davis, A. Herforth, and A. Kapuria. 2020. "The Food Systems Dashboard Is a New Tool to Inform Better Food Policy." *Nature Food* 1 (5): 243–246.

Frankema, E. 2014. "Africa and the Green Revolution: A Global Historical Perspective." *NJAS: Wageningen Journal of Life Sciences* 70–71 (1): 17–24.

FSCI (Food Systems Countdown Initiative). 2025. *The Food Systems Countdown Report 2024: Tracking Progress and Managing Interactions.* New York: Columbia University; Ithaca, NY: Cornell University; Rome: Food and Agriculture Organization of the United Nations (FAO); Geneva: Global Alliance for Improved Nutrition (GAIN).

Gollin, D. 2023. "Agricultural Productivity and Structural Transformation: Evidence and Questions for African Development." *Oxford Development Studies* 51 (4): 375–396.

Hardin, L. 2010. "CGIAR Centers and Their Role in the Green Revolution." In *The International Dimension of the American Society of Agronomy: Past and Future,* eds. B. Payne and J. Ryan, 19–22. Madison, WI: American Society of Agronomy.

Hayami, Y., and V.W. Ruttan. 1985. *Agricultural Development: An International Perspective,* 2nd ed. Baltimore: Johns Hopkins University Press.

Herforth, A., A. Venkat, Y. Bai, L. Costlow, C. Holleman, and W.A. Masters. 2022. "Methods and Options to Monitor the Cost and Affordability of a Healthy Diet Globally." Background paper for *The State of Food Security and Nutrition in the World 2022.* FAO Agricultural Development Economic Working Paper 22–03. FAO, Rome.

Ingram, J., and Thornton, P. 2022. "What Does Transforming Food Systems Actually Mean?" *Nature Food* 3: 881–882.

Lewis, W.A. 1954. "Economic Development with Unlimited Supplies of Labour." *The Manchester School* 22 (2): 139–191.

McMillan, M., and D. Rodrik. 2011. "Globalization, Structural Change, and Productivity Growth." NBER Working Paper 17143. National Bureau of Economic Research, Cambridge, MA.

Moscona, J., and K.A. Sastry. 2025. "Inappropriate Technology: Evidence from Global Agriculture." NBER Working Paper 33500. National Bureau of Economic Research, Cambridge, MA.

Pauw, K., O. Ecker, J. Thurlow, and A.R. Comstock. 2023. "Measuring Changes in Diet Deprivation: New Indicators and Methods." *Food Policy* 117: 102471.

Pingali, P.L. 2012. "Green Revolution: Impacts, Limits, and the Path Ahead." *PNAS* 109 (31): 12302–12308.

Reardon, T. 2015. "The Hidden Middle: The Quiet Revolution in the Midstream of Agrifood Value Chains in Developing Countries." *Oxford Review of Economic Policy* 31 (1): 45–63.

Rodrik, D. 2016. "Premature Deindustrialization." *Journal of Economic Growth* 21 (1): 1–33.

Rodrik, D., and J.E. Stiglitz. 2024. "A New Growth Strategy for Developing Nations." Essay prepared for the IEA-ERIA Project on the New Global Economic Order. Harvard University, Boston; Columbia University, New York.

Roessler, P., Y.I. Pengl, R. Marty, K.S. Titlow, and N. Van de Walle. 2022. "The Cash Crop Revolution, Colonialism and Economic Reorganization in Africa." *World Development* 158: 105934.

Rosenberg, T. 2014. "A Green Revolution, This Time for Africa." *New York Times*, April 9.

Ruggeri Laderchi, C., H. Lotze-Campen, F. DeClerck, et al. 2024. *The Economics of the Food System Transformation: Global Policy Report*. Food System Economics Commission (FSEC).

Solow, R.M. 1956. "A Contribution to the Theory of Economic Growth." *The Quarterly Journal of Economics* 70 (1): 65–94.

Stiglitz, J.E. 2018. "From Manufacturing-Led Export Growth to a Twenty-First Century Inclusive Growth Strategy: Explaining the Demise of a Successful Growth Model and What to Do About It." In *Inequality in the Developing World*, eds. C. Gradín, M. Leibbrandt, and F. Tarp, 284–318. Oxford, UK: Oxford University Press.

Suri, T., and C. Udry. 2022. "Agricultural Technology in Africa." *Journal of Economic Perspectives* 36 (1): 33–56.

Thurlow, J., B. Holtemeyer, S. Jiang, K. Pauw, and J. Randriamamonjy. 2025. "Measuring Agrifood Systems: New Indicators and Global Estimates." IFPRI Discussion Paper 2339. IFPRI, Washington, DC.

Timmer, P. 1998. "The Agricultural Transformation." In *Handbook of Development Economics*, vol. 1, 275–331. Cambridge, MA: Elsevier.

UN (United Nations). 2024. *Trade in Processed Food*. New York: United Nations Publications.

Willett, W., J. Rockström, B. Loken, M. Springmann, T. Lang, S. Vermeulen, and C.J. Murray. 2019. "Food in the Anthropocene: the EAT–Lancet Commission on Healthy Diets from Sustainable Food Systems." *Lancet* 393 (10170): 447–492.

Wollburg, P., T. Bentze, Y. Lu, C. Udry, and D. Gollin. 2024. "Crop Yields Fail to Rise in Smallholder Farming Systems in Sub-Saharan Africa." *PNAS* 121 (21): e2312519121.

SECTION II

Sustainability and Food Systems

Chapter 4

CLIMATE CHANGE
Understanding Impacts on Agrifood Systems and Evaluating Policy Options

Mark W. Rosegrant, Elizabeth Bryan, Timothy S. Thomas, and Keith Wiebe

Mark W. Rosegrant is a research fellow emeritus, IFPRI; **Elizabeth Bryan** is a research fellow, Natural Resources and Resilience Unit, IFPRI; **Timothy S. Thomas** and **Keith Wiebe** are senior research fellows, Foresight and Policy Modeling Unit, IFPRI.

Key messages

- As climate change became a central focus of food policy research, research expanded from projecting global and national impacts on agriculture to assessing risks to households and individuals and options for policy responses, and from a focus on mitigation toward actions that support both mitigation and adaptation.
- Model-based projections broadly agree on the key agricultural impacts of climate change: declining productivity growth and nutrient density of crops and reduced food security, with the greatest impacts at lower latitudes and in low-income countries.
- Research focused on identifying and reducing risks for vulnerable populations, particularly small-scale producers and women, identifies "hot spots" and effective adaptation strategies.
- Farm-level adaptations show promise for protecting livelihoods and food security but have not been adopted at scale. Research explores the barriers to greater impact of these innovations.
- Climate-smart agriculture research emphasizes synergies among adaptation, mitigation, and productivity goals at the farm level, and shows that climate-smart practices can reduce hunger.
- Evidence on disparities in climate impacts between high- and low-income countries and within countries, where impacts are greater on marginalized peoples, supports policies for climate justice.

Looking to the future, policy research on climate and food systems should aim to:

- **Improve identification of hotspots and targeting of resources** by combining yield projections from crop models with information on the local importance of crops and local indicators of vulnerability, including social inequalities.
- **Increase capacity for modeling** the frequency and impact of extreme climate events and better incorporate feedback mechanisms, impacts on labor productivity, economic growth, and nutrition, to allow comparison of the effectiveness of potential policy responses.
- **Identify and help to prioritize scalable adaptation strategies** that can be effective both at large and local scales.
- **Identify effective mechanisms to address climate injustice** and ways to make climate governance more representative of vulnerable communities, including women and poor populations.
- **Combine climate modeling and household survey analysis** at the country level to provide policy-relevant foresight and evidence to decision-makers as countries undergo major socioeconomic and environmental changes in coming years.
- **Inform the design and implementation of climate interventions and policies** that better address broader development goals such as food security, nutrition, and health.

Climate change is a major challenge of our time, with global and far-reaching—though often uncertain—effects on and from agriculture and food systems. Rising temperatures and changing weather patterns are already increasing the risk of hunger, malnutrition, and food insecurity among the world's poorest and most vulnerable populations. Natural disasters such as droughts and flooding are becoming more frequent and intense, while climate change makes arable land and freshwater scarcer and agricultural productivity harder to increase (Parry et al. 2009). New or more damaging pests, pathogens, and weeds will challenge farmers. Agriculture is also a major contributor to greenhouse gas (GHG) emissions. Thus, policies and investments for the agrifood sector must support both the goals of reducing GHG emissions and sequestering carbon (Mbow et al. 2019; Smith et al. 2018) and increasing farmers' resilience through adaptation to climate change and income growth.

However, as it becomes less likely that the world will meet its 2015 Paris Agreement goal of keeping the global temperature increase below 1.5°C, the risks to agriculture and difficulty of adaptation are growing rapidly, especially in low-income countries.

Climate change has rapidly shifted from a peripheral concern to a central focus of food policy research. Early research was primarily model based, with a strong focus on projecting potential global and national impacts of climate change on agrifood systems and examining adaptation policies and priorities in low- and middle-income countries (LMICs) and GHG mitigation policies in wealthier countries. This research has now expanded to encompass a wider array of methods, with a strong focus on assessing the risks and vulnerabilities faced by diverse households and individuals and their options for responding to climate shocks and stressors.

This chapter reviews the evolution of research on climate change, food security, and food systems and reflects on IFPRI's major contributions to understanding climate change impacts and identifying promising policies and investments for mitigation and adaptation. Looking ahead, we explore promising new research directions, including advancing understanding of climate risks and vulnerabilities to support policy prioritization that will facilitate low-emission development pathways and avoid maladaptation—that is, actions that increase vulnerability to climate change or result in adverse outcomes or diminished welfare for some populations.

Evolution of research

Models to assess global, regional, and national impacts of climate change

Several types of models are used to assess the impacts of climate change on agricultural production, food prices, and food security, including econometric models and simulation models. These all show broad areas of agreement on the key impacts of climate change, although specific quantitative results can vary substantially.

Econometric models estimate the impact of climate change on crop yields based on an analysis of the relationship between historical precipitation and temperature, on the one hand, and crop yields or total production on the other hand. Some models extend this analysis to consider impact on total factor productivity (TFP). Econometric analysis of TFP provides a more comprehensive measure of impact than crop yields, as it captures interactions between changes in outputs and inputs, including not only land but also capital, labor, and technology, throughout the entire sector (see, for example, Lobell, Bänziger et al. 2011). Other research considers the impact of climate on land values or farm net revenues (such as Mendelsohn et al. 1994, 1996).

Simulation models include both process-based crop models and global economic models.[1] *Process-based crop models* model physiological processes of crop growth and incorporate environmental factors such as temperature, water, and light, as well as soil parameters, cultivar choices, and agronomic management practices, to estimate the impact of climate variables (and therefore climate change) and other factors on crop yields (Feng et al. 2023). Prominent crop models include DSSAT (Jones et al. 2003) and the APSIM series (Keating et al. 2003), both developed by universities and research organizations.

Global economic models that address climate change impacts, including partial and computable general equilibrium models (CGEs), combine economic models of the agriculture sector (or multiple sectors, in the case of CGEs) with crop models and climate system models known as general circulation models (GCMs).

GCMs are numerical models representing physical processes in the atmosphere, ocean, cryosphere, and land surface, and are used to simulate changes in temperature, precipitation, and other variables over time. They are the most advanced tools currently available for simulating the response of the global climate system to increasing GHG concentrations. Global economic models also employ different assumptions about Shared Socioeconomic Pathways (SSPs), which represent alternative futures for economic and population growth, with implied Representative Concentration Pathways (RCPs), which represent potential GHG emission levels in the atmosphere and the subsequent increase in solar energy that would be absorbed (radiative forcing).

Global economic models can capture the economic feedback effects generated by climate shocks and estimate their impacts, not only on production but also on food demand, prices, and thus food security. For example, IFPRI's International Model for Policy Analysis of Agricultural Commodities and Trade (IMPACT)[2] is a system of models that links information from climate models, crop simulation models, and water

1 See De Salvo et al. (2013) and Feng et al. (2023) for detailed descriptions of these types of models.

2 www.ifpri.org/project/ifpri-impact-model/

models with a core global, partial equilibrium, multimarket model focused on the agriculture sector (see Robinson et al. 2024). It links to the GLOBE CGE model to capture whole-economy impacts of climate change. IMPACT relies on GCMs produced by ISIMIP (Inter-Sectoral Impact Model Intercomparison Project) and based on CMIP (Coupled Model Intercomparison Project) data created to inform the reports of the International Panel on Climate Change (IPCC).

Finally, it is important to note that IMPACT and similar modeling systems consider the impacts of temperatures and precipitation that affect crop productivity, but generally do not yet account for climate variability, extreme events (droughts, floods, and heat waves), sea level rise, or shifting pest and disease patterns. As a result, the aggregate modeled impacts summarized in the next section are likely conservative.

Modeling results: Impact of climate change

Across all types of models, most analyses have focused on four major crops: maize, rice, soybeans, and wheat (although IMPACT includes a much larger set of crop and livestock commodities, including many that are important in LMICs). This reflects the importance of these crops, which provide half of total global calorie availability. The models show broad areas of agreement on the key impacts of climate change (De Salvo et al. 2013; Feng et al. 2023).

Table 4.1 summarizes the findings on the impact of climate change on crop yields, commodity prices, and TFP from modeling studies, concentrating on relatively comprehensive meta-analyses and reviews of research studies that summarize the results from many analyses, as well as two seminal individual papers on global historical yields and TFP effects that address impacts in LMICs (Lobell, Bänziger et al. 2011 and Ortiz-Bobea et al. 2021, respectively). The estimated impacts vary across models, due to differences in the models themselves and differences in assumptions on the severity of climate change, projections of economic and agricultural growth, and underlying parameters. For each cited paper in Table 4.1 that covers multiple studies, there are also differences across individual studies, resulting in a range of projections around the mean values shown.

Crop yields and total factor productivity

There is substantial agreement among these studies that climate change is already having an appreciable negative impact on crop yields and TFP, and that it will continue to affect these key outcomes negatively to 2050 and

TABLE 4.1 Impact of climate change on crop yields, prices, and agricultural total factor productivity (TFP), mean results

				Impacts of climate change				
				Without adaptation (%)			With adaptation (%)	
Type of model	Year	Global/Region	Crop	Yield	Price	TFP	Yield	Citation
Crop model	2050	Global	Maize, Rice, Soybeans, Wheat	−11 (mean) −6.2 (median)			−4.6 (mean) −1.6 (median)	Hasegawa et al. (2022) Meta-analysis, 202 studies
Crop model	Change from 1998 to 2084	Global	Maize	−24.1				Jägermeyr et al. (2021) Results from 12 crop models and 5 GCMs
			Soybeans	−2.1				
			Rice	1.7				
			Wheat	17.5				
Crop model	2050	Africa	Wheat	−17				Knox et al. (2012) Review of 52 studies
			Sorghum	−15				
			Millet	−10				
			Maize	−5				
		South Asia	Maize	−16				
			Sorghum	−11				
Econometric model * Percent of studies that find negative impacts of climate change on crop yields.	2050	Low latitudes		−86*				Huang and Sim (2016) Review of 130 studies
		Lower-middle latitudes		−71*				
		Higher-middle latitudes		−15*				
		Upper latitudes		0*				

Table 4.1 continued

				Impacts of climate change				
				Without adaptation (%)			With adaptation (%)	
Type of model	Year	Global/Region	Crop	Yield	Price	TFP	Yield	Citation
Crop models, econometric regression, and field-warming experiments	Future impact per each 1°C increase in mean global temperature	Global	Maize	–7.4				
			Wheat	–6.0				Zhao et al. (2017)
			Rice	–3.2				Multiple studies
			Soybean	–3.1				
Econometric model	1980 to 2008	Global	Maize	–3.8	18.9			Lobell, Bänziger, et al. (2011)[a]
			Wheat	–5.5	6.4			
Econometric model	1961 to 2020	Global				–21.0		Ortiz-Bobea et al. (2021)[b]
		Africa				–26 to –34		
		LAC				–26 to –34		
		North America				–15		
		Europe				–10		
Global economic model	2050	Global	Coarse grains (mainly maize)	–12.5	11.8			Nelson et al (2014)
			Wheat	–9.9	15.9			Mean results from 10 global economic models
			Rice	–9.3	13.9			

Note: (a) Other studies of this type include Miao, Khanna, and Huang (2015); Schlenker and Roberts (2009); Lobell, Schlender, and Costa-Roberts (2011); and Choi and Helmberger (1993). (b) Other studies of this type include Liang et al. (2017) and Xiang et al. (2022).

beyond. The reduced crop yields and TFP due to climate change raise crop prices and reduce food consumption, worsening food security (see below). The negative impacts tend to be greatest for maize, although not in all studies. Maize is highly sensitive to temperature fluctuations, especially during critical growth stages, and has a relatively high water requirement, making it vulnerable to drought conditions. The estimated mean global impacts of climate change on crop yields are generally in the range of –5 percent to –15 percent in 2050 compared to a no-climate change scenario. Although this may seem relatively small, these reductions can cost 5 to 10 years of crop yield growth for major commodities. As shown in Table 4.1, the impacts of climate change vary by region. Nearly all studies that assess regional differences find that losses are larger at lower latitudes, mainly in low-income countries that can least afford or adapt to these impacts.

Nutrient effects

Recent findings show that climate change also affects the nutrient content of crops. According to the IPCC, while increased atmospheric CO_2 relative to current levels is projected to boost crop yields (CO_2 fertilization), it is also expected to significantly decrease the nutritional value of crops, particularly by lowering protein and key micronutrient density (the available nutrient content per unit of grain or other commodity) in staple grains like wheat and rice (Mbow et al. 2019). For example, a meta-study by Myers et al. (2014) found that wheat grown at 546–586 ppm atmospheric CO_2, compared with wheat grown at ambient levels, has 5.9–12.7 percent less protein, 5.2–7.5 percent less iron, and 3.7–6.5 percent less zinc (Mbow et al. 2019). Projections from IMPACT of per capita availability of nutrients in 2050 show that technological advances, market responses, and CO_2 fertilization will all increase crop and livestock production and therefore global availability of dietary protein, iron, and zinc (Beach et al. 2019). However, these potential benefits are significantly offset by the adverse impacts of climate change on both agricultural productivity and nutrient density. A study that compared projections based on two different datasets found that both predict global dietary reductions in protein, iron, and zinc relative to levels associated with current CO_2 emission levels, although the size of the declines differ (–4.1 percent, –2.8 percent, and –2.5 percent, respectively, for one dataset; –2.9 percent, –3.6 percent, and –3.4 percent, respectively, for the other) (Beach et al. 2019). The combined effects of elevated atmospheric CO_2, climate change, and associated nutrient reductions are projected to slow growth in nutrient availability for protein, zinc, and iron (Beach et al. 2019). Low-income nations already facing high

BOX 4.1 Modeling to inform policies for improved food security and climate change adaptation in the Philippines

From 2014 to 2016, IFPRI and the CGIAR Research Programs on Climate Change, Agriculture, and Food Security (CCAFS) and Policies, Institutions, and Markets (PIM) partnered with the Philippines' National Economic and Development Authority (NEDA), the country's top socioeconomic planning body, to assess the impacts of climate change on agriculture and food security and to examine policies and investments for adaptation. NEDA conducts macroeconomic forecasting, policy analysis, and research, playing a crucial role in shaping policies and investments for the agriculture sector and the broader economy to address climate change challenges. The project engaged key NEDA officials early on, ensuring that outputs from analysis and modeled scenarios were relevant to policies, plans, and investments. The modeling identified investments that can improve food security under climate change as well as policies that worsened outcomes, including rice import quotas. These outputs informed the preparation of the Philippine Development Plan 2017–2022 and the 2019 Rice Tariffication Law. In 2016, training sessions were provided, focused on using IMPACT results to inform policies on investments, food security, and malnutrition. NEDA organized three sessions for 63 government staff (22 men, 41 women) to build regional capacity in modeling methodologies.

Source: CGIAR (2018).

rates of nutrient deficiencies are likely to bear the greatest burden. Regardless of the baseline scenario used to project future consumption patterns, rising atmospheric CO_2 levels will hinder progress in reducing global nutrient deficiencies. In fact, the nutrient impacts offset any benefits of CO_2 fertilization on the availability of protein, iron, and zinc (Beach et al. 2019). These findings are consistent with another study, which found that the nutrient densities of protein, iron, and zinc of major crops are likely to decline by 3–17 percent if the world reaches 550 ppm atmospheric CO_2, resulting in an additional 175 million people who are zinc-deficient and 122 million who are protein-deficient worldwide in 2050, if dietary patterns do not change (Smith and Meyers 2018).

Food security

Climate change also reduces food security. The climate-change-related reductions in agricultural yields described above will be tempered, but not fully offset, by expansion in crop areas as prices rise. The resulting net decline in agricultural production decreases food availability and further raises prices, leading to lower per capita incomes and greater hunger. By 2050, world commodity prices are projected to be 16–18 percent higher as a result of climate change, as compared with a no-climate-change scenario, making it more difficult for low-income consumers to afford healthy diets (Sulser et al. 2021). Estimates from IFPRI's IMPACT model show that climate change will slow global progress toward ending hunger, leaving 78 million more people (a 19 percent increase) chronically hungry by 2050 compared to a no-climate-change scenario (Sulser et al. 2021). More than half of this increase is expected to occur in sub-Saharan Africa.

Investments in climate adaptation

The IMPACT model was used to evaluate the costs of adapting to climate change through investments in agriculture (Sulser et al. 2021). Offsetting the projected 78 million-person increase in hunger requires an increase in annual funding for international agricultural research from US$1.62 billion to $2.77 billion between 2015 and 2050. Additional investments of $12.7 billion in water and $10.8 billion in infrastructure per year are also required, totaling $24.7 billion annually to address climate-driven hunger impacts. Other modeling studies suggest even higher costs for adaptation or mitigation to alleviate hunger; the diverse estimates reflect differences in study objectives, sectors covered, climate considerations, and methodologies (Fan et al. 2018; Mason-D'Croz et al. 2019; Rosegrant et al. 2022). For example, Germany's Center for Global Development (ZEF) and the Food and Agriculture Organization of the United Nations (FAO) estimate that between $39 billion and $50 billion per year are needed to end hunger by 2030 (ZEF and FAO 2020). However, another study suggests that an investment of $33 billion annually could not only end hunger but also double small-scale producer incomes by 2030 and keep agricultural GHG emissions aligned with the Paris Agreement targets (Laborde et al. 2020).

The impact of adaptation and mitigation investments could be enhanced, and the prospects for meeting global sustainability and development goals in 2030 increased by improvements in supporting policies and investments. These enabling conditions and specific policy approaches—including improved value chains, finance, extension, gender-responsive policies and investments, social protection, implementation of smart subsidies, and agroecological and landscape approaches—are discussed in other chapters of this report.

Assessing the vulnerability of households and individuals to climate change and options to respond

Another body of policy research examines household and individual vulnerability to climate change and explores mechanisms to increase the resilience of the poor, women, and other vulnerable populations in the face of short- and long-term climate change impacts (see also Chapters 6 and 13).

Shift toward adaptation

The impacts of climate change on agriculture are devastating for the livelihoods and well-being of small-scale producers in LMICs, who are reliant on natural resources and have limited capacity to adapt (Adger et al. 2003; Easterling et al. 2007; Morton 2007). While early efforts to address climate change emphasized the need for mitigation of GHG emissions, the IPCC's 2007 assessment report included a focus on adaptation. The report highlighted the inevitability of changes in climate that were "baked in" due to historic emissions and, thus, the urgent need for adaptation (IPCC 2007; Mertz et al. 2009; Solomon et al. 2009).

Agricultural adaptation is now a key priority in many low-income countries to protect the livelihoods and employment of large shares of the population and to reduce the risk of worsening food insecurity and hunger (Bizikova et al. 2015; UNEP 2024). In the 2010s, low-income countries began to engage in adaptation planning in a more coordinated way through the National Adaptation Plan process, recognizing that they were likely to experience the worst impacts of climate change. The process, which was established in 2010 and reinforced by the Paris Agreement in 2015 (Huq et al. 2004; Reid and Huq 2008), helped 171 countries (87 percent of parties to the UNFCCC) put adaptation plans in place by 2024, although the effectiveness of these plans is mixed and implementation often lags (UNEP 2024). Broadly speaking, key adaptation strategies emphasized in many of these plans include adjustment of planting dates, diversification of production, selection of resilient and/or stress tolerant crop varieties and animal species, improved irrigation and water management, adoption of precision agriculture and other resource conserving strategies, and expansion of climate information services.

Early policy research introduced several important concepts, including vulnerability, exposure, sensitivity, and adaptive capacity, that have guided the study and measurement of adaptation (Adger 2006; Bryan et al. 2000; Smit et al. 1999, 2001). This work identified human adaptation as a process of adjustment to long-term changes in climate (often in anticipation of these changes), as distinct from coping with short-term climate variability, in order to minimize risks and take advantage of opportunities (Smithers and Smit 1997), while recognizing that effectively coping with climate shocks in the short-term may facilitate adaptation to long-term climate changes (Burton 1997). It also described the complexity of adaptation processes, which require actions at multiple scales and by multiple actors, as well as the varied forms of adaptation responses, which can be spontaneous or planned, anticipatory or reactive, and incremental or transformative (Kates et al. 2012; Smit et al. 2001).

Identifying hot spots

As climate change has become a reality in many parts of the world, the critical role of smallholder farmers in LMICs in adapting to climate change has been widely recognized, given that smallholders (those with less than 2 ha of land) produce 30–34 percent of the global food supply (Ricciardi et al. 2018). Research focused on identifying and reducing risks, building resilience capacities, and coping and adaptation strategies for smallholders began to emerge in the early 2000s. Papers explored vulnerability to climate change and identified "hotspots" based on the level of exposure to climate change, sensitivity of agricultural production systems, communities, households, and individuals to its negative impacts, and adaptive capacity of farm households and regions (Deressa and Hassan 2009; Gbetibouo and Hassan 2005; Gbetibouo, Ringler, and Hassan 2010; Kurukulasuriya et al. 2006). More recent analyses of hotspots integrate gender, recognizing that entrenched inequalities limit women's adaptive capacity (Koo et al. 2022; Lecoutere et al. 2023).

Farm-level adaptation

Research also documented and classified autonomous farm-level adaptation to long-term climate change (that is, spontaneous actions taken by farmers) (Bryant et al. 2000; Smit 1996). This work looked at the types of strategies and innovations adopted by farmers, including new technologies (such as irrigation and new crop varieties) and management strategies (such as soil and water conservation measures, changing planting dates, or adjusting inputs to deal with climate-related challenges), and diversification on and off farm (Below et al. 2010; Jagnani et al. 2021).

Autonomous adaptation measures initiated at the farm level are important but largely incremental (as opposed to systematic and transformational). They are thus insufficient to address the challenges posed by climate change at scale without support from policies and investments to accelerate and deepen adaptation and ensure participation of many stakeholders (Adger et al. 2005; Burton and Lim 2005; Smit and Skinner 2002; Thornton et al. 2018). Research has explored the determinants of farm-level adaptation in order to identify both constraints and entry points for interventions to facilitate greater uptake (Bryan et al. 2009, 2013b; Deressa et al. 2009a; Hassan and Nhemachena 2008; Silvestri et al. 2012). The key determinants identified include access to credit and climate information, land tenure security, and wealth status. Knowledge of these can inform policies and programs for managing climate risks and facilitating agricultural sector adaptation. Examples include the development of climate information services (Ziervogel et al. 2010) and insurance and bundled financial products (see Chapter 10) (Kramer 2023; Ndegwa et al. 2020), as well as social protection programs (see Chapter 11) (Tenzing 2019). IFPRI made an important contribution to this household-level research with a focus on how smallholder farmers' perceptions of long-term climate change affect their adoption of adaptation strategies (Gbetibouo et al. 2010; Deressa et al. 2011; Maddison 2007).

Emergence of climate-smart agriculture

Research has emphasized the potential synergies between adaptation and mitigation goals and actions and encouraged more coordinated efforts to achieve both goals, as well as greater productivity (Ayers and Huq 2009; Rosenzweig and Tubiello 2007). While the early promise of carbon markets as an opportunity for low-income countries to be paid for contributing to mitigation goals in the agriculture sector has not produced the expected boon (Lipper and Zilberman 2018), this "win-win" option did increase interest in finding climate solutions that offer benefits for adaptation, mitigation, and smallholder farmer livelihoods (Bryan et al. 2010). In addition, the Paris Agreement requires countries to identify voluntary emissions reductions in the context of long-term low-emission development strategies, thus balancing mitigation and development goals (Waisman et al. 2019). This search for synergies led to the development of the climate-smart agriculture (CSA) framework, which facilitates evaluation of agricultural practices and climate change strategies across three goals—productivity (and profits), adaptation, and mitigation (FAO 2010; Lipper and Zilberman 2018). For example, in Kenya, a study identified several farm-level practices that support all three goals; these practices include integrated soil fertility management, improved livestock feeding, and irrigation and soil and water conservation in arid areas (Bryan et al. 2013a). Modeling-based approaches to assess the effectiveness of CSA practices have shown that widespread adoption of these practices at the global level can contribute to all three goals, while increasing food accessibility and reducing hunger (De Pinto, Cenacchi et al. 2020). However, CSA is not a one-size-fits-all solution. Priorities for CSA vary based on the policy context, agroecological conditions, and farming systems among other factors, and different stakeholders often have different preferences regarding CSA outcomes (Khatri-Chhetri et al. 2017; Mwongera et al. 2017). The CSA framework has also been criticized as a set of purely technical solutions that ignore power dynamics that lead to inequitable distribution of benefits (Rosén et al. 2017; Taylor 2017).

Climate justice

Climate justice discussions arose from recognition of the divide between high-income countries, which bear most of the responsibility for climate change, and those mostly low-income nations that are experiencing the worst impacts. Concerns initially focused on the imperative for wealthier countries to reduce GHG emissions and support adaptation investments (Hossain et al. 2021), leading to the establishment of a Loss and Damage Fund at COP27 in 2022 (although the fund is still grossly underfunded) (see also Chapter 18).

More recently, greater attention is being paid to the unequal impacts of climate change—and of climate interventions—within countries and communities due to gender and other social differences (Bryan et al. 2024; FAO 2024). Entrenched inequalities in agrifood systems influence the distribution of climate change impacts in ways that increase the vulnerability of already marginalized groups, such as smallholder producers and other value chain actors, the poor, women, and Indigenous Peoples. Many of those affected by structural inequalities, who lack access to resources and services, are also least able to adapt to climate change impacts (Bryan et al. 2024). Even within the same household, individuals have different needs, preferences, and capacities to respond effectively to climate shocks and stressors (Kristjanson et al. 2017). These differences are driven by gender as well as other social identities, such as ethnicity, caste, and class (Djoudi et al. 2016) (see Box 4.2). Ensuring that resources and services are targeted to the most vulnerable groups (that is, through social protection programs and climate information and financial services); facilitating group-based approaches; addressing structural inequalities, such as women's unequal access to land or restrictive social norms; and ensuring that all stakeholders participate in policy processes are some approaches to addressing inequalities through climate action (Bryan et al. 2024). Implementing these measures also requires policymakers, practitioners, and other stakeholders to have the capacity to design and implement equitable policies and programs.

BOX 4.2 Gender in climate change research

Women and men experience and respond to climate change impacts differently. Over the past decade, research by IFPRI and others has advanced our understanding of the linkages between gender and climate change, particularly gender and climate-smart agriculture (CSA) (Gardezi et al. 2021). This work has explored gendered differences in vulnerability (exposure, sensitivity, adaptive capacity) and adoption of climate change responses (and the ability to negotiate for preferred responses), as well as outcomes of climate change and response choices for well-being (Bryan et al. 2017; Kristjanson et al. 2017). Gender-differentiated impacts of climate change depend on many factors, including how climate change affects gendered asset dynamics, labor allocation, and control over income. Men and women tend to own different assets, in line with sociocultural norms, and draw on these assets in different ways in response to climate and other types of shocks (Quisumbing et al. 2018; Rakib and Matz 2016). Whose assets are affected can depend on a variety of factors, including the type of shock and the context. Extreme climate events also affect labor allocation in agrifood systems. These events cause both men's and women's labor contributions to decline, particularly under heat stress, but the declines are larger among men. This leads to a relative increase in women's agricultural labor burden under increasingly hazardous working conditions that have implications for their well-being (Lee et al. 2021; Nico and Azzarri 2024). In addition, *The Unjust Climate*, an FAO report to which IFPRI contributed, shows that women-headed households experience greater income losses as a result of extreme climate events, and women farmers suffer greater crop losses despite adopting climate-smart strategies at rates similar to men (FAO 2024). The report suggests that gender-responsive agricultural extension services are needed to enhance the effectiveness of women's adaptation efforts (see Chapter 8).

Our understanding of women's and men's differing experiences with climate shocks and stressors, access to climate information, and awareness and adoption of CSA practices is enhanced by collection and analysis of intrahousehold data (Bryan et al. 2021; Twyman et al. 2014). Findings from Bangladesh, Kenya, Senegal, and Uganda show that while both men and women are adjusting their farming practices in response to climate change, women tend to be less aware of CSA practices and less likely to adopt them. However, when women have access to climate information and are informed about CSA practices, they are just as likely or, in some cases, more likely to adopt these practices, particularly when they pertain to their livelihood roles. Research from India, Kenya, and Uganda also shows that targeting women with information, including through innovative approaches such as video-based extension, can be an effective way to increase their awareness and adoption of CSA practices (see Chapter 8) (Barooah et al. 2023; Kato et al. 2023; Ndegwa et al. 2023). However, other constraints, such as lack of access to labor or capital, can limit adoption of some practices despite awareness and interest.

Women's empowerment is also an important vehicle for increasing climate resilience and managing climate risks (Bryan et al. 2024; Huyer et al. 2021). For example, a study from Bangladesh shows that women's input into agricultural decisions and participation in groups increases crop diversification away from rice and toward more fruit and vegetable production, with positive implications for both climate resilience and dietary diversity (De Pinto, Seymour et al. 2020). Women also play key roles in maintaining food security and nutrition under climate change, including through their roles along nutrition-sensitive value chains (such as dairy products) (Kane et al. 2024) and in minimizing food safety risks, which are increasing with climate change (Brown 2018).

Looking forward: Improving understanding of climate risks and vulnerabilities

Better identification of hotspots

Some critical consequences of climate change's impacts on agriculture are not captured by the global averages from models, because they will be most damaging in specific locations ("hotspots") or during extreme events, (Jägermeyr et al. 2021; Rosegrant et al. 2024). In addition to reducing productivity, climate change can also amplify agriculture's environmental impacts, including GHG emissions, water use and scarcity, soil erosion, and biodiversity loss, increase nitrogen and phosphorus pollution as a result of increased fertilizer use, and cause more frequent and severe pest outbreaks, leading to higher pesticide use and pollution (Yang et al. 2024).

Crop modeling helps identify locations that are particularly vulnerable to higher yield losses for any given crop under climate change (Jalloh et al. 2013; Hachigonta et al. 2013; Nelson et al. 2010; Thomas 2024; Thomas et al. 2018; Waithaka et al. 2013). Combining these data with information on the local importance of the crop can systematically identify hotspots, a critical step to assist planners in pinpointing investments geographically for the development of climate-resilient cultivars and to support farmers in growing crops better suited to emerging climate patterns. Identifying where these areas of high exposure and sensitivity overlap with populations that are more susceptible to harm—based on their wealth status, gender inequality, and degree of malnutrition, among other factors—can improve targeting of resources to the most vulnerable.

Advancing models to better understand climate change impacts

Uncertainty about the future climate—due to differences among climate models and emissions scenarios—and increased variability and intensity of precipitation due to higher temperatures make it reasonable to expect increasingly frequent extreme climate events. However, current models are inconsistent in projecting increased variability, and modeling the impact of extreme events on food security and agricultural productivity is challenging, given the current limitations of biophysical models. As computer processing speeds increase, doors are opening to allow better modeling of the changes in frequency of extreme events (Thomas, Robertson et al. 2022; Thomas, Schlosser et al. 2022), which will make it possible to compare the effectiveness of various interventions, not simply in raising mean yields but also in mitigating the impact of such climate shocks. Nonetheless, much work remains to be done in this area. Some recent efforts have quantified feedback mechanisms that have not been previously incorporated into most bioeconomic models, including measures of livestock productivity losses (Hertel and de Lima 2020; Thornton et al. 2022). However, aquaculture still needs to be incorporated into bioeconomic models of climate change impact.

An important area for further research is the impact of climate change on labor productivity losses due to heat stress. While substantial work is being done on heat stress in non-agricultural settings, little is known about the impact in the agriculture sector, where heat stress is likely worse because most work is done outdoors (Hertel and de Lima 2020). One recent study addresses this gap, revealing that some regions worldwide are already experiencing significant declines in agricultural workers' ability to perform physical labor due to heat exposure during critical crop-growing periods. The resulting declines in worker productivity will exacerbate challenges by increasing the demand for labor, reducing output, and driving up food prices, which will place additional strain on vulnerable populations (Nelson et al. 2024). Further research is needed to understand these impacts, as well as study of potential adaptation strategies, which could include shifting to crops that thrive in cooler seasons, increasing mechanization, and providing more access to shade, drinking water, and body-cooling measures (Nelson et al. 2024).

Substantial work has been done on the direct climate impacts on economic growth. One study shows that economic productivity is nonlinear with temperatures across all countries, peaking at an annual average of 13°C and declining sharply at higher temperatures (Burke et al. 2015). This relationship is consistent globally and evident in both agricultural and non-agricultural activities across developed and developing nations. Other work finds that higher temperatures significantly reduce economic growth in poor countries, lowering growth rates in addition to output levels, and broadly affect agriculture, industry, and political stability (Dell et al. 2012).[3] These findings highlight climate's role in economic development and, like the results on TFP and agricultural yields, point to substantial negative impacts on poorer nations. This work has not generally been incorporated into bioeconomic modeling, despite the important implications for household food security. Additional work on the relationship between economic growth under climate change in agricultural growth and general economic growth would be valuable. Likewise, in addition to the work described above on the decline of nutrient availability in foods due to the CO_2 fertilization effect, more research is needed on the implications for nutrition outcomes and the identification of interventions that could alleviate negative impacts of climate change on nutrition.

Challenges for adaptation

While the adverse impacts of climate change on agricultural production are many and severe, there are also many opportunities for adaptation that should be taken up by researchers and policymakers. Some of these adaptations occur across geographic scales—shifting maize, for example, to cooler locations, changing planting dates, or identifying cultivars more suited to the future climate. Others can be very local, such as barns with cooling for livestock or mechanization to replace agricultural labor. However, there is little evidence that adaptation measures, even if they are locally effective, are sufficient in the aggregate. Using comprehensive panel data across diverse geographies and a wide range of outcomes—including mortality, agricultural productivity, crime, conflict, economic output, and damage from flooding and tropical cyclones—a recent study finds limited evidence of effective adaptation (Burke et al. 2024). Among 21 outcomes studied, 6 show statistically significant

3 See also Barrage (2020); Burke, Davis, and Diffenbaugh (2018); Casey, Fried, and Goode (2022, 2023); Diffenbaugh and Burke (2019); Donadelli et al. (2017); Golosov et al. (2014); Hassler et al. (2019); Letta and Tol (2019); and Nordhaus and Boyer (2003) for the whole economy.

declines in sensitivity to climate change, 5 show increased sensitivity, and the others show no significant change. This suggests that the adaptation measures studied in the paper have not significantly reduced overall climate impacts. To reduce future damage, it is crucial to identify scalable, effective adaptation strategies (Burke et al. 2024).

Looking ahead, the impacts of climate change will be compounded by other major changes. In many countries, population growth is increasing demand for food and often for agricultural land to produce that food. This is likely to create pressure to expand agriculture into forests or other natural areas, leading to further increases in GHG emissions and possibly to losses in biodiversity, ecosystem services, and tourism. As populations grow and change, agricultural labor may move to other sectors, either seasonally or permanently. In response to this pressure, farms may consolidate, further freeing labor for agricultural processing or shifting it to manufacturing or services. Within households, changes in income sources can lead to reallocation of labor, with implications for the time requirements of both women and youth. Combining climate modeling and household survey analysis allows us to address these multidimensional analytical challenges. Country-level modeling enables researchers to project shifts among sectors, and impacts on food security, poverty, and gross domestic product (GDP). Providing this foresight to decision-makers can inform policymaking for adaptation and mitigation as well as food security.

Increasing the resilience of the poor, women, and other vulnerable populations

The vulnerability of marginalized populations can be expected to intensify as the impacts of climate change become more severe, unless policies, investments, and actions adequately address the disproportionate impacts of climate change on these groups. To ensure that climate action yields more equitable outcomes, while accelerating progress toward adaptation and low-emission development pathways, more research is needed on effective mechanisms to address climate injustice. Finding ways to make climate governance at multiple scales more representative and inclusive of the perspectives and knowledge of vulnerable communities will be essential. For example, women's leadership in climate governance and the inclusion of grassroots women's organizations in climate decision-making processes can ensure that policies, investments, and interventions reflect women's diverse needs and priorities (Morgan et al. 2024). Frameworks such as the Women's Empowerment in Agrifood Systems Governance (WEAGov) can be used to measure the extent of women's inclusion and leadership in climate governance (see Chapter 6) (Ragasa et al. 2022). Collective, group-based approaches are powerful vehicles for strengthening women's empowerment and building resilience to climate change (Cabot Venton et al. 2021), including by increasing access to public entitlements (Kumar et al. 2021) and ensuring that agricultural information reaches women (Alvi et al. 2021) (see Chapters 8 and 14). But more research is needed on the scope for these platforms to accelerate climate action while reducing inequalities and on the factors that determine the effectiveness of more inclusive climate governance.

To avoid maladaptation, climate action to increase resilience must go hand-in-hand with addressing inequalities and facilitating empowerment of vulnerable groups. Action must include targeted interventions that build the resilience capacities of the most vulnerable people to respond to current and future climate-related challenges, including displacement and migration (see Chapter 13). Promising interventions include those that reduce climate risks, such as crop insurance and financial products tailored to smallholder producers (see Chapter 10), including women, and social protection programs that also promote adaptation and mitigation objectives (see Chapter 11) (Hidrobo et al. 2024; Timu and Kramer 2021). However, more research is needed on the effectiveness of these interventions in increasing resilience and the extent to which they can reduce inequalities and facilitate empowerment. More research is also needed on mechanisms to ensure that climate finance for adaptation efforts, such as loss and damage funds, and mitigation actions that reach and benefit the most vulnerable agrifood system actors, are not diverted by powerful interests (see Chapters 15 and 18). Policy research must also move beyond a focus on agricultural production to include building resilience along entire agrifood value chains, especially for micro, small, and medium enterprises that play important roles in supply chains and can contribute to low-emission development (see Chapter 7) (Liverpool-Tasie and Parkhi 2020).

Last, while the evidence on inequalities in climate change impacts, adaptive capacities, and policy processes has grown considerably over the past decade, much of this evidence comes from local case studies. The FAO's *Unjust Climate* report has begun to quantify the differential impacts of climate change across LMICs—but key gaps remain (FAO 2024). For example, many analyses have been done at the level of the household head, ignoring intrahousehold differences in climate change impacts. More individual-level data are needed for monitoring adaptation progress among the most vulnerable groups and for more effective targeting of interventions.

Conclusions

Climate change will slow progress on food and nutrition security, with impacts varying by location, time, and scale. To counter the impact of climate change, investments in increasing productivity must be a high priority. As an example, the African Union recently approved the 10-year Strategy and Action Plan for the Comprehensive Africa Agriculture Development Programme (CAADP), reaffirming the commitments of member states to use 10 percent of public spending to finance agriculture with the goal of achieving 6 percent annual growth of agricultural GDP (see Chapter 19). Follow-through and focusing on productivity investments not only by the African Union but also by donors, international agencies, and governments around the world will be essential.

To increase the effectiveness of larger investments in productivity and to achieve complementary objectives, a number of additional actions should be taken. Greater effort is needed to integrate climate considerations into agricultural and food policies, national planning, and budgets, while also designing climate interventions that support gender equality and nutrition. Many adaptation and mitigation options address both climate and food security constraints, such as risk, uncertainty, imperfect markets, and lack of credit or insurance. For instance, redirecting subsidies for fertilizer, water, and energy into productivity-enhancing investments could drive agricultural growth and reduce GHG emissions (see Chapters 3 and 18) (Rosegrant et al. 2022). However, to be effective, climate solutions must be tailored to local contexts through inclusive planning processes that integrate local knowledge, values, and diverse perspectives. This has implications for the ways in which research is conducted (that is, in partnership with local research institutes and taking into consideration the priorities of diverse stakeholders) and policy decisions are made. Nature-based climate solutions, such as agroforestry, conservation agriculture, and natural water storage, can both contribute to reducing GHG emissions and also increase availability of safe water resources, make air safer to breathe, and increase food security.

Greater attention is being given to the unequal impacts of climate change—and climate interventions—due to poverty, gender, and other social differences. It is also recognized that climate interventions risk exacerbating these inequalities if they do not address resource gaps and prioritize solutions that meet the needs of the most vulnerable. More effective interventions would integrate technical solutions with enabling and empowering measures and approaches, such as expanding information and financial services, promoting inclusive governance arrangements, and addressing constraints to women's participation.

Future work must go beyond measuring and predicting biophysical and socioeconomic impacts to integrate adaptation and mitigation into development agendas and prioritize actions that advance development goals including food security, better nutrition, and gender equality. To prevent maladaptation, more research is needed to identify and assess interventions that enhance nutritional outcomes and build resilience in vulnerable groups, especially women and the poor. This will require better data collection, monitoring, and evaluation efforts to track progress toward these outcomes as climate actions are scaled up.

References

Adger, W.N. 2006. "Vulnerability." *Global Environmental Change* 16 (3): 268–281.

Adger, W.N., N.W. Arnell, and E.L. Tompkins. 2005. "Successful Adaptation to Climate Change Across Scales." *Global Environmental Change* 15 (2): 77–86.

Adger, W.N., S. Huq, K. Brown, D. Conway, and M. Hulme. 2003. "Adaptation to Climate Change in the Developing World." *Progress in Development Studies* 3 (3): 179–195.

Alvi, M., P. Barooah, S. Gupta, and S. Saini. 2021. "Women's Access to Agriculture Extension Amidst COVID-19: Insights from Gujarat, India and Dang, Nepal." *Agricultural Systems* 188: 103035.

Ayers, J.M., and S. Huq. 2009. "The Value of Linking Mitigation and Adaptation: A Case Study of Bangladesh." *Environmental Management* 43: 753–764.

Barooah, P., M. Alvi, C. Ringler, and V. Pathak. 2023. "Gender, Agriculture Policies, and Climate-Smart Agriculture in India." *Agricultural Systems* 212: 103751.

Barrage, L. 2020. "Optimal Dynamic Carbon Taxes in a Climate–Economy Model with Distortionary Fiscal Policy." *Review of Economic Studies* 87 (1): 1–39.

Beach, R.H., T.B. Sulser, A. Crimmins, et al. 2019. "Combining the Effects of Increased Atmospheric Carbon Dioxide on Protein, Iron, and Zinc Availability and Projected Climate Change on Global Diets: A Modelling Study." *Lancet Planetary Health* 3: e307–e317.

Below, T., A. Artner, R. Siebert, and S. Sieber. 2010. "Micro-Level Practices to Adapt to Climate Change for African Small-Scale Farmers: A Review of Selected Literature." IFPRI Discussion Paper 953. IFPRI, Washington, DC.

Bizikova, L., J.E. Parry, J. Karami, et al. 2015. "Review of Key Initiatives and Approaches to Adaptation Planning at the National Level in Semi-Arid Areas." *Regional Environmental Change* 15: 837–850.

Brown, L.R. 2018. *Aflatoxins in Food and Feed: Impacts, Risks, and Management Strategies.* GCAN Policy Note 9. Washington, DC: IFPRI.

Bryan, E., W. Akpalu, C. Ringler, and M. Yesuf. 2010. "Global Carbon Markets: Opportunities for Sub-Saharan Africa in Agriculture and Forestry." *Climate and Development* 2 (4): 309–331.

Bryan, E., M. Alvi, S. Huyer, and C. Ringler. 2024. "Addressing Gender Inequalities and Strengthening Women's Agency to Create More Climate Resilient and Sustainable Food Systems." *Global Food Security* 40: 100731.

Bryan, E., T. Deressa, G. Gbetibouo, and C. Ringler. 2009. "Adaptation to Climate Change in Ethiopia and South Africa: Options and Constraints." *Environmental Science and Policy* 12 (4): 413–426.

Bryan, E., E. Kato, and Q. Bernier. 2021. "Gender Differences in Awareness and Adoption of Climate-Smart Agriculture Practices in Bangladesh." In *Gender, Climate Change and Livelihoods: Vulnerabilities and Adaptations*, eds. J. Eastin and K. Dupuy, 123–142. Wallingford, UK: CABI.

Bryan, E., C. Ringler, B. Okoba, J. Koo, M. Herrero, and S. Silvestri. 2013a. "Can Agriculture Support Climate Change Adaptation, Greenhouse Gas Mitigation and Rural Livelihoods? Insights from Kenya." *Climatic Change* 118 (2): 151–165.

Bryan, E., C. Ringler, B. Okoba, C. Roncoli, S. Silvestri, and M. Herrero. 2013b. "Adapting Agriculture to Climate Change in Kenya: Household Strategies and Determinants." *Journal of Environmental Management* 114: 26–35.

Bryan, E., S. Theis, J. Choufani, A. De Pinto, R. Meinzen-Dick, and C. Ringler. 2017. "Gender-Sensitive, Climate-Smart Agriculture for Improved Nutrition in Africa South of the Sahara." In *2017 Annual Trends and Outlook Report (ATOR): A Thriving Agricultural Sector in a Changing Climate: Meeting Malabo Declaration Goals Through Climate-Smart Agriculture*, eds. A. De Pinto and J.M. Ulimwengu, 114–135. Washington, DC: IFPRI.

Bryant, C.R., B. Smit, M. Brklacich, T.R. Johnston, J. Smithers, Q. Chiotti, and B. Singh. 2000. "Adaptation in Canadian Agriculture to Climatic Variability and Change." In *Societal Adaptation to Climate Variability and Change*, eds. S.M. Kane and G.W. Yohe, 181–201. Dordrecht, Netherlands: Springer.

Burke, M., W.M. Davis, and N.S. Diffenbaugh. 2018. "Large Potential Reduction in Economic Damages Under UN Mitigation Targets." *Nature* 557: 549–553.

Burke, M., S.M. Hsiang, and E. Miguel. 2015. "Global Non-Linear Effect of Temperature on Economic Production." *Nature* 527: 235–239.

Burke, M., M. Zahid, M.C.M. Martins, et al. 2024. "Are We Adapting to Climate Change?" NBER Working Paper 32985. National Bureau of Economic Research, Cambridge, MA.

Burton, I. 1997. "Vulnerability and Adaptive Response in the Context of Climate and Climate Change." *Climatic Change* 36: 185–196.

Burton, I., and B. Lim. 2005. "Achieving Adequate Adaptation in Agriculture." *Climatic Change* 70: 191–200.

Cabot Venton, C., S.A. Prillaman, and J. Kim. 2021. "Building Resilience Through Self Help Groups: Evidence Review." In *The Resilience Evaluation, Analysis, and Learning Award Conference Proceedings*. Washington, DC: Food Security and Nutrition Network.

Casey, G., S. Fried, and E. Goode. 2022. "Projecting the Impacts of Temperature Change: The Role of Macroeconomic Dynamics." Paper presented at the International Monetary Fund Economic Review Conference, Rabat, Morocco, June 23.

Casey, G., S. Fried, and E. Goode. 2023. "Projecting the Impact of Rising Temperatures: The Role of Macroeconomic Dynamics." *IMF Economic Review* 71: 688–718.

CGIAR. CGIAR Research Program on Climate Change, Agriculture and Food Security. 2018. "Influencing Policies for Improved Food Security and Adaptation to Climate Change in the Philippines- CCAFS and PIM Inputs." In *Climate Change, Agriculture and Food Security Annual Report 2018*. Outcome Impact Case Report. Montpellier, France.

Choi, J.-S., and P. G. Helmberger. 1993. "How Sensitive Are Crop Yields to Price Changes and Farm Programs?" *Journal of Agricultural and Applied Economics* 25: 237–244.

Cho, S., and B. McCarl. 2017. "Climate Change Influences on Crop Mix Shifts in the United States." *Scientific Reports* 7: 40845.

Dell, M., B.F. Jones, and B.A. Olken. 2012. "Temperature Shocks and Economic Growth: Evidence from the Last Half Century." *American Economic Journal: Macroeconomics* 4 (3): 66–95.

De Pinto, A., N. Cenacchi, H.Y. Kwon, J. Koo, and S. Dunston. 2020. "Climate Smart Agriculture and Global Food-Crop Production. *PLoS ONE* 15 (4): e0231764.

De Pinto, A., G. Seymour, E. Bryan, and P. Bhandari. 2020. "Women's Empowerment and Farmland Allocations in Bangladesh: Evidence of a Possible Pathway to Crop Diversification." *Climatic Change* 163 (2): 1025–1043.

Deressa, T.T., and R.M. Hassan. 2009. "Economic Impact of Climate Change on Crop Production in Ethiopia: Evidence from Cross-Section Measures." *Journal of African Economies* 18 (4): 529–554.

Deressa, T.T., R.M. Hassan, and C. Ringler. 2011. "Perception of and Adaptation to Climate Change by Farmers in the Nile Basin of Ethiopia." *Journal of Agricultural Science* 149 (1): 23–31.

Deressa, T.T., R.M. Hassan, C. Ringler, T. Alemu, and M. Yesuf. 2009. "Determinants of Farmers' Choice of Adaptation Methods to Climate Change in the Nile Basin of Ethiopia." *Global Environmental Change* 19: 248–255.

De Salvo, M., D. Begalli, and G. Signorello. 2013. "Measuring the Effect of Climate Change on Agriculture: A Literature Review of Analytical Models." *Journal of Development and Agricultural Economics* 5 (12): 499–509.

Diffenbaugh, N.S., and M. Burke. 2019. "Global Warming Has Increased Global Economic Inequality." *PNAS* 116 (20): 9808–9813.

Djoudi, H., B. Locatelli, C. Vaast, et al. 2016. "Beyond Dichotomies: Gender and Intersecting Inequalities in Climate Change Studies" *Ambio 45* (Suppl 3): 248–262.

Donadelli, M., M. Jüppner, M. Riedel, and C. Schlag. 2017. "Temperature Shocks and Welfare Costs." *Journal of Economic Dynamics and Control* 82: 331–355.

Easterling, W.E., P.K. Aggarwal, P. Batima, et al. 2007. "Food, Fibre and Forest Products." In *Climate Change 2007: Impacts, Adaptation and Vulnerability*. Working Group II Contribution to the Fourth Assessment Report of the Intergovernmental Panel on Climate Change, eds. M.L. Parry, O.F. Canziani, J.P. Palutikof, P.J. van der Linden, and C.E. Hanson, 273–313. Cambridge, UK: Cambridge University Press.

Fan, S., D. Headey, D. Laborde, D. Mason-D'Croz, C. Rue, T.B. Sulser, and K. Wiebe. 2018. *Quantifying the Cost and Benefits of Ending Hunger and Undernutrition: Examining the Differences Among Alternative Approaches*. IFPRI Issue Brief. Washington, DC: IFPRI.

FAO (Food and Agriculture Organization of the United Nations). 2010. *"Climate-Smart" Agriculture: Policies, Practices and Financing for Food Security, Adaptation and Mitigation*. Rome.

FAO. 2024. *The Unjust Climate: Measuring the Impacts of Climate Change on Rural Poor, Women and Youth*. Rome.

Feng, X., H. Tian, J. Cong, and C. Zhao. 2023. "A Method Review of the Climate Change Impact on Crop Yield." *Frontiers in Forests and Global Change* 6: 1198186.

Gardezi, M., S. Michael, R. Stock, S. Vij, A. Ogunyiola, and A. Ishtiaque. 2021. Prioritizing Climate-Smart Agriculture: An Organizational and Temporal Review." *WIREs Climate Change* 13 (2): e755.

Gbetibouo, G.A., and R.M. Hassan. 2005. "Measuring the Economic Impact of Climate Change on Major South African Field Crops: A Ricardian approach. *Global and Planetary Change* 47: 143–152.

Gbetibouo, G.A., R.M. Hassan, and C. Ringler. 2010. "Modelling Farmers' Adaptation Strategies for Climate Change and Variability: The Case of the Limpopo Basin, South Africa." *Agrekon* 49 (2): 217–234.

Gbetibouo, G.A., C. Ringler, and R.M. Hassan. 2010. "Vulnerability of the South African Farming Sector to Climate Change and Variability: An Indicator Approach." *Natural Resources Forum* 34 (3): 175–187.

Golosov, M., J. Hassler, P. Krusell, and A. Tsyvinski. 2014. "Optimal Taxes on Fossil Fuel in General Equilibrium." *Econometrica* 82 (1): 41–88.

Hachigonta, S., G. Nelson, T.S. Thomas, and L.M. Sibanda, eds. 2013. *Southern African Agriculture and Climate Change: A Comprehensive Analysis.* IFPRI Research Monograph. Washington, DC: IFPRI.

Hasegawa, T., H. Wakatsuki, H. Ju, et al. 2022. "Summary Results from Different Types of Models: A Global Dataset for the Projected Impacts of Climate Change on Four Major Crops." *Scientific Data* 9: 58.

Hassan, R., and C. Nhemachena. 2008. "Determinants of African Farmers' Strategies for Adapting to Climate Change: Multinomial Choice Analysis." *African Journal of Agricultural and Resource Economics* 2 (1): 83–104.

Hassler, J., P. Krusell, C. Olovsson, and M. Reiter. 2019. "Integrated Assessment in a Multi-Region World with Multiple Energy Sources and Endogenous Technical Change Preliminary." 2019 Meeting Papers 809. Society for Economic Dynamics, Minneapolis.

Hertel, T.W., and C.Z. de Lima. 2020. "Viewpoint: Climate Impacts on Agriculture: Searching for Keys under the Streetlight."*Food Policy* 95: 101954.

Hidrobo, M., V. Mueller, E. Bryan, Z. Nesbitt-Ahmed, P. Laderach, and S. Roy. 2024. *Leveraging Social Assistance to Strengthen Women's and Girls' Climate Resilience: What Is the Potential, and What Are Promising Program Designs?* CGIAR Initiative on Fragility, Conflict, and Migration Policy Brief. Washington, DC: IFPRI.

Hossain, F., D. Falzon, M.F. Rahman, and S. Huq. 2021. "Toward Climate Justice: Making the Polluters Pay for Loss and Damage." In *Principles of Justice and Real-World Climate Politics*, eds. S. Kenehan and C. Katz, 149–170. London: Rowman & Littlefield.

Huang, K., and N. Sim. 2016. "Why Do Econometric Studies Disagree on the Effect of Warming on Agricultural Output? A Meta-Analysis." School of Economics and Public Policy Working Papers 2016-04. University of Adelaide, School of Economics and Public Policy, Adelaide, Australia.

Huq, S., H. Reid, M. Konate, A. Rahman, Y. Sokona, and F. Crick. 2004. "Mainstreaming Adaptation to Climate Change in Least Developed Countries (LDCs)." *Climate Policy* 4 (1): 25–43.

Huyer, S., T. Gumucio, K. Tavenner, et al. 2021. "From Vulnerability to Agency in Climate Adaptation and Mitigation." In *Advancing Gender Equality Through Agriculture and Environmental Research: Past, Present, and Future*, 261–294. Washington, DC: IFPRI.

IPCC (Intergovernmental Panel on Climate Change). 2007. *Climate Change 2007: Impacts, Adaptation and Vulnerability, Contribution of Working Group II to the Fourth Assessment Report of the IPCC.* Cambridge, UK: Cambridge University Press.

Jägermeyr, J., C. Müller, A.C. Ruane, et al. 2021. "Climate Impacts on Global Agriculture Emerge Earlier in New Generation of Climate and Crop Models." *Nature Food* 2 (11): 873–885.

Jagnani, M., C.B. Barrett, Y. Liu, and L. You. 2021. "Within-Season Producer Response to Warmer Temperatures: Defensive Investments by Kenyan Farmers." *The Economic Journal* 131 (633): 392–419.

Jalloh, A., G.C. Nelson, T.S. Thomas, R. Zougmoré, and H. Roy-Macauley, eds. 2013. *West African Agriculture and Climate Change: A Comprehensive Analysis.* IFPRI Research Monograph. Washington, DC: IFPRI.

Jones, J.W., G. Hoogenboom, C.H. Porter, et al. 2003. "The DSSAT Cropping System Model." *European Journal of Agronomy* 18: 235–265.

Kane, P.A., M.B. Barry, S. Eissler, et al. 2024. "Strengthening Women's Empowerment, Climate Resilience, and Nutrition Along the Goat Value Chain in Senegal: A Qualitative Study." IFPRI Discussion Paper 2274. IFPRI, Washington, DC.

Kates, R.W., W.R. Travis, and T.J. Wilbanks. 2012. "Transformational Adaptation When Incremental Adaptations to Climate Change are Insufficient." *PNAS* 109: 7156–7161.

Kato, E., P.B. Rwamigisa, H. Kamusiime, F. Sufian, and C. Ringler. 2023. "Do Informational Videos Impact Awareness, Knowledge, and Adoption of Climate-Smart Agricultural Practices? Insights from Rural Uganda." In *Reaching Smallholder Women with Information Services and Resilience Strategies to Respond to Climate Change*, eds. C. Ringler, M.F. Alvi, R. Birner, et al., 44–47. Washington, DC: IFPRI.

Keating, B.A., P.S. Carberry, G.L. Hammer, et al. 2003. "An Overview of APSIM, a Model Designed for Farming Systems Simulation." *European Journal of Agronomy* 18: 267–288.

Khatri-Chhetri, A., P.K. Aggarwal, P.K. Joshi, and S. Vyas. 2017. "Farmers' Prioritization of Climate-Smart Agriculture (CSA) Technologies." *Agricultural Systems* 151: 184–191.

Knox, J., T. Hess, A. Daccache, and T. Wheeler. 2012. "Climate Change Impacts on Crop Productivity in Africa and South Asia." *Research Letters* 7 (034032): 1–8.

Koo, J., C. Azzarri, A. Mishra, et al. 2022. "Effectively Targeting Climate Investments: A Methodology for Mapping Climate–Agriculture–Gender Inequality Hotspots." CGIAR GENDER Platform Working Paper #005. CGIAR GENDER Platform, Nairobi.

Kramer, B. 2023. "Climate Insurance: Opportunities for Improving Agricultural Risk Management in Kenya." In *Food Systems Transformation in Kenya*, eds. C. Breisinger, M. Keenan, J. Mbuthia, and J. Njuki, 285–304. Washington, DC: IFPRI.

Kristjanson, P., E. Bryan, Q. Bernier, et al. 2017. "Addressing Gender in Agricultural Research for Development in the Face of a Changing Climate: Where Are We and Where Should We Be Going?" *International Journal of Agricultural Sustainability* 15 (5): 482–500.

Kumar, N., K. Raghunathan, A. Arrieta, J. Amir, and S. Pandey. 2021. "The Power of the Collective Empowers Women: Evidence from Self-Help Groups in India." *World Development* 146: 105579.

Kurukulasuriya, P., R. Mendelsohn, R. Hassan, et al. 2006. "Will African Agriculture Survive Climate Change?" *World Bank Economic Review* 20 (3): 367–388.

Laborde, D., M. Parent, and C. Smaller. 2020. *Ending Hunger, Increasing Incomes, and Protecting the Climate: What Would It Cost Donors?* Winnipeg: International Institute for Sustainable Development; Washington, DC: IFPRI.

Lecoutere, E., A. Mishra, N. Singaraju, et al. 2023. "Where Women in Agri-Food Systems Are at Highest Climate Risk: A Methodology for Mapping Climate–Agriculture–Gender Inequality Hotspots." *Frontiers in Sustainable Food Systems* 7: 1197809.

Lee, Y., B. Haile, G. Seymour, and C. Azzarri. 2021. "The Heat Never Bothered Me Anyway: Gender-Specific Response of Agricultural Labor to Climatic Shocks in Tanzania." *Applied Economic Perspectives and Policy* 43 (2): 732–749.

Letta, M., and R. Tol. 2019. "Weather, Climate and Total Factor Productivity." *Environmental and Resource Economics* 73 (1): 283–305.

Liang, X., Y. Wu, R.G. Chambers, et al. 2017. "Determining Climate Effects on US Total Agricultural Productivity." *PNAS* 114 (12): E2285–E2292.

Lipper, L., and D. Zilberman. 2018. "A Short History of the Evolution of the Climate Smart Agriculture Approach and Its Links to Climate Change and Sustainable Agriculture Debates." In *Climate Smart Agriculture: Natural Resource Management and Policy*, vol. 52, eds. L. Lipper, N. McCarthy, D. Zilberman, S. Asfaw, and G. Branca, 13–30. Cham, Switzerland: Springer.

Liverpool-Tasie, L.S.O., and C.M. Parkhi. 2020. "Climate Risk and Technology Adoption in the Midstream of Crop Value Chains: Evidence from Nigerian Maize Traders." *Journal of Agricultural Economics* 72 (1): 158–179.

Lobell, D.B., M. Bänziger, C. Magorokosho, and B. Vivek. 2011. "Nonlinear Heat Effects on African Maize as Evidenced by Historical Yield Trials." *Nature Climate Change* 1: 42–45.

Lobell, D.B., W. Schlender, and J. Costa-Roberts. 2011. "Climate Trends and Global Crop Production Since 1980." *Science* 333 (6042): 616–620.

Maddison, D. 2007. "The Perception of and Adaptation to Climate Change in Africa." Policy Research Working Paper 4308. World Bank, Washington, DC.

Mason-D'Croz, D., T.B. Sulser, K. Wiebe, et al. 2019. "Agricultural Investments and Hunger in Africa Modeling Potential Contributions to SDG2 – Zero Hunger." *World Development* 116: 38–53.

Mbow, C., C. Rosenzweig, L.G. Barioni, et al. 2019. "Food Security." In *Climate Change and Land: An IPCC Special Report on Climate Change, Desertification, Land Degradation, Sustainable Land Management, Food Security, and Greenhouse Gas Fluxes in Terrestrial Ecosystems*, eds. P.R. Shukla, J. Skea, E. Calvo Buendia, et al., 437–550. Cambridge, UK: Cambridge University Press.

Mendelsohn, R., W. Nordhaus, and D. Shaw. 1994. "The Impact of Global Warming on Agriculture: A Ricardian Analysis." *American Economic Review* 84: 753–771.

Mendelsohn, R., W. Nordhaus, and D. Shaw. 1996. "Climate Impacts on Aggregate Farm Value: Accounting for Adaptation." *Agricultural and Forest Meteorology* 80: 55–66.

Mertz, O., K. Halsnæs, J.E. Olesen, et al. 2009. "Adaptation to Climate Change in Developing Countries." *Environmental Management* 43: 743–752.

Miao, R., M. Khanna, and H. Huang. 2015. "Responsiveness of Crop Yield and Acreage to Prices and Climate." *American Journal of Agricultural Economics* 98 (1): 191–211.

Morgan, M.Y., E. Bryan, and M. Elias. 2024. "Women's Leadership in Climate-Resilient Agrifood Systems: Defining a Future Research Agenda." *Environmental Research: Climate* 3: 023001.

Morton, J.F. 2007. "The Impact of Climate Change on Smallholder and Subsistence Agriculture." *PNAS* 104 (50): 19680–19685.

Mwongera, C., K.M. Shikuku, J. Twyman, et al. 2017. "Climate Smart Agriculture Rapid Appraisal (CSA-RA): A Tool for Prioritizing Context-Specific Climate Smart Agriculture Technologies." *Agricultural Systems* 151: 192–203.

Myers, S.S., A. Zanobetti, I. Kloog, et al. 2014. "Increasing CO_2 Threatens Human Nutrition." *Nature* 510: 139–142.

Ndegwa, M.K., C. Ringler, F.N. Muteti, E. Kato, and E. Bryan. 2023. "Can Participatory Video-Based Extension Increase Awareness and Knowledge of Climate Adaptation Practices? Insights from Rural Kenya." In *Reaching Smallholder Women with Information Services and Resilience Strategies to Respond to Climate Change*, eds. C. Ringler, M.F. Alvi, R. Birner, et al., 40–43. Washington, DC: IFPRI.

Ndegwa, M.K., A. Shee, C.G. Turvey, and L. You. 2020. "Uptake of Insurance-Embedded Credit in Presence of Credit Rationing: Evidence from a Randomized Controlled Trial in Kenya." *Agricultural Finance Review* 80 (5): 745–766.

Nelson, G.C., M.W. Rosegrant, A. Palazzo, et al. 2010. *Food Security, Farming, and Climate Change to 2050: Scenarios, Results, Policy Options*. Washington, DC: IFPRI.

Nelson, G.C., D. van der Mensbrugghe, H. Ahammad, et al. 2014. "Agriculture and Climate Change in Global Scenarios: Why Don't the Models Agree." *Agricultural Economics* 45 (1): 85–101.

Nelson, G.C., J. Vanos, G. Havenith, O. Jay, K.L. Ebi, and R.J. Hijmans. 2024. "Global Reductions in Manual Agricultural Work Capacity Due to Climate Change." *Global Change Biology* 30: e17142.

Nico, G., and C. Azzarri. 2024. "Climate Change and Sex-Specific Labor Intensity: An Empirical Analysis in Africa." *Global Food Security* 42: 100799.

Nordhaus, W.D., and J. Boyer. 2003. *Warming the World: Economic Models of Global Warming.* Cambridge, MA: MIT Press.

Ortiz-Bobea, A., T.R. Ault, C.M. Carrillo, R.G. Chambers, and D.B. Lobell. 2021. "Anthropogenic Climate Change Has Slowed Global Agricultural Productivity Growth." *Nature Climate Change* 11: 306–312.

Parry, M., A. Evans, M.W. Rosegrant, and T. Wheeler. 2009. *Climate Change and Hunger: Responding to the Challenge.* Rome: World Food Programme.

Quisumbing, A.R., N. Kumar, and J. Behrman. 2018. "Do Shocks Affect Men's and Women's Assets Differently? Evidence from Bangladesh and Uganda." Development Policy Review 36 (1): 3–34.

Ragasa, C., J. Kyle, P. Kristjanson, and S. Eissler. 2022. "Conceptualizing Women's Empowerment in Agrifood Systems Governance: A New Framework." IFPRI Discussion Paper 2153. IFPRI, Washington, DC.

Rakib, M., and J.A. Matz. 2016. "The Impact of Shocks on Gender-Differentiated Asset Dynamics in Bangladesh." *Journal of Development Studies* 52 (3): 377–395.

Reid, H., and S. Huq. 2008. *International and National Mechanisms and Politics of Adaptation: An Agenda for Reform.* New York: United Nations Development Programme.

Ricciardi, V., N. Ramankutty, Z. Mehrabi, L. Jarvis, and B. Chookolingo. 2018. "How Much of the World's Food Do Smallholders Produce?" *Global Food Security* 17: 64–72.

Robinson, S., S. Dunston, A. Mishra, et al. 2024. "The International Model for Policy Analysis of Agricultural Commodities and Trade (IMPACT): Model Documentation for Version 3.6." Modeling Systems Technical Paper 1. IFPRI, Washington, DC.

Rosegrant, M.W., T.B. Sulser, S. Dunston, et al. 2024. "Food and Nutrition Security Under Changing Climate and Socioeconomic Conditions." *Global Food Security* 41: 100755.

Rosegrant, M.W., T.B. Sulser, and K. Wiebe. 2022. "Global Investment Gap in Agricultural Research and Innovation to Meet Sustainable Development Goals for Hunger and Paris Agreement Climate Change Mitigation." *Frontiers in Sustainable Food Systems* 6: 965767.

Rosén, L., L.O. Naess, A. Nightingale, and J. Thompson. 2017. "'Triple Wins' or 'Triple Faults'? Analysing the Equity Implications of Policy Discourses on Climate-Smart Agriculture (CSA)." *Journal of Peasant Studies* 45 (1): 150–174.

Rosenzweig, C., and F.N. Tubiello. 2007. "Adaptation and Mitigation Strategies in Agriculture: An Analysis of Potential Synergies." *Mitigation and Adaption Strategies for Global Change* 12: 855–873.

Schlenker, W., and M.J. Roberts. 2009. "Nonlinear Temperature Effects Indicate Severe Damages to US Crop Yields Under Climate Change." *PNAS* 106 (37): 15594–15598.

Silvestri, S., E. Bryan, C. Ringler, M. Herrero, and B. Okoba. 2012. "Climate Change Perceptions and Adaptation of Agro-Pastoral Communities in Kenya." *Regional Environmental Change* 12 (4): 791–802.

Smit, B., I. Burton, R. Klein, et al. 1999. "The Science of Adaptation: A Framework for Assessment." *Mitigation and Adaptation Strategies for Global Change* 4: 199–213.

Smit, B., O. Pilifosova, I. Burton, et al. 2001. "Adaptation to Climate Change in the Context of Sustainable Development and Equity." In *Climate Change 2001: Impacts, Adaptation, and Vulnerability,* Contribution of Working Group II to the Third Assessment Report of the Intergovernmental Panel on Climate Change, 877–912. Cambridge, UK: Cambridge University Press.

Smit, B., D. McNabb, and J. Smithers. 1996. "Agricultural Adaptation to Climatic Variation." *Climatic Change* 33: 7–29.

Smit, B., and M.W. Skinner. 2002. "Adaptation Options in Agriculture to Climate Change: A Typology." *Mitigation and Adaptation Strategies for Global Change* 7: 85–114.

Smith, P., J. Nkem, K. Calvin, et al. 2018. "Interlinkages Between Desertification, Land Degradation, Food Security and GHG fluxes: Synergies, Trade-Offs and Integrated Response Options." In *Special Report on Climate Change and Land*, 551–672. Geneva: Intergovernmental Panel on Climate Change.

Smithers, J., and B. Smit. 1997. "Human Adaptation to Climatic Variability and Change." *Global Environmental Change* 7 (2): 129–146.

Solomon, S., G. Plattner, R. Knutti, and P. Friedlingstein. 2009. "Irreversible Climate Change Due to Carbon Dioxide Emissions." *PNAS* 106 (6): 1704–1709.

Sulser, T., K.D. Wiebe, S. Dunston, et al. 2021. *Climate Change and Hunger: Estimating Costs of Adaptation in the Agrifood System.* Food Policy Report. Washington, DC: IFPRI.

Taylor, M. 2017. "Climate-Smart Agriculture: What Is It Good For?" *Journal of Peasant Studies* 45 (1): 89–107.

Thomas, T.S. 2024. "The Impact of Climate Change on Agriculture." In *ReSAKKS Annual Trends and Outlook Report 2024: Advancing the Climate and Bioeconomy Agenda in Africa for Resilient and Sustainable Agrifood Systems*, eds. G. Tadesse, K. Glatzel, and M. Savadogo, 64–77. Kigali: AKADEMIYA2063; Washington, DC: IFPRI.

Thomas, T.S., V. Nazareth, and R. Folledo. 2018. "A Biophysical Approach to Modeling Alternative Agricultural Futures Under Climate Change." In *The Future of Philippine Agriculture Under a Changing Climate: Policies, Investments and Scenarios*, eds. M.W. Rosegrant and M.A. Sombilla, 377–449. Singapore: ISEAS Publishing.

Thomas, T.S., R.D. Robertson, K. Strzepek, and C. Arndt. 2022. "Extreme Events and Production Shocks for Key Crops in Southern Africa Under Climate Change." *Frontiers in Climate* 4: 787582.

Thomas, T.S., C.A. Schlosser, K. Strzepek, R.D. Robertson, and C. Arndt. 2022. "Using a Large Climate Ensemble to Assess the Frequency and Intensity of Future Extreme Climate Events in Southern Africa." *Frontiers in Climate* 4: 787721.

Thornton, P.K., P. Kristjanson, W. Förch, C. Barahona, L. Cramer, and S. Pradhan. 2018. "Is Agricultural Adaptation to Global Change in Lower-Income Countries on Track to Meet the Future Food Production Challenge?" *Global Environmental Change* 52: 37–48.

Thornton, P., G. Nelson, D. Mayberry, and M. Herrero. 2022. "Impacts of Heat Stress on Global Cattle Production During the 21st Century: A Modelling Study." *Lancet Planet Health* 6: e192–e201.

Timu, A.G., and B. Kramer. 2021. "Gender-Inclusive, Responsive, and Transformative Agricultural Insurance: A Literature Review." CCAFS Working Paper 417. CGIAR Research Program on Climate Change, Agriculture and Food Security (CCAFS), Wageningen, Netherlands.

Twyman, J., M. Green, Q. Bernier, et al. 2014. "Gender and Climate Change Perceptions, Adaptation Strategies, and Information Needs Preliminary Results from Four Sites in Africa." CCAFS Working Paper 83. CGIAR Research Program on CCAFS, Copenhagen.

UNEP (United Nations Environment Programme). 2024. *Adaptation Gap Report 2024: Come Hell and High Water—As Fires and Floods Hit the Poor Hardest, It Is Time for the World to Step Up Adaptation Actions*. Nairobi.

Waisman, H., C. Bataille, H. Winkler, et al. 2019. "A Pathway Design Framework for National Low Greenhouse Gas Emission Development Strategies." *Nature Climate Change* 9: 261–268.

Waithaka, M., G. Nelson, T.S. Thomas, and M. Kyotalimye, eds. 2013. *East African Agriculture and Climate Change: A Comprehensive Analysis*. IFPRI Research Monograph. Washington, DC: IFPRI.

Xiang, T., T.H. Malik, J.W. Hou, and J. Ma. 2022. "The Impact of Climate Change on Agricultural Total Factor Productivity: A Cross-Country Panel Data Analysis, 1961–2013." *Agriculture* 12 (12): 2123.

Yang, Y., D. Tilman, P. Smith, et al. 2024. "Climate Change Exacerbates the Environmental Impacts of Agriculture." *Science* 385 (671): 3747.

ZEF (Center for Development Research) and FAO. 2020. *Investment Costs and Policy Action Opportunities for Reaching a World Without Hunger (SDG 2)*. Bonn, Germany.

Zhao, C., B. Liu, S. Piao, et al. 2017. "Temperature Increase Reduces Global Yields of Major Crops in Four Independent Estimates." *PNAS* 114 (35): 9326–9331.

Chapter 5

ENVIRONMENTAL SUSTAINABILITY
The Intersection of Agrifood Systems with Ecosystem Health

Claudia Ringler, Wei Zhang, Ephraim Nkonya, Kris Wyckhuys, Sarobidy Rakotonarivo, and Ruth Meinzen-Dick

Claudia Ringler is director, Natural Resources and Resilience Unit (NRR), IFPRI; **Wei Zhang** is a senior research fellow, NRR, IFPRI; **Ephraim Nkonya** is a non-resident fellow, IFPRI; **Kris Wyckhuys** is with Chrysalis Consulting, Danang, Viet Nam, the Institute for Plant Protection, China Academy of Agricultural Sciences (CAAS), Beijing, and the School of the Environment, University of Queensland, Saint Lucia, Australia; **Sarobidy Rakotonarivo** is senior research fellow at the École Supérieure des Sciences Agronomiques at the University of Antananarivo, Madagascar; and **Ruth Meinzen-Dick** is a research fellow emeritus, IFPRI.

Key messages

- Unsustainable food systems are both a cause and consequence of environmental degradation. Despite global attention to climate change, biodiversity loss, and natural resource degradation, progress in understanding and addressing the problem has been slow at best.
- Over the last 50 years, research has evolved amid growing recognition of environmental degradation, the importance of ecosystem services and biodiversity, and the role of property rights, institutions, and governance.
- In response, researchers have offered insights on balancing agricultural productivity with environmental sustainability, as well as the role of governance and collaborative management.
- Innovative approaches, such as behavior-based methods that use games to simulate resource use decisions, have also catalyzed community action.
- More recent systems approaches aid in understanding complex relationships and can support integrated landscape approaches for development and conservation.

Looking to the future, research for sustainability should:

- **Strengthen systems thinking within and across food and linked systems.** Increasing sustainability requires considering not only food value chains but also diets, health, and energy systems, among others, in designing policies.
- **Commit to transdisciplinary research processes.** Research must actively include stakeholders who manage resources, such as farmers, in identifying and assessing interventions.
- **Strengthen political economy and policy process research.** Achieving wider implementation of tested interventions will require research on obstacles to adopting policies and practices for sustainability.
- **Apply an environmental justice lens in agrifood systems research.** Research on women's empowerment, energy use, and water insecurity highlights the critical importance of strengthening outcomes for those most disenfranchised by environmental degradation.
- **Improve the incorporation of feedback in systems models.** Better accounting for feedback loops within current models, such as the impacts of food systems on climate, is essential for understanding and acting on a range of environment–agrifood system interactions.
- **Accelerate game-changing innovations.** Emerging technologies can revolutionize how food is grown, but research on innovation constraints is needed to inform policies that support adoption of more sustainable technologies and practices at scale.

Climate change and biodiversity loss are arguably the greatest environmental challenges facing humanity today, and unsustainable agrifood systems are both a key cause and consequence of this rampant environmental degradation. Agrifood systems are directly linked to climate change, freshwater depletion and pollution, deforestation, and soil degradation and desertification (Figure 5.1). Following the 1992 Earth Summit, which drew global attention to unsustainable resource use, three of these critical challenges—biodiversity loss, climate change, and desertification—were taken up by United Nations conventions that established mutually agreed-upon goals and periodic global meetings to review progress. However, improvements in environmental sustainability have been uneven at best, with continued rapid degradation that poses a growing risk to all life on Earth. This chapter

FIGURE 5.1 The environmental impacts of agrifood production

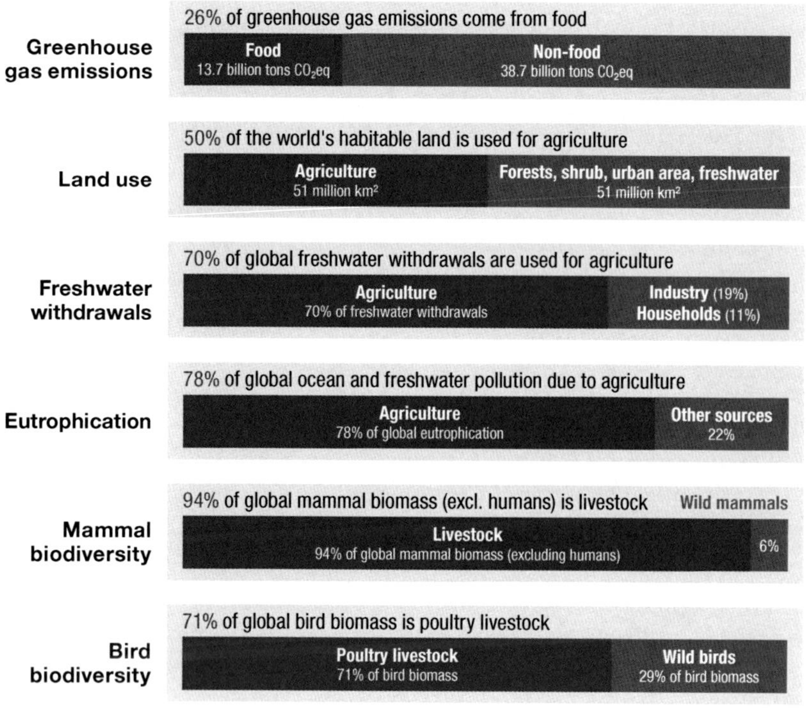

Source: Ritchie et al. (2022), based on data from Poore and Nemecek (2018); FAO (2011); and Bar-On et al. (2018).
Note: CO_2eq = carbon dioxide equivalent; km = kilometers.

reviews how key environmental challenges intersect with agrifood systems and describes research contributions toward understanding and addressing these challenges over the past 50 years, and then explores some future directions for environmental sustainability research.

The beginnings: The Green Revolution

Agricultural productivity and overall food security increased dramatically with Green Revolution technologies and practices introduced in the 1960s and 1970s. These achievements were accompanied by rapid increases in environmental degradation, greater dependence on inputs, and progressive loss of social-ecological resilience as a result of more expansive, genetically uniform monocultures that relied on greater use of chemical inputs and water control (Conway and Barbie 1988; Pingali 2012). Environmental impacts were neglected or, when acknowledged, were accepted as a necessary byproduct, to be mitigated as feasible. While these environmental externalities were neglected, the increased agricultural productivity did reduce the need to convert additional pasture and forest lands to agricultural production, conserving several million hectares of land (Stevenson et al. 2013). Yield improvements—and the associated land sparing—are recognized as an implicit benefit to environmental sustainability.

CGIAR contributed substantially to the technical advances of the Green Revolution (Evenson and Gollin 2003; Pingali 2012), and CGIAR research continues to be closely identified with its tenets, even as the organization has broadened its scope to food systems research (CGIAR Systems Organization 2021). When IFPRI was established in 1975 and then joined CGIAR in 1980, the benefits of the Green Revolution were being reaped largely in the form of lower levels of hunger due to increased yields of key staple crops (see Chapter 3).

IFPRI's initial remit was (1) to research the enabling policy environment for Green Revolution technologies and practices that would strengthen their impact on food security, (2) to assess trends and factors that could adversely affect the world food situation, and (3) to assess linkages between agricultural and broader economic development (see Chapter 17, Box 17.1) (Farrer 2000).

Even though IFPRI's mandate did not originally include environmental sustainability, the Institute did early on examine broader impacts of incentives, trade, and other policies and institutions affecting agrifood systems. In particular, researchers at IFPRI and elsewhere began to analyze the impacts of subsidies and other policy measures on agricultural productivity and the environment. Key environmental challenges included imbalanced use of fertilizers and pesticides as well as growing water scarcity. Research suggested changes that would better balance productivity with environmental goals, particularly for the intensively cultivated rice and wheat areas of Asia. Examples include research on irrigation water use in the Philippines (Rosegrant 1985), on the sustainability of irrigation (Svendsen 1987), on fertilizer subsidies in Bangladesh (Ahmed 1987), and on food subsidies in Mexico (Lustig 1986).

Policies and institutions for environmental sustainability under Agenda 21

As environmental impacts from human development worsened, including from agricultural intensification, the United Nations organized the Earth Summit in Rio de Janeiro in 1992, with the goal of putting sustainable development on the global policy agenda. The resulting Agenda 21 aimed to increase food production, reduce poverty levels, and improve the state of critical ecosystems (Desai and Ringler 2013). Another outcome of the Earth Summit was the establishment of three conventions to address particularly vexing environmental challenges, namely climate change, biodiversity loss, and desertification.

Just ahead of the Earth Summit, the second external program review of IFPRI had recommended, among other things, a research focus on environmental policies (CGIAR TAC 1991). The report noted that national agricultural policies tended to favor short-term food production over longer-term natural resource conservation, and that economic assessments of environmental externalities were lacking. Advances in environmental research linked to agrifood systems in the 1990s included work on water systems, soil health, tropical forests, and pest management.

Research on the future of water and other environmental resources

Recognition of the fundamental role of water in agricultural production systems, including in transforming low-production systems into high-production systems, led to substantial new research on optimizing agricultural water use in the context of hydrological systems, with attention to the role of institutions and governance arrangements as well as foresight studies (Meinzen-Dick 2014; Rosegrant et al. 2000, 2002). With water quantity quasi-finite and most withdrawn freshwater resources used in agriculture, understanding water futures and their implications for food security garnered growing attention. In 2000, the *World Water Vision* report (Cosgrove and Rijsberman 2000) described a global water crisis and called for systemic shifts and behavioral changes toward making water "everybody's business." This report was followed by *Water for Food, Water for Life: Comprehensive Assessment of Water Management in Agriculture*, led by the International Water Management Institute, which undertook a deep exploration of agricultural water futures (Molden 2007). A recent re-assessment of this work (Seijger et al. 2025) noted that most of its projected water productivity gains were too optimistic and had not materialized. This was mainly because expected photosynthesis breakthroughs had not occurred, and other interventions to address water scarcity had not taken place. Instead, agricultural production areas were expanded, freshwater resources dwindled, and environmental degradation continued.

Another milestone was the Millennium Ecosystem Assessment (MEA 2005), a scientific assessment of the world's ecosystems and their services that provides the basis for conserving and using ecosystems sustainably. The report systemized ecosystem services, which include provisioning services, such as food and water; regulating services, such as climate or flood regulation; supporting services, such as soil formation; and cultural services. More than 1,300 experts contributed to this report, including IFPRI and many other CGIAR researchers.

All these assessments highlighted the need for policies and institutions designed to address the environmental externalities of agricultural production. In this context, IFPRI researchers introduced some of the first global modeling tools to examine the impact of alternative agricultural futures and underlying water needs and key ecosystem services (Ringler et al. 2010; Rosegrant and Cai 2002; Rosegrant and Ringler 1999), such as food and water security, and related trade-offs across development goals. This work provided the natural resource basis for the development of IFPRI's well-known global modeling tools on the climate–water–food nexus (see Chapter 4 for more about these models).

Microeconomic research on policies and incentives for improved sustainability

In terms of national policies and institutions to counteract water scarcity and pollution related to agrifood systems, research focused on: market instruments, such as the benefits of water markets (Rosegrant and Binswanger 1994); the role of governance, including water-user associations (Cernea and Meinzen-Dick 1994); the potential of farmer-managed irrigation systems (Kolavalli and Brewer 1999); multiple-use water systems (Meinzen-Dick and Bakker 1999); and paying farmers to use less water (Ringler 2007). Most of these research areas have since been incorporated into the mainstream of water policy and practice.

Another body of research focused on soil fertility and sustainable land management in degraded, drier, or less-favored environments (for example, Scherr and Hazell 1994). Research on sustainable land management and poverty across Africa, Asia, and Latin America advanced the concept of "development domains." These were defined as geographic areas with similar characteristics that determine the comparative advantage of different rural livelihood strategies. The concept of domains provided a useful framework to better target policy interventions based on a combination of biophysical and socioeconomic criteria (Chamberlin et al. 2006; Pender et al. 2006), and also laid the foundation for later analysis of the cost of land degradation and improvement. This analysis (Nkonya et al. 2016) presented novel methodological approaches to using publicly available data to assess the economic costs of land degradation, including costs related to loss of ecosystem services and associated impacts on human well-being. Results suggest that the studied investments in land restoration have an average return-on-investment ratio of 5:1.

Research on property rights and natural resource institutions

Research by IFPRI and partner organizations also reflected the growing recognition of the role of property rights, institutions, and good governance for ensuring sustainable use of water and other natural resources (Baland and Platteau 1996; Bruns and Meinzen-Dick 2000; Ostrom 1990; see also Chapter 6). A growing body of research evidence on the effectiveness of communities in managing resources, along with broad recognition of the failure of top-down solutions and blueprints, led to policies devolving management of forests, fisheries, rangelands, and irrigation systems from the state to user groups (Colfer and Bryon 2001; Meinzen-Dick et al. 2001; Vermillion 1997). However, rapid devolution did not always lead to sustainable or equitable management of these common pool resources, highlighting the need for effective governance institutions (Meinzen-Dick et al. 2002; Ostrom 2001) and participatory research that involves communities.

In response to growing demand from CGIAR Centers and other partners for research on the institutional foundations of sustainable management practices, IFPRI led the formation of CGIAR's systemwide program on Collective Action and Property Rights (CAPRi) in 1995. CAPRi brought attention to the important institutional foundations of many sustainable natural resource management practices, such as soil erosion control and community-led forest management, that have both large spatial scales and long time horizons. Specifically, the spatial scale requires coordination among communities and with government, while the timeframe highlights the importance of property rights in creating incentives and authority for investments and ongoing management. CAPRi's work also increased interest in the critical role of local communities in improving sustainability. Partnerships with nongovernmental organizations (NGOs) and other development and conservation actors helped identify ways to strengthen local institutions, such as in India (Box 5.1), and interest grew in community-based conservation of biodiversity and natural resources, including groundwater. While studying factors affecting local community action, this work also recognized the importance of government policies and coordination, especially at higher levels, and the need for adaptive collaborative management of resources (Colfer and Prabhu 2023).

BOX 5.1 The Promise of Commons initiative in India: A success story

Lessons from CAPRi engagements, including trainings for nongovernmental organizations (NGOs) and government actors, in India on the key role of institutions for collective action and property rights in advancing environmental sustainability resonated with the Foundation for Ecological Security (FES), a local NGO working on ecological restoration and improved livelihoods. This led to long-term collaboration between FES and CGIAR Centers in strengthening community-based institutions and the Promise of Commons initiative. This initiative supports communities in securing their rights to the commons, such as common grazing lands or forests, facilitates collective action through village institutions, and links these institutions to government programs providing additional resources to restore degraded lands. These approaches, informed by IFPRI and other CGIAR Centers, have now been applied in more than 48,000 communities across 14 states in India (FES 2023). A recent impact assessment of FES work in Rajasthan state found that it facilitated significant increases in tree cover as well as tree and shrub diversity and reduced forest encroachment, likely as a result of institutional strengthening (Hughes et al. 2024). This support for community-led management is particularly important given the large benefit streams from ecosystem services emanating from India's common lands, including both direct income-generating services as well as services supporting ecological resilience (Guo et al. 2024; Sandhu et al. 2023).

Behavioral and systemic approaches

Since 2000, there has also been an increasing focus on gendered behavioral systems—that is, the norms and rules governing how women and men behave in different societies and environments—which are explored in more depth in Chapters 6 and 14 of this book. There was particular interest in women's role in environmental stewardship, the sustainable use and management of natural resources, and technology adoption, on the one hand. On the other hand, some researchers explored how women are affected differently from men by environmental degradation, such as deforestation and water depletion, given their important roles in the collection of firewood and domestic water supply (for example, Agarwal 2013; Meinzen-Dick et al. 2014). Additional major conceptual advances were made through work on women's rights to land and other natural resources and on the linkage between soil health and gender (Doss and Meinzen-Dick 2020; Meinzen-Dick et al. 2019; Zhang et al. 2021).

In recent years, behavioral experiments—which use games to study how people behave in different situations—have examined the dynamics of management of common pool resources and environmental services (Cárdenas and Ostrom 2004). These experiments have contributed to a better understanding of the role of group characteristics (including gender composition), individual and group behaviors, and approaches to resolving social dilemmas around resource management and overcoming coordination and collective action problems in agri-environmental settings (Bell and Zhang 2016; Bell et al. 2016). Both experimental games and mixed-methods approaches can provide insights into how farmers and other users of natural resources respond to pro-environmental financial incentives and how communities select and implement rules to improve resource governance and encourage restoration. This same research also led to new insights, including that games originally designed to measure the propensity for communities to cooperate (Cárdenas and Ostrom 2004) can also be used as an intervention to strengthen collective action (Cárdenas and Carpenter 2005; Janssen 2023). Innovative adaptation of games, paired with debriefings of participants, has provided experiential learning about natural resource management and institutional capacity development and has catalyzed actions to enhance self-governance (ElDidi et al. 2024; Falk et al. 2023).

A systems approach has also been used to understand the relationship between landscape complexity and ecosystem services provision.[1] This work recognizes that biodiversity and ecosystem services underpin the stability and sustainability of agriculture and that both agriculture and many of these key ecological functions operate at the landscape scale. Policy research examined the potential of integrated landscape approaches for development and conservation (Estrada-Carmona et al. 2024; Waeber et al. 2023) and advanced the use of integrated land use and ecosystem service modeling to inform decision-making (Li et al. 2021). Affiliated systems research includes assessment of the climate change mitigation benefits of nature-based crop protection, the impacts of agricultural development pathways on biodiversity, and the economic valuation of land use and natural systems to inform policy and investment decisions.[2]

1 This work includes Chaplin-Kramer et al. (2019), Karp et al. (2018), Zhang, Kato, et al. (2018), and Zhang, Lu, et al. (2018).

2 On nature based crop protection, see Wyckhuys, Furlong, et al. (2022) and Wyckhuys, Zhang, et al. (2022); on biodiversity, see Flachsbarth et al. (2015); and on economic valuation, see Huang et al. (2018), Sandhu et al. (2023), and Hettiarachchi et al. (2023).

Research needs for a sustainable future

Poverty and food insecurity cannot be reduced sustainably without addressing the converging crises of climate change, biodiversity loss, and land degradation.[3] And while the need for action has been established, there is no consensus on where best to intervene, or how (Chappell and LaValle 2011; Phalan et al. 2011). Given this immense set of challenges, what direction should agrifood systems policy research take to best address environmental sustainability? Looking ahead, we see six main paths for future policy research and action.

Strengthen systems thinking in policy research within and across food and linked environmental systems

Environmental outcomes from agrifood system interventions are complex, uncertain, variable over time, and affected by multiple interlocking natural and human systems. Assessing such outcomes requires systems thinking and robust evaluation strategies that include transdisciplinary approaches. Assessment of multiple, heterogeneous effects and multi-objective impact measures, as well as models or research designs that adequately account for behavior change at various scales, will be essential. The Economics of Ecosystems and Biodiversity for Agriculture and Food Evaluation Framework (TEEB 2018), to which IFPRI contributed, made the case for comprehensively measuring food system externalities, including positive and negative changes in ecosystem services and across all stages of the value chain. This work also provides the analytical framework for true cost accounting, the theme of two recent *State of Food and Agriculture* reports (FAO 2023, 2024).

Increasing the environmental sustainability of food systems, moreover, requires going beyond agricultural production systems. Water pollution, greenhouse gas emissions, and other forms of environmental degradation are also caused by food transportation, processing, and preparation (Ringler et al. 2018; Tassou et al. 2009).

Interest in addressing the role of dietary choices in environmental degradation has gained a lot of traction and is reflected in Sustainable Development Goal 12, which calls for sustainable consumption as well as production. The growing importance of environmental sustainability of diets is also reflected in the concept of "sustainable healthy diets" (FAO and WHO 2019), which requires balancing environmental and nutrition trade-offs and establishing

---

3 For discussion on this topic, see WFPF (2025), Berry et al. (2015), Lal (2009), Wheeler and von Braun (2013), and van der Sluijs and Vaage (2016).

coherent enabling and policy frameworks. This, in turn, requires stronger linkages between agrifood systems and health-policy actors for coordinated action in "win-win" areas, where improved nutrition and health interventions can also directly improve environmental outcomes (Clark et al. 2019). An example is the reduction of sugar consumption, which has both large health and large environmental benefits (Lee et al. 2020; Popkin and Ng 2021). Agrifood systems researchers should also work more closely with providers of alternative energy solutions to reduce the excessive reliance of "clean" energy on biofuel production. Biofuels are often derived from crops that require large amounts of land and water and chemical inputs, and as a result, their large-scale development has increased global food prices (Pimentel et al. 2009).

Double down on transdisciplinary research processes

Interdisciplinary research has identified policies, technologies, and institutions that can reduce the environmental footprint of agrifood systems. However, their implementation requires transdisciplinary approaches that actively engage farmer, pastoralist, fisher, and forest-dependent communities that are key stewards of our natural resource base (Enqvist et al. 2018), along with other key actors from government, civil society, and the private sector, to jointly identify and apply solutions. These transdisciplinary research approaches need to be applied to the holistic evaluation of agrifood systems' externalities, positive and negative, along with a comprehensive assessment of interventions across scales to identify and avoid any negative externalities or increased inequities.

The CGIAR 2022–2024 Research Portfolio created opportunities to deepen and enhance cross-Center collaborative research through the establishment of the Research Initiatives on Agroecological Transitions, Low-Emission Food Systems, Nature-Positive Solutions, and NEXUS Gains. Under these CGIAR Initiatives, transdisciplinary research in "living labs" and other collaborative and participatory action research processes combined biophysical and social science research to support actor groups in developing their own understanding and assessment of sustainability and crafting their own pathways toward sustainability. Living labs are place-based environments or spaces where innovation takes place in people's real-world settings with the goal of taking innovation out of formal, controlled "laboratory" environments. Principles of the Living Labs for People developed by CGIAR include co-production, gender equality and social inclusion, good governance, and institutional sustainability to advance existing and novel innovation processes (Habermann et al. 2023). In southeastern China, a living lab has emerged from various actor groups in Qingshan village aligning themselves with national policies on low-emission food systems transition and green growth. Low-carbon practices considered or adopted include composting, reducing emissions from waste burning, and producing rice more sustainably (Habermann et al. 2023). In Madhya Pradesh, India, visioning exercises in the agroecology living landscape (Bergamini et al. 2023) involving farmers, various government departments, the private sector, civil society, and researchers identified not only the technical innovations sought from agroecological innovations but also the relevant business models, policies, and behavior changes (Singh and Freed 2024). These approaches build on the tradition of participatory research on resource management with more systematic approaches to citizen science and involve a range of actors in identifying problems and solutions.

Strengthen political economy and policy process research

While sustainability-focused research has increasingly incorporated institutional and behavioral economics, limited implementation of tested interventions points to the need for a much stronger political economy focus and policy process research to address why policies are (or are not) adopted to drive improvement in environmental outcomes (see Chapter 15). This research should also improve the knowledge base on the enabling policy environment for "best-bet" environmental solutions as well as for removing hurdles to the adoption or diffusion of effective approaches. In particular, greater attention is needed to identify what will motivate different types of actors—from global and national political leaders to men and women in communities—to adopt policies and practices that account for long-term costs and benefits, including both those that are visible and externalities imposed on others.

Infuse an environmental justice lens in agrifood systems research

Environmental sustainability of agrifood systems cannot be achieved without environmental justice (see Chapter 4). This includes procedural and epistemic justice to ensure that those most disenfranchised by environmental degradation have a voice, that their knowledge systems are valued, and that they are involved in the development of solutions (Grafton et al. 2025). Work on women's empowerment (see Chapter 14), energy and water insecurity, and transdisciplinary research under the CGIAR Research Initiatives and new Science Programs and Accelerators provides a basis for strengthening attention to environmental justice.

Improve the incorporation of feedback loops in systems models

Most modeling systems, including those developed by CGIAR Centers, assess the impacts of climate change on food systems or the impacts of food systems on water depletion, without adequately incorporating feedback loops, such as the impacts of food systems on climate or the impact of long-term groundwater over-extraction on food security. Given the continued rapid degradation of environmental systems, more assessments of the direct impact of extreme water and land degradation on food security and livelihoods are needed, as well as assessments of how such increased degradation may constrain the options for jointly improving food security and biodiversity futures. Examples of areas needing more thorough assessment include tipping points of environmental change, such as peak glacier melt and peak groundwater depletion, which are expected to dramatically affect food production levels in South Asia, among other regions (Immerzeel et al. 2010), and ice mass losses that affect rice production in coastal lowland areas (Chen et al. 2012). Another example is the Atlantic Meridional Overturning Circulation collapse—potentially causing substantial cooling in the northern hemisphere paired with further warming and/or drying in parts of the southern hemisphere—which would make food production more challenging at a global level (Alcala 2021).

Accelerate game-changing innovations

The Green Revolution technologies were game-changing innovations with lasting food security benefits. Over the span of a few decades, tractors and combine harvesters replaced oxcarts, horse-drawn plows, and threshing boards, while high-yielding varieties and agrochemicals led to huge increases in primary productivity from the 1960s to 1980s. By drastically transforming established structures, these disruptive innovations created new markets, changed the very nature of agrifood production, and ultimately overtook well-established technologies.

In a similar way, emerging technologies such as modern forms of genome editing, alternative proteins, artificial intelligence, robotics, solar-powered irrigation and farming, and nature-based solutions such as agroecology can truly revolutionize the way in which food is grown and should be applied together (see Chapter 17) (Herrero et al. 2020). In particular, CRISPR/Cas9 is an advanced tool for gene editing that can dramatically improve crop yields, quality, and disease resistance (Aman Mohammadi et al. 2023). Artificial intelligence, similarly, is a breakthrough innovation that can benefit all components of food systems (Ben Ayed and Hanana 2021). Alternative proteins are another breakthrough that can dramatically reduce the environmental footprint of animal-source foods, but more research and policy engagement are required (Pingali et al. 2023). Moreover, new biodiversity-driven measures such as biological control, microbial soil inoculants, or science-based diversification tactics can be key enablers of a "greener revolution" (Snapp et al. 2010). The adoption of all of these game-changing innovations at scale can ensure food security for a swelling global population without disregarding human and planetary health. But just as with the Green Revolution, it is essential to examine emerging environmental and equity impacts of these technologies, rather than assuming that they provide technological panaceas.

Conclusions

The past 50 years have seen an expansion of research on sustainability in natural resource management within agriculture and in food systems more broadly, accompanied by advances in models, in metrics, and in understanding of the complexity and interconnectedness of these systems. Research has advanced from accepting environmental degradation as a byproduct to improving food security to identifying policies, institutions, and technologies that enhance food security without adversely affecting ecosystem health. Policy research has, furthermore, highlighted the need to involve resource users who are aware of the complexity of the socio-ecological conditions in development of policies and incentives for resource use.

Despite all of these advances, environmental crises of climate change, land degradation, and biodiversity loss are continuing to outpace our levels of action. To achieve food security now and in the future will require systems thinking that combines knowledge from multiple disciplines and resource users. It will also require careful assessment of political economy contexts and other factors that motivate actors at all levels from the local to the global to adopt technologies and practices that can sustain and enhance the resources that provide our food and livelihoods.

References

Agarwal, B. 2013. *Gender and Green Governance: The Political Economy of Women's Presence Within and Beyond Community Forestry.* Oxford, UK: Oxford University Press.

Ahmed, R. 1987. "Structure and Dynamics of Fertilizer Subsidy: The Case of Bangladesh." *Food Policy* 12 (1): 63–75.

Alcala, A. 2021. "Connecting the State of Meridional Overturning Circulation to Human Global Food Security: The Consequences of a Redistribution of Ecosystems in Response to Weakening." *Life: The Excitement of Biology* 11(3): 61–91.

Aman Mohammadi, M., M.R. Maximiano, S.M. Hosseini, and O.L. Franco. 2023. "CRISPR-Cas Engineering in Food Science and Sustainable Agriculture: Recent Advancements and Applications." *Bioprocess Biosystems Engineering* 46: 483–497.

Baland, J-M., and J-P. Platteau. 1996. *Halting the Degradation of Natural Resources: Is There a Role for Rural Communities?* Oxford, UK: FAO and Clarendon Press.

Bar-On, Y.M., R. Phillips, and R. Milo. 2018. "The Biomass Distribution on Earth." *PNAS* 115 (25): 6506–6511.

Bell, A. and W. Zhang. 2016. "Payments Discourage Coordination in Ecosystem Services Provision: Evidence from Behavioral Experiments in Southeast Asia." *Environmental Research Letters* 11: 114024.

Bell, A., W. Zhang, and K. Nou. 2016. "Pesticide Use and Cooperative Management of Natural Enemy Habitat in a Framed Field Experiment." *Agricultural Systems* 143: 1–13.

Ben Ayed, R., and M. Hanana. 2021. "Artificial Intelligence to Improve the Food and Agriculture Sector. *Journal of Food Quality* 1: 5584754.

Bergamini, N., A. Navarrete-Cruz, B. Triomphe, et al. 2023. *Vision to Action for Agroecological Transitions in the Living Landscapes: A Cross-Country Progress Report on Methodological Approaches.* Montpellier, France: CGIAR.

Berry, E.M., S. Dernini, B. Burlingame, A. Meybeck, and P. Conforti. 2015. "Food Security and Sustainability: Can One Exist without the Other? *Public Health Nutrition*18 (13): 2293–2302.

Bruns, B.R., and R. Meinzen-Dick, eds. 2000. *Negotiating Water Rights.* London: Intermediate Technology Press.

Cárdenas, J.C., and E. Ostrom. 2004. "What Do People Bring into the Game? Experiments in the Field about Cooperation in the Commons." *Agricultural Systems* 82 (3): 307–326.

Cárdenas, J.C., and J.P. Carpenter. 2005. "Three Themes on Field Experiments and Economic Development." In *Field Experiments in Economics (Research in Experimental Economics,* Vol. 10), eds. G.W. Harrison, J. Carpenter, and J.A. List, 71–123. Bingley, UK: Emerald Group.

Cernea, M.M., and R.S. Meinzen-Dick. 1994. "Design for Water User Associations: Organisational Characteristics." *ODI Irrigation Management Network Paper* 30 (17).

CGIAR System Organization. 2021. *CGIAR 2030 Research and Innovation Strategy: Transforming Food, Land, and Water Systems in a Climate Crisis*. Montpellier, France.

CGIAR TAC (Technical Advisory Committee). 1991. *Report of the Second External Programme Review of the International Food Policy Research Institute (IFPRI)*. Rome: TAC Secretariat, FAO.

Chamberlin, J., J.L. Pender, and B. Yu. 2006. "Development Domains for Ethiopia: Capturing the Geographical Context of Smallholder Development Options." DSGD Discussion Paper No. 43; EPTD Discussion Paper No. 159. IFPRI, Washington, DC.

Chaplin-Kramer, R., M. O'Rourke, N. Schellhorn, et al. 2019. "Measuring What Matters: Actionable Information for Conservation Biocontrol in Multifunctional Landscapes." *Frontiers in Sustainable Food Systems* 3: 60.

Chappell, M.J., and L.A. LaValle. 2011. "Food Security and Biodiversity: Can We Have Both? An Agroecological Analysis." *Agriculture and Human Values* 28: 3–26.

Chen, C.C., B. McCarl, and C.C. Chang. 2012. "Climate Change, Sea Level Rise and Rice: Global Market Implications." *Climatic Change* 110: 543–560.

Clark, M.A., M. Springmann, J. Hill, and D. Tilman. 2019. "Multiple Health and Environmental Impacts of Foods." *PNAS* 116 (46): 23357–23362.

Colfer, C.J.P., and Y. Byron, eds. 2001. *People Managing Forests: The Links between Human Well-Being and Sustainability*. Washington, DC: Resources for the Future.

Colfer, C.J.P., and R. Prabhu. 2023. *Responding to Environmental Issues through Adaptive Collaborative Management: From Forest Communities to Global Actors*. London: Routledge.

Conway, G.R., and E.B. Barbie. 1988. "After the Green Revolution: Sustainable and Equitable Agricultural Development." *Futures* 20 (6): 651–670.

Cosgrove, W.J., and F.R. Rijsberman. 2000. *World Water Vision: Making Water Everybody's Business*. London: Earthscan.

Desai, N., and C. Ringler. 2013. "Green Economy: Sustainable and Growing, but also Food Secure?" In *2012 Global Food Policy Report*, 28–37. Washington, DC: IFPRI.

Doss, C., and R. Meinzen-Dick. 2020. "Land Tenure Security for Women: A Conceptual Framework." *Land Use Policy* 99: 105080.

ElDidi, H., W. Zhang, I. Blackmore, et al. 2024. "Getting Ahead of the Game: Experiential Learning for Groundwater Governance in Ethiopia." *International Journal of the Commons* 18 (1): 66–81.

Enqvist, J.P., S. West, V.A. Masterson, L.J. Haider, U. Svedin, and M. Tengö. 2018. "Stewardship as a Boundary Object for Sustainability Research: Linking Care, Knowledge and Agency." *Landscape and Urban Planning* 179: 17–37.

Estrada-Carmona, N., R. Carmenta, J. Reed, et al. 2024. "Reconciling Conservation and Development Requires Enhanced Integration and Broader Aims: A Cross-Continental Assessment of Landscape Approaches." *One Earth* 7 (10): 1858–1873.

Evenson R.E., and D. Gollin. 2003. "Assessing the Impact of the Green Revolution, 1960 to 2000." *Science* 300 (5620): 758–762.

Falk, T., W. Zhang, R. Meinzen-Dick, R. Sanil, P. Priyadarshini, I. Soliev, and L. Bartels. 2023. "Games for Experiential Learning: Triggering Collective Changes in Commons Management." *Ecology and Society* 28 (1): 30.

FAO (Food and Agriculture Organization of the United Nations). 2011. *The State of the World's Land and Water Resources for Food and Agriculture: Managing Systems at Risk*. London: Earthscan; Rome: FAO.

FAO. 2023. *The State of Food and Agriculture 2023: Revealing the True Cost of Food to Transform Agrifood Systems*. Rome.

FAO. 2024. *The State of Food and Agriculture 2024: Value-Driven Transformation of Agrifood Systems*. Rome.

FAO and WHO (World Health Organization). 2019. *Sustainable Healthy Diets: Guiding Principles*. Rome.

Farrer, C. 2000. *IFPRI's First 10 Years*. Washington, DC: IFPRI.

FES (Foundation for Ecological Security). 2023. *Annual Report 2022–2023*. Anand, Gujarat, India.

Flachsbarth, I., B. Willaarts, H. Xie, G. Pitois, N.D. Mueller, C. Ringler, and A. Garrido. 2015. "The Role of Latin America's Land and Water Resources for Global Food Security: Environmental Trade-Offs of Future Food Production Pathways." *PLoS ONE* 10 (1): e0116733.

Grafton, R.Q., S. Fanaian, J. Horne, et al. 2025. "Rethinking Responses to the World's Water Crises." *Nature Sustainability* 8: 11–21.

Guo, Z., H. Sharma, M. Jadav, et al. 2024. "Measuring Above-Ground Carbon Stock Using Spatial Analysis and the InVEST Model: Application in the Thoria Watershed, India." *Environmental Research Communications* 6 (11): 115036.

Habermann, B., R. Nehring, W. Zhang, et al. 2023. "A Conceptual Framework of Living Labs for People for Sustainable Food Systems." IFPRI Discussion Paper 2227. IFPRI, Washington, DC.

Herrero, M., P.K. Thornton, D. Mason-D'Croz, et al. 2020. "Innovation Can Accelerate the Transition towards a Sustainable Food System." *Nature Food* 1 (5): 266–272.

Hettiarachchi, U., W. Zhang, T.T. Pham, K. Davis, and C. Fadda. 2023. "Ecosystem Services May Provide Large Economic Values in Kenya and Vietnam: A Value Transfer Application Based on Results from a Systematic Literature Review." IFPRI Discussion Paper 2228, IFPRI, Washington, DC.

Huang, J., K. Zhou, W. Zhang, et al. 2018. "Uncovering the Economic Value of Natural Enemies and the True Cost of Chemical Insecticides to Cotton Farmers in China." *Environmental Research Letters* 13: 064027.

Hughes, K., H. Sharma, P. Priyadarshini, T. Vagen, L.A. Winowiecki, and R.S. Meinzen-Dick. 2024. "Integrating Earth Observation, Biophysical, and Survey Data to Evaluate the Ecological Impacts of a Common Land Protection and Restoration Intervention in Rajasthan, India." *Humanities and Social Sciences Communications* 11: 1535.

Immerzeel, W.W., L.P.H. Van Beek, and M.F.P. Bierkens. 2010. "Climate Change Will Affect the Asian Water Towers." *Science* 328: 1382–1385.

Janssen, M., T. Falk, R. Meinzen-Dick, and B. Vollan. 2023. "Using Games for Social Learning to Promote Self-Governance." *Current Opinion in Environmental Sustainability* 62: 101289

Karp, D.S., R. Chaplin-Kramer, T.D. Meehan, et al. 2018. "Crop Pests and Predators Exhibit Inconsistent Responses to Surrounding Landscape Composition." *PNAS*: 115 (33): E7863–E7870.

Kolavalli, S., and J.D. Brewer. 1999. "Facilitating User Participation in Irrigation Management." *Irrigation and Drainage Systems* 13: 249–273.

Lal, R. 2009. "Soil Degradation as a Reason for Inadequate Human Nutrition." *Food Security* 1: 45–57.

Lee, J.Y., R.L. Naylor, A.J. Figueroa, and S.M. Gorelick. 2020. "Water-Food-Energy Challenges in India: Political Economy of the Sugar Industry." *Environmental Research Letters* 15 (8): 084020.

Li, M., Z. Guo, and W. Zhang. 2021. "Balancing Food Security and Environmental Sustainability by Optimizing Seasonal-Spatial Crop Production in Bangladesh." *Environmental Research Letters* 16: 074046.

Lustig, N. 1986. "Food Subsidy Programs in Mexico." IFPRI Working Papers on Food Subsidies. IFPRI, Washington, DC.

MEA (Millennium Ecosystem Assessment). 2005. *Ecosystems and Human Well-Being: Synthesis.* Washington, DC: Island Press.

Meinzen-Dick, R.S. 2014. "Property Rights and Sustainable Irrigation: A Developing Country Perspective." *Agricultural Water Management* 145: 23–31.

Meinzen-Dick, R.S., and M. Bakker. 1999. "Irrigation Systems as Multiple-Use Commons: Water Use in Kirindi Oya, Sri Lanka." *Agriculture and Human Values* 16 (3): 281–293.

Meinzen-Dick, R.S., A. Knox, and M. Di Gregorio, eds. 2001. *Collective Action, Property Rights, and Devolution of Natural Resource Management: Exchange of Knowledge and Implications for Policy.* Feldafing, Germany: Zentralstelle für Ernährung und Landwirtschaft.

Meinzen-Dick, R.S., C. Kovarik, and A. Quisumbing. 2014. "Gender and Sustainability." *Annual Review of Environment and Resources* 39: 29–55.

Meinzen-Dick, R.S., A. Quisumbing, C. Doss, and S. Theis. 2019. "Women's Land Rights as a Pathway to Poverty Reduction: Framework and Review of Available Evidence." *Agricultural Systems* 172: 72–82.

Meinzen-Dick, R.S., K. V. Raju, and A. Gulati. 2002. "What Affects Organization and Collective Action for Managing Resources? Evidence from Canal Irrigation Systems in India." *World Development* 30 (4): 649–666.

Molden, D., ed. 2007. *Water for Food, Water for Life: A Comprehensive Assessment of Water Management in Agriculture.* London: Earthscan; Colombo: International Water Management Institute.

Nkonya, E., A. Mirzabaev, and J. von Braun, eds. 2016. *Economics of Land Degradation and Improvement: A Global Assessment for Sustainable Development.* Cham, Switzerland: Springer.

Ostrom, E. 1990. *Governing the Commons: The Evolution of Institutions for Collective Action.* Cambridge, UK: Cambridge University Press.

Ostrom, E. 2001. "Decentralization and Development: The New Panacea." In *Challenges to Democracy. Political Studies Association Yearbook*, eds. K. Dowding, J. Hughes, and H. Margetts, 237–256. London: Palgrave Macmillan.

Pender, J., F. Place, and S. Ehui. 2006. "Strategies for Sustainable Land Management in the East African Highlands: Conclusions and Implications." In *Strategies for Sustainable Land Management in the East African Highlands*, eds. J. Pender, F. Place, and S. Ehui, 377–416. Washington, DC: IFPRI.

Phalan, B., M. Onial, A. Balmford, and R.E. Green. 2011. "Reconciling Food Production and Biodiversity Conservation: Land Sharing and Land Sparing Compared." *Science* 333: 1289–1291.

Pimentel, D., A. Marklein, M.A. Toth, et al. 2009. "Food Versus Biofuels: Environmental and Economic Costs." *Human Ecology* 37: 1–12.

Pingali, P. 2012. "Green Revolution: Impacts, Limits, and the Path Ahead." *PNAS* 109 (3)1: 12302–12308.

Pingali, P., J. Boiteau, A. Choudhry, and A. Hall. 2023. "Making Meat and Milk from Plants: A Review of Plant-Based Food for Human and Planetary Health." *World Development* 170: 106316.

Poore, J., and T. Nemecek. 2018. "Reducing Food's Environmental Impacts through Producers and Consumers." *Science* 360 (6392): 987–992.

Popkin, B.M, and S.W. Ng. 2021. "Sugar-Sweetened Beverage Taxes: Lessons to Date and the Future of Taxation." *PLoS Med* 18 (1): e1003412.

Ringler, C. 2007. "The Potential Role of Economic Instruments in River Basin Management: Increased Economic Efficiency through Paying Farmers to Use Less Water." *Ökologisches Wirtschaften* 1: 35–38

Ringler, C., A. Biswas, and S.A. Cline. eds. 2010. *Global Change: Impact on Water and Food Security.* Heilberg, Germany: Springer.

Ringler, C., J. Choufani, C. Chase, M. McCartney, J. Mateo-Sagasta, D. Mekonnen, and C. Dickens. 2018. *Meeting the Nutrition and Water Targets of the Sustainable Development Goals: Achieving Progress through Linked Interventions.* WLE Research for Development (R4D) Learning Series 7. Colombo, Sri Lanka: CGIAR Research Program on Water, Land and Ecosystems; Washington, DC: World Bank.

Ritchie, H., P. Rosado, and M. Roser. 2022."Environmental Impacts of Food Production." Our World in Data. https://ourworldindata.org/environmental-impacts-of-food

Rosegrant, M.W. 1985. "Efficiency and Equity Impacts of Alternative Water Allocation Methods in Diversion Irrigation Systems in the Philippines." Paper presented at the 1985 Annual Conference of the Australian Agricultural Economics Society, University of New England, Armidale, NSW, February 12–14.

Rosegrant, M.W., and H. Binswanger. 1994. "Markets in Tradable Water Rights: Potential for Efficiency Gains in Developing Country Water Resource Allocation." *World Development* 22 (11): 1613–1625.

Rosegrant, M.W., and X. Cai. 2002. "Global Water Demand and Supply Projections: Part 2. Results and Prospects to 2025." *Water International* 27 (2): 170–182.

Rosegrant, M.W., X. Cai, and S. Cline. 2002. *World Water and Food to 2025: Dealing with Scarcity.* Washington, DC: IFPRI.

Rosegrant, M.W., and C. Ringler. 1999. *World Water Vision Scenarios: Consequences for Food Supply, Demand, Trade, and Food Security: Results from the IMPACT implementation of the World Water Vision Scenarios* Washington, DC: IFPRI.

Rosegrant, M.W., C. Ringler, D.C. McKinney, X. Cai, A. Keller, and G. Donoso. 2000. "Integrated Economic-Hydrologic Water Modeling at the Basin Scale: The Maipo River Basin." *Agricultural Economics* 24 (1): 33–46.

Sandhu, H., W. Zhang, R.S. Meinzen-Dick, et al. 2023. "Valuing Ecosystem Services Provided by Land Commons in India: Implications for Research and Policy." *Environmental Research Letters* 18 (1): 013001.

Scherr, S.J., and P. Hazell. 1994. "Sustainable Agricultural Development Strategies in Fragile Lands." EPTD Discussion Paper No. 1. IFPRI, Washington, DC.

Seijger, C., A. Urfels, M. Christoforidou, P. Hellegers, G. Borghuis, S. Langan, and G. van Halsema. 2025. "More Food, But Less Land and Water for Nature: Why Agricultural Productivity Gains Did Not Materialize." *Agricultural Water Management* 307: 109229.

Singh, S., and S. Freed. 2024. *WP5: Understanding and Influencing Agency and Behaviour Change in India: Activity 5.1 Country Brief.* Agroecology Initiative Brief. Montpellier, France: CGIAR System Organization.

Snapp, S.S., M.J. Blackie, R.A. Gilbert, R. Bezner-Kerr, and G.Y. Kanyama-Phiri. 2010. "Biodiversity Can Support a Greener Revolution in Africa." *PNAS* 107 (48): 20840–20845.

Stevenson, J.R., N. Villoria, D. Byerlee, T. Kelley, and M. Maredia. 2013. "Green Revolution Research Saved an Estimated 18 to 27 Million Hectares from Being Brought into Agricultural Production." *PNAS* 110 (21): 8363–8368.

Svendsen, M.S. 1987. "Sustainability in Irrigated Agriculture." IIMI Working Paper No. 4. International Irrigation Management Institute, Kandy, Sri Lanka.

Tassou, S.A., G. De-Lille, and Y.T. Ge. 2009. "Food Transport Refrigeration: Approaches to Reduce Energy Consumption and Environmental Impacts of Road Transport." *Applied Thermal Engineering* 29 (8–9): 1467–1477.

TEEB (The Economics of Ecosystems and Biodiversity). 2018. *TEEB for Agriculture & Food: Scientific and Economic Foundations.* Geneva: UN Environment Programme.

van der Sluijs, J.P., and N.S. Vaage. 2016. "Pollinators and Global Food Security: the Need for Holistic Global Stewardship." *Food Ethics* 1: 75–91.

Vermillion, D.L. 1997. *Impacts of Irrigation Management Transfer: A Review of the Evidence.* Research Report 11.Kandy, Sri Lanka: International Irrigation Management Institute.

Waeber, P.O., A. Fellay, C.A. Garcia, et al. 2023. "Structuring the Complexity of Integrated Landscape Approaches into Selectable, Scalable, and Measurable Attributes." *Environmental, Science & Policy* 147 (2023): 67–77.

Wheeler, T., and J. von Braun. 2013. "Climate Change Impacts on Global Food Security." *Science* 341 (6145): 508–513.

WFPF (World Food Prize Foundation). 2025. "Hunger's Tipping point: An Urgent Call to Transform Food and Nutrition Security," WFPF Laureate Letter 2025. Des Moines, IA.

Wyckhuys, K.A.G., M.J. Furlong, W. Zhang, and G.C. Yubak Dhoj. 2022. "Carbon Benefits of Enlisting Nature for Crop Protection." *Nature Food* 3: 299–301.

Wyckhuys, K.A.G., W. Zhang, Y. Colmenarez, E. Simelton, B.O. Sander, and Y. Lu. 2022. "Tri-trophic Defenses as a Central Pivot of Low-Emission, Pest-Suppressive Farming Systems." *Current Opinion in Environmental Sustainability* 58: 101208.

Zhang, W., M. Elias, R. Meinzen-Dick, K. Swallow, D. Walker, C.C. Hernandez, and E. Nkonya. 2021. "Soil Health and Gender: Why and How to Identify the Linkages." *International Journal of Agricultural Sustainability* 19 (3/4): 269–287.

Zhang, W., E. Kato, F. Bianchi, P. Bhandary, G. Gort, and W. van der Werf. 2018. "Farmers' Perceptions of Crop Pest Severity in Nigeria Are Associated with Landscape, Agronomic and Socio-Economic Factors." *Agriculture, Ecosystems and Environment* 259: 159–167.

Zhang, W., Y.H. Lu, W. van der Werf, et al. 2018. "Multidecadal, County-Level Analysis of the Effects of Land Use, Bt Cotton, and Weather on Cotton Pests in China." *PNAS* 115 (33): E7700–E7709.

Chapter 6

TENURE
Policy Research on Resources, Rights, and Equity

Ruth Meinzen-Dick, Isabel Lambrecht, Frank Place, Uchendu Eugene Chigbu, Iliana Monterroso, and Diana Suhardiman

Ruth Meinzen-Dick is a research fellow emeritus, IFPRI; **Isabel Lambrecht** is a research fellow, Development Strategies and Governance Unit, IFPRI; **Frank Place** is a consultant, CGIAR; **Uchendu Eugene Chigbu** is a professor of Land Administration, Institute for Land, Livelihood and Housing/Department of Land and Spatial Siciences, Namibia University of Science and Technology; **Iliana Monterroso** is co-lead of the Mexico and Central America Initiative, Climate and Land Use Alliance and associate researcher, CIFOR-ICRAF; **Diana Suhardiman** is director of the Royal Netherlands Institute of Southeast Asian and Caribbean Studies/KITLV and special professor of Natural Resource Governance Climate and Equity, Cultural Anthropology and Development Studies, Faculty of Social Sciences, Leiden University.

Key messages

- Secure tenure of land and natural resources is critical to ensuring the equitable, efficient, and sustainable production of food and resilient rural livelihoods.
- Over the past 50 years, research on land tenure and property rights has evolved substantially, deepening our understanding of the importance of resource tenure and of the range of tenure forms across societies, resources, and time.
- Tenure-related studies applying a diversity of disciplines and methods have shown the variety of tenure arrangements, relationships between tenure and behavior, and the effects of reforms.
- Research has highlighted the complex linkages between tenure security and governance, provided insights into the effects of tenure formalization and individualization, and contributed to devolution programs that transfer rights and responsibility over natural resources from the state to user groups.

- Both research and social movements continue to stress the importance of equity of property rights, including gender equality as well as the rights of Indigenous Peoples, minority ethnic groups, and other marginalized local communities, over natural resources.

Moving forward, research on land tenure and property rights should examine key decisive and structural factors that impede tenure security:

- **Rights for the socially marginalized.** Identify how to overcome impediments to tenure security for ethnic minorities, women, youth, and other marginalized groups, including through the use of disaggregated data to assess indicators, patterns of tenure (in)security and its consequences, and changes over time.
- **Rights and tenure across resources.** Determine the best ways to secure rights for water, minerals, and ecosystem services, as well as tenure across resource systems, including through research on existing mechanisms for collective action, promising innovations to overcome injustices, and new ways to conceptualize and operationalize tenure that stress linkages across resources.
- **Tenure under structural transformation of economies.** Examine how structural transformations, such as urbanization and livelihood diversification, affect tenure and its governance to identify threats and opportunities for equitable transitions, particularly in peri-urban areas and urban food systems.
- **Research to support action.** Identify opportunities for reciprocal collaboration and partnerships, including across disciplines, to ensure that research addresses key issues, yields comprehensive information, and is shared with relevant audiences.

Secure tenure of land and natural resources is critical for ensuring equitable, efficient, and sustainable production of food and resilient rural livelihoods. Research on the importance of tenure of land and related natural resources has a long history in development research and policy. Over the past 50 years, IFPRI and other researchers and practitioners have engaged with diverse field contexts to expand our understanding of the different dimensions of tenure and of its role in food policy, rural development, land stewardship, natural resource governance, empowerment, and equity across different social groups.

In this chapter, we review foundational concepts and key lessons from research on tenure, before identifying priorities for further study, policy, and practice. Our focus is on low- and middle-income countries (LMICs).

Foundational concepts

Tenure refers to the institutional relationships, defined through law or custom, related to natural resources. It determines who can use the resource, for how long, and under what conditions (FAO 2002), including the rights and obligations of the holders of these resources (Bruce 1998). Land tenure has received the most attention, but tenure concepts also apply to a wider set of resources including water, forests and trees, and marine and mineral resources (Bruns and Meinzen-Dick 2000; Fortmann and Bruce 1988; Meinzen-Dick and Pradhan 2002). *Property rights* over resources are enforceable claims to the benefits from the resources.

Tenure security refers to the likelihood that the resource holder's rights will be upheld (Sjaastad and Bromley 2000), including the breadth, duration (time span), and assurance (provability) of resource rights, and freedom from imposition or interference by outside sources (Place et al. 1994). Tenure security enables rights holders to reap the benefits of their investments, either through use or upon transfer to another holder (Chigbu et al. 2017; Place et al. 1994). Resource tenure systems influence decisions about investments (Chigbu et al. 2021; Robinson 2005); those without secure tenure over a resource such as land or trees have less incentive to make long-term investments in the resource because they are not assured that they will reap the returns. For those without long-term tenure, such as renters or women living in contexts where land is considered to belong primarily to men, making long-term investments may be prohibited. In these cases, actions such as planting trees or constructing buildings would be construed as placing too much claim on a resource (Fortmann and Bruce 1988; Fortmann and Rocheleau 1985; Lambrecht and Asare 2016).

Rights do not necessarily imply full ownership and sole authority to use or dispose of a resource. Different individuals, families, groups, and even the state often hold overlapping rights, and the many combinations of these rights are commonly referred to as "bundles of rights." Rights are often categorized as *use rights*, which include the right to access, withdraw from, or exploit a resource for social and economic benefits, and *control* or *decision-making rights*, including the right to manage, exclude, and alienate (sell, lease, or subdivide) a resource (Meinzen-Dick et al. 2004; Schlager and Ostrom 1992). The complex interrelations between these rights and their holders—which together constitute a "web of interests"—underscore that resource rights are fundamentally a social construct (Arnold 2007; Hodgson 2004; Meinzen-Dick and Mwangi 2009). It is therefore not only ownership—holding all rights—that matters, but also use and management rights, which influence and affect the sustainability of the resource and the distribution of benefits.

Tenure governance

Rights over resources derive from many overlapping legal frameworks, a situation known as legal pluralism. While this has been widely recognized in the legal and anthropological literature, one of IFPRI's contributions to understanding tenure has been to bring legal pluralism into economic analyses of policy (Meinzen-Dick and Pradhan 2002). To recognize rights in practice requires considering not only state-issued titles but also claims based on international treaties and law (particularly for Indigenous Peoples), religious law and practices, customary law, rules deriving from projects (for example, specifying who can take products from a forest), and rules made by user groups (especially for water, forests, and rangelands) (see Box 6.1) (Meinzen-Dick and Pradhan 2002). Customary systems consist of the norms, customs, traditions, and practices of a given community, all of which have been inherited from ancestors or developed locally and accepted, reinterpreted, developed, and enforced by local ethnic or Indigenous groups to govern local resource access and use (Alden Wily 2011).

The coexistence of these types of law does not mean that all are equal: each is only as strong as the institution that stands behind it, and that strength is contextual. State law is often strongest in urban areas, while the ability of the state to enforce rights or regulations in remote areas is often limited, and customary institutions may be stronger (Meinzen-Dick and Pradhan 2002). The response of citizens to these laws affects the success or failure of their implementation. The coexistence of customary, religious, and statutory arrangements is a prevalent feature of land tenure (von Benda-Beckmann and Turner 2018). For example, in Africa as much as 70 percent of the continent's land is considered customary collective property, but less than 10 percent of this customary property is formally recognized in law (RRI 2023). This implies that the majority of the land under customary systems depends on social recognition.

Importance of tenure security and governance

The complex linkages between tenure security and governance are illustrated in Figure 6.1, a framework used by the IFPRI-led CGIAR System-wide

BOX 6.1 Indigenous tenure governance

Collective rights are essential to the tenure governance systems of Indigenous Peoples (ILO Convention 169, Part II, Arts. 13–19). The concept of territory, rather than land, is often used in Indigenous claims over resources. *Territory* refers to timber and nontimber forest products, as well as other ecosystem goods and services generated at the territorial level, including intangible ones such as carbon sequestration and traditional knowledge. Territorial concepts are contrary to the compartmentalization of rights, an approach that is promoted through the formalization of land rights, among other means (Monterosso, Cronkleton, and Larson 2019).

Indigenous commons refers both to resources claimed or held in possession by Indigenous Peoples at the territorial level, as well as the set of rules, including customary or traditional institutional arrangements, that promote collective action and define resource access, use, and control (Monterosso, Cronkleton, and Larson 2019). The state may or may not be recognized by these systems. Although governments claim ownership over water bodies and subterranean oil and mineral resources, Indigenous Peoples by custom consider these resources to be an integral part of their territories. Globally, communities and Indigenous Peoples are estimated to hold as much as 65 percent of all available land under customary tenure, but only 18 percent of this land has been formally recognized either as owned or designated for use by these groups (RRI 2023).

Program on Collective Action and Property Rights (CAPRi) to demonstrate the importance of tenure and governance institutions (Knox et al. 2002; Meinzen-Dick and Di Gregorio 2004). Time is mapped on the horizontal axis, from short term (within a season) to long term (decades or generations), and space on the vertical axis, spanning from a farm plot to a community, nation, or even global scale. This relational matrix has been widely used to consider whether tenure security or governance arrangements are likely to be important for the successful adoption of particular agricultural or natural resource management practices.

Practices such as planting a single plot with annual crops that give returns within a season are relatively straightforward, but those with longer time horizons or larger spatial scale, such as enhancing soil carbon or agroforestry, are more likely to be affected by tenure. Integrated pest management provides returns within a season but must be implemented across multiple farms. Thus, tenure security may not be a constraint, but coordination with other farmers is required. Practices like terracing, canal irrigation, watershed management,

FIGURE 6.1 Role of tenure security and coordination in agricultural and natural resource management practices

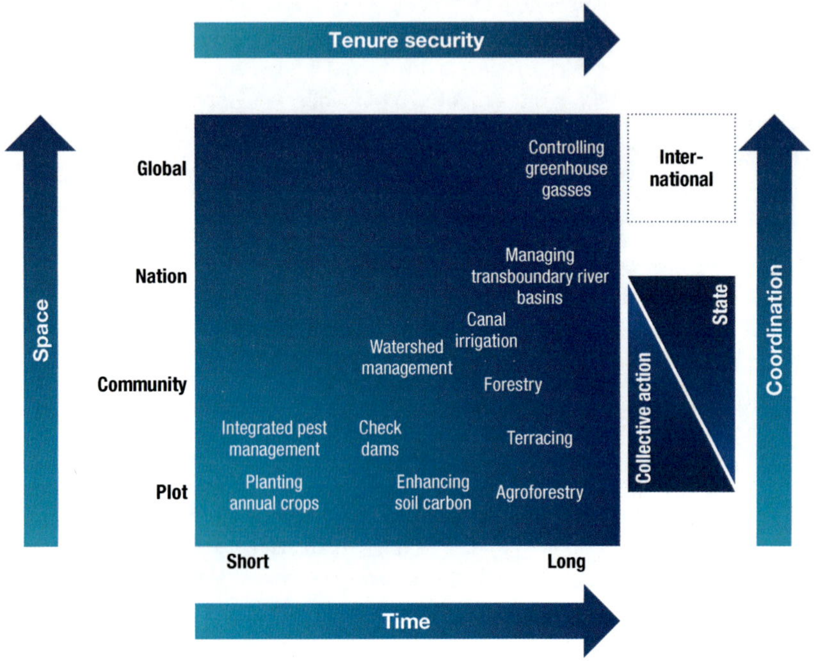

Source: Modified from Meinzen-Dick and Di Gregorio (2004).

Note: Placement of particular practices on this graph is illustrative. Where any particular practice (such as canal irrigation) would be placed depends on the investment required and scale relative to individual farms and communities.

or forestry are likely to require attention to coordination as well as to tenure. Coordination functions can be conducted through collective action, especially at more localized levels, or by the state, especially at higher levels, or through some form of co-management. Because many of the practices associated with sustainable resource management have long time horizons, large spatial scale, or both, tenure security and coordination are key for sustainability.

Tenure research: Evolution and impacts

Research on land tenure and property rights, along with policies and practices, has evolved substantially over the past half-century. Tenure-related studies have benefited from a diversity of disciplines and methods, which have shown the variety of tenure arrangements, relationships between tenure and behavior, and the effects of reforms. Many impact assessments have emphasized causal findings from quantitative studies, such as randomized allocation or regression discontinuity of pilot programs. Yet, broader observational studies and qualitative methods have been equally key to understanding the nature of tenure and how it affects behavior.[1] Ethnographic and qualitative research at IFPRI has identified the diversity of local concepts, tenure arrangements, and changes over time; these, in turn, have helped to shape survey questions that are meaningful in local contexts, capturing nuances of gendered tenure, and also have some comparability across sites (Awanyo 2009; Lambrecht et al. 2023; Pradhan et al. 2019). There is an iterative process linking conceptual and empirical work, with conceptual frameworks providing guidance on what to look for or how to interpret empirical findings, while findings from the field challenge and inform frameworks for understanding tenure arrangements. For example, the application of a conceptual framework commissioned by the Research Consortium on Women's Land Rights (Doss and Meinzen-Dick 2020) showed that forest and land reforms in Indonesia, Peru, and Uganda led to important progress for women's tenure security in those countries (Monterroso, Larson et al. 2019).

Formalization and individualization of land tenure

The evolutionary theory of land rights posits that with growing population pressures and market integration, land rights will evolve toward individualization and demand will increase for more formal tenure (Alchian and Demsetz 1973; Boserup 1965; Platteau 1996). Evidence from studies in Africa and Asia supports this theory, with findings showing that population increases led to the conversion of land to agriculture and individualization of rights under customary tenure systems (Otsuka and Place 2001). In these cases, the loss of forests and woodlands was partially offset by investments in tree crops and other agroforestry systems, which were facilitated by evolving tenure systems. Numerous land reforms have been piloted or scaled up to enhance individualization and formalization of property rights. Advocates have argued that individualization of land rights incentivizes production, while formalization provides greater assurance to rights holders, permits land to be used as collateral for credit, and allows rights holders to realize gains in land value by selling or leasing the land (De Soto 2000; Feder and Feeny 1991). At the end of the 1970s and early 1980s, property rights reforms in China and Viet Nam, moving from collective to individual land rights, were indeed associated with large

1 For example, see de la O Campos et al. (2015), Kilic et al. (2020), Lambrecht et al. (2023), and Lambrecht and Asare (2016).

increases in agricultural productivity and widespread reduction in rural poverty (Rozell and Swinnen 2004). As a result, China's reforms have been heralded as "the biggest antipoverty program the world has ever seen" (McMillan 2003, 94).

Inevitably, there are limitations to such reforms, and in some contexts, they come with important trade-offs between equity and efficiency. Agricultural decollectivization efforts across former Soviet states were not equally successful, demonstrating the importance of the macroeconomic and enabling environment for materializing gains of individualization (see Chapter 20 on Central Asia) (Dudwick et al. 2007; Rozelle and Swinnen 2004). In Latin America, the formalization of rights has been central to discussions of agrarian reform and, more recently, in debates on the nationalization of subsoil resources and on regulation of water systems (Monterroso, Cronkleton, and Larson 2019). Some contend that applying Western legal concepts of formalization and privatization has negatively affected rights holders, including through elite capture benefiting the rich and powerful, the exclusion of women and Indigenous groups, and disruption of customary governance and land management systems.[2] The African Union Framework and Guidelines on Land Policy, for instance, notes the need to decolonize Western-centric land policy to improve Africa-focused tenure, based on local realities in the continent (AU et al. 2010). In Nicaragua and the Philippines, conflicts emerged from the formalization of Indigenous claims over their ancestral lands (Larson et al. 2015).

While it is beyond the scope of this chapter to summarize the growing literature on the impacts of land formalization, systematic reviews of quantitative and qualitative research offer insights into the effects of tenure interventions on productivity, income, investments, and well-being. A systematic review of 20 quantitative and 9 qualitative studies on land tenure certification interventions found substantial effects on productivity and income. These effects resulted from increased tenure security, rather than credit access; and impacts were stronger in Latin America and Asia than in Africa. Despite these positive effects, the qualitative studies revealed concerns over displacement, perceived insecurity, social conflict, and gender inequality (Dwyer 2015; Lawry et al. 2017). Another systematic review of 36 quantitative and 23 qualitative studies found strong evidence of investment effects from land tenure security. However, the evidence was less clear on productivity and income

2 For example, see Benjaminsen et al. (2009), Cousins (2007), Meinzen-Dick and Mwangi (2009), Mwangi (2007), and Sjaastad and Bromley (2000).

effects, which may be related to the time necessary for tenure security to lead to these outcomes or the need for more support beyond tenure reforms (Higgins et al. 2018). The qualitative studies found that formalization programs did not always lead to stronger perceptions of tenure security, as we describe in the case of Myanmar in Box 6.2. In addition, certain groups were excluded, particularly women and poorer households, which was related, in some cases, to discrimination against women by the local land administration and social or financial barriers to women using the land productively.

BOX 6.2 Tenure insecurity and ambiguity of registered land in Myanmar

Land rights in Myanmar have been shaped by different regimes ruling the country, which has resulted in layers of revoked and active laws that sometimes contradict one another (Lambrecht et al. 2024; Mark 2016). Moreover, certain parts of the country are subject to the dual authority of the central government and an ethnic political organization, thus necessitating farmers to navigate—at times opportunistically—competing legal systems and spaces (Suhardiman et al. 2021).

Under Myanmar's constitution, all land and natural resources are considered state property. Before 2012, individuals were not legally permitted to buy or sell land, even though purchases and sales did occur in practice (Boutry and Allaverdian 2017). The 2012 Farmland Law made purchases, sales, or other forms of transfers of agricultural land parcels possible, provided the owner had a land use certificate for the parcel. Subsequently, Myanmar embarked on a hasty and incomplete effort to register agricultural parcels.

The process of land registration and certification did not unambiguously safeguard tenure security and sometimes stirred land disputes (Boutry and Allaverdian 2017; Oberndorf 2012; Thein et al. 2018). Furthermore, when combined with the Vacant, Fallow and Virgin Land Law, the land certification program provided a renewed legal foundation for widespread land expropriations, often from ethnic minorities or in conflict areas (Suhardiman et al. 2021). Nationwide, farmers held land use certificates for approximately 78 percent of all agricultural parcels. However, in the Hills and Mountain zone, home to most of the ethnic minority groups, only 41 percent of all agricultural parcels had land use certificates (Lambrecht et al. 2024).

IFPRI's mixed-method research further exposed striking differences between de jure and de facto property rights within households, showing the importance of understanding legal pluralism in the implementation of reforms. Although at least 80 percent of documented parcels were reported as registered in the name of one household member (usually a man), perceived transfer rights and decision-making over the land were much more often jointly held by men and women (Figure B6.1) (Lambrecht et al. 2023). Qualitative research showed that respondents usually considered property as jointly owned and prioritized household harmony over individual legal authority (Lambrecht et al. 2023). Future land policy should promote awareness about joint land titling while also implementing supportive measures to facilitate access to services linked to land documentation, and it should ensure tenure security for all ethnic groups.

FIGURE B6.1 Land use rights (share of parcels)

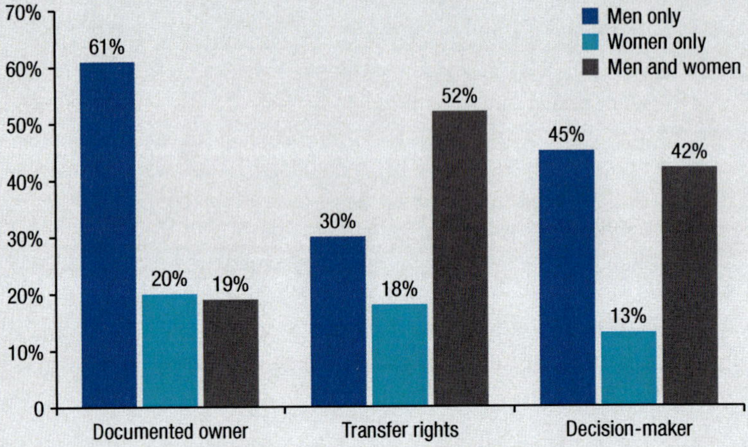

Source: Lambrecht et al. (2023), estimates based on the 2015 Myanmar Poverty and Living Conditions Survey.

In a review of 117 studies estimating the impact of land tenure interventions on a range of human well-being outcomes, 82 percent found at least one positive outcome and 14 percent at least one negative outcome. Additionally, 73 percent of studies estimating environmental effects found positive outcomes and 15 percent negative outcomes (Tseng et al. 2020).

Common property

While much of the economic literature and policy attention has centered on tenure in smallholder agriculture, there has also been considerable focus on the management of *common property*, that is, resources that are held and managed collectively. IFPRI and CAPRi research has contributed to this prmimarily by studying water, rangelands, and forests, as well as by synthesizing lessons across different types of resources.[3] As noted in Figure 6.1, many natural resource management processes take place above the level of an individual farm. Some activities, such as landscape-level terracing to control water flows, can be accomplished through coordination across individual farms, but many others, such as those involving forests, rangelands, or water resources, are frequently managed as common property. In these cases, the use and management rights are vested in a collective, whether a social group, such as a tribe, or a formal organization, such as a forest users' group or water users' association.

Tenure security for these commons depends on how common property is recognized externally, as well as the effectiveness of governance arrangements within the groups in providing assurance that the resource will be shared equitably among members and over time. Organizations such as the International Association for the Study of the Commons and the CAPRi program fostered learning about tenure security across different resources. Studies of water rights highlighted the variable nature of property rights and the importance of identifying many different bases for claims on resources (Bruns and Meinzen-Dick 2000; Meinzen-Dick and Pradhan 2002), principles that also apply to land-based resources. Holding property together was identified as a way to strengthen the collective management of the commons (Coward 1990).

Growing evidence on effective community management of commons (Baland et al. 2007; Ostrom 2015), combined with studies on fiscal constraints and the deficiencies of state-managed common pool resources in the 1990s, led to a range of "devolution" programs that aimed to transfer responsibility over natural resources from the state to user groups. These programs, which spanned different sectors, were variously referred to as irrigation management transfer, joint forest management, fisheries co-management, and devolution of rangeland management. IFPRI contributed to studies of such reforms in irrigation, and the CAPRi program convened comparative analysis across resources (Meinzen-Dick et al. 1999). Many such programs sought to transfer responsibilities for resource management or maintenance from the government to communities, without the corresponding rights. Evidence of

3 On water, see Bruns and Meinzen-Dick (2005), Bruns et al. (2005), Meinzen-Dick and Pradhan (2002), and Zwarteveen and Meinzen-Dick (2001). On rangelands, see McCarthy (2004), Goodhue and McCarthy (2009), and Mwangi (2009). On forests, see Mwangi et al. (2011), and Tachibana and Adhikari (2009). On synthesized lessons, see Meinzen-Dick et al. (1997), Meinzen-Dick and Di Gregorio (2004), Meinzen-Dick et al. (2006), and Otsuka and Place (2001).

the importance of property rights held by communities led to the devolution of these rights from the state to communities (or, in many cases, restitution of rights customarily held by communities, which had been claimed by the state) (Meinzen-Dick et al. 1999). Such devolution programs had mixed success in improving environmental outcomes, highlighting the need to go beyond panaceas for resource governance (Meinzen-Dick 2007; Ostrom 2007) and engage more with the complex processes of adaptive co-management, whereby communities, government agencies, and others with interests in the resource collaboratively plan, negotiate, evaluate, and adapt plans as necessary (see also Chapter 15) (Colfer et al. 2022).

In Latin America, even as land administration projects prioritized the formalization of individual rights and the expanded colonization of large forest lands, social mobilization by Indigenous Peoples and Afro-descendant communities kept important land regions under collective tenure (Monterroso, Cronkleton, and Larson 2019). A large proportion of national territory has been formally recognized as either designated for or owned by Indigenous groups and local communities in Mexico (52 percent), Peru (35 percent), Colombia (34 percent), Brazil (23 percent), and Nicaragua (28 percent) (RRI 2023). Reforms in land and forest policy triggered changes in regulations, which redefined tenure rights to increase community access to Indigenous and traditional commons, especially forested regions, that had previously been claimed by governments (Ortega 2004). The modes of transfer and types of rights granted vary, from collective land and territorial titling to co-management schemes and concession contracts that recognize lesser or temporary rights (RRI 2023).

Equity

While issues of efficiency of resource use and investment have been prominent in the literature on private property, and sustainability of resource management prominent in the literature on the commons, there has been major interest in equity of property rights across both bodies of work. In addition to gender equity in resource rights, a key aspect of equity relates to the rights over natural resources of Indigenous Peoples, minority ethnic groups, and other marginalized local communities over natural resources. Social equity concerns, historical justice, and potential equity–efficiency trade-offs were key considerations in the political economy of Asian land reforms and property rights reforms initiated in former communist regimes (Otsuka 2007; Swinnen 1999).

Distributional issues of land policy reached another critical point after large-scale biofuel production and the 2007–2008 world food price crisis triggered interest in land acquisitions for agriculture. Multinational companies and sovereign wealth funds of countries dependent on large-scale food imports, as well as national elites who saw opportunities for profit (Borras et al. 2010, 2011; Jayne et al. 2016; von Braun and Meinzen-Dick 2009), sought to purchase or lease large tracts of land formally under government ownership. These outside investors—whether international or national elites—often promised to invest in improving the productivity of seemingly unused land with corresponding water resources. In most cases, the land targeted was under customary tenure and used by local communities in a variety of ways. Protests by civil society organizations and studies of large-scale land deals indicated that in many cases, the investors did not engage in full consultations and obtain the consent of communities, or they failed to carry through with their promised investments and did not ultimately yield higher productivity or profits than previous land uses.[4] Indigenous Peoples, ethnic minorities, and women were particularly likely to have been bypassed in any negotiations, thereby losing their customary property rights.

IFPRI was among the first organizations to call for an internationally enforceable code of conduct on large-scale land acquisitions in 2009 (von Braun and Meinzen-Dick 2009). The UN Special Rapporteur for the Right to Food issued a set of core principles and measures to address the human rights challenges of such large-scale land acquisitions (De Schutter 2009). IFPRI and more than 40 other organizations participated in a collaborative research project led by the International Land Coalition, which brought together perspectives from research, civil society, and intergovernmental organizations involved in land policy (Anseeuw 2012). This contributed to several international consultations convened by the Committee on Food Security and other organizations, which utimately led to the development of the Voluntary Guidelines on the Responsible Governance of Tenure (VGGTs) (FAO 2012) and the Guiding Principles on Large Scale Land Based Investments in Africa (AU et al. 2014). Waning of the global food price crisis and improvements in land governance in a number of countries reduced outside demand for land acquisitions for food and biofuels after 2013. However, because the VGGTs remain voluntary, with limited enforcement, large-scale land acquisitions such as for palm oil production, carbon offsets, and green energy production have continued with inadequate consultation of local communities, and remain a serious threat to tenure security, especially for community lands (IPES-Food 2024; Lay et al. 2024).

4 Examples include Behrman et al. (2012), Cotula et al. (2009), Nnoko-Mewanu (2016), Schoneveld et al. (2011); and Wisborg (2014).

INDIGENOUS PEOPLES

A key aspect of equity relates to the resource rights of Indigenous Peoples, minority ethnic groups, and other marginalized local communities. The 2007 UN Declaration on the Rights of Indigenous Peoples states that the collective right "to the lands, territories and resources which they have traditionally owned, occupied or otherwise used or acquired constitute(s) the minimum standards for the survival, dignity and well-being of the indigenous peoples of the world" and requires the free, prior, and informed consent (FPIC) of Indigenous Peoples for any actions that would affect their lands, territories, or rights. Nevertheless, because the majority of Indigenous lands and territories are not formally recognized by the state (RRI 2023), and the customs of ethnic minorities differ from majority ethnic groups in their country, these resources are particularly vulnerable to expropriation, and the process of obtaining formal recognition of collective lands is more challenging than for companies to acquire formal rights over such lands (Notess et al. 2018).

WOMEN

Providing property rights at the household level is not enough to ensure key development outcomes: women's rights to land and other assets are important for many of these outcomes, including women's empowerment, technology adoption, government services, consumption, and food security.[5] Yet women often face barriers to tenure security, including from both customary and official norms and practices, such as inheritance regimes.[6] IFPRI research has played an important part in highlighting limitations on women's land and resource rights, and it has also been critical in countering misleading narratives—such as that women own only 1, 2, or 5 percent of the land in LMICs (Doss et al. 2014; Doss et al. 2015)—and in providing more nuanced understandings of women's land rights across the life cycle (Lambrecht 2016; Pradhan et al. 2019).

While many formalization programs excluded women and marginalized groups (Askew and Odgaard 2019; Joireman 2008; Lastarria-Cornhiel 1997), several governments leveraged their formalization reforms to enhance equity, especially gender equality, and redress gender-discriminatory customs (Aldashev et al. 2012; Deere and Leon 2003). Well-known examples include Rwanda's land tenure regularization and Ethiopia's first-level land registration,

---

5 For the range of this work, see Agarwal (1994), Doss 2006; Haddad et al. (1997), Meinzen-Dick et al. (2014), Meinzen-Dick et al. (2019), Panda and Agarwal (2005), Peterman et al. (2011), and Quisumbing and Maluccio (2003).

6 This work includes Lastarria-Cornhiel (1997), Meinzen-Dick et al. (1997), Andersson Djurfeldt (2020), and Morgan et al. (2023).

which endeavored to include women in the registration process, such as through ensuring joint registration of husband's and wife's names, along with reforms to the family code that gave women stronger rights.

Impact assessments of Rwanda's pilot program showed mixed results: legally married women had their rights documented and secured, but those in informal marriages experienced an erosion of their rights. Investment in soil conservation increased, especially by female-headed households who previously had been the most tenure insecure (Ayalew Ali et al. 2014; Lawry et al. 2023). Five years after the completion of Rwanda's registration process, however, researchers found an increase in informal land transfers and reductions in equity gains in land rights (Ayalew Ali et al. 2021). These findings indicate significant challenges in sustaining long-term objectives if they are not aligned with local incentives and customs.

In Ethiopia, studies found near-universal registration of land among both male- and female-headed households. Beyond having registered rights, knowledge of land rights affected the adoption of long-term investments in the land (Deininger et al. 2008; Kumar and Quisumbing 2015). Female landholders were more likely to rent out land if they had a certificate and thus had more opportunities to earn income from the land (Holden et al. 2011).

Because legislative reforms alone are often insufficient to secure women's land rights in practice, programs such as community-based legal aid are needed to increase women's knowledge of land rights. IFPRI research showed that the effectiveness of such programs is conditioned by social context (Mueller et al. 2018). In many countries, women's rights to land remain highly contested, especially in cases of divorce or widowhood.

YOUTH

Land rights for youth is a key issue in countries with high rates of population growth. As people live longer, land inheritance is delayed, especially as compared to decades ago, and unclaimed land is difficult to find outside of sparsely populated areas that are typically unfavorable for cultivation. Increasing land values make it difficult for young people to buy land, and as a result, many youth migrate out of rural areas (Amare et al. 2024; Kosec et al. 2018). As this challenge has received more attention, there has been keen interest in understanding how rental markets—which require much less capital—can provide youth with access to land, while still protecting the rights of the elderly, such as widows who depend on the land for their livelihoods. Studies in Africa show rental markets can transfer land from land-rich households to land-poor households, but the extent to which youth benefit is mixed, depending on transaction costs and the security of tenure for land owners (Deininger et al. 2017; Jin and Jayne 2013; Ricker-Gilbert and Chamberlin 2018). However, providing long-term security for tenants remain an issue that must be addressed to ensure investments in land's long-term sustainability.

Looking ahead

Given the social, economic, and political importance of land and resource tenure, there is a continuing need for research on tenure formalization, management of the commons, efficiency, and equity, among other topics. As IFPRI's work advances, there is a need to address persistent and emerging questions that can provide research-informed policy solutions for reducing poverty and hunger in developing countries. A fundamental research question remains: What are decisive and structural factors that impede tenure security? This leads to four more specific questions: How can we overcome the impediments to tenure security for the socially marginalized? What are the best ways to secure tenure beyond land, and across other resources? How can tenure be governed in an environment undergoing a structural transformation of economies and/or governance modalities, such as decentralization? Last, what opportunities exist for research to support action? We highlight select issues that address these questions and merit additional attention.

Securing rights for the socially marginalized

While research has shown the importance of tenure security for ethnic minorities, women, and youth—as well as highlighting the many normative and structural factors that restrict their control over land and other resources—there is an ongoing need for evidence on how to overcome these barriers and provide secure rights for these groups. For women and youth, this requires attention to inheritance systems as well as land. Closer attention to personal and structural conditions, both formal and informal, can improve our understanding of additional transformative approaches needed to accompany legal tenure reforms (Morgan et al. 2023). The inclusion of a module on property rights, documentation, and tenure security for land and dwellings in the Women's Empowerment Metric for National Statistical Systems (WEMNS) developed by IFPRI will allow for disaggregating these measures by gender, age, and ethnicity, as well as linking them to agricultural or other data on outcomes as this module is rolled out in nationally representative surveys (Seymour et al. 2024). Such data will be valuable for identifying indicators, patterns of land tenure (in)security and its consequences, as well as changes over time or in response to reforms.

Rights and tenure across resources

Water rights. Water rights can be subject to many types of competing interests: rivers that span countries and regions can put upstream users in conflict with downstream users, aquifers can set current users against future users, and wherever water scarcity is felt, farmers compete with other users. Climate change increases the difficulty in predicting and securing water supplies. These competing interests in water create challenges for governance, from the local to the global level, to develop and enforce policies and regulations, with continual review and revision necessary as water resources change over time (see also Chapter 5). The unique characteristics of water as fugitive, shared, and vital to life make it problematic to apply conventional (land-based) concepts of property rights, but there is growing policy interest in how to secure water tenure as a basis for both equity and investment (FAO 2002). This calls for detailed research documenting existing water tenure and collective action mechanisms, as well as action research with governments and communities to adapt water tenure to strengthen security and equity (Hodgson et al. 2024).

Mineral rights. Mineral rights have often been claimed by the state or separated from land rights. Power differentials between mining companies and communities, along with environmental degradation, cause conflicts in many countries. Such tensions are increasing as demand and support for the transition to renewable energy lead to more extraction of key minerals. Pressure to secure access to deposits of these often rare minerals has also sparked concerns over new waves of land displacement. Many of these deposits are found in Indigenous Peoples' territories, with a global inventory of energy transition minerals finding that 54 percent of projects are located on or near Indigenous Peoples' lands (Owen et al. 2022). Political economy research on the processes of negotiation and implementation (or cancellation) of such mining projects can help address the power dynamics in negotiations and FPIC processes, as well as identify promising innovations to overcome injustices.

Rights to ecosystem services. Climate change and biodiversity loss have turned attention to ecosystem services. The rising use of payment for ecosystem services (PES) as an incentive mechanism to encourage carbon sequestration or biodiversity conservation creates both opportunities and risks, especially for communities in forests or areas of high biodiversity. Legal frameworks are needed to determine how such renumerations can be shared. Currently, communities without recognized land ownership are often bypassed by PES mechanisms; those who do participate often must forfeit their local management rights to comply with external regulations. A review of 31 LMIC countries found that carbon rights are explicitly defined in 9 countries, but only 3 recognize community rights to carbon on collective lands (RRI 2021). Collaborative research on mechanisms to recognize and reward local stewardship of commons and other lands with high ecosystem services (Sandhu et al. 2023), as is being conducted in India under the Promise of Commons initiative, can help shape new PES mechanisms that are more equitable and effective (Lawry 2024).

Tenure across resource systems. Although most tenure-related reforms have been carried out separately for agricultural lands, forests, rangelands, marine resources, fisheries, and water, the biophysical and social interconnections among these systems call for better understanding of tenure across resources. Concepts of community territorial governance and landscape management may help overcome resource compartmentalization. Collaborative research with Indigenous groups may provide new ways to conceptualize and operationalize tenure that stress linkages across resources—and users.

Tenure under structural transformation of economies

While agricultural land and associated natural resources continue to be central to rural livelihoods, urbanization and diversification of livelihoods change tenure relationships. Alternative employment opportunities can reduce dependence on farmland and resources. Rising land values near cities and roads can create opportunities for those with secure tenure to profit, but they also create potential for "land grabs" and displacement for those without secure, documented tenure and knowledge of the formal legal system. Empirical research on these processes is needed to identify threats and opportunities for equitable transitions, particularly in peri-urban areas and in urban food systems.

Opportunities for research to support action

Research has made many valuable contributions to tenure policy, programs, and investments, as well as to action by civil society. Key examples include research by IFPRI and others on the importance of women's land rights leading to advocacy by women's organizations and more policy provisions for women's land rights; research on large-scale land acquisitions harming communities with customary rights that informs processes leading to the formulation and implementation of international and national policies on land acquisitions; and research on the importance of collective tenure leading to innovations in securing pastoralists' and communities' tenure (Knight et al. 2012; Sulle 2021). Some of these achievements were enabled by planned coordination and collaboration among researchers and policymakers, donors, and civil society groups, such as at the annual Land Conference hosted by the World Bank, the International Land Coalition, and the African Union Land Policy Center. Given the central role of resource tenure in agricultural performance, environmental management, gender equality, and livelihoods for youth, reciprocal communication and ongoing partnerships are needed between researchers and practitioners to ensure that research addresses key issues and is shared with relevant audiences. Partnerships with governments, donors, project implementers, and civil society can increase the relevance and uptake of research findings, but researchers need to remain reflexive about their positionality in the research.

Greater collaboration between different disciplines would yield more-comprehensive information for decision-makers. These efforts should combine qualitative studies with analyses based on surveys, remote sensing, and administrative data, with samples suitable for causal impact assessments, as well as legal analysis and political economy research to address power issues (see Chapter 15).

Competition over land and water resources is a source of conflict, and conflict can increase tenure insecurity by displacing populations and destroying official records as well as social histories of tenure (Ratner et al. 2013). Thus, including tenure assessments and addressing resource conflicts can play an important role in action research on fragile and conflict-affected contexts (see Chapter 13) (Bruch et al. 2016; Ratner et al. 2017).

The past 50 years of research have deepened our understanding of the importance of resource tenure and broadened our understanding of the range of tenure forms across societies, resources, and time. The fundamental importance of resource tenure for economic development, environmental sustainability, equity, and identity calls for continued research and innovation in this field.

References

Agarwal, B. 1994. *A Field of One's Own: Gender and Land Rights in South Asia*. Cambridge, UK: Cambridge University Press.

Alchian, A., and H. Demsetz. 1973. "The Property Right Paradigm." *Journal of Economic History* 33 (1): 16–27.

Aldashev, G., I. Chaara, J-P. Platteau, and Z. Wahhaj. 2012. "Using the Law to Change the Custom." *Journal of Development Economics* 97 (2): 182–200.

Alden Wily, L. 2011. *The Tragedy of Public Lands: Understanding the Fate of the Commons under Global Commercial Pressure*. Rome: International Land Coalition.

Amare, M., H. Ghebru, G. Mavrotas, and A. Ogunniyi. 2024. "The Role of Land Inheritance in Youth Migration and Employment Choices: Evidence from Rural Nigeria." *European Journal of Development Research* 36 (1): 135–160.

Andersson Djurfeldt, A. 2020. "Gendered Land Rights, Legal Reform and Social Norms in the Context of Land Fragmentation: A Review of the Literature for Kenya, Rwanda and Uganda." *Land Use Policy* 90: 104305.

Anseeuw, W., L. Alden Wily, L. Cotula, and M. Taylor. 2012. *Land Rights and the Rush for Land. Findings of the Global Commercial Pressures on Land Research Project.* Rome: International Land Coalition.

Arnold, C.A. 2007. "The Reconstitution of Property: Property as a Web of Interests." *Harvard Environmental Law Review* 26 (2).

Askew, K., and R. Odgaard. 2019. "Deeds and Misdeeds: Land Titling and Women's Rights in Tanzania." *New Left Review* 118: 68–85.

AU (African Union), AfBD (African Development Bank), and UNECA (Economic Commission for Africa). 2010. *Framework and Guidelines on Land Policy in Africa.* Addis Ababa: AUC-ECA-AfDB Consortium.

AU, AfDB, and UNECA. 2014. *Guiding Principles on Large Scale Land Based Investments in Africa.* Addis Ababa: UNECA.

Awanyo, L. 2009. "Exploring the Connections: Land Tenure, Social Identities, and Agrobiodiversity Practices in Ghana." *Geografiska Annaler: Series B, Human Geography* 91 (2): 137–155.

Ayalew Ali, D., K. Deininger, and M. Goldstein. 2014. "Environmental and Gender Impacts of Land Tenure Regularization in Africa: Pilot Evidence from Rwanda." *Journal of Development Economics* 110: 262–275.

Ayalew Ali, D., K. Deininger, G. Mahofa, and R. Nyakulama. 2021. "Sustaining Land Registration Benefits by Addressing the Challenges of Reversion to Informality in Rwanda." *Land Use Policy* 110: 104317.

Baland, J-M., J-P. Platteau, and M. Olson. 2007. *Halting Degradation of Natural Resources: Is There a Role for Rural Communities?* Rome: Food and Agriculture Organization of the United Nations.

Behrman, J., R. Meinzen-Dick, and A. Quisumbing. 2012. "The Gender Implications of Large-Scale Land Deals." *Journal of Peasant Studies* 39 (1): 49–79.

Benjaminsen, T.A., S. Holden, C. Lund, and E. Sjaastad. 2009. "Formalisation of Land Rights: Some Empirical Evidence from Mali, Niger and South Africa." *Land Use Policy* 26 (1): 28–35.

Borras, S.M., R. Hall, I. Scoones, B. White, and W. Wolford. 2011. "Towards a Better Understanding of Global Land Grabbing: An Editorial Introduction." *Journal of Peasant Studies* 38 (2): 209–216.

Borras, S.M., P. McMichael, and I. Scoones. 2010. "The Politics of Biofuels, Land and Agrarian Change: Editors' Introduction." *Journal of Peasant Studies* 37 (4): 575–592.

Boserup, E. 1965. *The Conditions of Agricultural Growth: The Economics of Agrarian Change under Population Pressure*. London: George Allen and Unwin.

Boutry, M., and C. Allaverdian. 2017. *Land Tenure in Rural Lowland Myanmar: From Historical Perspectives to Contemporary Realities in the Dry Zone and the Delta*. Of Lives and Land Myanmar Series. Nogent sur Marnes, France: GRET.

Bruce, J.W. 1988. *Review of Tenure Terminology*. Tenure Brief 1. Madison, WI: University of Wisconsin Land Tenure Center.

Bruch, C., C. Muffett, S. Nichols, eds. 2016. *Governance, Natural Resources and Post-Conflict Peacebuilding*. New York: Routledge.

Bruns, B., and R. Meinzen-Dick, eds. 2000. *Negotiating Water Rights*. London: Intermediate Technology Press.

Bruns, B., C. Ringler, and R.S. Meinzen-Dick, eds. 2005. *Water Rights Reform: Lessons for Institutional Design*. Washington, DC: IFPRI.

Chigbu, U.E., M.R. Mabakeng, and A. Chilombo. 2021. "Strengthening Tenure and Resource Rights for Land Restoration." UNCCD Global Land Outlook Working Paper. United Nations Convention to Combat Desertification (UNCCD), Bonn, Germany.

Chigbu, U.E., A. Schopf, W.T. de Vries, F. Masum, S. Mabikke, D. Antonio, and J. Espinoza. 2017. "Combining Land-Use Planning and Tenure Security: A Tenure Responsive Land-Use Planning Approach for Developing Countries." *Journal of Environmental Planning and Management* 60 (9): 1622–1639.

Colfer, C.J.P., R. Prabhu, and A.M. Larson, eds. 2022. *Adaptive Collaborative Management in Forest Landscapes: Villagers, Bureaucrats and Civil Society*. Abingdon, UK: Taylor & Francis.

Cotula, L., S. Vermeulen, R. Leonard, and J. Keeley, eds. 2009. *Land Grab or Development Opportunity? Agricultural Investment and International Land Deals in Africa*. London: International Institute for Environment and Development.

Cousins, B. 2007. "More Than Socially Embedded: The Distinctive Character of 'Communal Tenure' Regimes in South Africa and Its Implications for Land Policy." *Journal of Agrarian Change* 7 (3): 281–315.

Coward, E. 1990. "Property Rights and Network Order: The Case of Irrigation Works in the Western Himalayas." *Human Organization* 49 (1): 78–88.

Deere, C.D., and M. Leon. 2003. "The Gender Asset Gap: Land in Latin America." *World Development* 31 (6): 925–947.

Deininger, K., D. Ayalew Ali, S. Holden, and J. Zevenbergen. 2008. "Rural Land Certification in Ethiopia: Process, Initial Impact, and Implications for Other African Countries." *World Development* 36 (10): 1786–1812.

Deininger, K., S. Savastano, and F. Xia. 2017. "Smallholders' Land Access in Sub-Saharan Africa: A New Landscape?" *Food Policy* 67: 78–92.

de la O Campos, A.P., N. Warring, C. Brunelli, C. Doss, and C. Kieran. 2015. *Gender and Land Statistics: Recent Developments in FAO's Gender and Land Rights Database.* Rome: FAO.

De Schutter, O. 2009. *Large-Scale Land Acquisitions and Leases: A Set of Core Principles and Measures to Address the Human Rights Challenge.* Geneva: Office of the United Nations High Commissioner for Human Rights.

De Soto, H. 2000. *The Mystery of Capital: Why Capitalism Triumphs in the West and Fails Everywhere Else.* New York: Basic Books.

Doss, C. 2006. "The Effects of Intrahousehold Property Ownership on Expenditure Patterns in Ghana." *Journal of African Economies* 15 (1): 149–180.

Doss, C., C. Kovarik, A. Peterman, A. Quisumbing, and M. van den Bold. 2015. "Gender Inequalities in Ownership and Control of Land in Africa: Myth and Reality." *Agricultural Economics* 46 (3): 403–434.

Doss, C., and R.S. Meinzen-Dick. 2020. "Land Tenure Security for Women: A Conceptual Framework." *Land Use Policy* 99: 105080.

Doss, C., R.S. Meinzen-Dick, and A. Bomuhangi. 2014. "Who Owns the Land? Perspectives from Rural Ugandans and Implications for Large-Scale Land Acquisitions." *Feminist Economics* 20 (1): 76–100.

Dudwick, N., K. Fock, and D. Sedik. 2007. "Land Reform and Farm Restructuring in Transition Countries: The Experience of Bulgaria, Moldova, Azerbaijan, and Kazakhstan." World Bank Working Papers. World Bank, Washington, DC.

Dwyer, M.B. 2015. "The Formalization Fix? Land Titling, Land Concessions and the Politics of Spatial Transparency in Cambodia." *Journal of Peasant Studies* 42 (5): 903–928.

FAO (Food and Agriculture Organization of the United Nations). 2002. *Land Tenure and Rural Development.* FAO Land Tenure Series 3. Rome.

FAO. 2012. *Voluntary Guidelines on the Responsible Governance of Tenure of Land Fisheries and Forests in the Context of National Food Security.* Rome.

Feder, G., and D. Feeny. 1991. "Land Tenure and Property Rights: Theory and Implications for Development Policy." *World Bank Economic Review* 5 (1): 135–153.

Fortmann, L., and J.W. Bruce. 1988. *Whose Trees? Proprietary Dimensions of Forestry.* Boulder, CO: Westview Press.

Fortmann, L., and D. Rocheleau. 1985. *Women and Agroforestry: Four Myths and Three Case Studies.* Nairobi: International Centre for Research on Agroforestry (ICRAF).

Goodhue, R.E., and N. McCarthy. 2009. "Traditional Property Rights, Common Property, and Mobility in Semi-Arid African Pastoralist Systems." *Environment and Development Economics* 14 (1): 29–50.

Haddad, L.J., J. Hoddinott, and H. Alderman, eds. 1997. *Intrahousehold Resource Allocation in Developing Countries: Models, Methods, and Policy.* Baltimore: Johns Hopkins University.

Higgins, D., T. Balint, H. Liversage, and P. Winters. 2018. "Investigating the Impacts of Increased Rural Land Tenure Security: A Systematic Review of the Evidence." *Journal of Rural Studies* 61: 34–62.

Hodgson, S. 2004. *Land and Water—the Rights Interface.* FAO Legal Papers Online 36. Rome: FAO.

Hodgson, S., A.-K. Rothe, S. Espinosa Flor, J. Troell, and B. Kiersch. 2024. *Water Tenure Assessment.* Rome: FAO.

Holden, S.T., K. Deininger, and H. Ghebru. 2011. "Tenure Insecurity, Gender, Low-Cost Land Certification and Land Rental Market Participation in Ethiopia." *Journal of Development Studies* 47 (1): 31–47.

IPES-Food (International Panel of Experts on Sustainable Food Systems). 2024. *Land Squeeze: What Is Driving Unprecedented Pressures on Global Farmland and What Can Be Done to Achieve Equitable Access to Land?*

Jayne, T.S., J. Chamberlin, L. Traub, et al. 2016. "Africa's Changing Farm Size Distribution Patterns: The Rise of Medium-Scale Farms." *Agricultural Economics* 47 (S1): 197–214.

Jin, S., and T.S. Jayne. 2013. "Land Rental Markets in Kenya: Implications for Efficiency, Equity, Household Income, and Poverty." *Land Economics* 89 (2): 246–271.

Joireman, S.F. 2008. "The Mystery of Capital Formation in Sub-Saharan Africa: Women, Property Rights and Customary Law." *World Development* 36 (7): 1233–1246.

Kilic, T., C. Doss, and C. Kieran. 2020. "Measuring Ownership, Control, and Use of Assets." World Bank Policy Research Working Paper No. 8146. World Bank, Washington, DC.

Knight, R.S., J. Adoko, T. Auma, A. Kaba, A. Salomao, S. Siakor, and I. Tankar. 2012. *Protecting Community Lands and Resources: Evidence from Liberia, Mozambique and Uganda.* Washington, DC: Namati.

Knox, A., R. Meinzen-Dick, and P. Hazell. 2002. "Property Rights, Collective Action, and Technologies for Natural Resource Management: A Conceptual Framework." In *Innovation in Natural Resource Management: The Role of Property Rights and Collective Action in Developing Countries*, eds. R. Meinzen-Dick, A. Knox, F. Place, and B. Swallow, 12–44. Baltimore: Johns Hopkins University.

Kosec, K., H. Ghebru, B. Holtmeyer, V. Mueller, and E. Schmidt. 2018. "The Effect of Land Access on Youth Employment and Migration Decisions: Evidence from Rural Ethiopia." *American Journal of Agricultural Economics* 100 (3): 931–954.

Kumar, N., and A.R. Quisumbing. 2015. "Policy Reform Toward Gender Equality in Ethiopia: Little by Little the Egg Begins to Walk." *World Development* 67: 406–423.

Lambrecht, I.B. 2016. "'As a Husband I Will Love, Lead, and Provide.' Gendered Access to Land in Ghana." *World Development* 88: 188–200.

Lambrecht, I.B., and S. Asare. 2016. "The Complexity of Local Tenure Systems: A Smallholders' Perspective on Tenure in Ghana." *Land Use Policy* 58: 251–263.

Lambrecht, I.B., B. Belton, P. Fang, B. Minten, and P.T. Naing. 2024. "Agricultural Land: Inequality and Insecurity." In *Myanmar's Agrifood System: Historical Development, Recent Shocks, Future Opportunities.* Washington, DC: IFPRI.

Lambrecht, I.B., K. Mahrt, N.L.K. Synt, H.E. Win, and K.Z. Win. 2023. "Gender Gaps in Land Rights: Explaining Different Measures and Why Households Differ in Myanmar." *Agricultural Economics* 54 (5): 728–741.

Larson, A.M., P.J. Cronkleton, and J.M. Pulhin. 2015. "Formalizing Indigenous Commons: The Role of 'Authority' in the Formation of Territories in Nicaragua, Bolivia, and the Philippines." *World Development* 70: 228–238.

Lastarria-Cornhiel, S.1997. "Impact of Privatization on Gender and Property Rights in Africa." *World Development* 25 (8): 1317–1333.

Lawry, S. 2024. "Community Forest Stewards: Managers, Not Labourers." *Forest News*, July 17.

Lawry, S., R. McLain, M. Rugadya, G. Alvarado, and T. Heidenrich. 2023. *Land Tenure Reform in Sub-Saharan Africa: Interventions in Benin, Ethiopia, Rwanda, and Zimbabwe*, 1st ed. London: Routledge.

Lawry, S., C. Samii, R. Hall, A. Leopold, D. Hornby, and F. Mtero. 2017. "The Impact of Land Property Rights Interventions on Investment and Agricultural Productivity in Developing Countries: A Systematic Review." *Journal of Development Effectiveness* 9 (1): 61–81.

Lay, J., W. Anseeuw, S. Eckert, I. Flachsbarth, C. Kubitza, K. Nolte, and M. Gigeret. 2024. *Taking Stock of the Global Land Rush: Few Development Benefits, Many Human and Environmental Risks. Analytical Report III.* Bern: University of Bern.

Mark, S. 2016. "Are the Odds of Justice 'Stacked' Against Them? Challenges and Opportunities for Securing Land Claims by Smallholder Farmers in Myanmar." *Critical Asian Studies* 48 (3): 443–460.

McCarthy, N. 2004. *Managing Resources in Erratic Environments: An Analysis of Pastoralist Systems in Ethiopia, Niger and Burkina Faso.* Research Report 135. Washington, DC: IFPRI.

McMillan, J. 2003. *Reinventing the Bazaar: A Natural History of Markets*, 1st ed. New York: W.W. Norton.

Meinzen-Dick, R.S. 2007. "Beyond Panaceas in Water Institutions." *PNAS* 104 (39): 15200–15205.

Meinzen-Dick, R.S., and M. Di Gregorio, eds. 2004. *Collective Action and Property Rights for Sustainable Development*. 2020 Focus Brief 11. Washington, DC: IFPRI.

Meinzen-Dick, R.S., L.R. Brown, H.S. Feldstein, and A.R. Quisumbing. 1997. "Gender, Property Rights, and Natural Resources." *World Development* 25 (8): 1303–1315.

Meinzen-Dick, R.S., N. Johnson, A.R. Quisumbing, et al. 2014. "The Gender Asset Gap and Its Implications for Agricultural and Rural Development." In *Gender in Agriculture and Food Security: Closing the Knowledge Gap*, eds. A.R. Quisumbing, R. Meinzen-Dick, T.L. Raney, A. Croppenstedt, J.A. Behrman, and A. Peterman, 91–115. Dordrecht, Netherlands: Springer and FAO.

Meinzen-Dick, R.S., A. Knox, and M. Di Gregorio, eds. 1999. *Collective Action, Property Rights, and Devolution of Natural Resource Management: Exchange of Knowledge and Implications for Policy*. Feldafing, Germany: Zentralstelle für Ernährung und Landwirtschaft.

Meinzen-Dick, R.S., and E. Mwangi. 2009. "Cutting the Web of Interests." 26 (1): 36–43.

Meinzen-Dick, R.S., E. Mwangi, and S. Dohrn. 2006. *Securing the Commons*. CAPRi Policy Brief 4. Washington, DC: IFPRI.

Meinzen-Dick, R.S., and R. Pradhan. 2002. "Legal Pluralism and Dynamic Property Rights." CAPRi Working Paper 22. IFPRI, Washington, DC.

Meinzen-Dick, R., R. Pradhan, and M. Di Gregorio. 2004. "Understanding Property Rights." In *Collective Action and Property Rights for Sustainable Development*, eds. R.S. Meinzen-Dick and M. Di Gregorio. 2020 Focus Brief 11. Washington, DC: IFPRI.

Meinzen-Dick, R.S., A. Quisumbing, C. Doss, and S. Theis. 2019. "Women's Land Rights as a Pathway to Poverty Reduction: Framework and Review of Available Evidence." *Agricultural Systems* 172: 72–82.

Morgan, M., A.M. Larson, S. Trautman, E. Garner, M. Elias, and R. Meinzen-Dick. 2023. *Gender Transformative Approaches to Strengthen Women's Land and Resource Rights*. Bogor, Indonesia: Centre for International Forestry Research (CIFOR); Nairobi: World Agroforestry (ICRAF) and International Fund for Agricultural Development (IFAD).

Monterroso, I., P. Cronkleton, and A.M. Larson. 2019. "Commons, Indigenous Rights, and Governance." In *Routledge Handbook of the Study of the Commons*, eds. B. Hudson, J. Rosenbloom, and D. Cole, 1st ed., 376–339. New York: Routledge.

Monterroso, I., A.M. Larson, N. Liswanti, and Z. Cruz-Berga. 2019. *Mobilizing Change for Women within Collective Tenure Regimes*. Washington, DC: Research Consortium on Women's Land Rights.

Mueller, V., L. Billings, T. Mogues, A. Peterman, and A. Wineman. 2018. "Filling the Legal Void? Impacts of a Community-Based Legal Aid Program on Women's Land-Related Knowledge, Attitudes, and Practices." *Oxford Development Studies* 46 (4): 453–469.

Mwangi, E. 2007. "Subdividing the Commons: Distributional Conflict in the Transition from Collective to Individual Property Rights in Kenya's Maasailand." *World Development* 35 (5): 815–834.

Mwangi, E. 2009. "Property Rights and Governance of Africa's Rangelands: A Policy Overview." *Natural Resources Forum* 33 (2): 160–170.

Mwangi, E., R.S. Meinzen-Dick, and Y. Sun. 2011. "Gender and Sustainable Forest Management in East Africa and Latin America." *Ecology and Society* 16 (1): 17.

Nnoko-Mewanu, J.M. 2016. "Who Is Not at the Table: Land Deal Negotiations in Southwestern Cameroon." PhD Thesis, Iowa State University, Ames.

Notess, L., P.G. Veit, I. Monterroso, et al. 2018. *The Scramble for Land Rights: Reducing Inequity between Communities and Companies*. Washington, DC: World Resources Institute.

Oberndorf, R. 2012. *Legal Review of Recently Enacted Farmland Law and Vacant, Fallow and Virgin Lands Management Law Improving the Legal and Policy Frameworks Relating to Land Management in Myanmar*. Yangon, Myanmar: Land Core Group, Food Security Working Group.

Ortega, R.R. 2004. *Models for Recognizing Indigenous Land Rights in Latin America*. Biodiversity Series 99. Washington, DC: World Bank.

Ostrom, E. 2007. "A Diagnostic Approach for Going Beyond Panaceas." *PNAS* 104 (39): 15181–15187.

Ostrom, E. 2015. *Governing the Commons: The Evolution of Institutions for Collective Action*. Cambridge, UK: Cambridge University Press.

Otsuka, K. 2007. "Efficiency and Equity Effects of Land Markets." In *Handbook of Agricultural Economics*, vol. 3, eds. R. Evanson and P. Pingali, 2671–2703. Amsterdam: Elsevier.

Otsuka, K., and F. Place, eds. 2001. *Land Tenure and Natural Resource Management: A Comparative Study of Agrarian Communities in Asia and Africa*. Baltimore: Johns Hopkins University Press.

Owen, J.R., A.M. Lechner, J. Harris, R. Zhang, and É. Lèbre. 2022. "Energy Transition Minerals and Their Intersection with Land-Connected Peoples." *Nature Sustainability* 6 (2): 203–211.

Panda, P., and B. Agarwal. 2005. "Marital Violence, Human Development and Women's Property Status in India." *World Development* 33 (5): 823–850.

Peterman, A., A. Quisumbing, J. Behrman, and E. Nkonya. 2011. "Understanding the Complexities Surrounding Gender Differences in Agricultural Productivity in Nigeria and Uganda." *Journal of Development Studies* 47 (10): 1482–1509.

Place, F., M. Roth, and P. Hazell. 1994. "Land Tenure Security and Agricultural Performance in Africa: Overview of Research Methodology." In *Searching for Land Tenure Security in Africa*, eds. J.W. Bruce and S.E. Migot-Adholla, 15–39. Washington, DC: World Bank.

Platteau, J.P. 1996. "The Evolutionary Theory of Land Rights as Applied to Sub-Saharan Africa: A Critical Assessment." *Development and Change* 27 (1): 29–86.

Pradhan, R., R. Meinzen-Dick, and S. Theis. 2019. "Property Rights, Intersectionality, and Women's Empowerment in Nepal." *Journal of Rural Studies* 70: 26–35.

Quisumbing, A.R., and J.A. Maluccio. 2003. "Resources at Marriage and Intrahousehold Allocation: Evidence from Bangladesh, Ethiopia, Indonesia, and South Africa." *Oxford Bulletin of Economics and Statistics* 65 (3): 283–327.

Ratner, B.D., R. Meinzen-Dick, J. Hellin, et al. 2017. "Addressing Conflict through Collective Action in Natural Resource Management." *International Journal of the Commons* 11 (2): 877–906.

Ratner, B.D., R. Meinzen-Dick, C. May, and E. Haglund. 2013. "Resource Conflict, Collective Action, and Resilience: An Analytical Framework." *International Journal of the Commons* 7 (1): 183–208.

Ricker-Gilbert, J., and J. Chamberlin. 2018. "Transaction Costs, Land Rental Markets, and Their Impact on Youth Access to Agriculture in Tanzania." *Land Economics* 94 (4): 541–555.

Robinson, E.J.Z. 2005."Reassessing the Interaction between Investment and Tenure Uncertainty." *Environment and Development Economics* 10 (2): 143–157.

Rozelle, S., and J.F.M. Swinnen. 2004. "Success and Failure of Reform: Insights from the Transition of Agriculture." *Journal of Economic Literature* 42 (2): 404–56.

RRI (Rights and Resources Initiative). 2021. *Status of Legal Recognition of Indigenous Peoples,' Local Communities' and Afro-Descendant Peoples' Rights to Carbon Stored in Tropical Lands and Forests*. Washington, DC.

RRI. 2023. *Who Owns the World's Land? Global State of Indigenous, Afro-Descendent, and Local Community Land Rights Recognition from 2015–2020*, 2nd ed. Washington, DC.

Sandhu, H., W. Zhang, R. Meinzen-Dick, et al. 2023. "Valuing Ecosystem Services Provided by Land Commons in India: Implications for Research and Policy." *Environmental Research Letters* 18 (1): 013001.

Schlager, E., and E. Ostrom. 1992. "Property-Rights Regimes and Natural Resources: A Conceptual Analysis." *Land Economics* 68 (3): 249–262.

Schoneveld, G.C., L.A. German, and E. Nutakor. 2011. "Land-Based Investments for Rural Development? A Grounded Analysis of the Local Impacts of Biofuel Feedstock Plantations in Ghana." *Ecology and Society* 16 (4): 10.

Seymour, G., J. Heckert, A. Quisumbing, et al. 2024. "Measuring Women's Empowerment in National Surveys: Development of the Women's Empowerment Metric for National Statistical Systems (WEMNS)." IFPRI Discussion Paper 2254. IFPRI, Washington, DC.

Sjaastad, E., and D.W. Bromley. 2000. "The Prejudices of Property Rights: On Individualism, Specificity, and Security in Property Regimes." *Development Policy Review* 18 (4): 365–389.

Suhardiman, D., J. Bright, and C. Palmano. 2021. "The Politics of Legal Pluralism in the Shaping of Spatial Power in Myanmar's Land Governance." *Journal of Peasant Studies* 48 (2): 411–435.

Sulle, E. 2021. *Qualitative Outcome Study: The Contribution of the CGIAR Research Program on Livestock to Land Use Planning Processes in Ethiopia, Kenya and Tanzania.* Nairobi: International Livestock Research Institute.

Swinnen, J.F.M. 1999. "The Political Economy of Land Reform Choices in Central and Eastern Europe." *Economics of Transition* 7 (3): 637–664.

Tachibana, T., and S. Adhikari. 2009. "Does Community-Based Management Improve Natural Resource Condition? Evidence from the Forests in Nepal." *Land Economics* 85 (1): 107–31.

Thein, U.S., J.C. Diepart, U.H. Moe, and C. Allaverdian. 2018. *Large-Scale Land Acquisitions for Agricultural Development in Myanmar: A Review of Past and Current Processes.* MRLG Thematic Study Series 9. Vientiane, Laos: Mekong Region Land Governance.

Tseng, T.-W.J., B.E. Robinson, M.F. Bellemare, et al. 2020. "Influence of Land Tenure Interventions on Human Well-Being and Environmental Outcomes." *Nature Sustainability* 4 (3): 242–251.

von Benda-Beckmann, K., and B. Turner. 2018. "Legal Pluralism, Social Theory, and the State." *Journal of Legal Pluralism and Unofficial Law* 50 (3): 255–274.

von Braun, J., and R. Meinzen-Dick. 2009. *'Land Grabbing' by Foreign Investors in Developing Countries: Risks and Opportunities.* IFPRI Policy Brief 13. Washington, DC: IFPRI.

Wisborg, P. 2014. "Transnational Land Deals and Gender Equality: Utilitarian and Human Rights Approaches." *Feminist Economics* 20 (1): 24–51.

Zwarteveen, M., and R. Meinzen-Dick. 2001. "Gender and Property Rights in the Commons: Examples of Water Rights in South Asia." *Agriculture and Human Values* 18 (1): 11–25.

Chapter 7

FOOD VALUE CHAINS
Transformations in Low- and Middle-Income Countries

Thomas Reardon, Bart Minten, Sudha Narayanan, and Johan Swinnen

Thomas Reardon is a university distinguished professor, Department of Agricultural, Food, and Resource Economics, Michigan State University, and a nonresident senior research fellow, IFPRI; **Bart Minten** is program leader, Myanmar Strategy Support Program, and a senior research fellow, Development Strategies and Governance Unit (DSG), IFPRI; **Sudha Narayanan** is a senior research fellow, DSG, IFPRI; and **Johan Swinnen** is director general, IFPRI.

Key messages

- In recent decades, food value chains (FVCs) in low- and middle-income countries (LMICs) have expanded dramatically, with major increases in the number of actors, volume and diversity of services and products, and geographic length.
- In the 1970s and 1980s, agrifood systems were dominated by traditional FVCs and parastatal companies. Beginning in the late 1980s, two FVC "revolutions" were sparked by market liberalization, including the dismantling of parastatals and price controls and opening to foreign direct investment; by public infrastructure investments; and by changing food consumption patterns.
- The "quiet revolution" was marked by the rapid expansion of micro, small, and medium enterprises (MSMEs) in the midstream of FVCs, including in wholesale, logistics, processing, and farm input marketing. Research focused on this expansion and the formation of spontaneous clusters of farms and enterprises, the central role of wholesale markets, and the eventual concentration and disintermediation of midstream MSMEs.
- The subsequent modern revolution was marked by the growth of large private enterprises, and concentration of FVC activities among fewer actors. Research has focused on the impacts of the rise of supermarkets, large processors, and agribusiness input firms on farmer technologies and incomes, consumers, and the logistics sector and on competition.

Looking ahead, priorities for researchers and policymakers should include:

- **Innovative FVC structures.** Spontaneous clusters of farms and MSMEs, new types of farm cooperatives, and outsource services are likely to grow in importance. Research should systematically consider where, how, and why these new models emerge and whether they are viable and able to meet market requirements, as well examining the role of policies in their success or failure.
- **Short- and long-term shocks to FVC reliance.** As climate, economic, conflict, and other shocks become more common, increasing the resilience of both individual value chain segments and whole value chains is essential. This will require a better understanding of the differential effect of shocks on FVCs and FVC actors, and of the enabling environment that can best foster private sector investment in resilience, including for farms.
- **Trends in market structures.** Several current trends are likely to continue shaping FVCs. As market concentration continues, research and policy must find an appropriate balance between competition and concentration. To meet growing demand and the need for food safety standards, public and private investments infrastructure, such as wholesale markets and cold chains, will be essential. The emergence of new technologies also calls for policy frameworks and investment to make the most of new opportunities.

Food value chains increasingly link farm production to input providers upstream and to often-distant consumers downstream, passing through a variety of intermediaries including traders, storage providers, processors, wholesalers, and retail enterprises. Over the past 50 years, the economic, demographic, and policy context of food value chains (FVCs) in low- and middle-income countries (LMICs) has changed enormously, with policy research evolving to address this changing context. FVCs have become much more important to the food economies of LMICs as they have become larger in both the volume and diversity of agrifood products. These changes have accompanied two broad phases in agrifood policy and two "revolutions" that have completely reshaped the structure of FVCs in LMICs (Reardon and Timmer 2007). Initial growth in FVCs was advanced by a proliferation of off-farm micro, small, and medium enterprises (MSMEs) and, more recently, by increased concentration in the sector as larger, modern firms expand their role. Evolving food systems policies have been closely linked to these structural changes. Policy debates and research have also evolved over this period, both influenced by and contributing to the FVC transformation, with policy research providing evidence, analysis, and recommendations to inform policy decisions.

This chapter discusses drivers and patterns in FVC transformation, along with policy research topics and contributions over the past 50 years, and provides a look at future research and policy needs. The chapter first outlines the growth and importance of FVCs to LMIC economies and employment, and then provides a brief overview of the broad transformations of FVCs since 1975, looking first at the pre-liberalization phase, then at the post-liberalization phase, and in the latter, the set of drivers that gave rise to the two FVC revolutions that took place in that phase—the "quiet revolution" and the "modern revolution." For each of these, it examines the patterns, policy issues, and research conducted by IFPRI and collaborating institutions. The final section looks ahead at the key trends that will continue to shape FVCs and the policy and research priorities for the coming years.

The chapter focuses mainly on transformation of domestic FVCs in LMICs, rather than the transformation of global value chains. However, we note that trade policies, foreign direct investment (FDI), and actors from global value chains, such as large food industry and agribusiness companies, have shaped the changes in domestic FVCs.

Expansion of domestic food value chains

Growth of FVCs in LMICs, and in their importance in LMIC economies, employment, and diets, has been rapid over the past 50 years.[1] In sub-Saharan Africa (SSA), rural-to-urban FVCs grew 800 percent between 1985 and 2010 (Haggblade 2011), and urban areas now account for about 60 percent of the region's food consumption (Reardon, Echeverría et al. 2019). In India, urban food expenditures increased threefold (in real terms) between 1991 and 2006, supplied by growth in domestic rural-to-urban FVCs (Minten, Reardon, and Vandeplas 2009).

Rural-to-urban FVCs are very important because urban consumers, such as in SSA now purchase nearly all of their food (per World Bank Living Standards Mearurement Study [LSMS] data), of which 80–90 percent comes

1 Growth of FVCs in LMICs has been notably faster than the similar transformation that occurred in the United States in the mid-20th century.

FIGURE 7.1 Agrifood sector value chains

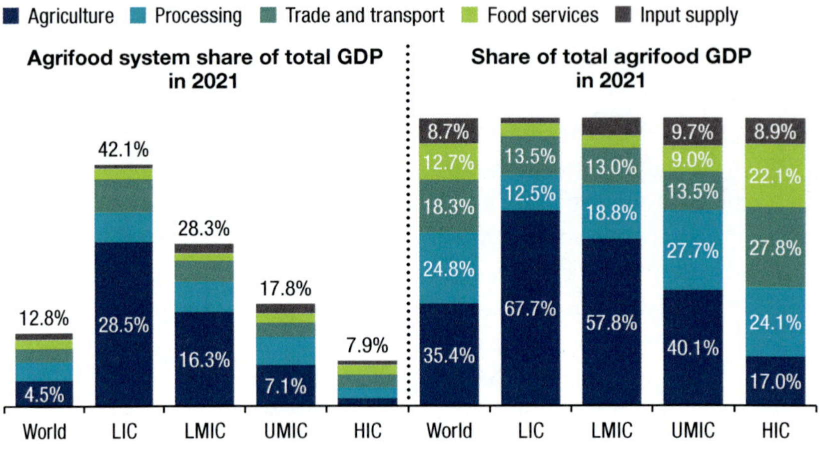

Source: Thurlow et al. 2025.

Note: LIC= low-income countries; LMIC = lower-middle-income countries; UMIC = upper-middle-income countries; and HIC = high-income countries.

via FVCs from rural areas and the rest from imports (Liverpool-Tasie et al. 2021). Rural-to-rural urban-to rural FVCs have also become very important in Asia, Latin America, and SSA. By the 2010s, around 60 percent of rural food consumption in SSA came from purchases, including in both unfavorable and favorable agroecological zones and in both poorer and richer countries (Dzanku et al. 2024). India's rural consumers now purchase about 80 percent of their food (Reardon, Mishra et al. 2020).

As a country's gross domestic product (GDP) increases, the agrifood system share of GDP falls. As shown in Figure 7.1, the agrifood sector now makes up about 42 percent of GDP in low-income countries (LICs) and only 7.9 percent in high-income countries (HICs) (Thurlow et al. 2025). However, as GDP rises, the share of off-farm segments within the agrifood sector also grows: today, the farm share is 67.7 percent for LICs but only 17 percent for HICs.

Related to this growth, rural and urban household employment data indicate that off-farm food systems employment is growing faster than agricultural employment across all country groups (Figure 7.2). A study (Dolislager et al. 2021) using LSMS survey data from SSA, Asia, and Latin America points to several key findings: for rural areas, food systems employment—in agrifood processing, wholesale, logistics, retail, and food service, mainly in MSMEs

FIGURE 7.2 Agrifood systems employment

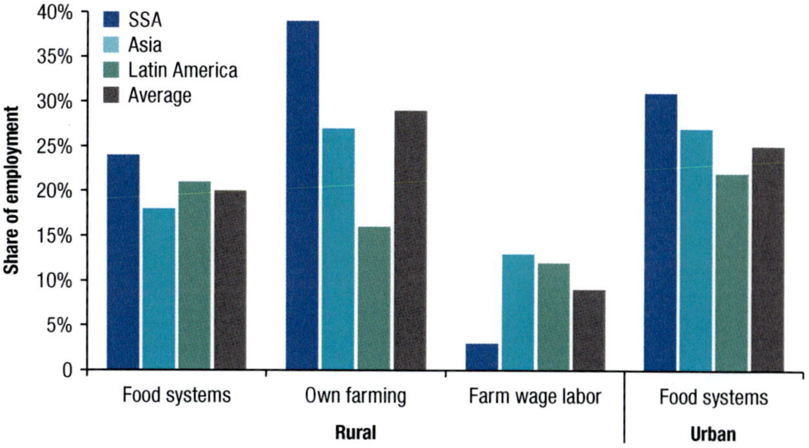

Source: Dolislager et al. 2021.

Note: Data cover 178,794 households with 460,654 individuals. sub-Saharan Africa (SSA) is represented by Ethiopia, Malawi, Niger, Nigeria, Tanzania, and Uganda; Asia by Bangladesh, Cambodia, Indonesia, and Nepal; and Latin America by Mexico, Nicaragua, and Peru.

as wage-based or self-employment—now accounts for 20 percent of rural employment (measured in full-time equivalents, or FTEs). The relatively small share in rural Asia (18 percent) indicates that processing and wholesale are more agglomerated in urban areas in Asia than in SSA. The average share of own-farming employment is just slightly higher than food systems employment, at 29 percent, and well above the share of employment as farm wage labor, which averages only 9 percent. The study also found that food systems employment is important in urban areas, averaging 25 percent of FTEs.

Over the past 50 years, there has been a profound transformation of FVCs in LMICs through two broad policy phases, the pre-liberalization (and pre-globalization) phase, roughly 1970–1990, and the post liberalization (and globalization) phase, roughly 1990 to the present.

During the pre-liberalization phase, FVCs were characterized structurally by a mix of the traditional stage of FVCs (dominated by micro enterprises and spatially short supply chains in a food economy with a limited urban share) and a parastatal segment (including large grain and oilseed processors, public distribution such as grain depots and subsidized "fair price" shops, and state input retailers) coupled with price controls and subsidies, particularly on grains.

FIGURE 7.3 The J-curve of concentration: Stages in FVC transformation

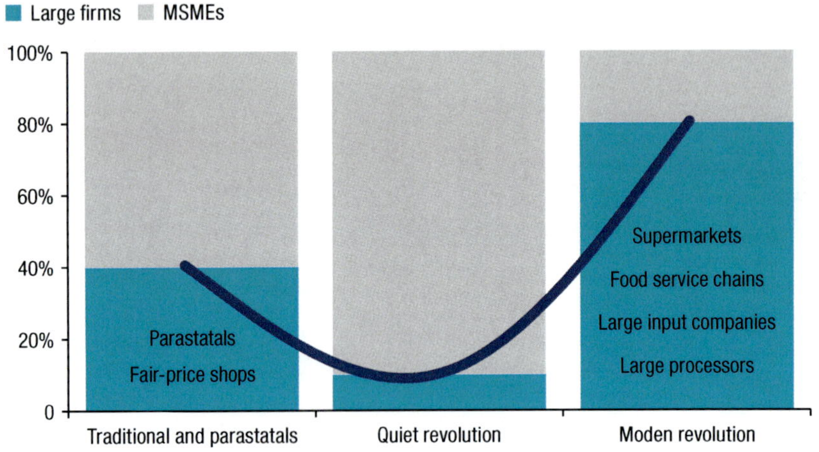

Source: Derived from Reardon (2015).
Note: MSMEs = micro, small, and medium enterprises.

During the post-liberalization phase, FVCs underwent two revolutions (Reardon and Timmer 2007). We discuss the drivers of these revolutions in detail later, but summarize their features here to set the stage:

1. A *quiet revolution*, featuring a rapid proliferation of MSMEs in the midstream segments of FVCs (processing, wholesale, and logistics) and agri-dealers in farm input value chains.[2]
2. A *modern revolution*, featuring the proliferation of supermarket chains, large processors, specialized wholesale firms, and large farm input firms (Reardon and Timmer 2007). These large firms were formed by the privatization of parastatals, by foreign direct investment (FDI), and by expansion of MSMEs.

The above two phases and two revolutions can be illustrated by a J-curve pattern of concentration (Figure 7.3) (Reardon 2015). The left side of the J is the pre-liberalization phase, when FVCs are moderately concentrated—a fragmented sector made up of traditional micro enterprises coexists with a concentrated parastatal sector. The quiet revolution in the hidden middle,

2 See Reardon, Chen et al. (2012) for the idea of the "quiet revolution" of MSMEs; and Reardon (2015) for the term "hidden middle."

post-liberalization, is in the middle of the J-curve, when parastatals have been eliminated, MSMEs proliferate, and the agrifood sector experiences de-concentration. The modern revolution is on the right side of the J-curve, when entry and formation of large firms concentrate the sector.

The timing of the phases and of the associated transformations of FVCs is approximate, as it differs (sometimes years earlier, sometimes years later) by region and country. As a rough approximation, SSA and South Asia are primarily in the middle segment of the J-curve at present, and Central and Eastern Europe and much of Southeast Asia and Latin America are at the early inflection of the third segment of the J-curve or well into it. We structure the rest of the chapter around these policy phases and FVC revolutions.

Pre-liberalization policy phase

Patterns and drivers

The pre-liberalization phase, in the decades preceding the 1990s, was characterized by a mix of traditional FVCs, dominated by micro enterprises and spatially short supply chains, and large-scale state-run agrifood firms (parastatals) combined with policies of price controls and subsidies. For governments, international credit was relatively abundant to support spending on farm and consumer subsidies, parastatals, and public infrastructure up until the structural adjustment era of the 1980s and 1990s (see Chapter 3).

For political economy reasons, governments wanted to ensure consistent and low-cost supplies of food to emerging cities and inputs to farmers. However, governments and many researchers believed that traditional FVCs had serious deficiencies, and in socialist countries, this assumption extended to market mechanisms more generally. It was assumed that the high transaction costs associated with numerous small and disbursed food distributors prevented traditional FVCs from supplying large volumes of food to cities or inputs to farmers. Moreover, the practices of traditional wholesalers were deemed exploitative, with accusations of price gouging harming both farmers and consumers. Given these perceived deficiencies of traditional FVCs and the desire of both governments and donors to ensure that pre- and post-farmgate supply chain contexts were favorable to the emerging Green Revolution in Asia and Latin America and to agricultural growth in LMICs more generally, governments adopted public goods solutions in the form of parastatals, regulations, and subsidies (see Chapters 3 and 17).

The parastatals included large processors that procured food grains and export crops from farmers, public distribution institutions such as grain depots, subsidized "fair price" shops that marketed grains to urban consumers, and state-run farm input firms that supplied inputs and credit to farms in an effort to address market imperfections (Swinnen and Maertens 2007). Support to farmers and consumers was bolstered by the development of public infrastructure to reduce marketing costs, including roads, electricity connections, market information systems, and public wholesale markets (Abbott 1967).

Research themes, policy critiques, and policy implications

For several decades prior to the 1990s, researchers questioned the widespread perception (the "conventional wisdom") that viewed traditional FVCs as anti-competitive, exploitative, and inefficient and that had justified the replacement of traditional FVCs with parastatals. Research in India contradicted this conventional wisdom, demonstrating a high degree of competition in paddy, rice, wheat, and sorghum markets (Lele 1971). In Peru, research showed that small traders were competitive and that their profit margins were commensurate with their high transaction costs and risk levels (Scott 1985). Later studies of traditional small and medium traders, which considered distance from markets, risks, and transaction costs in their analysis, such as an IFPRI study (Minten and Kyle 1999) in the former Zaire, likewise found that the profit margins of small traders largely reflected traders' high costs and risks, rather than excessive profits.

IFPRI researchers tested the market efficiency of traditional FVCs at the subnational (meso) level using tests of spatial market integration—that is, whether price signals and information are transmitted smoothly across markets, such that prices in different markets generally move together. A study in Bangladesh found that overall grain market integration was moderately high, with low integration apparent in less than 10 percent of markets, despite the conventional wisdom that these markets are inefficient (Goletti et al. 1995).

IFPRI researchers also conducted assessments of the cost-effectiveness of subsidies and public retail food distribution programs, which served as the FVC for many urban residents, in targeting poor consumers (see Chapter 11). Extensive research contributions by IFPRI in Egypt were influential in prompting the government to rethink and refine its food distribution systems, particularly in targeting and improving the efficiency of food subsidies (Ahmed and Bouis 2002; Alderman and von Braun 1986).

Post-liberalization policy phase: Drivers of transformation

We first describe the policy and nonpolicy drivers of the two FVC revolutions in this post-liberalization phase, and then discuss the revolution in some depth. Liberalization was introduced in the late 1980s in early-reforming countries and in the 1990s and 2000s in late-reforming countries. The reform was driven by untenable fiscal and trade balance deficits created by the existing public systems, external pressures such as conditionality imposed by the International Monetary Fund and World Bank through structural adjustment programs (see Chapter 3), as well as accumulating evidence that while parastatal systems were inefficient,[3] the domestic private sector was not necessarily inefficient or exploitative (Reardon and Timmer 2007).

Liberalization

Domestic market liberalization was important to both the proliferation of MSMEs and the formation and entry of large private enterprises (LPEs) in FVCs. Liberalization policies included reduction and elimination of price controls and subsidies and a major shift away from parastatals and cheap food policies as countries moved toward liberalized, pro-private-sector food policies (Timmer et al. 1983). A key component of these reforms was the privatization, scaleback, and elimination of parastatals, including state food retail chains, grain procurement systems, farm input depots, and government agrarian banks.

Policies liberalizing foreign trade and FDI introduced in the 1990s and 2000s particularly facilitated the modern revolution through (1) the elimination or reduction of FDI restrictions preventing foreign modern LPEs from entering LMICs; (2) privatization of parastatals, often by selling them to domestic or foreign LPEs; and (3) liberalization of product trade internally and externally.

IFPRI research informed the shift to liberalization in FVCs, with notable success in rice policies. For example in Bangladesh in the late 1990s, IFPRI studies (for example, del Ninno and Dorosh 2001) emphasized the role of private traders in stabilizing rice markets during severe floods and highlighted the negative consequences of government market interventions. This encouraged a shift toward market-driven solutions that included a reduction in government controls, an increased role for private traders, and promotion of free trade, changes that helped Bangladesh better manage the rice market during natural disasters, ultimately leading to more resilient food systems.

3 On Africa, see Delgado and Mellor (1984) and Christensen and Witucki (1982).

As the shift into the post-liberalization phase proceeded, a rich literature on political economy factors emerged that identified country-specific paths to liberalization. This work lies beyond the scope of this chapter but includes studies on the drivers and transition paths from agrifood socialism to liberalization in China, Viet Nam, Central and Eastern Europe, and the former Soviet Union (Swinnen and Rozelle 2006) and on the transition away from parastatals in Asia and Africa (Kherallah et al. 2002; Rashid et al. 2008).

Infrastructure investments

Public infrastructure investments, particularly the development of wholesale markets, roads, and electricity systems, underpinned both FVC revolutions. First, wholesale market investments were made by municipal, state, and national governments, as well as by the private sector. Wholesale markets provide several crucial advantages: they created agglomeration economies that could be served by burgeoning wholesale and logistics MSMEs in the quiet revolution (Reardon, Liverpool-Tasie, and Minten 2021) and later could supply supermarket chains in the modern revolution, as occurred with horticultural FVCs in Central America (Berdegué et al. 2005). Additionally, they formed important nodes linking rural towns to secondary cities and to primary cities through lengthening rural-to-urban FVCs, such as with fish FVCs in Nigeria (see Liverpool-Tasie et al. 2024). Wholesale markets also served to mitigate transaction costs and risk for buyers (including urban food firms, MSMEs, and large firms alike) and for sellers (farmers, rural traders, and processors). For example, public investment in these markets was shown to promote development of the teff FVC in Ethiopia, as well as spurring technological change among teff farmers (see Box 7.1) (Minten, Tamru et al. 2016). Finally, these markets serve as determinants of food safety in the supply of perishable products (Fafchamps et al. 2008). However, while the development of wholesale markets was particularly crucial during the quiet revolution, their significance has tended to decrease in the modern stage as supermarkets increasingly source directly from processors and farmers (Reardon and Timmer 2012).

Second, the development of roads facilitated the lengthening of FVCs. In Bangladesh, for example, feeder roads into rural areas supported the rapid development of the aquaculture FVC, which, in just 10 years, experienced a tripling of midstream MSMEs serving the domestic urban market (Hernández et al. 2018). Research in Ethiopia demonstrated the centrality of highway improvements coupled with MSME investments in larger trucks in reducing transport costs and expanding the teff FVC serving Addis Ababa (Minten, Tamru et al. 2016).

BOX 7.1 The quiet revolution in the Ethiopian teff value chain

Research on the Ethiopian teff value chain, conducted by IFPRI and collaborators from Ethiopia's national agricultural research system, illustrates the developments on the demand and supply side of processed and prepared grains. Traditionally, consumers have bought teff as a grain, cleaned it themselves, had it custom milled, and then prepared enjera at home. Although these practices are still maintained in rural areas and small towns, they have changed in Addis Ababa since the early 2000s. As custom milling and home cleaning of grain declined, consumers began to buy teff flour or prepared enjera, driving a nearly 50 percent increase in teff mills, enjera-making enterprises, and retail outlets in the city's neighborhoods. Wholesale marketing of teff has also surged. Research shows that the marketed surplus of teff increased rapidly over the past decade, and significantly more trade was reported on average in these markets over time. The reported number of trucks increased over the decade by some 70–80 percent across lean and peak periods. Growth in the teff value chain is exceeding urban population growth in Ethiopia, indicating higher consumption levels in the cities, more trade between rural areas that may pass through urban wholesale markets, and growth of transportation by trucks.

Source: Minten, Tamru et al. (2016).

Third, the public and private provision of electricity has been crucial to both FVC revolutions. For example, the expansion of the electricity grid played a key role in facilitating the rapid proliferation of cold storage MSMEs in Agra, India, in the 2000s (Reardon, Chen et al. 2012), as did the diffusion of private generators in Bihar (Minten, Reardon et al. 2014).

Rapid growth in urban and rural food demand and markets

Changes in food demand have contributed to—and been driven by—the proliferation of MSMEs in the quiet revolution and of large processors and supermarkets in the modern revolution. When urban areas accounted for a small share of total food consumption before the 1990s, their "demand pull" effect on domestic FVCs was small. With the huge increase in the size and share of the urban population in the past three decades, urban food consumption now accounts for the majority of national food consumption in LMICs.[4] These

4 For an example in Africa, see Liverpool-Tasie et al. (2021); for India, see Reardon, Mishra et al. (2022).

changes deeply affected the growth and composition of FVCs, incentivizing the quiet revolution and the modern revolution, both of which occurred earliest in urban areas.

Rural areas have likewise seen an enormous increase in the share and volume of purchased foods in consumption, with similar impacts on FVCs. In contrast to urban areas, however, the quiet revolution has continued to dominate rural food economies in the post-liberalization period, with the modern revolution emerging later in rural towns, such as supermarket chains spreading into rural towns in Chile (Faiguenbaum et al. 2002) and Indonesia (Reardon, Stamoulis, and Pingali 2007), and e-commerce extending into rural towns in China (Guo et al. 2022).

Change in composition of food demand

Two important changes have occurred in diets, driven on the demand side by increasing incomes and employment outside the home (and thus opportunity cost of time to prepare and process food at home), and on the supply side, by a rapid increase in the supply and accessibility of a greater diversity of foods.

First, starting roughly in the 1990s, diets diversified—the share of grains in overall food consumption declined in both urban and rural areas, while demand increased for animal-source foods, fruits and vegetables, and processed foods (see Chapter 12).[5] By the 2010s, grains made up only one-quarter of food consumption in India (Reardon and Minten 2011) and just over one-third of food expenditures in Ethiopia (Worku et al. 2017).

Second, starting in the 1980s and 1990s, urban and rural households began buying processed foods, spurring growth of processing MSMEs and LPEs, which made processed foods more accessible and further promoted processed food consumption. For example, urban, and eventually rural households, gradually shifted away from growing or buying raw ingredients, such as raw grain, and processing and preparing them at home, instead buying raw grain and having it custom milled by a local MSME, then shifting to buying milled flour and processed foods (such as jarred sauces, bread, noodles, and so on). This final shift tended to entail first buying non-ultra-processed foods such as bread and then adding ultra-processed foods, such as sodas and cookies, to the diet, as well as consuming more meals away from home in food-service MSMEs.[6]

---

5 For Africa, see Reardon, Liverpool-Tasie et al. (2024); for Asia, see Pingali (2007).

6 For Africa, see Reardon, Tschirley et al. (2021); on food service MSMEs in Tanzania, see Sauer et al. (2021).

Post-liberalization: The quiet revolution

General patterns and research findings

In the quiet revolution, the MSME sector responded strongly to liberalization, better infrastructure, and the increase in demand for nongrain and processed foods. Detailed survey-based research conducted by IFPRI, Michigan State University (MSU), and collaborators from LMIC national agriculture research systems over 20 years has shown that MSMEs have been very dynamic in the midstream segments of FVCs, such as processing, wholesale, logistics, and farm input retailers (agri-dealers), since 1990.

Through investments and the formation of new enterprises, MSMEs have supplied the huge increase in volumes of these products to rural and urban areas (Reardon, Liverpool-Tasie et al. 2024). In Africa, MSME wholesalers supply horticultural and animal-source products to expanding cities as well as to rural purchasers, with studies documenting this supply for milk and vegetables marketed in Addis Ababa (Minten, Habte et al. 2020; Minten, Mohammed, and Tamru 2020); tomatoes in Tanzania, where the number of tomato traders increased sixfold in three decades (Reardon 2024); aquaculture fish sent to Dhaka and other towns in Bangladesh (Hernández et al. 2018); and aquaculture and capture fish supplied from spontaneous clusters of fish capture and farming operations to consumers throughout Nigeria (Liverpool-Tasie et al. 2024).

FVCs lengthened considerably as the urban share of LMIC populations and national food economies grew quickly (see Liverpool-Tasie et al. 2021; Reardon et al. 2020) and induced a large increase in third-party logistics, for example, in the Nigerian maize FVC (Liverpool-Tasie, et al. 2021). While in some cases, such as green leafy vegetables, domestic FVCs remain spatially short or medium in length, many of the FVCs developed by MSMEs are spatially long, such as for tomatoes and fish in Nigeria.[7]

The lengthening of food supply chains from rural to urban areas was facilitated by substantial public investments in roads and growth of the wholesale sector, including (1) an increase in wholesale markets, such as in China (Huang et al. 2007), India (Minten and Reardon 2011), Ethiopia (Minten, Tamru et al. 2016), and Nigeria (Liverpool-Tasie et al. 2021); and (2) the proliferation of urban wholesalers and rural traders, for example increased entry of traders into the market in Madagascar (Barrett 1997).

7 On vegetables in Nigeria, see Liverpool-Tasie et al. (2023, 2024).

BOX 7.2 The quiet revolution in potato value chains in India

In India, government R&D of potato varieties and extension services provided for the new varieties were complemented by subsidies for tubewells and cold storage, as well as major investments in road improvements, the power grid, and communications networks. Research by IFPRI and Michigan State University in Bihar and Agra looked at the impact on potato value chains from farmer to consumer. Following the government investments, growing urban demand for vegetables was met by a significant shift from food grain to potato production in agricultural zones near Agra and in Bihar as well as rapid investment by many MSMEs in refrigerated cold storage. Farmers flocked to the cold storage firms, with the share of farmers using the storage sites in the Agra study area rising from 1 percent in 1999 to 99 percent by 2009. Wholesalers, mainly from Delhi, came and bought directly from local traders or farmers at the cold storage enterprise sites. Investments in cold storage, which de-seasonalized the potato supply, as well as in the new varieties and infrastructure, reduced consumer prices for potatoes, fueling an increase in consumer demand and further growth in the sector.

Source: Reardon, Chen et al. (2012) for Agra; Minten, Reardon et al. (2014) for Bihar.

The rapid spread of MSMEs has not only responded to but also fueled increases in demand for nongrain products. For example, the rapid development of cold storage MSMEs has had a bidirectional impact on potato demand in India (see Box 7.2). Income and population increases in India's large cities in the 1990s and 2000s led to a major increase in the demand for vegetables in general and potatoes in particular. The development of the potato value chain and cold storage options lowered prices for consumers, further increasing demand.

MSMEs have also responded to the growing demand for processed foods. Grain millers and food service MSMEs spread rapidly, such as in urban Africa, where an IFPRI study documented the increase in food vendors selling prepared millet and rice dishes in Ouagadougou.[8]

Despite the evidence, this midstream dynamism has not been widely perceived or adequately acknowledged in the policy debate, as governments, donors, and researchers have too often maintained negative views of traders as

8 On grain millers in Africa, see Reardon, Tschirley et al. (2021); on Ouagadougou, see Reardon, Thiombiano, and Delgado (1989).

largely exploitative actors, rather than as dynamic investors. Midstream enterprises were also thought to be few in number, leading to a dearth of input and midstream value chain services. This idea is captured by the term "the missing middle," often heard in international development forums and from donors, governments, and nongovernmental organizations (NGOs). Yet, in view of the sector's dynamism, the midstream is more accurately termed the "hidden middle" (Reardon 2015). As a consequence of this misperception, donors, NGOs, governments, and even international agricultural research centers still promote direct actions and investments to create midstream activity, rather than supporting and leveraging existing local, grassroots innovations.

Policy research and recommendations

Highlights from the policy research literature on the quiet revolution in MSMEs, which continue to be relevant to today's FVCs, include the role of spontaneous clusters, wholesale markets, disintermediation in FVCs, and continued debate about exploitation of farmers in FVCs.

SPONTANEOUS CLUSTERS OF FARMS AND MSMEs

A large body of research in the past 20 years, including important collaborative work of IFPRI with MSU, has focused on the emergence and rapid development of "spontaneous clusters" of farms and off-farm MSMEs. This work has served to balance the public debate that has mainly focused on "managed" clusters, such as agro-parks and special economic zones promoted by governments, donors, NGOs, and large "anchor" companies. Spontaneous clusters, in contrast, grow up organically through the co-location and evolution of MSMEs—primarily those supplying farm inputs and midstream processing of agricultural products—alongside farms (Belton et al. 2024; Reardon, Liverpool-Tasie et al. 2024).

Spontaneous clusters of farms and MSMEs have commonly formed near water sources and highways that connect farming areas to city markets. Off-farm MSMEs providing services to farmers in these clusters have helped to attract more farmers, and rising numbers of farms increase demand for their services. Public investments in roads, electrification, wholesale markets, and certain aspects of governance, such as control of banditry and conflict, favor both the functioning of MSMEs as individual firms and the emergence of spontaneous clusters. This has been shown, for example, in the proliferation of modern cold storage in Bihar and Uttar Pradesh in India and in Bangladesh in the 2000s and 2010s (Minten, Reardon et al. 2014; Reardon, Chen et al. 2012) and for milk chilling centers in Uganda (Van Campenhout et al. 2021) and India

(Burkitbayeva et al. 2023). While government investments in infrastructure have provided important enabling conditions for these clusters, these investments usually were not intentionally designed to support the clusters.

In the Ethiopian Rift Valley, IFPRI conducted research on spontaneous clusters of vegetable farmers and off-farm MSMEs (Minten, Mohammed et al. 2020). Vegetable farmers began production at this site because of good growing conditions (especially easy access to water) and highway links to the growing Addis Ababa market. Wholesalers, logistics firms, and input suppliers as well as agricultural service providers (such as well-digging outfits) moved into the area, creating an agglomeration of services within a decade that supported further expansion of vegetable farming, which now supplies a large share of the vegetables sold in Addis Ababa. Research on aquaculture clusters in Bangladesh and Myanmar also produced evidence of rapid, spontaneous "symbiotic" co-location and proliferation of farms, wholesalers, input suppliers, and transporters.[9]

Research findings also suggest an important link between lengthening of FVCs, the proliferation of MSMEs in those longer chains, and development of spontaneous clusters. For example, a large spontaneous cluster of tomato farmers and related off-farm actors in several states in northern Nigeria sends tomatoes some 500 to 1,000 kilometers to central and southern Nigeria (Liverpool-Tasie et al. 2023).

Finally, research has found that spontaneous clusters can promote farmers' adoption of technology by reducing their transaction costs and thus creating incentives for farmers to commercialize. This was shown, for example, for aquaculture clusters in Bangladesh (Hu et al. 2019).

WHOLESALE MARKETS AND MSME TRADER DYNAMICS

Looking at the role of wholesale markets, IFPRI and collaborators first assessed the impact of government restrictions on direct sales by farmers to these markets. Governments such as India's had restricted sales in wholesale markets to licensed traders prior to liberalization to facilitate monitoring and control of trader margins. IFPRI studies in the 2000s informed the reform of India's 1963 Agricultural Produce Marketing Act, which regulated the marketing of agricultural products in government wholesale markets (Joshi et al. 2004). The research showed that policies to restrict entry into agricultural trade are neither necessary nor useful for ensuring fair prices, and that governments

9 On Bangladesh, see Hernandez et al. (2018) and Hu et al. (2019); on Myanmar, see Belton et al. (2018).

should focus instead on facilitating entry and competition among traders and processors. This finding was supported by detailed trader surveys in Benin, Madagascar, and Malawi showing that marketing costs are nearly proportional to transaction size and that larger traders do not have significantly different profit rates than smaller traders (Fafchamps 2005).

Second, IFPRI research recommended technological and institutional innovations to upgrade wholesale markets and other commodity trading places. One such innovation in which IFPRI research played a crucial role was the Ethiopian Commodity Exchange (Gabre-Madhin 2012).

THE CONTINUING DEBATE ON EXPLOITATION

Research also continued to debunk some tenets of conventional wisdom. For many decades, trader–farmer interactions were depicted in terms of traders' "tying" credit to farmers' agreement to sell them their output, in what has often been viewed as an exploitative relationship (for example, Harriss-White 2008). Recent work shows a decrease in such tied output–credit arrangements, at least in output markets. These arrangements become less advantageous to farmers once they have better access to roads (and thus to competing traders), to cash from rural nonfarm employment, and to digital information that allows them to compare offers. This decrease has been observed, for example, in rice value chains in Bangladesh, China, India, and Viet Nam (Reardon, Chen et al. 2014).

Recent research also shows that informal relations between MSME traders and processors and small farmers can benefit farmers by helping to resolve idiosyncratic market failures. Recent literature on relational contracts (Macchiavello et al. 2022) suggests that these informal arrangements can include provision of transport, credit, and information for small farmers, who primarily rely on upstream and downstream MSMEs for inputs, services, and output sales (Liverpool-Tasie et al. 2020). Research on the Rwanda coffee FVC has provided an example of these relationships (Macchiavello and Morjaria 2021).

CONCENTRATION AND DISINTERMEDIATION IN THE MSME SECTOR

First, as noted above in the J-curve of concentration, after the parastatal era there was a proliferation of MSMEs, followed by increasing concentration as LPEs enterd FVCs and many MSMEs either exited or were acquired by LPEs. In Zimbabwe, for example, the grain processing sector de-concentrated in the early post-liberalization period in the 1990s as MSMEs entered and parastatal plants were sold or eliminated, and then re-concentrated in the 2010s as LPEs acquired processing MSMEs (Reardon, Tschirley et al. 2021; Rubey 1995). In

China, the rapid shuttering of small rice mills, accompanied by a concentration and urbanization of rice mills, occurred in the 2000s (Reardon, Chen et al. 2014).

Second, IFPRI and MSU research identified "disintermediation" in the MSME trader sector in a number of countries. Disintermediation refers to a reduction in the number of midstream actors between consumers and upstream sellers in the value chain. To illustrate, it can involve a mill or urban wholesaler buying directly from farmers and bypassing the traditional link with the rural (village) trader, or a mill or urban wholesaler selling directly to urban retailers, bypassing the traditional link with a semi-wholesaler who makes the link to the city market. Disintermediation has been found to be closely linked to the rise of third-party logistics, which provides the crucial delivery link for the disintermediating wholesaler or mill. Examples of these phenomena are shown for rice in Bangladesh, China, and India (Reardon, Chen et al. 2014) and maize value chains in Myanmar (Belton et al. 2024).

Post-liberalization: The modern revolution

Conventional wisdom at the start of the post-liberalization phase

Conventional wisdom assumed that conditions in LMICs were unfavorable to investment in and development of modern LPEs. One tenet of this wisdom held that parastatals would crowd out LPEs, a somewhat circular argument given that the rationale for establishing parastatals was that new LPEs would not emerge in LMICs and traditional, local LPEs were too inefficient to meet the needs of growing cities. In addition, it was thought that both foreign LPEs and private domestic investors would consider the market conditions inadequate to create or invest in LPEs. This tenet likewise reflected a circular argument, since during the pre-liberalization phase, strong strictures were placed on FDI to protect both small and large local firms. Last, it was argued that supermarkets would not spread because demand conditions were unfavorable in LMICs and a "supermarket revolution" was considered impossible (for example, see Goldman 1974).

Conventional wisdom also held that, in contrast to the parastatals, LPEs do not aim to provide public goods, and thus domestic LPEs were exploitative and speculative, as would be any foreign LPEs established in LMICs. This assumption, combined with a protectionist perspective and fears of foreign LPEs outcompeting local firms of all sizes, led to FDI regulation that initially slowed investments by foreign LPEs.

Events and research proved much of this conventional wisdom wrong, as we discuss next.

General patterns

The immediate effects of FDI liberalization reforms were huge inflows of FDI from supermarket chains, large processors, large logistics and wholesale distribution firms (often to service FDI food industry firms), and large farm input companies (Reardon and Timmer 2007). For example, Brazil's relaxation of FDI constraints spurred large investments by dairy multinationals such as Nestlé and Parmalat (Farina 2002). Similar experiences have been documented in Poland (Dries and Swinnen 2004).

Historically, FDI had been mainly "vertical" to create export bases, such as tea plantations, but the new wave of FDI was mainly "horizontal," intended to provide services, domestic markets for, and inputs to local FVCs (Awokuse and Reardon 2018). FDI initially flowed from HICs, such as multinational retailers Carrefour and Walmart, which established stores in Brazil; and then from emerging market countries, such as Korean supermarket chains that enter China.

As these LPEs entered the market, they competed with existing firms. New LPEs often acquired legacy LPEs, such as old family firms, and bought up or outcompeted MSMEs over time. In addition, sales of parastatals to domestic firms created LPEs. For example, various milling parastatals were sold to a Tanzania-based company, Bakhresa, allowing it to grow into a regional multinational firm (Reardon, Tschirley et al. 2021). This concentration is shown in the right side of the J-curve in Figure 7.3.

Policy research and recommendations from the modern revolution

Research has examined the development of the modern revolution, both geographically and temporally, and has explored the impacts of policy, the role of restructuring, and effects on nutrition and inclusion, among other topics.

IMPACTS OF POLICY ON THE MODERN REVOLUTION

As LPEs emerged upstream (large input companies), midstream (processors), and downstream (food service chains, supermarkets), research looked at how liberalization of the domestic sector and FDI influenced their expansion. One strand of research focused on the entry of foreign export firms and development of domestic firms and cooperatives that sold to those firms and exported directly, and examined their effects on farmers' cash cropping and food crop productivity.[10] Another strand of research focused on the entry of foreign multinational firms marketing domestically, such as sugar and dairy processing multinationals in Central and Eastern Europe (Dries and Swinnen 2004; Gow and Swinnen 1998).

RISE OF THIRD-PARTY LOGISTICS AND MODERN-SECTOR DISINTERMEDIATION

Geographic lengthening of both domestic and export FVCs induced a further large increase in third-party logistics enterprises. An important symbiosis developed between supermarkets and large processors and these logistics enterprises, as well as with specialized wholesalers, as has been shown in Asia and Latin America (Reardon, Henson, and Berdegué 2007) and China (Hu et al. 2004).

However, the rise of supermarkets and large processors also sidelined some actors and institutions that had been important in the quiet revolution. For example, the modernization of supermarket procurement systems has reduced the role of traditional wholesale markets and MSME traders, as supermarkets increasingly rely on direct procurement from farms and processors. This modern-sector disintermediation has been documented in Central America (Berdegué et al. 2005) and in Central and Eastern Europe (Reardon and Swinnen 2004).

FVCs, MARKET POWER, AND REGULATIONS

Recent consolidation in agribusiness, food processing, and especially retailing have triggered discussions about the impact of market power. Prior to liberalization, concerns about excessive concentration in FVCs focused on the dominant position of state companies and food processors and related impacts on competition. With price and trade liberalization, and more recently the modern revolution, this discussion has now shifted to implications of concentration in trading, input markets, and the retail sector (Deconinck 2020).

Overall, the welfare effects of concentration in FVCs are more complex, both theoretically and empirically, than often presented (Swinnen and Vandeplas 2010). Concentration may benefit consumers and producers for several reasons, including scale economies in production or R&D, reductions in transaction costs, and countervailing power in value chains. Interest in countervailing power, in particular, has grown as concentration in the retail sector has increased the sector's power in FVCs, which may offset the market power of food processors and traders.

---

10 For an example in Guatemala, see von Braun et al. (1989).

Poor consumers may welcome large retailers in their communities since these stores can give them access to cheaper food (and products in general). Growing retail concentration may also benefit consumers in the policy arena. When farmer and agribusiness interests (or consumer and large retailers' interests) are aligned, the size advantage of large companies in collective action and the numbers advantage of farmers and consumer groups can affect policies.

Many complaints of market power problems have come from farmers and smaller food processing companies. However, in this case also, the situation is more complex than it seems. For producers, large retailers may enhance, rather than reduce, competition in traditional markets, where trading is controlled by traditional middlemen. Moreover, markets characterized by vertical coordination and interlinked contracts within FVCs may enhance the bargaining power of farmers, and large downstream firms may have less incentive to wield market power to reduce prices paid to their suppliers (Kuijpers and Swinnen 2016; Sexton 2013; Swinnen and Vandeplas 2011). In all of these cases, there is a trade-off between concentration and both the efficiency effects of the relationships in value chains and their distributional effects. Empirical studies find mixed evidence on the relationship between concentration in FVCs and efficiency and rent distribution (Maertens et al. 2012; Minten et al. 2009).

Relatedly, market power may be an essential, but temporary, component of innovation in FVCs (Zilberman et al. 2019). Creators of new products can take advantage of market power with both upstream and downstream components (buyers and suppliers) of FVCs. This market power can be reflected in contracting terms and the nature of vertical coordination. However, in a dynamic environment, this market power may not last, so there is a trade-off between static and dynamic efficiency.

EFFECTS OF SUPERMARKETS ON NUTRITION

In conventional wisdom, supermarkets were thought to have several drawbacks. First, it was thought that introduction of supermarkets food prices would have impacts on LMIC consumers and be unaffordable for some. This idea has been challenged for a number of LMICs (Minten and Reardon 2008), including India (Minten et al. 2010). Modern retail at first may price its products high when it serves a niche market, but it eventually uses its economies of scale in procurement and marketing to lower prices to gain market share beyond the middle class by including the poor.

Second, supermarkets were thought to drive purchases of ultra-processed food. This belief has been disproved, for example, in Kenya, where researchers found no significant difference between the range of foods, especially processed foods, sold by small retailers and supermarket chains alike (Demmler et al. 2018).

Third, it was thought that supermarkets do not sell healthy foods such as fruits and vegetables. While this is the case in the early stages of supermarket penetration, over time, supermarkets shift into these products and even reduce their cost to poor consumers (Minten and Reardon 2008).

INCLUSION OR EXCLUSION

There has been debate over whether (and under what conditions) the modern transformation of the food industry has led to inclusion of small farmers, rural laborers, and small processors and of women and youth, or whether it has led to exclusion of these groups. In general, research tells a mixed story (Reardon, Barrett et al. 2009). Many case studies provide evidence of substantial inclusion, such as small farmers in the Polish dairy sector (Dries and Swinnen 2004); in export horticulture in Senegal (Maertens and Swinnen 2009); for sugar in Central Europe (Gow and Swinnen 1998); and in Madagascar (Minten, Randrianarison et al. 2009). Some studies show that inclusion is conditional on rights to resources, as shown for women's ownership of sugarcane plots in Uganda (Ambler et al. 2022). A synthesis of research on women and inequality in FVCs in Africa and Asia finds that women often face significant constraints to FVC participation (Quisumbing et al. 2021).

Inclusion has often been a result of resource provision contracts implemented by large firms to resolve idiosyncratic market failures that affect small farms. Indeed, the spread of modern firms throughout FVCs has increased farm technology adoption, quality upgrading, and product differentiation, as well as net incomes, as these firms use contracts and private standards to incentivize farms and processors to supply the needed volumes, quality, and safety traits of products (Barrett et al. 2012; Swinnen 2007; Swinnen and Kuijpers 2019).

However, there are also cases where company standards imposed by large firms on farms have led to exclusion and even shuttering of small farms, as occurred in Brazil's dairy sector (Farina 2002). In view of these mixed results, the policy issue has been whether and how to assist small farms and other actors to upgrade to meet the requirements of large enterprises. This research parallels the work on relational contracts for the quiet revolution.

Looking forward

Dramatic changes have occurred in LMIC food systems over the past 50 years. In 1975, when IFPRI was founded, food systems were largely in the traditional stage, urban populations were small, and subsistence agriculture was the norm, with only pockets of commercialization and intensification. Looking forward to the coming decades, we expect that some current trends will persist and can envision the accompanying food policy challenges and opportunities, although unexpected changes and challenges are also likely. In this section, we look at the most rapidly occurring trends and related policy and research needs.

Innovative FVC structures

Three types of innovative FVC structures discussed in this chapter are likely to grow in importance and merit increased attention.

SPONTANEOUS CLUSTERS

Spontaneous clusters, already important to supplying domestic and some export markets in Asia and Africa, are likely to expand in size and number as domestic and intraregional demand for diverse non-grain products grows in the coming decades. Research and policy attention to many aspects of these clusters are needed (Belton et al. 2024; Reardon, Liverpool-Tasie et al. 2024): (1) the policy and nonpolicy drivers of their growth and constraints to their formation and development; (2) effects of these "agglomeration economies" on the growth and innovation of firms and farms, such as technology innovation among farms;[11] (3) their governance, institutions, and informal coordination and how these affect transaction costs within and between clusters and the urban markets;[12] (4) their influence on nutrition within clusters and in the markets they supply, such as their effect on the cost of nutrient-dense foods to consumers;[13] (5) their effects on inclusiveness, especially the dynamics of entry and growth of employment for youth and women; and (6) their effects on food safety and the environment, as well as the related impacts of institutional arrangements such as land and water access in the cluster.

FARM CLUSTERS OR NEW COOPERATIVES

Farm cooperatives are an age-old structure aggregating farmers to source inputs, receive extension, or provide market outputs. Traditional location-based and

---

11 For an example in Bangladesh, see Hu et al. (2019).

12 On Peru, see Escobal and Cavero (2012).

13 On fish in Malawi, see Muhonda et al. (2024).

state-established and -managed cooperatives have declined over the decades and been replaced by a new generation of cooperatives and managed clusters. These are set up by smaller groups of farms with the objectives of meeting market requirements, enforcing use of technologies, and meeting quality and safety standards. They are emerging and spreading in LMICs, supported by large food and input companies that are, for example, seeking group certification for global market access.[14]

Research has identified particular cases of these new cooperatives, but a more systematic understanding is needed of where, how, and why these clusters emerge and of whether they are viable or profitable and able to meet market requirements. Research should consider these questions as well as examining the role of policies in the success or failure of these new groups. One interesting case showed the rapid diffusion of these new cooperatives in Chile and Brazil in the 1990s, but most proved unprofitable and folded (Berdegue 2001; Farina 2002). While these new cooperatives are often seen as a panacea for farmers' problems, they frequently encounter issues with organizational sustainability and profitability (Narayanan et al. 2025).

AGRICULTURAL OUTSOURCE SERVICE ENTERPRISES

As important suppliers in many LMIC markets, these services help farmers reach both cost-sensitive commoditized product markets as well as quality- and safety-sensitive markets. Outsource services will likely continue to spread geographically and into new sectors, such as animal products. Importantly, they will continue to expand the use of locally innovative technologies, such as rice combines, drones, and fruit tree hormone treatment. As the sophistication and cost of new technologies, including equipment such as drones, will likely be unaffordable for individual small farmers, these farmers are expected to increasingly turn to outsource services instead. Research and policy attention is needed to build understanding of several aspects of these service enterprises (Awokuse et al. 2024; Belton et al. 2024): (1) the spread of these services and the "business ecosystem" of policy and nonpolicy enabling factors that condition their emergence and development; (2) technological changes (such as adoption of robotics) in these enterprises and their innovative organizational relations, for example, with agricultural input and equipment companies (such as their emerging contract relationships with multinational firms); (3) the evolving business models of these firms, including their relationships

14 On shrimp in Bangladesh, see Narayanan et al. (2025).

with farmer cooperatives; and (4) ways to optimize inclusiveness, such as rural employment and environmental impacts of these firms.

Short- and long-term shocks to FVC resilience

FVCs will increasingly need to cope with shocks, including extreme weather events, local and international conflicts, epidemics, rising cybercrime, and trade policy and other economic disruptions (see Chapters 13 and 16). Debate around building FVC resilience has largely focused on government policies, regulations, and public investments to strengthen resilience and to help farms and firms cope with shocks. While this focus remains important, the formulation and effectiveness of policies can be enhanced by partnering with both large and small private sector firms and by considering these firms' strategies and actions for managing risks and coping with shocks.

DIFFERENTIATED POLICIES

Shocks typically affect all segments of FVCs (de Brauw and Pacillo 2022) but often cause disproportionate harm to rural zones, which are least able to prepare for or cope with shocks. As a result, events such as weather shocks deepen poverty both directly, through effects on farm production, and indirectly, as FVC actors such as traders, processors, and retailers shift procurement away from areas deemed vulnerable to shocks (Reardon and Zilberman 2018). Shocks also have a disproportionate effect on FVCs in the transitional stage, which tend to be geographically long but also structurally fragmented (dominated by small firms), meaning they have less capacity than large firms to manage risks and recuperate from shocks (Reardon and Vos 2021). Policymakers will need to ensure that entire FVCs as well as individual segments (such as farmers) are resilient and adaptable, with policies that promote stability and the capacity to handle a variety of shocks (see Chapter 13). Both researchers and policymakers need a better understanding of the differential impacts of shocks by FVC type and segment, as well as in vulnerable zones, to better craft appropriate policy and investment strategies.

ENABLING ENVIRONMENT FOR PRIVATE SECTOR ADAPTATION

Research and policy should increasingly consider and promote an enabling environment that allows private sector enterprises, whether small or large, to pivot to new technologies and practices, as well as sourcing and marketing strategies, in the face of shocks. This increases the resilience of FVCs overall: for example, during the COVID-19 pandemic, firms changed their sourcing and marketing geographies, helped their suppliers to weather the shock, and rapidly adopted e-commerce and other digital solutions (Reardon and Swinnen 2020). Food industry and agribusiness firms generally undertake a wide range of actions to adapt to shocks, such as using third-party logistics firms as partners in reconfiguring supply chains to minimize damages (Reardon and Vos 2021).

PRIVATE SECTOR SUPPORT FOR FARM AND FVC RESILIENCE

Both large and small private sector FVC firms could increase their role in supporting small farmers' resilience in the face of shocks and even in mitigating farmers' contributions to shocks by helping to reduce their environmental footprint (Swinnen et al. 2024). While the policy debate around promoting adoption of climate-smart agriculture by farmers has focused on government policies, the importance of the private sector in enabling and encouraging farms to be shock resilient has been poorly understood and underappreciated. Going forward, research and policy should increasingly engage the private sector as partners in building farmers' resilience. Private sector firms have a business incentive to support resilience in order to ensure consistent supply from farmers (Swinnen et al. 2024). In past decades, firms have demonstrated their capacity to convert this incentive into concrete measures to assist farmers or to impose requirements on their suppliers by providing price premiums, information, loans, access to inputs, and other measures to help farmers meet the market's volume and quality requirements. This same power is being applied to help, and also require, farmers to stabilize their supplies to FVCs, such as by investing in small-scale irrigation.

COMPETITION POLICIES

While private sector influence can benefit FVC resilience, large firms can be expected to wield greater influence than smaller firms over their suppliers, which can endanger the fair distribution of income (Swinnen et al. 2024). Addressing this issue requires transparent and effective competition policies and regulations. This risk was recognized in earlier FVC innovations made by large firms and gave rise to legislation such as the European Union's Unfair Trade Practices (UTP) regulations. Public sector policy can complement private sector assistance by redirecting existing "non-green" subsidies to support better information and traceability infrastructure and by establishing compensation policies to address social concerns and political constraints arising throughout FVCs (Swinnen et al. 2024).

Opportunities and challenges related to changing market structures

FVC actors and the governments that support and regulate them will face both opportunities and challenges from a continuation of current trends in market structures, including ongoing concentration, a shift toward larger food markets in urban and rural areas, and the rapid growth in demand for diverse food products. Each of these has implications for forward-looking policy research and policies.

CONCENTRATION AND COMPETITION

As more countries move from the transitional stage to the modern stage of FVCs, wherein larger firms play a prominent role in inputs, processing, marketing, and retailing, more policy attention to competition and concentration in FVCs is needed. FVCs are generally becoming more concentrated, though at varying rates for different countries, products, and FVC segments. Policy debate has focused on how concentration affects the inclusion of small-scale firms and farms, consumer prices, and FVC innovation, with research showing that the results are mixed, depending on conditions. Notable benefits for both farmers and consumers can arise from the formation of large-scale firms, but reduced competition can also lead to negative impacts.

Different regulatory approaches are possible, including competition policies, the use of standards (either public regulations or private or NGO standards), or new types of regulations such as the UTP rules, which focus on the nature of transactions and contracting across the supply chain to ensure fair practices and efficient outcomes. Yet, farmers have often been unsatisfied with these competition regulations. Their demand for regulation often emerges when shocks trigger changes in vertical relationships and conditions. For example, recent food price volatility has contributed to farmer complaints about asymmetric price transmission within FVCs. Given the conceptual and empirical nuances and complexities, as well as the relative scarcity of rigorous studies on concentration and competition, especially in LMICs, future food policy research should give this area substantial attention, especially the specifics of how concentration and competition affect outcomes for producers and consumers.

NEW INVESTMENT IN PUBLIC INFRASTRUCTURE

Urbanization will continue to be a key factor driving transformation of FVCs and related policy challenges. Over the next 25 years, the urban population is projected to grow by 2.5 billion people, which will significantly contribute to FVC growth. Rural purchases of food have grown quickly as a share of consumption and total volume over the past several decades, and will continue to be a key driver of FVC growth in coming years. Diet diversification beyond basic grains will continue (see Chapter 12), with horticulture, animal feed, animal products, and oil seeds playing a central role in future FVCs, with potentially positive impacts on nutrition outcomes.

These rapidly growing, vast food markets will require increased investments in wholesale markets; transportation infrastructure, particularly to link rural and urban areas; electricity systems; and fuel provision. For consumers, these investments tend to make food more affordable by reducing transportation costs. For midstream enterprises—those we have termed the hidden middle, including wholesalers, third-party logistics firms, processors, and agri-dealer input firms—these investments create an enabling business environment and should be a major priority in the decades to come. Given the growing importance of these agents in shaping efficient, equitable, safe, and resilient food systems, further study of their roles, operations, and public investment needs is essential.

FOOD SAFETY AND AFFORDABILITY

Rapid development of perishable food FVCs will require increased attention to food safety. Researchers and policymakers should focus on ensuring the safe and affordable delivery of the growing quantity of traded perishable food, from both national and international sources, within an increasingly volatile environment. Meeting growing demand for fruits and vegetables, fish, dairy, and meat will also require large-scale private and public investment in cold chain infrastructure. Gaining a better understanding of the role of cold chains and how they can be designed to generate beneficial outcomes will be crucial for these settings, as has been done for dairy products in Africa (Van Campenhout et al. 2021) and potato value chains in Asia (Reardon, Chen et al. 2012).

NEW TECHNOLOGIES

The rise of new technologies—such as artificial intelligence, smartphones, drones, genetic improvements, and other innovations—combined with agricultural labor scarcity will accelerate transformation in all segments of FVCs, with demand for these technologies growing over time. Moreover, these new technologies have potential to greatly reduce the link between modernization, consolidation, and technology upgrading, allowing for "modernization without consolidation." Recent innovations include technologies and technology services, such as outsource services, which allow small farmers (and off-farm enterprises) to enjoy economies of scale, machine and agronomic upgrading, and information systems that previously were only accessible to medium and large firms. This evolution partly reflects the innovative technologies themselves, such as small-scale irrigation linked to meso-scale solar electrical grids. Policymakers will need to design frameworks, supported by relevant research, that facilitate the safe, ethical, and efficient adoption of these new technologies in ways that maximize benefits while minimizing the risks.

References

Abbott, J.C. 1967. "The Development of Marketing Institutions." In *Agricultural Development and Economic Growth*, eds. H.M. Southworth and B.F. Johnston, 364–399. Ithaca, NY, Cornell University Press.

Ahmed, A., and H.E. Bouis. 2002. "Weighing What's Practical: Proxy Means Tests for Targeting Food Subsidies in Egypt." *Food Policy* 27 (5–6): 519–540.

Alderman, H., and J. von Braun. 1986. "Egypt's Food Subsidy Policy: Lessons and Options." *Food Policy* 11 (3): 223–237.

Ambler, K., K.M. Jones, and M. O'Sullivan. 2022. "Empowering Women to Engage in Commercial Agriculture." In *AEA Papers and Proceedings*, vol. 112, 546–550. Nashville: American Economic Association.

Awokuse, T., and T. Reardon. 2018. "Agrifood Foreign Direct Investment and Waves of Globalization of Emerging Markets: Lessons for U.S. Firms." *Economic Review - Federal Reserve Bank of Kansas City* 75–96.

Barrett, C.B. 1997. "Food Market Liberalization and Trader Entry: Evidence from Madagascar." *World Development* 25 (5): 763–777.

Barrett, C.B., M.E. Bachke, M.F. Bellemare, H.C. Michelson, S. Narayanan, and T.F. Walker. 2012. "Smallholder Participation in Contract Farming: Comparative Evidence from Five Countries." *World Development* 40 (4): 715–730.

Belton, B., A. Cho, M. Hall, B. Minten, and T. Reardon. 2024. "Wholesalers and the Transformation of Myanmar's Maize Value Chains." *Applied Economic Perspectives and Policy* 47 (1): 125–153.

Belton, B., A. Hein, K. Htoo, L. Seng Kham, A. Sandar Phyoe, and T. Reardon. 2018. "The Emerging Quiet Revolution in Myanmar's Aquaculture Value Chain." *Aquaculture* 493: 384–394.

Berdegué, J.A., F. Balsevich, L. Flores, and T. Reardon. 2005. "Central American Supermarkets' Private Standards of Quality and Safety in Procurement of Fresh Fruits and Vegetables." *Food Policy* 30 (3): 254–269.

Burkitbayeva, S., E. Janssen, and J. Swinnen. 2023. "Hiding in Plain Sight: The Emergence of Modern Dairy Farms in India." *Journal of Agribusiness in Developing and Emerging Economies* 13 (2): 194–210.

Christensen, C., and L. Witucki. 1982. "Food Problems and Emerging Policy Responses in Sub-Saharan Africa." *American Journal of Agricultural Economics* 64 (5): 889–896.

Deconinck, K. 2020. "Concentration in Seed and Biotech Markets: Extent, Causes, and Impacts." *Annual Review of Resource Economics* 12 (1): 129–147.

Delgado, C.L., and J.W. Mellor. 1984. "A Structural View of Policy Issues in African Agricultural Development." *American Journal of Agricultural Economics* 66 (5): 665–670.

del Ninno, C., and P.A. Dorosh. 2001. "Averting a Food Crisis: Private Imports and Public Targeted Distribution in Bangladesh after the 1998 Flood." *Agricultural Economics* 25 (2–3): 337–346.

Demmler, K.M., O. Ecker, and M. Qaim. 2018. "Supermarket Shopping and Nutritional Outcomes: A Panel Data Analysis for Urban Kenya." *World Development* 102: 292–303.

Dolislager, M., T. Reardon, A. Arslan, L. Fox, S. Liverpool-Tasie, C. Sauer, and D. Tschirley. 2021. "Youth and Adult Agrifood System Employment in Developing Regions: Rural (Peri-urban to Hinterland) vs. Urban." *Journal of Development Studies* 57 (4): 571–593.

Dries, L., and J.F. Swinnen. 2004. "Foreign Direct Investment, Vertical Integration, and Local Suppliers: Evidence from the Polish Dairy Sector." *World Development* 32 (9): 1525–1544.

Dzanku, F.M., L.S.O. Liverpool-Tasie, and T. Reardon. 2024. "The Importance and Determinants of Purchases in Rural Food Consumption in Africa: Implications for Food Security Strategies." *Global Food Security* 40: 100739.

Escobal, J.A., and D. Cavero. 2012. "Transaction Costs, Institutional Arrangements and Inequality Outcomes: Potato Marketing by Small Producers in Rural Peru." *World Development* 40 (2): 329–341.

Fafchamps, M., E. Gabre-Madhin, and B. Minten. 2005. "Increasing Returns and Market Efficiency in Agricultural Trade." *Journal of Development Economics* 78 (2): 406–442.

Fafchamps, M., R.V. Hill, and B. Minten. 2008. "Quality Control in Nonstaple Food Markets: Evidence from India." *Agricultural Economics* 38 (3): 251–266.

Faiguenbaum, S., J.A. Berdegué, and T. Reardon. 2002. "The Rapid Rise of Supermarkets in Chile: Effects on Dairy, Vegetable, and Beef Chains." *Development Policy Review* 20 (4): 459–471.

Farina, E. 2002. "Consolidation, Multinationalization, and Competition in Brazil: Impacts on Horticulture and Dairy Product Systems." *Development Policy Review* 20 (4): 441–457.

Gabre-Madhin, E.Z. 2012. *A Market for Abdu: Creating a Commodity Exchange in Ethiopia*. IFPRI Occasional Essays Series. Washington, DC: IFPRI.

Goldman, A. 1974. "Outreach of Consumers and the Modernization of Urban Food Retailing in Developing Countries: Low Outreach Is an Important Barrier to the Establishment of a More Modern, More Economical Supermarket System in Developing Countries." *Journal of Marketing* 38 (4): 8–16.

Goletti, F., R. Ahmed, and N. Farid. 1995. "Structural Determinants of Market Integration: The Case of Rice Markets in Bangladesh." *The Developing Economies* 33 (2): 196–198.

Gow, H.R., and J.F. Swinnen. 1998. "How Foreign Direct Investment Has Stimulated Growth in the Central and Eastern European Agri-Food Sectors: Vertical Contracting and the Role of Private Enforcement Capital." Working Paper No. 18. World Bank Policy Research Group, Washington, DC.

Guo, J., S. Jin, J. Zhao, H. Wang, and F. Zhao. 2022. "Has COVID-19 Accelerated the E-Commerce of Agricultural Products? Evidence from Sales Data of E-stores in China." *Food Policy* 112: 102377.

Haggblade, S. 2011. "Modernizing African Agribusiness: Reflections for the Future." *Journal of Agribusiness in Developing and Emerging Economies* 1 (1): 10–30.

Harriss-White, B. 2008. "Introduction: India's Rainfed Agricultural Dystopia." *European Journal of Development Research* 20 (4): 549–561.

Hernández, R.A., B. Belton, T. Reardon, C. Hu, X. Zhang, and A. Ahmed. 2018. "The 'Quiet Revolution' in the Aquaculture Value Chain in Bangladesh." *Aquaculture* 493: 456–468.

Hu, D., T. Reardon, S. Rozelle, P. Timmer, and H. Wang. 2004. "The Emergence of Supermarkets with Chinese Characteristics: Challenges and Opportunities for China's Agricultural Development." *Development Policy Review* 22 (4): 557–586.

Hu, C., X. Zhang, T. Reardon, and R.A. Hernandez. 2019. "Value-Chain Clusters and Aquaculture Innovation in Bangladesh." *Food Policy* 83: 310–326.

Huang, J., S. Rozelle, X. Dong, Y. Wu, H. Zhi, X. Niu, and Z. Huang. 2007. *Agrifood Sector Studies China: Restructuring Agrifood Markets in China—The Horticulture Sector.* London: International Institute for Environment and Development.

Joshi, P.K., A. Gulati, P.S. Birthal, and L. Tewari. 2004. "Agriculture Diversification in South Asia: Patterns, Determinants and Policy Implications." *Economic and Political Weekly* 39 (24): 2457–2467.

Kherallah, M., C.L. Delgado, E.Z. Gabre-Madhin, N. Minot, and M. Johnson. 2002. *Reforming Agricultural Markets in Africa: Achievements and Challenges.* Baltimore: Johns Hopkins University Press for IFPRI.

Kuijpers, R., and J. Swinnen. 2016. "Value Chain and Technology Transfer to Agriculture in Developing and Emerging Economies." *American Journal of Agricultural Economics* 98 (5): 1403–1418.

Lele, U.J. 1971. *Food Grain Marketing in India: Private Performance and Public Policy.* Ithaca, NY: Cornell University Press.

Liverpool-Tasie, L.S.O., A. Wineman, S. Young, et al. 2020. "A Scoping Review of Market Links between Value Chain Actors and Small-Scale Producers in Developing Regions." *Nature Sustainability* 3: 799–808.

Liverpool-Tasie, L.S.O., A. Wineman, M.U. Amadi, et al. 2024. "Rapid Transformation in Aquatic Food Value Chains in Three Nigerian States." *Frontiers in Aquaculture* 3: 1302100.

Liverpool-Tasie, L.S.O, T. Reardon, and B. Belton. 2021. "'Essential Non-Essentials': COVID-19 Policy Missteps in Nigeria Rooted in Persistent Myths about African Food Value Chains." *Applied Economic Perspectives and Policy* 43 (1): 205–224.

Liverpool-Tasie, L.S.O., R. Ya'u, A. Ibrahim, et al. 2023. "Transformation of Vegetable Value Chains in Nigeria: Spatial and Seasonal Lengthening & Intensification." RSM2SNF Working Paper. Michigan State University, East Lansing.

Macchiavello, R., and A. Morjaria. 2021. "Competition and Relational Contracts in the Rwanda Coffee Chain." *Quarterly Journal of Economics* 136 (2): 1089–1143.

Macchiavello, R., T. Reardon, and T. Richards. 2022. "Empirical Industrial Organization Economics to Analyze Developing Country Food Value Chains." *Annual Review of Resource Economics* 14: 193–220.

Maertens, M., B. Minten, and J. Swinnen. 2012. "Modern Food Supply Chains and Development: Evidence from Horticulture Export Sectors in Sub-Saharan Africa." *Development Policy Review* 30 (4): 473–497.

Maertens, M., and J.F. Swinnen. 2009. "Trade, Standards, and Poverty: Evidence from Senegal." *World Development* 37 (1): 161–178.

Minten, B., and S. Kyle. 1999. "The Effect of Distance and Road Quality on Food Collection, Marketing Margins, and Traders' Wages: Evidence from the Former Zaire." *Journal of Development Economics* 60 (2): 467–495.

Minten, B., L. Randrianarison, and J.F. Swinnen. 2009. "Global Retail Chains and Poor Farmers: Evidence from Madagascar." *World Development* 37 (11): 1728–1741.

Minten, B., and T. Reardon. 2008. "Food Prices, Quality, and Quality's Pricing in Supermarkets vs. Traditional Markets in Developing Countries." *Review of Agricultural Economics* 30 (3): 480–490.

Minten, B., T. Reardon, K.M. Singh, and R. Sutradhar. 2014. "The New and Changing Roles of Cold Storages in the Potato Supply Chain in Bihar." *Economic and Political Weekly* 49 (52): 98–108.

Minten, B., Y. Habte, S. Tamru, and A. Tesfaye. 2020. "The Transforming Dairy Sector in Ethiopia." *PLoS ONE* 15 (8): e0237456.

Minten, B., B. Mohammed, and S. Tamru. 2020. "Emerging Medium-Scale Tenant Farming, Gig Economies, and the COVID-19 Disruption: The Case of Commercial Vegetable Clusters in Ethiopia." *European Journal of Development Research* 32: 1402–1429.

Minten, B., E. Tamru, E. Engida, and K. Tadesse. 2016. "Feeding Africa's Cities: The Case of the Supply Chain of Teff to Addis Ababa." *Economic Development and Cultural Change* 64 (2): 265–297.

Minten, B., T. Reardon, and A. Vandeplas. 2009. "Linking Urban Consumers and Rural Farmers in India: A Comparison of Traditional and Modern Food Supply Chains." IFPRI Discussion Paper 883. IFPRI, Washington, DC.

Muhonda, P., E. Rice, A. Bennett, L.S.O. Liverpool-Tasie, B. Belton, and E. Abaidoo. 2024. "Investigating the Inclusiveness of the Usipa Value Chain in Malawi." *World Development Perspectives* 33: 100552.

Narayanan, S., B. Belton, R. Kabir, A.Z. Sakil, A.H. Khan, and R. Hernandez. 2025. *The Cluster Panacea? An Evaluation of Three Interventions in Shrimp Value Chains in Bangladesh.* Washington, DC: IFPRI.

Pingali, P. 2007. "Westernization of Asian Diets and the Transformation of Food Systems: Implications for Research and Policy." *Food Policy* 32 (3): 281–298.

Quisumbing, A., J. Heckert, S. Faas, et al. 2021. "Women's Empowerment and Gender Equality in Agricultural Value Chains: Evidence from Four Countries in Asia and Africa." *Food Security* 13: 1101–1124.

Rashid, S., A. Gulati, and R.W. Cummings, Jr. 2008. *From Parastatals to Private Trade: Lessons from Asian Agriculture.* Baltimore: Johns Hopkins University Press for IFPRI.

Reardon, T. 2015. "The Hidden Middle: The Quiet Revolution in the Midstream of Agrifood Value Chains in Developing Countries." *Oxford Review of Economic Policy* 31 (1): 45–63.

Reardon, T. 2024. "Tomato Wholesale Markets in Tanzania: Important, Growing, Transforming – but Hidden from Debate." PowerPoint presentation at ICAE, Delhi, India, August 7.

Reardon, T., C.B. Barrett, J.A. Berdegué, and J.F, Swinnen. 2009. "Agrifood Industry Transformation and Farmers in Developing Countries." World Development 37 (11): 1717–1727.

Reardon, T., K.Z. Chen, B. Minten, and L. Adriano. 2012. *The Quiet Revolution in Staple Food Value Chains in Asia: Enter the Dragon, the Elephant, and the Tiger.* Mandaluyong, the Philippines: Asian Development Bank and IFPRI.

Reardon, T., K.Z. Chen, B. Minten, L. Adriano, T.A. Dao, J. Wang, and S. Das Gupta. 2014. "The Quiet Revolution in Asia's Rice Value Chains." *Annals of the New York Academy of Sciences* 1331: 106–118.

Reardon, T., R. Echeverría, J. Berdegué, B. Minten, S. Liverpool-Tasie, D. Tschirley, and D. Zilberman. 2019. "Rapid Transformation of Food Systems in Developing Regions: Highlighting the Role of Agricultural Research and Innovations." *Agricultural Systems* 172: 47–59.

Reardon, T., S. Henson, and J. Berdegué. 2007. "'Proactive Fast-Tracking' Diffusion of Supermarkets in Developing Countries: Implications for Market Institutions and Trade." *Journal of Economic Geography* 7 (4): 1–33.

Reardon, T., L.S.O. Liverpool-Tasie, B. Belton, M. Dolislager, B. Minten, B. Popkin, and R. Vos. 2024. "African Domestic Supply Booms in Value Chains of Fruits, Vegetables, and Animal Products Fueled by Spontaneous Clusters of SMEs." *Applied Economic Perspectives and Policy* 46 (2): 390–413.

Reardon, T., L.S.O. Liverpool-Tasie, and B. Minten. 2021. "Quiet Revolution by SMEs in the Midstream of Value Chains in Developing Regions: Wholesale Markets, Wholesalers, Logistics, and Processing." *Food Security* 13: 1577–1594.

Reardon, T., and B. Minten. 2011. "The Quiet Revolution in India's Food Supply Chains." IFPRI Discussion Paper 1115. IFPRI, Washington, DC.

Reardon, T., A. Mishra, C.S.R. Nuthalapati, M. Bellemare, and D. Zilberman. 2020. "COVID-19's Disruption of India's Transformed Food Supply Chains." *Economic & Political Weekly* 55 (18): 18–22.

Reardon, T., K. Stamoulis, and P. Pingali. 2007. "Rural Nonfarm Employment in Developing Countries in an Era of Globalization." *Agricultural Economics* 37 (S1): 173–183.

Reardon, T., and J.F.M. Swinnen. 2004. "Agrifood Sector Liberalization and the Rise of Supermarkets in Former State-Controlled Economies: A Comparative Overview." *Development Policy Review* 22 (5): 515–523.

Reardon, T., and J. Swinnen. 2020. "COVID-19 and Resilience Innovations in Food Supply Chains." *IFPRI Blog*, July 6.

Reardon, T., T. Thiombiano, and C. Delgado. 1989. "L'importance des Céréales Non-traditionnelles dans la Consommation des Riches et des Pauvres à Ouagadougou." *Economie Rurale* 190: 9–14.

Reardon, T., and C.P. Timmer. 2007. "Transformation of Markets for Agricultural Output in Developing Countries since 1950: How Has Thinking Changed?" In *Handbook of Agricultural Economics, Vol. 3: Agricultural Development: Farmers, Farm Production and Farm Markets*, eds. R.E. Evenson and P. Pingali, 2808–2855. Amsterdam: Elsevier.

Reardon, T. and C.P. Timmer. 2012. "The Economics of the Food System Revolution." *Annual Review of Resource Economics* 4: 225–264.

Reardon, T., D. Tschirley, L.S.O. Liverpool-Tasie, et al. 2021. "The Processed Food Revolution in African Food Systems and the Double Burden of Malnutrition." *Global Food Security* 28: 100466.

Reardon, T., and R. Vos. 2021. "Food Supply Chains: Business Resilience, Innovation, and Adaptation." In *2021 Global Food Policy Report: Transforming Food Systems after COVID-19*, 64–73. Washington, DC: IFPRI.

Reardon, T., and D. Zilberman. 2018. "Climate Smart Food Supply Chains in Developing Countries in an Era of Rapid Dual Change in Agrifood Systems and the Climate." In *Climate Smart Agriculture: Building Resilience to Climate Change*, eds. L. Lipper, N. McCarthy, D. Zilberman, S. Asfaw, and G. Branca, vol. 52, 335–351. Rome: Food and Agriculture Organization of the United Nations; Cham, Switzerland: Springer.

Rubey, L. 1995. "The Impact of Policy Reform on Small-Scale Agribusiness: A Case Study of Maize Processing in Zimbabwe." *African Rural Urban Studies* 2 (2–3): 93–119.

Sauer, C.M., T. Reardon, D. Tschirley, et al.. 2021. "Consumption of Processed Food & Food Away from Home in Big Cities, Small Towns, and Rural Areas of Tanzania." *Agricultural Economics* 52 (5): 749–770.

Scott, G.J. 1985. *Markets, Myths, and Middlemen: A Study of Potato Marketing in Central Peru*. Lima: International Potato Center (CIP).

Sexton, R.J. 2013. "Market Power, Misconceptions, and Modern Agricultural Markets." *American Journal of Agricultural Economics* 95 (2): 209–219.

Swinnen, J.F., ed. 2007. *Global Supply Chains, Standards and the Poor: How the Globalization of Food Systems and Standards Affects Rural Development and Poverty*. Wallingford, UK, and Cambridge, MA: CABI.

Swinnen, J.F., and R. Kuijpers. 2019. "Value Chain Innovations for Technology Transfer in Developing and Emerging Economies: Conceptual Issues, Typology, and Policy Implications." *Food Policy* 83: 298–309.

Swinnen, J.F., and M. Maertens. 2007. "Globalization, Privatization, and Vertical Coordination in Food Value Chains in Developing and Transition Countries." *Agricultural Economics* 37 (S1): 89–102.

Swinnen, J.F., L. Ronchi, and R. Reardon. 2024. "Harness Agrifood Value Chains to Help Farmers Be Climate Smart." *Science* 386 (6725): 974–977.

Swinnen, J.F., and S. Rozelle. 2006. *From Marx and Mao to the Market: The Economics and Politics of Agricultural Transition*. Oxford, UK: Oxford University Press.

Swinnen, J.F., and A. Vandeplas. 2010. "Market Power and Rents in Global Supply Chains." *Agricultural Economics* 41: 109–120.

Swinnen, J., and A. Vandeplas. 2011. "Rich Consumers and Poor Producers: Quality and Rent Distribution in Global Value Chains." *Journal of Globalization and Development* 2 (2): 2–2.

Thurlow, J., B. Holtemeyer, S. Jiang, K. Pauw, and J. Randriamamonjy. 2025. "Measuring Agrifood Systems: New Indicators and Estimates." IFPRI Discussion Paper 2339. IFPRI, Washington, DC.

Timmer, C.P., W.P. Falcon, and S.R. Pearson, eds. 1983. *Food Policy Analysis.* Baltimore: Johns Hopkins University Press and World Bank.

Van Campenhout, B., B. Minten, and J.F. Swinnen. 2021. "Leading the Way: FDI and Dairy Value Chain Upgrading in Uganda." *Agricultural Economics* 52 (4): 607–631.

von Braun, J., D. Hotchkiss, and M.D.C. Immink. 1989. *Nontraditional Export Crops in Guatemala: Effects on Production, Income, and Nutrition.* Research Report 73. Washington, DC: IFPRI.

Worku, I., M. Dereje, B. Minten, and K. Hirvonen. 2017. "Diet Transformation in Africa: The Case of Ethiopia." *Agricultural Economics* 48 (S1): 73–86.

SECTION III

Supporting Farmers

Chapter 8

AGRICULTURAL EXTENSION SERVICES
From Transfer of Technology to Facilitation for Innovation

Kristin Davis, Rikin Gandhi, Jawoo Koo, Berber Kramer, Alesha Miller, Jona Repishti, David J. Spielman, and Rasheed Sulaiman V.

Kristin Davis is a senior research fellow, Natural Resources and Resilience Unit (NRR), IFPRI; **Rikin Gandhi** is chief executive officer, Digital Green; **Jawoo Koo** is a senior research fellow, NRR, IFPRI; **Berber Kramer** is a senior research fellow, Markets, Trade, and Institutions Unit, IFPRI-Kenya; **Alesha Miller** is chief strategy officer, Digital Green; **Jona Repishti** is assistant director, Global Gender, Digital Green; **David J. Spielman** is director, Innovation Policy and Scaling Unit, IFPRI; and **Rasheed Sulaiman V.** is director, Centre for Research on Innovation and Science Policy (CRISP).

Key messages

- Agricultural extension is key to providing information, services, technologies, and skills. It impacts nutrition, health, and food security; poverty reduction, livelihoods, and jobs; climate adaptation and mitigation; gender equality, youth, and social inclusion; and environmental health and biodiversity.
- Delivery methods for agricultural extension services have changed significantly over the past 50 years. From a mainly public institution focused on in-person services, extension today employs many digital methods, including AI, that supplement traditional face-to-face approaches.
- Other key changes include the growing roles of the private sector and value chain approaches, as well as a broadening of extension's mandate from simply transferring productivity-enhancing technologies to facilitating agricultural innovation and brokering relationships and services as an intermediation agency.
- As extension becomes more pluralistic in methods and providers, a major challenge is measuring and monitoring performance and evaluating impacts to remain effective and efficient and to provide accountability to clientele and funders.

- IFPRI with other research and implementation partners has contributed to conceptualization, content and knowledge management and sharing, and impact evaluation of extension programs, for which experimental methods such as randomized controlled trials are increasingly common.

Looking ahead, further research is needed to adapt and improve extension within a more pluralistic approach using digital tools:

- **Adapting governance to pluralistic advisory systems** requires a better understanding of the appropriate roles of diverse public and private sector providers, how to coordinate and finance their contributions, and how to reach different user groups equitably, while advancing more sustainable agricultural practices.
- **Strengthening capacity** for extension services is needed at both the individual and organizational levels for problem solving, digital literacy, and adaptive management. Research should identify ways to effectively develop capacities for pluralism, to develop gender- and nutrition-sensitive strategies, and to use digital tools that are relevant and equitable.
- **Improving metrics, tracking, and data quality** is essential to evaluate impact in extension services. Frameworks and indicators as well as data collection to assess performance and outcomes need to be implemented at scale, with support from national or state governments. Data quality is critical for timely and localized advice.
- **Generating evidence to inform extension methods**, such as bundling of services with complementary products, is key to cost-effectiveness and relevance. Research must look beyond addressing informational constraints to technology adoption to more holistic solutions that address a wide range of barriers to uptake of social, institutional, and technical innovations.

Agrifood systems in many low- and middle-income countries (LMICs) rely on smallholder farmers and other small-scale value chain actors to produce, process, and market the agricultural goods and services that nourish and sustain entire nations. Agricultural extension and rural advisory services play a key role in these agrifood systems by supporting farmers' efforts to enhance productivity, strengthen resilience to shocks, and conserve the natural resource base on which these systems depend. These services encompass "the entire set of organizations that support and facilitate people engaged in agricultural production to solve problems and to obtain information, skills, and technologies to improve their livelihoods and well-being" (Birner et al. 2009, 342). For brevity, we refer to this as "extension" throughout the chapter.

Extension has long been considered a way to provide farmers with information on recommended inputs, technologies, and on-farm practices: which crop varieties to cultivate, how to apply fertilizer, and how to protect crops and livestock from pest and disease threats. But the reality of extension is far more complex and nuanced. Among the many functions of extension services today, they are expected to provide tailored educational services, linkages to value chain actors and opportunities, and technology testing and evaluation. Often, these services are also expected to support social programs, multisectoral initiatives, private sector investment schemes, and governmental administrative activities. Whether extension providers are called officers, agents, agrovets, village-based advisors, or agronauts, they are often the frontline agents of change in smallholder systems.

Innovations—both technical and institutional—have driven agrifood system change, facilitating an increase in productivity and greater capacity to cope with challenges and shocks. In part due to the success of the Green Revolution, in which research-center-developed improved crop varieties and practices such as fertilizer and irrigation increased yields, innovation continues to be viewed as a supply-driven, technical process in the context of LMIC agriculture (see Chapter 17). Governments, donors, and implementing partners have tended to develop and promote technological innovations to farmers through linear, top-down processes in a paradigm focused on research and technology without broader attention to sociocultural factors or systems thinking (Woltering 2024).

This chapter examines the global evolution of agricultural extension and rural advisory services over the past 50 years and the shift from a "transfer of technology" approach, characterized by the dissemination of information to farmers, to a more sophisticated "facilitation for innovation" paradigm. The chapter applies IFPRI's "best fit" conceptual framework, which was developed more than a decade ago to analyze extension services and systems and has since been widely used for extension diagnostics and reform efforts. The analysis focuses on exploring changes within the overall environment (*frame conditions*) in which extension operates, and how this affects the extension *characteristics* (governance, organizational capacity and management, and advisory methods and tools). In doing so, the chapter highlights key changes in paradigms and approaches, current and emerging trends, and directions for future research.

Looking back

During the past 50 years, the overall environment (or frame conditions) for extension has evolved significantly in response to societal shifts, which has affected extension characteristics. Throughout this evolution, IFPRI and partners have contributed to conceptualization, content and knowledge management and sharing, and evidence through impact evaluations.

As a formal institution, extension began in the 1850s as a supply-driven approach to transfer technologies from knowledge centers to farmers (Swanson et al. 1997) and was later crucial in disseminating innovations during the Green Revolution. Fifty years ago, extension was mainly a top-down, technology-focused, and publicly provided service that primarily transferred technology from research stations to farmers through extension. An example of this type of extension was the World Bank's training and visit approach, which was designed to help professionalize LMIC public extension services through a rigid system of regular staff training followed by visits to farmers (Anderson 2007).

In the mid-1990s, this transfer-of-technology paradigm faced growing criticism of the standardized models being promoted in many countries. Interest shifted to participatory methods (Scoones and Thompson 1994), farmer-to-farmer learning, and adult education and experiential learning approaches, such as farmer field schools (Waddington and White 2014). Rather than focusing on technologies to solve researcher-perceived problems, these "bottom-up" approaches emphasized developing solutions jointly with farmers while empowering them to problem-solve. They also focus on the institutional factors, social relationships, and broader farming and market systems that affect innovation within agriculture (see Chapter 17).

Around this time, the Neuchâtel Initiative began meeting annually to exchange experiences and views and to develop a common understanding of rural extension issues. The Initiative was initially an informal group of bilateral and multilateral agricultural development agencies and researchers, including IFPRI. In 2010, the Initiative was formalized as the Global Forum for Rural Advisory Services (GFRAS), which was supported by IFPRI and several donors. Fifteen years on, the Forum has engaged in policy reform in over 25 countries, raised awareness of the need for more professionalism in extension, and trained more than 3,000 people using the global "new extensionist"[1] learning modules (GFRAS 2020).

---

1 The New Extensionist is a global view of extension and advisory services (EAS) that reinvents and articulates its role in the rapidly changing rural context. It argues for an expanded role for EAS within agricultural innovation systems and development of new capacities at different levels to play this role. See Davis and Sulaiman (2014).

BOX 8.1 Agricultural innovation and the best-fit approach

Extension is a key component of a country's agricultural innovation system (Rajalahti et al. 2008). An innovation system consists of a set of interrelated agents, their interactions, and the institutions ("rules of the game") that condition their behavior in relation to the common objective of generating, diffusing, and using knowledge and/or technology (Spielman 2005). The innovation system encompasses the best-fit frame conditions and affects characteristics of advisory services. The first characteristic is *governance* structures, which are essentially the institutional setup for providing and financing services. The second is *organizational capacity and management*, which include capacity to provide advisory services and the ways that services are managed within governance structures. The third is *advisory methods* used by extension agents to interact with farmers. Together, these characteristics shape extension's performance, which not only leads to outcomes such as technical changes among farmers but also supports development of the organizational, managerial, and marketing capacities of farmers that lead to impact.

In the late 2000s, IFPRI scholars collaborated to develop a framework for designing and analyzing pluralistic extension services. This "best-fit" approach aims to help identify ways to provide and finance extension services that work best for different contexts (Birner et al. 2009) (Box 8.1). It distinguishes between extension characteristics of governance structures, organizational capacity and management, and advisory methods. Since its development, the framework has been used as the foundation for many studies and projects (Davis and Spielman 2017). IFPRI also contributed to extension thinking for the World Bank's *Agricultural Innovation Systems: An Investment Sourcebook* (World Bank 2012). The sourcebook is a comprehensive document on identifying, designing, and implementing investments, approaches, and complementary interventions to strengthen innovation systems and promote agricultural innovation and equitable growth. It is an important resource for the research community and nongovernmental organizations as well as the private sector, farmer organizations, and individuals with an interest in agricultural innovation.

Many extension scholars now put the focus of extension and advisory services on (1) brokering functions (Klerkx and Leeuwis 2009) that link farmers to various services, (2) communication for innovation (Leeuwis 2004),[2] and

2 This evolution can be seen in the retitling of the book from *Agricultural Extension* to the current title, *Communication for Rural Innovation: Rethinking Agricultural Extension.*

(3) scaling of support (Schut et al. 2022) for rural and agricultural actors in the innovation system. This more encompassing view of extension is known as "facilitation for innovation," which was initially documented in 1995, but has since grown and expanded (Engel 1995). Facilitation for innovation is about extension playing a broader role in agricultural innovation systems, engaging with a wider range of stakeholders beyond research and farmers, and using knowledge and resources essential for problem-solving. The facilitation process enables varied actors within the agricultural innovation system to share knowledge and engage in collective and coordinated action to enable innovation and impact at scale, with extension serving as an agency for intermediation among these diverse and pluralistic actors.

Evolution and current trends in extension

Frame conditions: The enabling environment

Changes in extension paradigms affect the frame conditions—that is, the environment that supports or deters implementation of policies and programs. A number of key shifts have occurred in recent years. As a result of the rapid growth in digital tools, the number of global digital advisory services increased nearly 10 times between 2009 and 2019 (Phatty-Jobe et al. 2020), leading many donors and governments to consider how they can use these tools to cost-effectively scale extension services. A shift toward understanding and supporting "farming as a business" has led to interventions that provide farmers with marketing skills or enable them to connect to global value chains (Arulmanikandan et al. 2023). Catholic Relief Services, for instance, developed the Skills for Marketing and Rural Transformation set of manuals that outlines steps and provides guidance for farmers to market their produce (CRS and MEAS 2016). Nutrition, while part of (mainly women-focused) extension programs in the 1970s, was recognized as an important topic for extension to tackle again in the 2000s. IFPRI's work on nutrition and extension includes evidence on how to provide nutrition-sensitive school meals (Gelli and Bliznashka 2024) and how adoption of small-scale irrigation and water harvesting techniques is linked to nutritional outcomes (Balana et al. 2024). There is also a growing focus on including women and youth as extension clientele (Babu et al. 2021; Franzel et al. 2021).

A major paradigm shift has occurred in response to climate change, biodiversity loss, and natural resource degradation, with attention now on extension that can support farmers in adapting to and mitigating climate change (see Chapter 5) (Sulaiman 2020). Extension and advisory platforms and international agencies, as well as farmers themselves, are increasingly focusing on environmental health and biodiversity, including through sustainable intensification, agroecological approaches (FAO 2022; Gliessman 2022; Sulaiman 2021), regenerative agriculture, and nature-based solutions.

For instance, the CGIAR Research Initiative Ukama Ustawi used various methods to disseminate climate-smart agriculture solutions—such as conservation agriculture, drought-tolerant maize varieties, crop rotation, and climate information—to help farmers in maize-based systems in East and Southern Africa sustainably intensify their agricultural production (CGIAR n.d.-a). The CGIAR Research Initiative on Nature-Positive Solutions worked with farmer groups in three continents to co-develop social, institutional, and technological community and agricultural innovations that include women and youth, provide livelihoods, and preserve the natural environment (CGIAR n.d.-b). Farmers around the globe are recognizing the shortfalls of industrial agriculture and turning to agroecological practices that work with nature to feed the planet (Thurow 2024).

As priorities shift, funding and policies help to provide an enabling environment to implement new paradigms. For instance, GFRAS received funding to support climate adaptation and mitigation (GFRAS n.d.-b) and agroecology (GFRAS 2023) in extension. The African Union has called for 30 percent of Africa's agricultural land to be managed sustainably in the Kampala Declaration on the Detailed Programme for the Development of Agriculture (Ulimwengu 2025). At the national level, for instance in Kenya, agroecological principles are being integrated into the constitution and various policies and laws (Awiti and Ndiwa 2024).

Governance and capacity

Pluralism is a major evolving factor in the governance of extension. In the 1970s, many LMICs solely employed public extension models. However, in response to liberalization policies and the perceived failure of government models in the 1980s, nongovernmental organizations, private companies, and farmer organizations increasingly began providing these services. Today we see many different public, private, and civil society actors providing and financing advisory services, a trend that also includes public–private partnerships (Bruce and Costa 2019) and multistakeholder platforms (Schut et al. 2019). Pluralism can lead to greater reach and enhanced focus on women, the environment, and financial sustainability, but it can also lead to coordination difficulties and duplicated efforts.

The growing role of value chain approaches and the private sector in extension services is arguably the biggest shift in recent decades (Zhou and Babu 2015). This demand-driven development perspective focuses on product characteristics, and the production process is determined by perceived consumer demand and communicated to producers through value chains. Through this value-chain-driven process, capital, technology, and advisories flow upstream to farmers, with private extension services playing an important role. The relative importance of this model varies by commodity, institutions, infrastructure, and level of development. Hybrid models are common, in which the public, private, and civil society sectors fund and provide extension. For instance, the Agricultural Technology Management Agency in India is a mechanism at the district level created to provide decentralized extension services through a semi-autonomous registered society under government control (Swanson and Rajalahti 2010). In Uganda, the government has begun a scale-up effort designed to enhance the reach of its public extension service by training some 32,000 private village agents (Scheer and Okelai 2019).

Additionally, farmers and farmer organizations, who are also part of the private sector, provide services today. For example, the National Smallholder Farmers Association of Malawi is a smallholder-owned membership organization that provides information to members through radio, newsletters, and trainings on topics such as farming methods, literacy, and business skills (Makwenda 2010). Also in Malawi, we see examples of local farmer extensionists—"lead farmers"—who provide a link between the public extension system and farming communities, helping to translate public messages and demonstrate practices for their fellow farmers (Ragasa 2020).

Some private sector models, particularly those focused on specific commodities, risk some pitfalls. Private providers may focus only on crops of interest to the company, rather than holistically helping farmers who often have diversified operations, or on high-potential geographies, farmers, and crops, thus potentially excluding marginalized groups and locations (Muyanga and Jayne 2008; Feder et al. 2011).

Private sector extension is too diverse to draw general conclusions about its impact. A book synthesizing global lessons of private sector extension concludes that those programs implemented as part of value chains are effective in their areas but have limited coverage (Babu and Zhou 2015). It finds that the private sector cannot be a substitute for public sector extension and recommends developing public–private partnerships to increase the efficiency and effectiveness of both public and private extension. Further, it is important to distinguish between extension from upstream actors such as input or service providers and downstream actors. Upstream providers will want to increase their market share without necessarily having farmers' interests at heart, while downstream actors have an interest in upgrading the quantity or quality of agricultural production, which is an objective more aligned with that of public extension services and farmers.

Decentralization of public services is another governance trend that is usually intended to provide more accountability to clientele (see Chapter 15). For instance, Kenya changed its constitution in 2010 to decentralize government services as part of a strategy for accelerating growth and increasing equality (World Bank 2015). However, decentralization has implications for organizational and management capacities of extension organizations. Many times, local-level extension offices and departments lack the individual and organizational capacity necessary for planning, monitoring, evaluating, and managing extension programs (World Bank 2019). Thus there is need for strengthening these capacities at the local level.

Other organizational and management capacity-building trends include focuses on adaptive management, leadership, alliance building and partnering, and mechanisms to incentivize performance (Davis and Sulaiman 2014). These organizational capacities are particularly important for addressing the crises and change seen today. In exploring extension's response to COVID-19, researchers noted that capacities for flexibility, collaboration, and partnerships were important for coping with change (Davis et al. 2021). In addition, to address risks and shocks and better link farmers to markets, further methods are needed to mobilize farmers into groups, strengthen their organizational and managerial capacities, and support community-managed services and community-based extension workers.

Using participatory approaches is a key capacity for extension staff and organizations. Many extension programs have been designed and implemented with little farmer participation (Lampach et al. 2021). Traditional extension approaches often deliver generic, regionwide advice, missing opportunities to provide farm-specific insights. These usually prioritize yield maximization, often based on controlled experimental data from test plots, rather than addressing the economic realities faced by small-scale farmers. Farmers do not necessarily aim to maximize yields; they need to consider both the costs and productivity of recommended practices and technologies. Risk-neutral farmers will want to maximize profits rather than yields, while risk-averse farmers, who face a trade-off between increasing profitability and reducing risk, may find that profitable investments expose them to more risk than they are willing to accept. The tendency of extension systems to recommend an innovation if it improves yields, without considering costs and risks, has become even more problematic given recent challenges related to climate change. Thus, the emphasis on yields has limited the effectiveness of extension programs because this emphasis does not correspond with farmers' actual preferences and undermines their trust, adoption rates, and overall results. It is critical to have skills to better understand farmers' needs and involve them in program design and evaluation to make extension relevant to farmers.

Digital tools

Extension methods are typically classified into individual (farm visits, phone calls), group (demonstrations, tours, trainings), and mass (bulletins, leaflets, radio, television) communications (van den Ban and Hawkins 1988). Since 1975, extension outreach methods have evolved substantially in the digital space, from radio to the rise of mobile phones, big data, and artificial intelligence (AI). Despite the digital evolution, radio remains an important tool for extension, providing information that is inclusive, accessible, and affordable. A relatively high percentage of rural dwellers, including women, own radios, allowing for extensive reach and social inclusion, and evidence shows farmers prefer listening to advisory messages rather than reading them (Gorman et al. 2018). With availability in local languages, radio programs have potential to reach vulnerable groups and to include feedback and monitoring through call-in (or text-in) numbers or discussion in local study groups. Radio's impacts can be maximized when it is integrated into wider agricultural extension programs and is locally relevant. EFarm Radio International, The Mediae Company, and BBC Media Action have worked to scale up radio programs by collaborating with existing national extension services and training radio stations to produce quality programs (Rao 2015).

Today, digital applications (apps), social media, videos, television, smartphones, and micro-projectors can all provide extension information. In Kenya, for instance, digital information and advisory services are rapidly penetrating agricultural systems, with 46 percent of farmers recently reporting their use (60 Decibels 2024). Digital extension tools abound in Kenya thanks to the importance of agriculture, high connectivity levels, use of mobile phones and mobile money, and data transparency (CTA 2019). In this rapidly growing space, some digital tools have been rigorously evaluated, while many have not.

Television has become a tool to deliver "edutainment"—educational content presented in an entertaining format. For instance, The Mediae Company

BOX 8.2 Democratizing and localizing extension through artificial intelligence

Artificial intelligence (AI) may reshape extension services by moving services to greater levels of personalization and site-specificity. Digital Green's Farmer.Chat AI assistant exemplifies the ways in which agentic AI can localize extension by offering increasingly precise and personalized support to smallholder farmers (Digital Green n.d.). This multilingual chatbot is accessible through text, voice, and images and can deliver advice in multiple underrepresented local languages such as Hausa, Kiswahili, and Odia. It can combine complex scientific research with dynamic weather and market data to help farmers make more informed decisions on crop management, pest control, and sales. For instance, Farmer.Chat helped Annred Gakii, a smallholder farmer in Kenya, treat a sick cow and provided tailored advice to her on crop diversification, leading to increased yields and reduced costs (Digital Green 2024). Drawing on more than 350,000 farmer queries, Farmer.Chat also highlights patterns in the challenges faced by farmers, enabling extension services to adapt to evolving needs and creating new pathways to co-design solutions with farming communities. Tools like Farmer.Chat also provide opportunities to explore and measure these effects at scale, thereby demonstrating not only the benefits from agentic AI services but also the heterogeneity in benefits across different farmers and contexts and AI's cost-effectiveness as a complement or substitute to traditional in-person extension services. Digital Green and IFPRI collaborate in several countries to examine these research topics.

has reached millions of smallholder farmers through its farm makeover reality programs "Shamba Shape Up" (in Kenya and Uganda) and "Munda Makeover" (in Zambia), as well as through a farming contest for young farmers, "Don't Lose the Plot?" (VOA 2018). Globally, the reach of digital tools is growing quickly, especially among youth (71 percent of young people use the internet) (UNICEF 2017). However, as with any extension method, constraints to access and participation of marginalized groups and areas must be considered.

The digital revolution has helped to democratize extension by allowing information and innovation to flow not just in one direction from researchers to farmers via extension, but in multiple directions to multiple actors and at a low cost to large audiences. Big data and AI, such as large language models, have great potential for delivering scientific information to farmers and extension staff and for responding to and aggregating information and requests from farmer. Coupling these tools with innovative delivery channels can help providers tailor advice to local contexts and allow them to increase their reach (Box 8.2).

IFPRI has contributed to extension methods through content and knowledge management and sharing as well as big data and open access policies that support extension. IFPRI co-founded the CGIAR Platform for Big Data in Agriculture and spearheaded CGIAR's work on open access and open data, making data comply with FAIR (findable, accessible, interoperable, and reusable) standards. IFPRI research underpins several initiatives on building open-source models and data frameworks to enhance AI-driven agricultural advisory services and support global digital public goods. The Institute is also leveraging its AI expertise to design a range of pilot studies exploring how AI can enhance agricultural advisory services. For example, IFPRI developed a pilot study in Uganda, led by the CGIAR Research Initiative on Digital Innovation, to help Farm Radio International process farmers' voice messages in a local Bantu language by using AI tools to automate transcription, reducing the need for manual effort (Mganga et al. 2023). However, while digital tools and advancements have facilitated much greater access to information, they have also created a digital divide affecting marginalized groups, who are also affected by elements of intersectionality (Gillwald and Partridge 2022) and geography (Saha et al. 2024). Despite promising adoption trends, digital extension providers struggle to reach and benefit populations, particularly women, with low literacy and limited digital skills, highlighting persistent gaps in accessibility for these groups (60 Decibels 2024).

A growing body of evidence suggests that not all digitally enabled extension approaches are equally effective in improving adoption, productivity, income, or welfare outcomes (Spielman et al. 2021). While digital tools can increase the effectiveness and decrease the costs of information delivery to small-scale farmers, with more personalization to fit marginalized farmers' unique conditions, the design attributes of such approaches matter, and it is important to consider these to make digital tools inclusive (Van Campenhout et al. 2020). While digital tools can provide tailored advice (Box 8.2), scholars still recommend the continued input of human experts (Tzachor et al. 2023). Combining virtual approaches with face-to-face interventions and multidirectional communication can make the provision of information and advisory services much more effective. For instance, Digital Green, working with the agriculture ministries of India, Ethiopia, and Kenya, created nearly 8,000 farmer-to-farmer videos in more than 50 languages, enabling extension agents to provide timely, locally relevant information to more farmers. Randomized

BOX 8.3 Bundling insurance and advisory services

Farmers face risks of weather extremes but often cannot afford crop insurance that could mitigate these risks (see Chapter 10). Picture-based insurance (PBI) delivers affordable, comprehensive, and easy-to-understand crop insurance. The PBI tool uses farmers' smartphone pictures to cheaply verify farmers' claims, making crop insurance more attractive and accessible to small-scale farmers (Ceballos et al. 2019). By relying on pictures showing crop characteristics, PBI minimizes the costs of loss verification and detects damage at the plot level. This instrument lends itself to natural synergies with site-specific, picture-based agro-advisories that recommend better (climate-smart) technologies and practices (IFPRI. n.d.). Yet, research on the impacts of picture-based advisories found that they did not have larger effects on farmer knowledge or practices than generic advisories; effects of advisories were driven by engagement with the advisories, regardless of whether these were generic or personalized (Ceballos et al. 2024).

controlled trials have shown that the community video approach leads to 18–22 percent increases in yields and up to a 24 percent increase in farmers' profits relative to traditional extension.

Additionally, many digital tools are not scaled at a significant level. Investments in digital tools for extension are generally low, with a focus on specific enterprises, and are primarily fueled by donors, with private investment lagging (CTA 2019). Some services have been discontinued when the implementing projects ended, perhaps due to the lack of digital infrastructure. For instance, a project may provide smartphones to village extension service providers to deliver advisories to farmers (who do not have access to smartphones or extension platforms themselves), but these smartphones may get damaged or stolen over time. Thus, although the project may set up a system for creating and disseminating content, the system may not be sustainable, and the content may stop reaching the end user if there is no plan to invest in and maintain the necessary digital infrastructure once the project ends. For AI, the challenges for scaling also include accuracy, data limitations, performance, and costs.

Methods discussed so far have focused mainly on the dissemination of information. However, information is just one extension service that farmers need to take up innovations. Bundling information with credit or insurance could strengthen both information dissemination and uptake of financial services (Box 8.3) (Timu et al. 2024).

BOX 8.4 Methods for evaluating extension

Evaluations of extension programs are systematic and objective assessments of projects, programs, or policies, and their results. Evaluation methods include performance indicators, formal surveys, participatory methods, public expenditure tracking surveys, and impact evaluations (which establish a causal relationship between an intervention and observed outcomes). Evaluations examine the relationship between activities and outputs of an extension program and intended outcomes and impacts. The relevance, efficiency, effectiveness, impact, and sustainability of extension can be measured using various indicators related to knowledge, attitudes, perceptions, practices, and downstream impacts such as farm productivity or profitability (Christoplos et al. 2012).

Effectiveness of extension and financing strategies

Globally, evidence on the effectiveness of agricultural extension programs remains uneven, with outcomes varying significantly by context, crop type, and target group. Generally, public research and extension programs have shown high rates of return, although conclusions may depend on methods used (Alston et al. 2000). Extension research and evaluation (see Box 8.4) are complicated by the diversity of delivery methods, which range from traditional in-person training to digital-enabled platforms, combined with the inherent complexity of agriculture and its socio-ecological setting. Methodological challenges such as selection bias, limited baseline data, and resource constraints further impede the ability to assess impacts rigorously. Not surprisingly, the existing evidence on the impacts of extension is mixed. A meta-analysis found that while some interventions achieved substantial yield increases, many showed limited or no effects (Ogundari 2022). Similarly, a review of 41 analyzed programs with rigorous evaluation methods found that only 12 delivered significant improvements in farmer yields or profits (J-PAL 2023). These findings highlight the need for better-designed, evidence-based extension strategies that account for contextual diversity and address systemic barriers to effectiveness.

IFPRI is known for its work on assessments of the design and impact of extension and advisory services in LMICs. This includes conceptual work (the best-fit framework), global measurements of extension systems (Davis et al. 2020; Davis, Makhija, and Spielman 2021), and in-depth reviews of the evolution of the practice and performance of extension systems across countries and programs (Nakasone and Torero 2016; Spielman et al. 2021). Other

IFPRI research covers topics such as digital innovations in extension services for a picture-based advisory system piloted in Haryana, India, and seven counties in Kenya (Ceballos et al. 2024); the "edutainment" reality TV show in Zambia; bottom-up approaches and decentralized extension structures to improve access to and quality of extension services in Malawi (Ragasa et al. 2024); digital extension strategies that consider men's and women's needs in Uganda (Lecoutere et al. 2023; Van Campenhout et al. 2023); and video-mediated extension approaches in Africa and South Asia (Abate et al. 2023; Ringler et al. 2023; Van Campenhout et al. 2020).

Evaluations of extension impact have evolved from straightforward adoption studies focused on uptake of technologies in relation to farmers' attributes to approaches that account for the frame conditions—that is, the enabling environment and agricultural innovation system. IFPRI and partners are increasingly using experimental methods, such as randomized controlled trials, that can identify the effects of extension alone by controlling for other factors. For instance, in Ethiopia, researchers assessed the impacts of video-mediated agricultural extension provision on farmers' adoption of improved agricultural technologies and practices. Results showed that video-mediated extension significantly increased uptake of recommended technologies and practices by improving extension access and farmer knowledge (Abate et al. 2023). Similarly, for IFPRI's picture-based advisory study in India and Kenya (Box 8.3), a comparison of users and matched non-users would have led to the conclusion that the advisories had an impact, but comparing outcomes with a randomized control group showed that the personalized advisories did not affect knowledge or practices more than generic advisories. Even in the case of a national television show, to which access cannot be restricted, it is possible to conduct a rigorous evaluation of the intervention using encouragement designs, which introduce exogenous variation in the probability of accessing an innovation (such as the television show) by randomizing an encouragement to access it (such as text reminders to watch the show), as IFPRI did for an edutainment assessment in Zambia (Aggarwal et al. 2024).

Randomized controlled trials are often seen as the gold standard because of their rigor and capacity to attribute observed differences between treatment and control groups to the intervention using straightforward data analysis methods. They can be complicated and expensive to deploy, but the major cost in these trials is often the implementation and monitoring of the intervention that is being evaluated. If implementers and researchers coordinate from the start and commit to a randomized rollout of the intervention, and there is no duplication of implementation or monitoring efforts, costs can be reduced substantially. Another advantage of randomized trials is that, while baseline data are helpful to have, this data is not absolutely required, given that there is no need to control for a wide range of potential confounding factors, unlike for many quasi-experimental methods. Baseline data are mainly used to test for balance between treatment and control groups, which can also be done using secondary or administrative data. Also, sample sizes in these trials can be smaller than sample sizes in well-designed quasi-experimental evaluations because control observations will be more comparable to the treated sample by design. As such, randomized trials may offer greater value for money than quasi-experimental methods.

Evidence on who benefits and to what extent from different types of extension is also important for determining how to finance it. Traditionally, extension—like many public goods—has been publicly financed. Farmers are willing to pay for digital information, but only small amounts, with lower willingness to pay among women than men (Hidrobo et al. 2022). As a result, financing extension through fee-based advisories or extension services could introduce inequities and high administrative costs relative to fees collected from farmers. More sustainable models for financing extension would involve vertical or horizontal coordination, whereby farmer groups, private sector, or other value chain actors with an interest in promoting their products and services or increasing the quantity or quality of agricultural produce fund advisory services for smallholder farmers. Public funding for extension will remain crucial, though, to avoid extension becoming primarily a marketing tool.

Future of extension: New research directions

Looking to 2050 and beyond, we can expect continued challenges to food and nutrition security and livelihoods amid rising demand for food and fuel (Gotor and Masso 2024). These ongoing issues will require the continued evolution of extension's governance, capacities, and methods. Further research is needed to understand how national extension systems can support smallholder farmers, other rural actors, and our food systems through 2050 and beyond. Over the past 50 years, extension has largely been supply-driven, focused on priorities of donors, researchers, and governments. Several national governments have made substantial effort to strengthening extension, including Brazil, China, Rwanda, and Uganda,[3] and IFPRI

3 See Bergamasco et al. (2016) on Brazil; Yonggong (1998) on China; Davis et al. (2010) on Ethiopia; MacNairn and Davis (2018) on Rwanda; and Rwamigisa et al. (2019) on Uganda.

has helped to evaluate several of these programs and to transfer lessons from one system to another (Ambler et al. 2016). However, we now need to better trace the pathway from policy and funding decisions to deployment and assessment of national extension services. Going forward, we need research on LMIC extension governance, organizational capacities and management, and methods, particularly digitally enabled methods, to help achieve sustainable impact at scale.

Adapting governance to pluralistic systems. Better understanding is needed of how governments can best facilitate innovation and coordinate and regulate pluralistic extension providers, including supporting extension in private value chains. Evidence is needed to determine how to best combine novel and traditional approaches and to coordinate multiple providers of information and advisory services. Research should focus on identifying which mechanisms work for effective coordination of pluralistic services, and how public, private, and public–private partnership extension models can become more financially sustainable, without losing focus on socially inclusive models that help farmers transition toward environmentally sustainable agricultural practices. Inclusion and environmental sustainability goals may not be in the primary interest of profit-maximizing private sector advisory providers, but they are important objectives for the public sector to incentivize. Relatedly, further research is needed on new models for extension, including on private sector and value-chain-based extension, as well as on hybrid models and models seeking to extend reach by supplementing public systems with lead farmers. These models should be evaluated for impact and approaches to scaling them up.

Building capacity for extension. Research on the best ways to enhance capacities in extension services at both the individual and organizational levels for problem solving, digital literacy, and adaptive management is needed. In addition, it is important to identify the capacities that public and private extension staff need to effectively use digital tools. Further research is also needed on organizational capacities to provide gender- and nutrition-sensitive adaptation and mitigation strategies and strategies to conserve natural resources. With the diverse demands now placed on extension, finding the right balance of skills and expertise within an extension organization and within a pluralistic system should be examined.

Improving metrics, tracking, and data. In terms of management, there is also a clear need for improved metrics to track and measure impact in extension delivery (USAID and Agrilinks 2019). Efforts have been made toward this goal, including the development of frameworks (Sulaiman et al. 2022) and indicators to assess performance and outcomes (Davis 2020). However, these have not been implemented at scale. For agricultural research, the Agricultural Science and Technology Indicators (ASTI) program provides open-source data on financial, human, and institutional resources in agricultural research systems (both public and private sector) across LMICs (ASTI n.d.). Extension proponents have discussed establishing an ASTI-style collection of metrics for extension systems. However, the complexity and pluralistic nature of extension, especially in decentralized systems, make the task of tracking resources daunting. It is critical to have national or state government ownership in such efforts: deeper engagement at the country or state level will be needed to develop a better understanding of countries' experiences with extension implementation and metrics generation and use, as well as potential opportunities for specific partners to collaborate with the public sector to facilitate cost and time efficiencies by incorporating digital tools. National pluralistic extension platforms that are part of GFRAS could be a possible option for national-level monitoring (GFRAS n.d.-a). Relatedly, data quality is an important consideration as extension relies increasingly on big data and digital advisories. Extension needs capacity to source, use, and evaluate data used in advisories, and feedback loops are critical to ensure that data underlying localized advisories on weather, markets, and practices are accurate.

Informing extension methods. To build the evidence base on methods, we need to explore how bundling of extension services with complementary products can help lower costs of distribution, reduce and manage risk, and increase technology adoption and agricultural incomes, while being responsive to different user groups, including women, youth, and those with limited literacy. Given the many barriers to technology adoption, addressing solely information constraints will not ensure uptake of innovation. We need holistic solutions that address risk and liquidity constraints, high transportation costs, and limited access to output markets, among other barriers. With the changing nature of extension, impact evaluations of bundled products and services that include advisories are needed to quantify the added value of the extension services and identify complementarities and synergies, if any, with the bundle components. If there are no clear synergies, then extension should not be included as a bundled product but instead offered separately to facilitate choice.

The appropriate roles of government and other actors are of interest here: some methods are controlled by public authorities (such as radio) and some by the private sector. This can affect efficiency, access, and quality of services. For digital extension methods, evidence is needed regarding what digital infrastructure can facilitate these services and how barriers to use can be reduced. Further questions remain about how AI can be harnessed to provide relevant tailored solutions that use human-centered design. Finally, research is needed on how extension outreach methods can be inclusive, reaching marginalized populations and providing high-quality content as extention's reach expands through the use of digital tools. Extension has adapted its governance, capacities, and methods in response to the changing environment or frame conditions. While many approaches have come and gone, extension will remain an important institution for supporting smallholders, including marginalized populations, in managing risk and improving their livelihoods and nutrition while protecting the environment. Over the next decades, we urgently need more evidence on how extension can more efficiently and effectively facilitate innovation to provide sustainable impact at scale.

References

Abate, G.T., T. Bernard, S. Makhija, and D.J. Spielman. 2023. "Accelerating Technical Change through ICT: Evidence from a Video-Mediated Extension Experiment in Ethiopia." *World Development* 161: 106089.

Aggarwal, S., F. Ceballos, B. Kramer, et al. 2024. "It's Showtime! Evaluating the Effect of TV Programming on Agricultural Productivity in Rural Zambia." AEA RCT Registry. September 12.

Alston, J.M., C. Chan-Kang, M.C. Marra, P.G. Pardey, and T. Wyatt. 2000. *A Meta-Analysis of Rates of Return to Agricultural R&D: Ex Pede Herculem?* IFPRI Research Report 113. Washington, DC: IFPRI.

Ambler, K., A. de Brauw, S. Godlonton, and A. Vanderkooy. 2016. "Adapting and Implementing Fomento in Africa: Learning through a Partners Workshop." *IFPRI Blog*, February 12.

Anderson, J. 2007. "Agricultural Advisory Services." Background Paper for *World Development Report 2008: Agriculture for Development*. World Bank, Washington, DC.

Arulmanikandan, B., P. Vaishnavi, and A.R. Jayan. 2023. "The Role of Agricultural Value Chain Extension in Facilitating Agricultural Transformation." In *Emerging Trends in Agricultural Extension Education*, eds. O. Prakash, A. Verma, M. Sultana, C. Patil N.D., and N. Khoisnam. Bhubaneswar, Odisha, India: S P Publishing.

ASTI (Agricultural Science & Technology Indicators). n.d. "ASTI Is Moving from IFPRI to FAO." Accessed March 2025. www.asti.cgiar.org/

Awiti, A.O., and A.M. Ndiwa. 2024. *Evaluating the Integration of Agroecological Principles into Kenya's Legal and Policy Framework*. Nairobi: CIFOR-ICRAF.

Babu, S., S. Franzel, K. Davis, and N. Srivastava. 2021. "Drivers of Youth Engagement in Agriculture. Insights from Guatemala, Niger, Nigeria, Rwanda, and Uganda." IFPRI Discussion Paper 2010. IFPRI, Washington, DC.

Babu, S.C., and Y. Zhou. 2015. "Private Sector Extension: Synthesis of the Case Studies." In *Knowledge Driven Development: Private Extension and Global Lessons*, eds. Y.Z. and S.C. Babu, 253–267. Cambridge, MA: Academic Press.

Balana, B., D.K. Mekonnen, T. Arega, et al. 2024. "Do Public Works Investments in Watershed Rehabilitation and Small-Scale Irrigation Improve Nutrition and Resilience? Evidence from Bureau for Humanitarian Assistance Interventions in Support of Ethiopia's Productive Safety Net Program." IFPRI Discussion Paper 2308. IFPRI, Washington, DC.

Bergamasco, S., R. Borsatto, and C. Thomson. 2016. "Evaluation of Extension Reforms in Brazil." Technical Report. Washington, DC: IFPRI; São Paulo, Brazil: UNICAMP.

Birner, R., K. Davis, J. Pender, et al. 2009. "From Best Practice to Best Fit: A Framework for Designing and Analyzing Pluralistic Agricultural Advisory Services Worldwide." *The Journal of Agricultural Education and Extension* 15 (4): 341–355.

Bruce, K., and H. Costa. 2019. "Enabling Environment for PPPs in Agricultural Extension Projects: Policy Imperatives for Impact." *Journal of Rural Studies* 70: 87–95.

Ceballos, F., A. Chugh, and B. Kramer. 2024. "Impacts of Personalized Picture-Based Crop Advisories: Experimental Evidence from India and Kenya." IFPRI Discussion Paper 2322. IFPRI, Washington, DC.

Ceballos, F., B. Kramer, and M. Robles. 2019. "The Feasibility of Picture-Based Insurance (PBI): Smartphone Pictures for Affordable Crop Insurance." *Development Engineering* 4: 100042.

CGIAR. n.d.-a. "CGIAR Research Initiative on Diversification in East and Southern Africa." Accessed March 2025. www.cgiar.org/initiative/diversification-in-esa/

CGIAR. n.d.-b. "CGIAR Research Initiative on Nature-Positive Solutions." Accessed March 2025. www.cgiar.org/initiative/nature-positive-solutions/?section=news-events

Christoplos, I., P. Sandison, and S. Chipeta. 2012. *Guide to Evaluating Rural Extension*. Lausanne, Switzerland: GFRAS.

CRS (Catholic Relief Services) and MEAS (Modernizing Extension and Advisory Services). 2016. *The Seven Steps of Marketing: A Smart Skills Manual*. Baltimore: Urbana, IL.

CTA (Centre for Agricultural and Rural Cooperation). 2019. *The Digitalisation of African Agriculture Report 2018-2019*. Wageningen, Netherlands: CTA/Dalberg Advisers.

Davis, K. 2020. "Comparison of National Extension Systems: Application of the Best-Fit Framework." In *Agricultural Extension: Global Status and Performance in Selected Countries*, eds. K. Davis, S.C. Babu, and C. Ragasa, 53–95. Washington, DC: IFPRI.

Davis, K., S. Makhija, and D.J. Spielman. 2021. *Agricultural Extension and Rural Advisory Services: What Have We Learned? What's Next?* Policies, Institutions, and Markets (PIM) Synthesis Brief. Washington, DC: PIM.

Davis, K., A. Snider, T. Archibald, B. Grove, and S. Babu. 2021. "Organizational Innovation in Times of Crises: The Case of Extension and Advisory Services." *Journal of International Agricultural and Extension Education* 28 (1): 6–14.

Davis, K., and D.J. Spielman. 2017. "Applying the Best-Fit Framework to Assess and Strengthen National Extension and Advisory Services." *Journal of Agricultural and Extension Education* 24 (3): 80–90.

Davis, K., and R.V. Sulaiman. 2014. "The 'New Extensionist': Roles and Capacities to Strengthen Extension and Advisory Services." *Journal of International Agricultural Education and Extension* 21 (3): 6–18.

Davis, K., B. Swanson, D. Amudavi, et al. 2010. "In-Depth Assessment of the Public Agricultural Extension System of Ethiopia and Recommendations for Improvement." IFPRI Discussion Paper 1041. IFPRI, Washington, DC.

Davis, K.E., S.C. Babu, and C. Ragasa. 2020. *Agricultural Extension: Global Status and Performance in Selected Countries*. Washington, DC: IFPRI.

Digital Green. 2024. "Farmer.Chat: Shaping Modern Agriculture for Youth in Kenya." *Insights* [blog]. December 4.

Digital Green. n.d. "Improving the Speed and Efficiency of Agricultural Extension with AI." Accessed March 2025. https://farmerchat.digitalgreen.org/

Engel, P. 1995. *Facilitating Innovation: An Action-Oriented Approach and Participatory Methodology to Improve Innovative Social Practice in Agriculture*. Wageningen, the Netherlands: Wageningen Agricultural University.

FAO (Food and Agriculture Organization of the United Nations). 2022. *Enabling Extension and Advisory Services to Promote Agroecology*. Rome.

Feder, G., R. Birner, and J.R. Anderson. 2011. "The Private Sector's Role in Agricultural Extension Systems: Potential and Limitations." *Journal of Agribusiness in Developing and Emerging Economies* 1 (1): 31–54.

Franzel, S., R. Miiro, N. Uwitonze, K. Davis, B. Luzobe, and R. Rurangwa. 2021. "Engaging Youth in Agribusiness through Private Sector Extension and Advisory Services: Lessons from Rwanda and Uganda." In *Proceedings of the Virtual Conference of the Association for International Agricultural and Extension Education, April 12-14*, ed. AIAEE Executive Board. College Station, TX: Association for International Agricultural and Extension Education.

Gelli, A., and L. Bliznashka. 2024. *Program Pathways for Sustainable, Nutrition Sensitive School Meals: An Updated Framework.* CGIAR Initiative on Fragility, Conflict, and Migration. Washington, DC: IFPRI.

GFRAS (Global Forum for Rural Advisory Services). n.d.-a. "Country Fora." Accessed March 2025. www.g-fras.org/en/community/country-fora.html

GFRAS. n.d.-b. *Extension Support for Climate Adaptation and Mitigation.* New Extensionist Learning Kit (NELK) Module. Accessed March 2025. https://nelk.g-fras.org/en/home-alias-alias/topic-1/38-extension-support-for-climate-adaptation-and-mitigation/preview

GFRAS. 2020. *Adding Value to Local Action: How GFRAS Promotes Advocacy and Leadership in Rural Advisory Services for Sustainable Development.* Lausanne, Switzerland.

GFRAS. 2023. *Transforming Rural Economies and Youth Livelihoods – Agroecology Toolkit.* Lausanne, Switzerland.

Gillwald, A., and A. Partridge. 2022. "Gendered Nature of Digital Inequality: Evidence for Policy Considerations." Background Paper. Research ICT Africa, Cape Town, South Africa.

Gliessman, S. 2022. "Can Agricultural Extension Be of Service to Agroecology?" *Agroecology and Sustainable Food Systems* 46 (7): 953–954.

Gorman, M., O. Keogh, P. Clarke, and P. Leonard. 2018. "The Use and Potential of Local Radio for Extension Services in the West of Ireland." *International Journal of Agricultural Extension* 6: 51–59.

Gotor, E., and C. Masso. 2024. *What Do We Know About the Future of the Environment and Biodiversity in Relation to Food System Transformation?* CGIAR Initiative on Foresight. Montpellier, France: CGIAR.

Hidrobo, M., G. Palloni, D.O. Gilligan, J.C. Aker, and N. Ledlie. 2022. "Paying for Digital Information: Assessing Farmers' Willingness to Pay for a Digital Agriculture and Nutrition Service in Ghana." *Economic Development and Cultural Change* 70 (4): 1367–1402.

IFPRI. n.d. "Picture-Based Crop Insurance (PBI)." Accessed March 2025. www.ifpri.org/project/pbinsurance/

J-PAL (Abdul Latif Jameel Poverty Action Lab). 2023. *Improving Agricultural Information and Extension Services to Increase Small-Scale Farmer Productivity.* J-PAL Policy Insights. Cambridge, MA.

Klerkx, L., and C. Leeuwis. 2009. "Shaping Collective Functions in Privatized Agricultural Knowledge and Information Systems: The Positioning and Embedding of a Network Broker in the Dutch Dairy Sector." *Journal of Agricultural Education and Extension* 15 (1): 81–105.

Lampach, N., N.-V. Phu, and N.-T. The. 2021. *The Effect of Agricultural Extension Programs on Technical Efficiency of Crop Farms in Low and Middle-Income Countries.* Amsterdam: SSRN, Elsevier.

Lecoutere, E., D.J. Spielman, and B. Van Campenhout. 2023. "Empowering Women through Targeting Information or Role Models: Evidence from an Experiment in Agricultural Extension in Uganda." *World Development* 167: 106240.

Leeuwis, C. 2004. *Communication for Rural Innovation: Rethinking Agricultural Extension*, 3rd ed. Hoboken, NJ: Wiley-Blackwell.

MacNairn, I., and K. Davis. 2018. *Rwanda: Desk Study of Extension and Advisory Services: Developing Local Extension Capacity (DLEC) Project.* San Francisco: Digital Green.

Makwenda, B. 2010. "How NASFAM Supports Smallholders in Malawi." In *Alternatives to Land Acquisitions: Agricultural Investment and Collaborative Business Models*, eds. L. Cotula and R. Leonard, 63–70. London and Edinburgh: International Institute for Environment and Development.

Mganga, N., E. Jones-Garcia, A.G. Monsalue, and J. Koo. 2023. *Longa: An Automated Speech Recognition Tool for Bantu Languages.* Digital Innovation Technical Report December 2023. Washington, DC: IFPRI.

Muyanga, M., and T.S. Jayne. 2008. "Private Agricultural Extension System in Kenya: Practice and Policy Lessons." *Journal of Agricultural Education and Extension* 14 (2): 111–124.

Nakasone, E., and M. Torero. 2016. "A Text Message Away: ICTs as a Tool to Improve Food Security." *Agricultural Economics* 47 (S1): 49–59.

Ogundari, K. 2022. "A Meta-Analysis of the Impact of Agricultural Extension Services." *China Agricultural Economic Review* 14 (2): 221–241.

Phatty-Jobe, A., A. Seth, and K. Norton. 2020. *Digital Agriculture Maps: 2020 State of the Sector in Low and Middle-Income Countries.* London: GSMA AgriTech Programme.

Ragasa, C. 2020. "Effectiveness of the Lead Farmer Approach in Agricultural Extension Service Provision: Nationally Representative Panel Data Analysis in Malawi." *Land Use Policy* 99: 104966.

Ragasa, C., C. Alvarez-Mingote, and P. McNamara. 2024. "Bottom-Up Approaches and Decentralized Extension Structures for Improving Access to and Quality of Extension Services and Technology Adoption: Multi-Level Analysis from Malawi." *European Journal of Development Research* 36: 1093–1146.

Rajalahti, R., W. Janssen, and E. Pehu. 2008. "Agricultural Innovation Systems: From Diagnostics Toward Operational Practices." Agriculture and Rural Development Discussion Paper 38. World Bank, Washington, DC.

Rao, S. 2015. *Using Radio in Agricultural Extension.* Note 18. GFRAS Good Practice Notes for Extension and Advisory Services. Lindau, Switzerland: GFRAS.

Ringler, C., M. Alvi, R. Birner, et al., eds. 2023. *Reaching Smallholder Women with Information Services and Resilience Strategies to Respond to Climate Change.* Washington, DC: IFPRI.

Rwamigisa, P.B., T. Kakuba, and S. Massawe. 2019. "Uganda's Single Spine Agricultural Extension System: What It Is, Progress and Early Lessons on Its Implementation." Unpublished document.

Saha, P., A.K. Prusty, and C. Nanda. 2024. "Extension Strategies for Bridging Gender Digital Divide." *Journal of Applied Biology & Biotechnology* 12 (4): 76–80.

Scheer, J., and J.A. Okelai. 2019. *Village Agent Model Study: Likely Effects on the Ugandan Agricultural Sector. Developing Local Extension Capacity Project.* Washington, DC: US Agency for International Development.

Schut, M., Kamanda, J., Gramzow, A., et al. 2019. "Innovation Platforms in Agricultural Research for Development: Ex-Ante Appraisal of the Purposes and Conditions Under Which Innovation Platforms Can Contribute to Agricultural Development Outcomes." *Experimental Agriculture* 55 (4): 575–596.

Schut, M., C. Leeuwis, M. Sartas, et al. 2022. "Scaling Readiness: Learnings from Applying a Novel Approach to Support Scaling of Food System Innovations." In *Root, Tuber and Banana Food System Innovations*, eds. G. Thiele, M. Friedmann, H. Campos, V. Polar, and J.W. Bentley. Cham, Switzerland: Springer Cham.

Scoones, I., and J. Thompson, eds. 1994. *Beyond Farmer First.* London: IT Publications.

60 Decibels. 2024. *Digital Agriculture Services in Kenya: The Farmer Perspective.* New York.

Spielman, D.J. 2005. *Innovation Systems Perspectives on Developing-Country Agriculture: A Critical Review.* Washington, DC: IFPRI.

Spielman, D.J., E. Lecoutere, S. Makhija, and B. Campenhout. 2021. "Information and Communications Technology (ICT) and Agricultural Extension in Developing Countries." *Annual Review of Resource Economics* 13: 177–201.

Sulaiman, V.R. 2020. "Agricultural Extension and Advisory Services in Support of Climate Change Adaptation and Mitigation: An Evidence Review." In *Strategies for Supporting Sustainable and Climate Smart Advisory Services: April 2020.* Zurich, Switzerland: Swiss Re Foundation; Lausanne, Switzerland: ISF.

Sulaiman, V.R. 2021. "The Importance of Agroecology." *CoSAI News& Blogs*, June 18.

Sulaiman, V.R., D. Chuluunbaatar, P. Djamen, C. Grovermann, and A. Holley. 2022. *Indicator Framework for National Extension and Advisory Service Systems: Metrics for Performance and Outcome Measurement.* Rome: FAO.

Swanson, B.E., R.P. Bentz, and A.J. Sofranko, eds. 1997. *Improving Agricultural Extension: A Reference Manual.* Rome: FAO.

Swanson, B.E., and R. Rajalahti. 2010. "Strengthening Agricultural Extension and Advisory Systems: Procedures for Assessing, Transforming, and Evaluating Extension Systems." Agriculture and Rural Development Discussion Paper 44. IFPRI, Washington, DC.

Thurow, R. 2024. *Against the Grain: How Farmers around the Globe Are Transforming Agriculture to Nourish the World and Heal the Planet.* Evanston, IL: Agate.

Timu, A.G., D. Manoti, A. Shee, and L. You. 2024. "Impacts of Gender-Inclusive Extension Approaches on Farmer Understanding and Willingness to Pay for Bundled Financial Services." *Current Research in Environmental Sustainability* 8: 100268.

Tzachor, A., M. Devare, C. Richards, et al. 2023. "Large Language Models and Agricultural Extension Services." *Nature Food* 4: 941–948.

Ulimwengu, J.M. 2025. "An Agenda for Agri-Food Transformation." Africa News Agency, January 30.

UNICEF. 2017. *The State of the World's Children 2017: Children in a Digital World.* New York.

USAID (US Agency for International Development) and Agrilinks. 2019. "Measuring Performance in Ag Extension: Common Metrics Framework," webinar.

Van Campenhout, B., E. Lecoutere, and D.J. Spielman. 2023. "Hiding or Pleasing: Spousal Disagreement among Ugandan Maize Farmers." *Journal of Development Studies* 59 (1): 39–57.

Van Campenhout, B., D.J. Spielman, and E. Lecoutere. 2020. "Information and Communication Technologies to Provide Agricultural Advice to Smallholder Farmers: Experimental Evidence from Uganda." *American Journal of Agricultural Economics* 103 (1): 317–337.

van den Ban, A.W., and S. Hawkins. 1988. *Agricultural Extension.* London: Longman Scientific & Technical.

VOA (Voice of America). 2018. "'Can You Dig It?' Africa Reality Show Draws Youth to Farming." February 13. Washington, DC.

Waddington, H., and H. White. 2014. *Farmer Field Schools: From Agricultural Extension to Adult Education.* 3ie Systematic Review Summary 1. London: International Initiative for Impact Evaluation (3ie).

Woltering, L. 2024. "Stuck on Scale: Rethinking Scaling and Systems Change for Agricultural Research for Development for Food Systems Transformation." PhD thesis, Wageningen University and Research, Wageningen, Netherlands.

World Bank. 2012. *Agricultural Innovation Systems: An Investment Sourcebook.* Washington, DC.

World Bank. 2015. "Kenya—Enhancing Decentralized Service Delivery Through Participatory Project: Summary Overview: Kenya Devolution." Washington, DC.

World Bank. 2019. "Kenya's Devolution." Brief. November 26.

Yonggong, L. 1998. "Institutional and Policy Reform of Rural Extension in China during the Transition towards a Market Economy." In *Training for Agricultural and Rural Development 1997-98*, ed. L.V. Crowder. Rome: FAO.

Zhou, Y., and S.C. Babu. 2015. *Knowledge Driven Development: Private Extension and Global Lessons*. Amsterdam: Elsevier.

Chapter 9

QUALITY SEEDS, IMPROVED VARIETIES

The Economics of Crop Genetic Improvement and Farmer Uptake

Berber Kramer and David J. Spielman

Berber Kramer is a senior research fellow, Markets, Trade, and Institutions Unit, IFPRI-Kenya; and **David J. Spielman** is director, Innovation Policy and Scaling Unit, IFPRI.

Key messages

- Crop genetic improvement has greatly enhanced agricultural productivity, food security, and economic opportunities for smallholder farmers globally. CGIAR has contributed significantly to efforts of national partners to develop high-yielding, stress-resistant, and nutritionally improved crop varieties.
- Early policy analyses, largely informed by the successes of the Green Revolution, documented substantial gains in agricultural productivity, income generation, and poverty alleviation. However, subsequent research highlighted persistent adoption challenges, including slow varietal turnover and seed replacement, which have been linked to diverse institutional constraints (such as weak extension systems), economic barriers (high seed prices, liquidity constraints), and behavioral factors (risk aversion, limited farmer knowledge).
- Initial crop improvement programs focused primarily on maximizing yields. Over time, programs have shifted toward prioritizing multidimensional traits, including tolerance to abiotic stresses (such as drought and heat), resistance to pests and diseases, enhanced nutrient use efficiency, nutritional quality (biofortification), and consumer- and market-preferred characteristics, to more effectively address challenges such as climate change and changing diets.

Looking to the future, research and policy recommendations should focus on:

- **Strategically bundling improved crop varieties with complementary inputs and services**—such as agricultural advisories, credit, insurance, fertilizer, and guaranteed market access—to help overcome multiple simultaneous adoption constraints. Developing evidence-based bundles, carefully calibrated to specific local contexts, offers a promising pathway to accelerating varietal adoption, turnover, and seed replacement.

- **Make significant investments in both formal and informal seed system infrastructure and regulatory reforms** to fully realize the potential of crop genetic innovation. Policy reforms should enhance seed quality assurance, improve intellectual property frameworks, streamline varietal registration and release processes, and encourage private sector investment.
- **Strengthen global consensus on the role of science—particularly precision genetics—in food systems.** This includes an urgent need for greater investment in rigorous impact assessments, using both predictive modeling and real-world evaluations to capture context-specific outcomes as well as broader impacts on economies.
- **Integrate systematic use of market intelligence to align crop improvement with climate change, urbanization, and shifting diets.** This will require an improved understanding of how crop genetic improvement programs use market intelligence, and what barriers they face in adopting more systematic and transdisciplinary processes for prioritization.

Crop improvement has long been a cornerstone of global efforts to enhance agricultural productivity, improve food security, and foster economic development. By conserving, sharing, and developing genetic materials that are used to develop high-yielding, stress-resistant, and more nutritious varieties, CGIAR genebanks, crop improvement programs, and seed system projects have contributed substantially to the efforts of national partners to deliver genetic gain—that is, genetic traits embodied in seeds and other types of planting material—to small-scale, resource-poor farmers. Together with national research programs, extension services, private seed companies, and women and men farmers' organizations, CGIAR has helped improve income, nutrition, and livelihood opportunities for both the rural and urban poor.

IFPRI researchers and the wider community of applied economists and policy analysts have consistently gravitated to studying crop improvement, although over time, the focus of policy research has evolved significantly. Early analyses by IFPRI and partners in the 1980s and 1990s provided evidence of the potential social and economic gains from public investment in plant breeding, supported by findings from field-based studies of varietal adoption patterns, trends, and impacts. While governments in low- and middle-income countries (LMICs) continue to invest in genetic innovation and seed systems to support smallholder farmers, concerns have emerged about persistently low and heterogeneous rates of adoption among farmers in many agricultural systems. As a result, more recent research has focused on alleviating institutional and behavioral constraints to adoption and introducing innovative mechanisms to accelerate adoption and realize productivity and welfare outcomes.

Policy research has also undergone a shift in focus from the original challenge of replacing landraces and local varieties with high-yielding improved varieties to an emphasis on (1) continuously replacing older improved varieties with more recently released varieties (*varietal turnover*), (2) encouraging farmers to cultivate high-quality seed irrespective of genetic change (*seed replacement*), and (3) moving from a singular focus on higher yields to targeting multiple traits, including abiotic stress tolerance, pest and disease resistance, improved use-efficiency for resources such as soil nutrients and water, higher micronutrient content, specific processing and preparation qualities, and other farmer-, consumer-, and market-facing characteristics. Combined, these shifts aim to make improved varieties more attractive to farmers, consumers, and other value chain actors, and also strengthen the contribution of crop varieties to addressing major food systems challenges such as climate change and dietary transitions.

In this chapter, we explore evidence on the contribution of crop genetic improvement to productivity, nutrition, environment, and poverty outcomes. We also discuss important areas for future research, such as documenting the empirical impacts and cost-effectiveness of innovations through which new varieties reach farmers; improving our understanding of how to accelerate varietal adoption, turnover, and quality seed use; and providing insights on the role of information, incentives, and policies in amplifying the impacts that crop genetic improvement may have on a range of development outcomes.

Evidence on crop improvement and farmer uptake

Identifying solution-oriented, evidence-based approaches to accelerate varietal adoption and turnover

Researchers have been interested in the economic and social dimensions of technological change in agriculture for decades. Seminal work in the late 1950s and early 1960s provided the iconic logistic growth functions, or "s-curves," that are still used today to describe the process by which new technologies are adopted by

FIGURE 9.1 The classic adoption curve

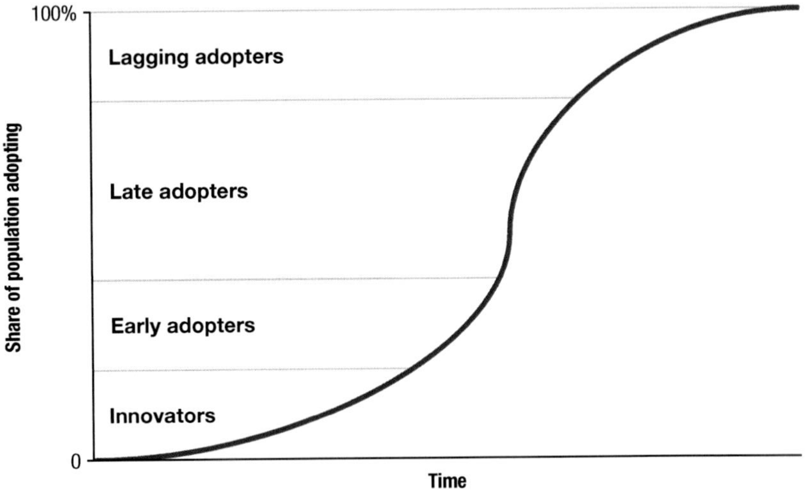

Source: Adapted from Rogers (1962).

different types of farmers over time (Figure 9.1) (Griliches 1957; Rogers 1962). Given the expected impacts of improved varieties on productivity and poverty, a major emphasis for policy researchers has been to produce evidence on how to speed up the rate at which genetic gains reach farmers' fields.

Early research focused on the *adoption* of improved varieties, which is typically defined as the sustained replacement of landraces or local varieties by varieties that were improved by breeding programs (Spielman and Smale 2017). Adoption involves a significant change in the genetic composition of varieties being cultivated by farmers. Over a period of about 20 years starting in the mid-to-late 1960s, the Green Revolution experience with semidwarf rice and wheat varieties in South and Southeast Asia provided much of the evidence on the nature and impact of the adoption process (Evenson and Gollin 2003; Foster and Rosenzweig 1995; Lipton and Longhurst 1989). IFPRI researchers and colleagues shaped many of the early studies on adoption—not just of high-yielding rice and wheat varieties, but also of other country–crop combinations such as maize in Southern Africa (Smale et al. 1995) and rice and millet in West Africa (Adesina and Zinnah 1995; Smale et al. 2018).

Despite early successes in increasing productivity and reducing hunger and poverty (Fan et al. 2008; Hazell and Ramasamy 1991), evidence from subsequent decades and regions tends to suggest low and variable adoption rates that are associated with a wide range of factors (Feder et al. 1985; Feder and Umali 1993). Particular emphasis was given to the slow rate of varietal *turnover*, that is, the replacement of first-generation varieties with more recently released varieties that sustain or enhance the original gains, such as pest and disease resistance (Brennan and Byerlee 1991; Heisey and Brennan 1991). Findings often pointed to adoption constraints such as the farmer's education level, landholding size, access to extension, or gender of the household head, and the need for more targeted efforts to reach the types of farm-households that are less likely to adopt new varieties.

Interest in accelerating adoption remains strong especially in Africa south of the Sahara, as concerns persist over slow or stagnant productivity growth in agriculture (Ray et al. 2012; Wollburg et al. 2024). But as many farmers in the Global South have become aware of improved varieties (for example, see Sheahan and Barrett 2017), policy emphasis is shifting slowly toward a combined approach: accelerating adoption where it is still lagging, and accelerating varietal turnover and seed replacement in most other places. Accelerating adoption will require a major shift in farmers' technology paradigm—the what, when, and how of their farming decisions, which are often shaped by the support provided by agricultural extension services, seed suppliers, starter kits, and fertilizer-seed-credit packages. Varietal turnover and seed replacement is often more continuous and incremental in nature, and more dependent on functional markets, institutions, and targeted policies.

Beyond their support to public plant breeding programs and extension services, governments often view seed subsidies as their primary policy instrument for promoting turnover and replacement. Subsidies address farmers' most obvious constraint—high sensitivity to seed prices (Mason and Ricker-Gilbert 2013; Mason and Smale 2013)—but often fail to address other important incentive issues and desirable outcomes, such as diversification in the crops that farmers produce (which contributes to soil health and dietary diversity in the home), unless intentionally designed to address these shortcomings (Spielman and Smale 2017). For example, in Malawi, introducing subsidies for legumes in addition to maize led to measurable improvements in diets (Matita et al. 2022).

Studies have also examined alternative explanations for low adoption, turnover, and replacement, looking, for example, at farmers' and consumers' preferred traits to inform product development priorities in breeding programs (Chowdhury et al. 2011; Oparinde et al. 2016), at how farmers obtain seed with these traits when markets are fragmented or absent (Lipper et al. 2010), and at how specific regulations or program interventions could increase accessibility and affordability (Smale 2005; Tripp and Louwaars 1997).[1]

Another branch of research builds on behavioral insights derived from exchanges between social psychology, economics, and public policy, and engages with seminal issues such as farmer heterogeneity (Suri 2011), learning processes (Foster and Rosenzweig 1995; Leathers and Smale 1991), risk and uncertainty (Rosenzweig and Binswanger 1993), and social institutions such as gender norms (Galiè et al. 2025). Studies on downside risk, learning dynamics, and imperfect markets for a wide range of crops and countries represent an important expansion of the evidence base and suggest ways to encourage adoption, turnover, and replacement by alleviating behavioral constraints (Gars and Ward 2019; Hanna et al. 2014).

One means of alleviating these constraints is by combining new varieties with other essential inputs and services. The rationale for bundling improved varieties with products such as advisories, finance, insurance, fertilizers, or market guarantees, is that farmers face multiple barriers to technology adoption, especially more marginalized farmers (Suri and Udry 2022) (see also Chapter 8). Marketing a new variety to risk-averse farmers, for example, may succeed only if it is accompanied by a small loan to address liquidity and credit constraints, a low-cost index insurance policy to address weather risk, digital advisories to provide easy access to technical and business insights, product quality assurance signals, or contracts to ensure postharvest market opportunities.[2] Similarly, providing information to women farmers or adjusting the manner in which the information is delivered to them may increase their capacity to make informed decision-making on, and their bargaining power over, which varieties to plant (Lecoutere et al. 2023; Kramer, Trachtman, and Zuze 2023).

The idea that improved varieties can be bundled with other products, services, and programs is not new. Packages of improved varieties, inorganic fertilizer, credit, extension services, and minimum support pricing were a hallmark of the Green Revolution. What is new, however, is the expansion of bundling options into entirely new realms of activity, as well as the commitment

---

1 Early research on these questions relied mainly on nonexperimental methods such as dichotomous choice models to establish correlations between adoption and its potential drivers; quasi-experimental methods requiring strong assumptions to (weakly) identify causal effects of innovations to increase adoption; or qualitative approaches that provide insufficient generalizability beyond small samples of farmers.

2 On credit, see Kramer et al. (2024); on insurance, see Boucher et al. (2024) and Ward et al. (2020); on digital advisories, see Fabregas et al. (2025, 2019); on product quality assurance signals, see Miehe, Swallow et al. (2023), Miehe, Van Campenhout et al. (2023), Bold et al. (2017), and Hsu and Wambugu (2022); and on contracts, see Arouna et al. (2021) and Michler and Wu (2020).

to generating rigorous evidence of what works and for whom (Barrett et al. 2020). While bundles may differ significantly by crop, country, market, or context, each one represents an opportunity for experimentation and learning, in which improved varieties contribute to the larger process of technical change and system transformation (see Chapter 8).

In fact, experiments that explore the impact of these bundles are providing actionable policy recommendations that address last-mile delivery problems for improved varieties. For example, bundling stress-tolerant varieties with index insurance products requires careful calibration of the index to the variety's performance under stress, such as moderate to severe drought during key stages of the plant's growth. The index should trigger payouts only once reaching a "tipping point" at which the stress-tolerance trait is overwhelmed, such as extreme drought that continues for several days beyond the plant's capacity to withstand drought (Ward et al. 2020; Lybbert and Carter 2014). Even then, farmers may not consider stress-tolerant varieties and insurance coverage to be complementary (Kramer, Waweru, and Malcarne 2023). Costing insurance requires not only an understanding of the probability that extreme events might occur, and the value of the potential losses, but also the discounts required to encourage smallholders to purchase coverage despite their lack of familiarity with index insurance (see also Chapter 10).

Creating an enabling environment to incentivize public and private innovation to accelerate varietal turnover

Experiments that have focused on addressing demand-side barriers to adoption tend to test solutions that increase demand for improved varieties made available through formal channels—primarily commercial seed markets—rather than through the nonmarket exchanges of seed on which many smallholder farmers rely. Thus, whereas innovations in bundled solutions to increase demand have advanced rapidly, broader seed systems innovations to strengthen both formal and informal channels are evolving more slowly. This slow pace has spurred an emphasis on upstream and midstream value chain development (Reardon and Timmer 2012), as well as lessons learned from private industry (Donovan et al. 2021; ter Steeg and Louwaars 2024), to design innovative approaches that address supply-side constraints to adoption, turnover, and replacement.

Seed systems innovations include a sequence of supply-side regulatory reforms and demand-side investments designed to attract private investment into the delivery of improved varieties and quality seeds to smallholder farmers, while they also draw attention to the economic trade-offs, unintended consequences, and operational challenges of such reforms and investments. Producing marketable seeds requires land, labor, irrigation, regulation, capital, expertise, and business and marketing experience to forecast and generate demand, all of which can be scarce in LMICs, thus introducing supply-side constraints in their seed industries. Lifting these constraints is important for firms and market regulators in order to ultimately make new varieties available to farmers and other value chain actors. Unfortunately, compared to investments in crop improvement, investments in seed production and other production-related aspects of the seed system have been relatively small.

Although many private companies in Asia have achieved notable successes in the seed industry, Africa's seed sector remains at a relatively nascent stage. Initiatives such as the Program for Africa's Seed Systems, led by AGRA (formerly the Alliance for a Green Revolution in Africa), and the Integrated Seed Sector Development project, led by Wageningen University & Research, have contributed to mobilizing attention and resources around the policy dimensions of seed sector development, especially where formal (commercial) and informal (farmer-based) seed systems intersect (Bagamba et al. 2023; Munyi and De Jonge 2015; Wellard et al. 2019). Similar efforts such as the Enabling the Business of Agriculture indicators, the African Seed Access Index, and the Access to Seeds Index, combined with CGIAR programs and projects, have also drawn attention to the need for a stronger enabling environment. Emphasis has been placed on early generation seed production and distribution, varietal registration and release procedures, seed quality assurance systems, intellectual property rights, seed subsidies, and international seed trade (Kuhlmann 2021; Spielman and Smale 2017).

Demonstrating how public investment in crop improvement can generate desirable development outcomes

IFPRI and its partners continue to conduct research that quantifies the social returns to improved variety development and delivery, justifies the necessary investment of scarce public resources, and unpacks the political economy factors that shape the policy discourse around crop improvement.

Research spanning several decades has helped to demonstrate how public spending on agricultural R&D—usually involving plant breeding programs as the primary funding vehicle—contributes to agricultural productivity growth (Alston and Pardey 2021; Alston et al. 1995, 2022; Griliches 1964; Hayami and Ruttan 1971). More recent evidence has helped to solidify the connection between crop improvement and productivity growth (Evenson and Gollin 2003), while other work has extended this analysis to link R&D and poverty outcomes (Fan et al. 2000; Fan and Hazell 2001). A recent ex ante impact assessment suggests that faster productivity growth due to crop improvement could increase economywide income by 2 to 15 times the current level of public R&D spending on food crops in LMICs, with the potential to reduce the population at risk of hunger by up to 11 million people (Wiebe et al. 2021).

This entire body of literature draws on a rich combination of analytical techniques and data sources, ranging from econometric analysis using national statistics on crop production, yields, and area (Fan et al. 2000; Fan and Hazell 2001), relatively straightforward economic surplus models (Fuglie and Echeverria 2024), structural models estimating the impacts of the Green Revolution (Gollin et al. 2021), and sophisticated global computational models (Evenson and Rosegrant 2003; Rosegrant et al. 2023). Models developed or improved upon by IFPRI, such as DREAM, RIAPA, IMPACT, and MIRAGRODEP, have played a critical role in the latter type of research, as has the Institute's statistical and econometric modeling (see also Chapter 4).

While many of these studies have withstood the test of time and remain seminal to our understanding of crop improvement's importance, the nature of the evidence is rapidly evolving, including the methods being applied to demonstrate impacts (Box 9.1). In particular, there is a move to use more rigorous experimental and quasi-experimental methods to establish causal relationships with greater credibility, precision, and context-specificity (Stevenson et al. 2023). For example, randomized controlled trials (RCTs) have become a cornerstone of evidence-based policy evaluation, including the evaluation of impacts from using improved varieties under farmers' conditions (for example, Emerick et al. 2016; Boucher et al. 2024).

Using RCTs to evaluate the impacts of crop improvement, as has been done by IFPR researchers for a variety of crops in Africa and Asia, has helped create the evidence needed to advance these traits into widespread use by farmers.[3] IFPRI-led studies also shed light on which incentives and institutions lead to changes in farmers' expectations, preferences, and behaviors, and ultimately shape their adoption, turnover, and replacement decisions in cost-effective and inclusive ways. Importantly, whether they rely on relatively simple economic surplus models or complex global trade ones, these RCT-based methods offer greater analytical resolution and nuance at the country, crop, and trait levels than modeling-based approaches, despite well-reasoned criticisms of the approach (Barrett and Carter 2020, 2010; Rosenzweig and Udry 2020).

3 Examples of RCT-based studies include Abate et al. (2018, 2023); Ahmed et al. (2021); Bernard et al. (2023); De Brauw et al. (2018, 2019); and Van Campenhout et al. (2021).

BOX 9.1 Methods for assessing the impacts of crop improvement

Research on crop improvement must consider whether the social returns justify investing scarce public resources in plant breeding. Below, we examine the variety of approaches used to address this question.

EX ANTE MODELING-BASED IMPACT ASSESSMENTS. Continuous ex ante assessments of crop improvement programs and other development interventions are essential to justifying and prioritizing scarce public resources. These assessments utilize economic models such as DREAM, RIAPA, IMPACT, MIRAGRODOP, and GLOBE, as well as biophysical models and data to enhance and calibrate these models. They conceptualize outputs from investments in breeding as "innovations" that increase productivity growth, and then quantify the returns to agricultural R&D and the contribution of crop improvement by comparing scenarios with and without adoption of improved varieties (that is, with higher versus lower rates of productivity growth). This method is often used to inform investments in agricultural R&D. An advantage is that the use of macroeconomic models facilitates comparison across regions, crops, and outcome indicators to identify where the greatest gains can be achieved, depending on the diverse objectives of donors and other decision-makers. However, the approach requires making strong assumptions about the effects of improved varieties on yield growth and about the relationship between yield growth and poverty reduction, food security, and nutrition, neither of which are always founded in empirical observations.

AGRONOMIC PERFORMANCE TRIALS. Field trials play a pivotal role in evaluating the real-world performance of new crop varieties. Conducted under controlled conditions in diverse agro-ecological settings, these trials measure agronomic traits such as yield, resistance to pests and diseases, and adaptation to local conditions. Data from these trials inform breeders, farmers, and policymakers about the practical implications of adopting new varieties and their comparative advantages over existing ones. However, performance under controlled conditions does not always translate into similar performance under farmers' practices. Under uncontrolled conditions, the decision to adopt a new variety could introduce behavioral responses that cannot be captured through this type of method.

ADOPTION STUDIES. Studies documenting the adoption of new crop varieties provide empirical evidence on the extent to which varieties are adopted, perceptions of those varieties, and reasons for adoption or non-adoption. These studies often aim to document changes in agricultural practices and productivity associated with adopting new varieties by comparing adopters and nonadopters. Given that the decision to adopt a new variety is likely correlated with unobservable farmer characteristics that also influence agricultural practices and productivity, findings from adoption studies should not be interpreted as causal evidence. Nonetheless, insights gained from these studies are critical for understanding the rate of adoption of improved varieties, farmers' perceptions of these varieties, and potential barriers to adoption, all of which can be used to design effective extension and outreach programs.

EMPIRICAL (QUASI-)EXPERIMENTAL EVALUATIONS. None of the methods described above show that investments in crop improvement have an impact on the ground. Rigorous evaluations employ robust methodologies such as randomized controlled trials (RCTs) or quasi-experimental designs to assess the causal impact of offering new crop varieties. By randomly assigning treatments that introduce exogenous variation in the probability of accessing and/or adopting the new variety, these evaluations measure the direct effects of adopting new varieties on agricultural productivity, income generation, food security and nutrition, gender equity, the ability to protect consumption or investments from climate shocks, and other indicators of impact. The use of RCTs in policy research is an extension of methods regularly employed by breeders and agronomists, wherein plants, rather than people, are compared across treatment and control groups in highly controlled environments. The main difference is that policy researchers observe people rather than plants, looking at the choices that individual farmers make and the benefits that can be achieved, given their own preferences, expectations, and constraints. Such evaluations provide credible evidence for policymakers and development practitioners to prioritize investments in crop improvement programs. Evidence from such evaluations could also be used to verify whether the impacts predicted ex ante by computational models did indeed occur and to refine model calibration with empirically observed adoption and impact estimates.

Looking to the future

Strengthening opportunities for impact through crop improvement and seed systems

Looking to the future, the cumulative evidence strongly suggests that a better understanding of the impact pathways leading from genetic improvement to broader developmental outcomes can potentially improve the efficiency and effectiveness of future policy and investment choices. This is particularly relevant when such choices aim to address large and complex agricultural transformation challenges beyond productivity growth or poverty reduction, such as better diets and nutrition, climate adaptation and mitigation, gender equity, social inclusion, and addressing environmental degradation and biodiversity loss.

This means moving beyond yield improvement as the sole indicator of breeding progress. Since it is well recognized that a new variety often will not succeed with farmers if its yield is not at least equal to the incumbent variety, the question becomes what other traits should be targeted in product profiles. Potential targets include abiotic stress-tolerance traits for drought, flood, heat, cold, and other adverse weather events induced by climate change; abiotic stress-resistance traits for existing or new pests and diseases that flourish in changing climatic conditions; and nutrition traits conferred by biofortification, that is, breeding for traits that enhance the micronutrient content of food staples (Bouis et al. 2024). National research centers and breeding programs have invested significantly in all of these areas with their CGIAR partners, with standout examples including CIMMYT's Drought Tolerant Maize for Africa, IRRI's Stress Tolerant Rice for Africa and South Asia, and HarvestPlus's work across CGIAR on biofortification, among many others.

Opportunities exist to target other impacts of crop improvement, such as addressing gender equality by targeting production or processing traits that reduce drudgery for women, or by focusing on crops traditionally grown and marketed by women. Some breeding programs have documented or emphasized women's and men's preferred varietal traits and incorporated gender into their participatory breeding and variety selection processes (Ashby and Polar 2019; Weltzien et al. 2019). Programs with explicit gender objectives have made considerable progress in advancing gender equity by supporting women's empowerment, addressing discriminatory gender norms, or creating social and economic opportunities for women (Galiè et al. 2025). However, less emphasis has been placed on identifying which supply- and demand-side interventions can equitably improve varietal adoption, turnover, and replacement (Kramer and Trachtman 2024). There is still a strong need for testing new product concepts and complementary solutions among diverse groups of individuals.

Breeding better crops and delivering them to farmers will never be a simple "silver bullet" solution to achieve productivity growth, poverty reduction, improved nutrition, climate resilience, gender and social equity, or environmental sustainability. Given the complexity of these challenges, no set of traits in a given variety of a single crop can address all of them simultaneously. Even with a proven "blockbuster" crop variety, it is impossible to simultaneously address the diverse needs of heterogeneous farmers who operate under different agroecological and socioeconomic conditions. Therefore, a more nuanced approach is needed that bundles together combinations of technological, technical, and social innovations to address multiple outcomes and achieve stronger and more equitable impacts (Barrett et al. 2020, 2022). Behavioral, social, and economic intelligence is key to identifying opportunities and constraints through which these sociotechnical bundles can help improve adoption and strengthen impacts.

Future priorities for policy and policy research

Looking forward, ensuring that improved varieties and quality seed have a sizable impact on development outcomes in the run-up to 2050 will require several actions.

Build a stronger global consensus around the role of science and technology in our food systems. From the very basic tenets of Mendelian genetics to the role of vaccines in society, science is under siege in many communities. For decades, IFPRI and partners have explored the social and economic impacts of improved genetics in agriculture, helping to demonstrate that science is critical to ending global hunger, malnutrition, and poverty. As the pace of scientific achievement increases, precision genetics will likely play a role in agriculture, as it already does in the healthcare and pharmaceutical sectors, among many others. IFPRI-led studies on genetically modified crops, gene editing, marker-assisted selection, and other applications of genetics and molecular biology to agriculture in the Global South have contributed to ongoing debates around scientific evidence,[4] complementing other independent and influential sources on the role of precision genetics in our global food system (Qaim 2016). Choices on which technologies to move forward should and will be science-based even when public perception suggests otherwise (Paarlberg 2023).

- Evidence is essential to mobilizing science to both safeguard and improve human and environmental health through crop improvement programs and seed sector development.

4 This work includes Falck-Zepeda and Zambrano (2011), Omamo and Von Grebmer (2005), Pixley et al. (2022, 2019), and Smale et al. (2009).

Use market intelligence systematically in crop improvement programs and seed sector development. This is especially important in light of major global challenges related to climate change, dietary transitions, and rapidly evolving food systems. Market intelligence brings together strategic information on future crops, market segments, trait preferences, and client needs to inform decisions and prioritize investments in breeding pipelines and seed systems. It can offer analytical insights that shape the entire product design process from the selection of desired traits to placement in an appropriate market segment. Such information can reorient investments in crop improvement to address climate change and associated shifts in what types of varieties can be grown where, as well as to accommodate changes in dietary preferences in a more urbanized and marketized world, where most people purchase their food rather than grow it themselves.

- IFPRI's economic policy models and its data assets from large-scale household surveys can help inform this prioritization process by providing ex ante projections of potential returns on investment in a given pipeline, or by providing more granular subnational estimates of indicators for various impact areas, such as poverty, food security, dietary diversity, malnutrition, women's empowerment, and resilience to climate change.

Employ IFPRI's behavioral research to provide insights on how to encourage the use of market intelligence. Systematic processes for gathering and using market data remain rare, with evidence showing that less than 45 percent of surveyed breeding professionals use structured methods (Rice et al. 2024). Power dynamics and attitudes further limit meaningful contributions from social scientists (Cullen et al. 2023). Without high-quality market data or cross-disciplinary collaboration, decision-makers risk relying on imprecise or biased information, hindering effective prioritization of breeding investments, and limiting impactful investments in crop improvement. An online experiment suggests that breeding experts, including natural and social scientists involved in plant breeding, may over-extrapolate from low-quality data and may be biased to prefer cheaper, lower-quality information from disciplines that are more familiar to them (Trachtman et al. 2024). This research highlights the need to strengthen awareness of data quality in market research for impactful R&D prioritization and to quantify the gains from increasing such awareness, expanding data availability to prioritize investments, and investing in the systematic use of such information.

Increase investment in assessing the impact of improved varieties and quality seed. This should include complementing ex ante modeling with rigorous empirical measurement of impacts and cost-effectiveness using randomized evaluations that can capture context-specific granularity and shed light on institutional and behavioral mechanisms, such as adoption rates, crowding-out or crowding-in of other inputs, and impacts for diverse groups of farmers, who are otherwise elusive in a modeling framework due to unobservable preferences, risks, expectations, or other factors that are difficult to discern.

- Recent CGIAR efforts to strengthen impact assessment of crop improvement through coordinated trials across a wide range of settings (Kilonzi et al. 2025) will be instrumental, as are ongoing studies to evaluate the impact of various socio-technical bundles.

Link ex ante modeling with ex post evaluations to refine policy guidance for governments and their development partners. For instance, models used for ex ante impact assessment could be updated with robust and externally valid empirical estimates of adoption of new improved varieties under realistic seed system scenarios, as well as their potential effects on productivity. Conversely, general equilibrium effects should be considered more extensively, especially in labor and other factor markets, beyond the immediate first-order outcomes of a given intervention, which are captured through randomized evaluations.

- Using findings from ex post evaluations to estimate general equilibrium effects is an important area of further research (Rosenzweig and Udry 2020; Taylor and Filipski 2014).

Improve the policy, regulatory, and investment landscape. Significant changes are needed in the landscape that shapes the performance of crop improvement programs and seed sector development efforts. Of particular importance are the international agreements and national policies governing the exchange of genetic resources for crop improvement purposes that affect both innovation and entrepreneurship. This includes designing and implementing intellectual property rights regimes that address both breeders' and farmers' rights to plant genetic resources and the gains from innovation, and national varietal registration and release processes and mutual recognition of released varieties through regional regulatory harmonization to increase access to genetic gain across borders (Bishaw et al. 2025), as well as the role of genebanks in conserving genetic resources for future generations (Halewood et al. 2012; Koo et al. 2003; Mekonnen and Spielman 2021). At a more local level, this includes addressing the production and equitable distribution of early generation seed to facilitate commercial firms, farmer-based enterprises, and seed sector development, and the design of credible seed quality assurance systems.

- IFPRI's role in this area is to inform efforts to improve policy, regulatory, and investment landscapes, and to evaluate costs and benefits of doing so.

Conclusion

Research by IFPRI and the wider policy research community highlights the continued importance of plant breeding, crop improvement, and seed sector development. Improved varieties, varietal turnover, and seed replacement will remain key drivers of technical change among smallholder farmers, with benefits that include lower food prices for workers, income and livelihood opportunities for rural farm and nonfarm households, and low-cost inputs for firms in the service and industrial sectors. A rich body of work demonstrates the value of investing in genetic innovation and delivering these innovations to farmers' fields. However, the technology adoption paradox continues to frustrate many observers and decision-makers: with so much invested in agricultural R&D, why are the rates of improved variety adoption, variety turnover, and seed replacement so low and variable across crops and countries in the Global South? The answer can be found in the specificities of crops, people, and places, though that should not prevent investment in generating evidence-based recommendations to affect policy. Key areas for future consideration include building a stronger global consensus around the role of science and technology in our food systems, applying data-driven market intelligence to prioritize breeding, and advancing dramatic policy, regulatory, and investment reforms to reshape the performance of crop improvement programs and seed sector development efforts.

References

Abate, G.T., T. Bernard, A. de Brauw, and N. Minot. 2018. "The Impact of the Use of New Technologies on Farmers' Wheat Yield in Ethiopia: Evidence from a Randomized Control Trial." *Agricultural Economics* 49 (4): 409–421.

Abate, G.T., T. Bernard, S. Makhija, and D.J. Spielman. 2023. "Accelerating Technical Change Through ICT: Evidence from a Video-Mediated Extension Experiment in Ethiopia." *World Development* 161: 106089.

Adesina, A.A., and M.M. Zinnah. 1993. "Technology Characteristics, Farmers' Perceptions and Adoption Decisions: A Tobit Model Application in Sierra Leone." *Agricultural Economics* 9 (4): 297–311.

Ahmed, A.U., J. Hoddinott, N. Abedin, and N.Hossain. 2021. "The Impacts of GM Foods: Results from a Randomized Controlled Trial of Bt Eggplant in Bangladesh." *American Journal of Agricultural Economics* 103 (4): 1186–1206.

Alston, J.M., G.W. Norton, and P.G. Pardey. 1995. *Science Under Scarcity*. Ithaca, NY: Cornell University Press.

Alston, J.M., and P.G. Pardey. 2021. "The Economics of Agricultural Innovation," *Handbook of Agricultural Economics* 5: 3895–3980.

Alston, J.M., P.G. Pardey, and X. Rao. 2022. "Payoffs to a Half Century of CGIAR Research." *American Journal of Agricultural Economics* 104 (2): 502–529.

Arouna, A., J.D. Michler, and J.C. Lokossou. 2021. "Contract Farming and Rural Transformation: Evidence from a Field Experiment in Benin." *Journal of Development Economics* 151: 102626.

Ashby, J.A., and V. Polar. 2019. "The Implications of Gender Relations for Modern Approaches to Crop Improvement and Plant Breeding." In *Gender, Agriculture and Agrarian Transformations*, ed. C.E. Sachs, 11–34. London: Routledge.

Bagamba, F., P.R. Ntakyo, G. Otim, D.J. Spielman, and B. Van Campenhout. 2023. "Policy and Performance in Uganda's Seed Sector: Opportunities and Challenges." *Development Policy Review* 41 (3): 12665.

Barrett, C.B., T. Benton, K.A. Cooper et al. 2020. "Bundling Innovations to Transform Agri-Food Systems." *Nature Sustainability* 3 (12): 974–976.

Barrett, C.B., T.G. Benton, J. Fanzo, et al. 2022. *Socio-Technical Innovation Bundles for Agri-Food Systems Transformation*. Cham, Switzerland: Springer Nature.

Barrett, C.B., and M.R. Carter. 2010. "The Power and Pitfalls of Experiments in Development Economics: Some Non-Random Reflections." *Applied Economic Perspectives and Policy* 32 (4): 515–548.

Barrett, C.B., and M.R. Carter. 2020. "Finding Our Balance? Revisiting the Randomization Revolution in Development Economics Ten Years Further On." *World Development* 127: 104789.

Bernard, T., S. Lambert, K. Macours, and M. Vinez. 2023. "Impact of Small Farmers' Access to Improved Seeds and Deforestation in DR Congo." *Nature Communications* 14 (1): 1603.

Bishaw, Z., Y. A.Yigezu, and G. Berhane. 2025. *Where Do We Stand After Ten Years of Harmonized Seed Trade Regulations in COMESA? Case Studies from Egypt and Sudan.* Policy Brief. CGIAR Initiative on Fragility to Resilience in Central and West Asia and North Africa. Montpelier: CGIAR.

Bold, T., K.C. Kaizzi, J. Svensson, and D. Yanagizawa-Drott. 2017. "Lemon Technologies and Adoption: Measurement, Theory and Evidence from Agricultural Markets in Uganda." *The Quarterly Journal of Economics* 132 (3): 1055–1100.

Bold, T., S. Ghisolfi, F. Nsonzi, and J. Svensson. 2022. "Market Access and Quality Upgrading: Evidence from Four Field Experiments." *American Economic Review* 112 (8): 2518–2552.

Boucher, S.R., M.R. Carter, J.E. Flatnes, T.J. Lybbert, J.G. Malacarne, P.P. Mareyna, and L.A. Paul. 2024. "Bundling Genetic and Financial Technologies for More Resilient and Productive Small-Scale Farmers in Africa." *The Economic Journal* 134 (662): 2321–2350.

Bouis, H., J. Foley, K. Lividini, et al. 2024. "Biofortification: Future Challenges for a Newly Emerging Technology to Improve Nutrition Security Sustainably." *Current Developments in Nutrition* 8 (12): 104478.

Brennan, J.P., and D. Byerlee. 1991. "The Rate of Crop Varietal Replacement on Farms: Measures and Empirical Results for Wheat." *Plant Varieties and Seeds* 4: 99–106.

Chowdhury, S., J.V. Meenakshi, K.I. Tomlins, and C. Owori . 2011. "Are Consumers in Developing Countries Willing to Pay More for Micronutrient-Dense Biofortified Foods? Evidence from a Field Experiment in Uganda." *American Journal of Agricultural Economics* 93 (1): 83–97.

Cullen, B., A. Snyder, D. Rubin, and H.A. Tufan. 2023. "'They Think We Are Delaying Their Outputs.' The Challenges of Interdisciplinary Research: Understanding Power Dynamics Between Social and Biophysical Scientists in International Crop Breeding Teams." *Frontiers in Sustainable Food Systems* 7: 1250709.

De Brauw, A., P. Eozenou, D.O. Gilligan, C. Hotz, N. Kumar, and J.V. Meenakshi. 2018. "Biofortification, Crop Adoption and Health Information: Impact Pathways in Mozambique and Uganda." *American Journal of Agricultural Economics* 100 (3): 906–930.

De Brauw, A., M. Moursi, and A.B. Munhaua. 2019. "Vitamin A Intakes Remain Higher Among Intervention Participants 3 Years After a Biofortification Intervention in Mozambique." *British Journal of Nutrition* 122 (1): 1175–1181.

Donovan, J., P. Rutsaert, D.J. Spielman, K.M. Shikuku, and M. Demont. 2021. "Seed Value Chain Development in the Global South: Key Issues and New Directions for Public Breeding Programs." *Outlook on Agriculture* 50 (4): 366–377.

Emerick, K., A. De Janvry, E. Sadoulet, and M.H. Dar. 2016. "Technological Innovations, Downside Risk, and the Modernization of Agriculture." *American Economic Review* 106 (6): 1537–1561.

Evenson, R.E., and D. Gollin, eds. 2003. *Crop Variety Improvement and its Effect on Productivity: The Impact of International Agricultural Research.* Wallingford, UK: CAB International.

Evenson, R.E., and M. Rosegrant. 2003. "The Economic Consequences of Crop Genetic Improvement Programmes." In *Crop Variety Improvement and Its Effect on Productivity: The Impact of International Agricultural Research*, eds. R. E. Evenson, D. Gollin, 473–497. Wallingford, UK: CAB International.

Fabregas, R., M. Kremer, M. Lowes, R. On, and G. Zane. 2025. "Digital Information Provision and Behavior Change: Lessons from Six Experiments in East Africa." *American Economic Journal: Applied Economics* 17 (1): 527–566.

Fabregas, R., M. Kremer, and F. Schilbach. 2019. "Realizing the Potential of Digital Development: The Case of Agricultural Advice." *Science* 366 (6471): 3038.

Falck-Zepeda, J.B., and P. Zambrano. 2011. "Socio-Economic Considerations in Biosafety and Biotechnology Decision Making: The Cartagena Protocol and National Biosafety Frameworks." *Review of Policy Research* 28 (2): 171–195.

Fan, S., A. Gulati, and S. Thorat. 2008. "Investment, Subsidies, and Pro-Poor Growth in Rural India." *Agricultural Economics* 39 (2):163–170.

Fan, S., and P.B.R. Hazell. 2001. "Returns to Public Investments in the Less-Favored Areas of India and China." *American Journal of Agricultural Economics* 83 (5): 1217–1222.

Fan, S., P.B.R. Hazell, and S. Thorat. 2000."Government Spending, Growth and Poverty in Rural India." *American Journal of Agricultural Economics* 82 (4): 1038–1051.

Feder, G., and D.L. Umali. 1993. "The Adoption of Agricultural Innovations: A Review." *Technological Forecasting and Social Change* 43 (3–4): 215–239.

Feder, G., R.E. Just, and D. Zilberman. 1985. Adoption of Agricultural Innovations in Developing Countries: A Survey," *Economic Development and Cultural Change* 33 (2): 255–298.

Foster, A.D., and M.R. Rosenzweig. 1995. "Learning by Doing and Learning from Others: Human Capital and Technical Change in Agriculture." *Journal of Political Economy* 103 (6): 1176–1209.

Fuglie, K.O., and R.G. Echeverria. 2024. "The Economic Impact of CGIAR-Related Crop Technologies on Agricultural Productivity in Developing Countries, 1961–2020." *World Development* 176 (2024): 106523.

Galiè, A., B. Kramer, D.J. Spielman, N. Kawarazuka, A.M. Rietveld, and S. Aju. 2025. "Inclusive and Gender-Transformative Seed Systems: Concepts and Applications." *Agricultural Systems* 226: 104320.

Gars, J., and P.S. Ward. 2019. "Can Differences in Individual Learning Explain Patterns of Technology Adoption? Evidence on Heterogeneous Learning Patterns and Hybrid Rice Adoption in Bihar, India." *World Development* 115 : 178–189.

Gollin, D., W. Hansen, and A.M. Wingender. 2021. "Two Blades of Grass: The Impact of the Green Revolution," *Journal of Political Economy* 129 (8): 2344–2384.

Griliches, Z. 1957. "Hybrid Corn: An Exploration in the Economics of Technological Change." *Econometrica* 25 (4): 501–522.

Griliches, Z. 1964. "Research Expenditures, Education and the Aggregate Agricultural Production Function." *American Economic Review* 54 (6): 961–974.

Halewood, M., I.L. Noriega, and S. Louafi. 2012. *Crop Genetic Resources as a Global Commons: Challenges in International Law and Governance.* Routledge.

Hanna, R., S. Mullainathan, and J. Schwartzstein. 2014. "Learning through Noticing: Theory and Evidence from a Field Experiment." *The Quarterly Journal of Economics* 129 (3): 1311–1353.

Hayami, Y., and V.W. Ruttan. 1971. *Agricultural Development: An International Perspective.* Baltimore: Johns Hopkins University Press.

Hazell, P.B.R, and C. Ramasamy, eds. 1991. *The Green Revolution Reconsidered: The Impact of High-Yielding Rice Varieties in South India.* Baltimore: Johns Hopkins University Press for IFPRI.

Heisey, P.W., and J. Brennan. 1991. "An Analytical Model of Farmers' Demand for Replacement Seed." *American Journal of Agricultural Economics* 73 (4): 1044–1052.

Hsu, E., and A. Wambugu. 2022. "Can Informed Buyers Improve Goods Quality? Experimental Evidence from Crop Seeds." G^2LM|LIC Working Paper 74, Gender, Growth, and Labour Markets in Low Income Countries Programme, IZA Institute of Labor Economics, Bonn, Germany.

Kilonzi, E., B. Kramer, and C. Trachtman. 2025. "Workshop: Assessing On-the-Ground Impacts of Genetic Innovation in Agriculture." *IFPRI Blog*, January 15.

Koo, B., P.G. Pardey, and B.D. Wright. 2003. "The Economic Costs of Conserving Genetic Resources at the CGIAR Centers." *Agricultural Economics* 29 (3): 287–297.

Kramer, B., S. Pattnaik, P.S. Ward, and Y. Xu. 2024. *Impacts of an Innovative Credit+ Insurance Bundle for Marginalized Farmers: Evidence from a Cluster Randomized Trial in Odisha, India.* Washington, DC: IFPRI.

Kramer, B., and C. Trachtman. 2024. "Gender Dynamics in Seed Systems: An Integrative Review of Seed Promotion Interventions in Africa." *Food Security* 16 (1): 19–45.

Kramer, B., C. Trachtman, and L. Zuze. 2023. "Information as a Source of Empowerment: The Role of Climate Information Services," AICCRA Working Paper 17, Accelerating Impacts of CGIAR Climate Research for Africa, CGIAR, Montpellier, France.

Kramer, B., C. Waweru, and J.G. Malacarne. 2023. "Summing the Parts: How Does 'Bundling' Affect Willingness-to-Pay for Seeds and Insurance in a Sample of Kenyan Farmers?" Conference Paper, IFPRI, Washington, DC.

Kuhlmann, K. 2021. "Mapping Inclusive Law and Regulation: A Comparative Agenda for Trade and development." *African Journal of International Economic Law* 2: 48–87.

Leathers, H.D., and M. Smale. 1991. "A Bayesian Approach to Explaining Sequential Adoption of Components of a Technological Package." *American Journal of Agricultural Economics* 73 (3): 734–742.

Lecoutere, E., D.J. Spielman, and B. Van Campenhout. 2023. "Empowering Women Through Targeting Information or Role Models: Evidence from an Experiment in Agricultural Extension in Uganda." *World Development* 167 (2023): 106240.

Lipper, L., C.L. Anderson, and T.J. Dalton, eds. 2010. *Seed Trade in Rural Markets Implications for Crop Diversity and Agricultural Development*. London: FAO and Earthscan.

Lipton, M., and R. Longhurst. 1989. *New Seeds and Poor People*. London: Routledge.

Lybbert, T.J., and M.R. Carter. 2014. "Bundling Drought Tolerance and Index Insurance to Reduce Rural Household Vulnerability." In *Sustainable Economic Development: Resources, Environment, and Institutions*, eds. A. Balisacan, U. Chakrovarty, and M. Ravago, 401–414. Oxford: Academic Press.

Mason, N.M., and J. Ricker-Gilbert. 2013. "Disrupting Demand for Commercial Seed: Input Subsidies in Malawi and Zambia." *World Development* 45: 75–91.

Mason, N.M., and M. Smale. 2013. "Impacts of Subsidized Hybrid Seed on Indicators of Economic Well-Being Among Smallholder Maize Growers in Zambia." *Agricultural Economics* 44 (6): 659–670.

Matita, M., L. Chiwaula, E.W. Chirwa, J. Mazalale, and H. Walls. "Subsidizing Improved Legume Seeds for Increased Household Dietary Diversity: Evidence from Malawi's Farm Input Subsidy Programme with Implications for Addressing Malnutrition in All Its Forms." *Food Policy* 113 (2022): 102309.

Mekonnen, D.K., and D.J. Spielman. 2021. :Changing Patterns in Genebank Acquisitions of Crop Genetic Materials: An Analysis of Global Policy Drivers and Potential Consequences." *Food Policy* 105: 102161.

Michler, J.D., and S.Y. Wu,. 2020. "Relational Contracts in Agriculture: Theory and Evidence." *Annual Review of Resource Economics* 12 (1): 111–127.

Miehe, C., R. Sparrow, D.J. Spielman, and B. Van Campenhout. 2023. "The (Perceived) Quality of Agricultural Technology and Its Adoption: Experimental Evidence from Uganda," IFPRI Discussion Paper 02918, IFPRI, Washington, DC.

Miehe, C., B. Van Campenhout, L. Nabwire, R. Sparrow, and D.J. Spielman. 2023. "Miracle Seeds: Biased Expectations, Complementary Input Use, and the Dynamics of Smallholder Technology Adoption," IFPRI Discussion Paper 02183, IFPRI, Washington, DC.

Munyi, P., and B. De Jonge. 2015. "Seed Systems Support in Kenya: Consideration for an Integrated Seed Sector Development Approach." *Journal of Sustainable Development* 8 (2): 161–173.

Omamo, S.W., and K. von Grebmer, eds. 2005. *Biotechnology, Agriculture, and Food Security in Southern Africa*. Washington, DC: IFPRI.

Oparinde, A., A. Banerji, E. Birol, and P. Ilona. 2016. "Information and Consumer Willingness to Pay for Biofortified Yellow Cassava: Evidence from Experimental Auctions in Nigeria." *Agricultural Economics* 47 (2): 215–233.

Pixley, K.V., J.B. Falck-Zepeda, K.E. Giller, et al. 2019. "Genome Editing, Gene Drives, and Synthetic Biology: Will They Contribute to Disease-Resistant Crops, and Who Will Benefit? *Annual Review of Phytopathology* 57 (1): 165–188.

Pixley, K.V., J.B. Falck-Zepeda, R.L. Paarlberg, et al. 2022. "Genome-Edited Crops for Improved Food Security of Smallholder Farmers." *Nature Genetics* 54 (4): 364–367.

Qaim, M. 2016. *Genetically Modified Crops and Agricultural Development*. London: Palgrave Macmillan.

Ray, D.K., N. Ramankutty, N.D. Mueller, P.C. West, and J.A. Foley. 2012. "Recent Patterns of Crop Yield Growth and Stagnation," *Nature Communications* 3 (1): 1293.

Reardon, T., and C.P. Timmer. 2012. "The Economics of the Food System Revolution." *Annual Review of Resource Economics* 4 (1): 225–264.

Rice, B., B. Kramer, and C. Trachtman. 2024. *Barriers to Using Market Intelligence in Plant Breeding: Evidence from a Survey of Breeding Professionals*. Market Intelligence Brief Series no. 21. Montpellier: CGIAR.

Rogers, E.M. 1962. *Diffusion of Innovations*. New York: Free Press.

Rosegrant, M.W., B. Wong, T.B. Sulser, N. Dubosse, and T.J. Lybbert. 2023. "Benefit–Cost Analysis of Increased Funding for Agricultural Research and Development in the Global South." *Journal of Benefit-Cost Analysis* 14 (S1): 181–205.

Rosenzweig, M.R., and H.P. Binswanger. 1993. "Wealth, Weather Risk and the Composition and Profitability of Agricultural Investments." *Economic Journal* 103 (416): 56–78.

Rosenzweig, M.R., and C. Udry. 2020. "External Validity in a Stochastic World: Evidence from Low-Income Countries." *Review of Economic Studies* 87 (1): 343–381.

Sheahan, M., and C.B. Barrett. 2017. "Ten Striking Facts about Agricultural Input Use in Sub-Saharan Africa." *Food Policy* 67: 12–25.

Smale, M., ed. 2005. *Valuing Crop Biodiversity: On-Farm Genetic Resources and Economic Change.* Wallingford, UK: CABI Publishing.

Smale, M., A. Assima, A. Kergna, V. Thériault, and E. Weltzien. 2018. "Farm Family Effects of Adopting Improved and Hybrid Sorghum Seed in the Sudan Savanna of West Africa." *Food Policy* 74: 162–171.

Smale, M., P.W. Heisey, and H.D. Leathers. 1995. "Maize of the Ancestors and Modern Varieties: The Microeconomics of High-Yielding Variety Adoption in Malawi." *Economic Development and Cultural Change* 43 (2): 351–368.

Smale, M., P. Zambrano, G. Gruère, et al. 2009. *Measuring the Economic Impacts of Transgenic Crops in Developing Agriculture during the First Decade: Approaches, Findings, and Future Directions.* Washington, DC: IFPRI.

Spielman, D.J., and M. Smale. 2017. "Policy Options to Accelerate Varietal Change among Smallholder Farmers in South Asia and Africa South of the Sahara." IFPRI Discussion Paper 1666. Washington, DC: IFPRI.

Stevenson, J.R., K. Macours, and D. Gollin. 2023. "The Rigor Revolution: New Standards of Evidence for Impact Assessment of International Agricultural Research." *Annual Review of Resource Economics* 15 (1): 495–515.

Suri, T. 2011. "Selection and Comparative Advantage In Technology Adoption." *Econometrica* 79 (1): 159–209.

Suri, T., and C. Udry. 2022. "Agricultural Technology in Africa." *Journal of Economic Perspectives* 36 (1): 33–56.

Taylor, J.E., and Filipski, M.J. 2014. *Beyond Experiments in Development Economics: Local Economy-wide Impact Evaluation.* Oxford: Oxford University Press.

ter Steeg, E.M.S., and Louwaars, N.P., 2024. "Market-Driven Transitions in the Vegetable Seed Sector in Sub-Saharan Africa," *Agricultural Systems* 221: 104113.

Trachtman, C., B. Kramer, and J. do Nascimento. 2025. "Market Information and R&D investment under Ambiguity: A Framed Artefactual Experiment with Plant Breeding Professionals." IFPRI Discussion Paper 2314. IFPRI, Washington, DC.

Tripp, R., and N. Louwaars. 1997. "Seed Regulation: Choices on the Road to Reform." *Food Policy* 22 (5): 433–446.

Van Campenhout, B. 2021. "The Role of Information in Agricultural Technology Adoption: Experimental Evidence from Rice Farmers in Uganda." *Economic Development and Cultural Change* 69 (3): 1239–1272.

Van Campenhout, B., D.J. Spielman, and E. Lecoutere. 2021. "Information and Communication Technologies to Provide Agricultural Advice to Smallholder Farmers: Experimental Evidence from Uganda." *American Journal of Agricultural Economics* 103 (1): 317–337.

Ward, P.S., S. Makhija, and D.J. Spielman. 2020. "Drought-Tolerant Rice, Weather Index Insurance, and Comprehensive Risk Management for Smallholders: Evidence from a Multi-Year Field Experiment in India." *Australian Journal of Agricultural and Resource Economics* 64 (2): 421–454.

Wellard, K., J. Osando, H. Odame, et al., 2019. *End of Program Evaluation for AGRA Africa's Seed Systems Programs Evaluation: Evaluation Report.* Nairobi: AGRA.

Weltzien, E., F. Rattunde, A. Christinck, K. Isaacs, and J. Ashby. 2019. "Gender and Farmer Preferences for Varietal Traits: Evidence and Issues for Crop Improvement." *Plant Breeding Reviews* 43: 243–278.

Wiebe, K., T.B. Sulser, S. Dunston, M.W. Rosegrant, K. Fuglie, D. Willenbockel, and G.C. Nelson. 2021. "Modeling Impacts of Faster Productivity Growth to Inform the CGIAR Initiative on Crops to End Hunger." *PLoS One* 16 (4): e0249994.

Wollburg, P., T. Bentze, Y. Lu, C. Udry, and D. Gollin. 2024. "Crop Yields Fail to Rise in Smallholder Farming Systems in Sub-Saharan Africa." *PNAS* 121 (21): e2312519121.

Chapter 10

AGRICULTURAL INSURANCE
Policies and Programs for Reducing Farmer Risk

Francisco Ceballos, Peter Hazell, Ruth Hill, and Berber Kramer

Francisco Ceballos is a research fellow, Markets, Trade, and Institutions Unit (MTI), IFPRI; **Peter Hazell** is a research fellow emeritus, IFPRI; **Ruth Hill** is director, MTI, IFPRI; and **Berber Kramer** is a senior research fellow, MTI, IFPRI.

Key messages

- Farm households face numerous risks that can discourage investments and trap them in poverty. Insurance should be a useful tool to reduce this risk, but agricultural risks are inherently difficult to insure against, and farmers' insurance preferences and capacity to pay are diverse.
- Many governments provided subsidized multi-peril crop insurance in the 1970s, but these programs were largely abandoned given their high cost for governments, the inappropriate incentives they created for farmers and lenders, and in many cases because they had also become corrupt and subject to political interference.
- Index-based insurance, based on rainfall or weather measurements at local weather stations, remote sensing, or area-yield assessments, was pioneered in the 1990s. Although these products showed promise, demand was low, especially among smallholders.
- Research identified many factors associated with low demand, including affordability, gender, limited education, and lack of trust; but most prominent was the finding that basis risk—the difference between insurance payouts and actual losses—is likely to undermine uptake.
- Evidence shows that when farmers are confident in being covered by insurance, they substantially increase farm investments, including in higher-return activities.
- Recent focus has been on designing and testing new insurance products that are potentially more attractive to farmers. Many rely on innovative

technologies to reduce basis risk, while others are testing "bundles" of products, insurance education, and easier payment schedules to increase attractiveness. Nevertheless, these products continue to leave many vulnerable farmers without insurance.

Looking to the future, research on agricultural insurance should:

- **Enhance the value of insurance for farmers**, recognizing that it may only be possible to insure part of the risk farmers face with market instruments alone. This may require introducing more holistic insurance instruments that tackle multiple risks at once, designing optimal bundles of insurance with other products and technologies (ideally risk-reducing ones such as improved seeds or seasonal weather forecasting), or designing insurance that complements existing ways of managing risk (such as informal transfers from family and friends during crisis times).
- **Integrate insurance with agricultural digital tools** to increase access for underserved populations, reduce costs, and contribute to the scaling of insurance products. Use of artifical intelligence, for instance through image recognition or large language models, can help improve index accuracy, insurance understanding, and client-insurer interactions.
- **Better integrate the design and development of agricultural insurance** with alternative risk management options available to farm households and with other forms of public assistance, such as safety net programs and emergency relief.
- **Provide empirical cost-benefit analyses of different types of insurance subsidies** to better inform policy choices. Research should analyze different insurance models and compare them to alternative ways of reducing and managing risk.

Farm households face a host of market and production risks that can threaten their livelihoods and food security, lead to loss of capital assets and reduced ability to repay debt and, over time, trap them in poverty (Carter and Barrett 2006; Dercon and Hoddinott 2004). Moreover, since new technologies or farming practices that could increase long-term farm productivity and household welfare are often more risky than established technologies and practices, farmers can be discouraged from adopting or investing in them. In principle, insurance against production risks should be a useful aid for farm households by transferring part of their risk to the insurance market, but unfortunately many agricultural risks are inherently difficult to insure against.

The insurance supply problem

Insurance works best for risks that are known (allowing it to be appropriately priced), unlikely to hit all people insured at once (allowing risks to be pooled), and which cannot be influenced by those insured through neglectful or deliberate mismanagement (the so-called *moral hazard* problem). Life, medical, and auto insurance typically meet these requirements and are widely available from private insurers, but apart from some specific perils like hail, most agricultural production and market risks do not meet them. For instance, data for pricing risks (time-series data on individual farm or plot-level yields) are mostly unavailable, and climate change is increasing the frequency and severity of future production losses in ways that are difficult to predict. At the same time, hazards that impact crop yields and livestock mortality tend to affect many farmers in one location simultaneously (that is, the risk is *covariate*), making it difficult to pool risks. Market or price risks pose a similar covariate problem, worsened by the fact that they are affected by political and economic forces that are inherently difficult to predict for insurance purposes. Additionally, assessing agricultural losses can be cost-prohibitive due to short timeframes around harvest and the remoteness of many farms in low- and middle-income countries (LMICs). Such constraints can also reduce the ability of insurers to monitor farmers' management practices to limit moral hazard problems.

The diversity of farmers' insurance needs also makes agricultural insurance markets challenging to sustain. Differences in farmers' risk exposure require insurers to pool farmers facing similar risks into groups who can be offered the same contracts. However, for any given contract, only those farmers facing average or higher than average levels of risk tend to purchase the insurance while lower risk farmers opt out (the *adverse selection* problem). The insurer must then raise the premium rate to reflect the higher average risk level of the remaining pool of insured farmers, which in turn can lead to more lower-risk farmers opting out of the pool. The subsequent downward spiral can eventually undermine the financial viability of the insurance contract.

For these reasons, private agricultural insurance products that have reached scale without any subsidies have been largely limited to specific perils like hail and frost (Hazell and Timu 2024). The few attempts to provide comprehensive coverage for crop and livestock losses on a commercial basis either failed or only reached small niche markets, such as insuring commercial farmers for high-value products or tree crops.

The absence of adequate insurance against agricultural losses has long concerned governments, and increasingly so as climate change increases both the likelihood and severity of climate-related losses (see Box 10.1 on rates of

BOX 10.1 Regional distribution of insured crop farms, coverage, and vulnerable households (percent)

Region	Insured farms 2020	Premiums collected 2020	Value of insurance coverage 2020	Insured farms as share of all farms 2015	Total premium collected as share of agricultural GDP 2017	Share of total population vulnerable and exposed to natural hazards 2021
Asia	98.1	92.7	96.8	61.7	0.60	20.4
China & India	95.0	91.2	96.3	74.9		
Other	3.1	1.5	0.5	9.7		
Africa	1.5	0.2	0.1	7.6	0.08	36.1
LAC	0.4	7.1	3.1	4.8	0.45	15.6
Total	100.0	100.0	100.0	53.5		

Source: Calculated from GIZ (2021), Swiss Re (2019), Lowder et al. (2016), and Hill et al. (2024).

Note: These are approximate figures, as there is variation in the years for which data are available both between and within columns, and not all the underlying data sources provide full country representation.

Across Asia, Africa, and Latin America and the Caribbean, 53.5 percent of all farms are insured, but 95 percent of all the insured farms are located in China and India, where they benefit from heavily subsidized national programs. Elsewhere, very few farms are insured and insurance covers just a tiny fraction of farmers' exposure, as indicated by very low ratios of premiums collected to agricultural GDP, and very large shares of the population remain exposed to natural hazards with little capacity to cope with these threats.

coverage). This concern has led to many attempts both to provide publicly supported insurance programs to protect farmers against production losses and to intervene in agricultural markets to help reduce price risks for farmers.

We limit our review to the development of agricultural insurance programs and to recent innovations that are helping to make insurance more commercially viable, while also flagging IFPRI's contributions to these developments.

Initial government responses

Given the limited development of private agricultural insurance markets, many governments have sought to support farmers by offering multi-peril crop insurance (MPCI) programs, which provide comprehensive coverage against a wide array of production risks on an indemnity basis. One of the earliest of these was the Federal Crop Insurance Corporation program established in the United States in 1938. By the 1970s, there were many such programs around the world, which often tied the insurance to credit by underwriting farm loans provided by state-owned agricultural development banks. Given the high cost of providing such broad coverage, substantial subsidies were needed to make insurance premiums affordable to most farmers and achieve the desired scale of coverage, and by the late 1970s, several billion dollars were spent globally each year on public subsidies for MPCI programs.

To assess the value of MPCI programs, IFPRI and the Inter-American Institute for Cooperation on Agriculture (IICA) cosponsored a set of case studies of insurance programs in Australia, Brazil, Japan, Mexico, Panama, and the United States, and organized an international conference on crop insurance and agricultural credit in 1982 (Hazell et al. 1986; Hazell and Valdés 1984). They concluded that the MPCI programs were very expensive for governments yet failed to achieve their primary purposes. These programs faced serious moral hazard problems, encouraged farmers to take excessive risks, undermined good insurance and bank lending practices, and in many cases had become corrupt and subject to political interference. The findings attracted the attention of several national and international aid agencies and contributed to the phasing out or revamping of most MPCI programs by the end of the century.

The need for effective agricultural insurance remained, and IFPRI, among others, began research in the early 1990s on the potential of index-based insurance (IBI) products to fill the gap. By basing insurance payouts on an objectively measured index for a geographic area, such as readings from a local weather station, satellite imagery, or average crop yields in the area (area-yield insurance), IBI should reduce the cost of providing insurance by eliminating the need for assessing farm-level losses and reducing incentives for engaging in moral hazard (since payouts are determined by an index beyond the farmer's control, instead of farm-level losses). Several countries, including India, had already established area-yield insurance schemes (Dandekar 1976), and new possibilities were seen for rainfall-based indexes using local weather station data (Miranda 1991). A growing literature, including research conducted by IFPRI together with the World Bank and several universities on the feasibility of weather-based IBI, saw sufficient potential to warrant its further development (for example, Hazell 1992; Miranda 1991; Skees et al. 1999.) Several weather-based IBI programs were pioneered around the world in the late 1990s, and many more were launched soon after, especially once remote-sensing data and crop modeling techniques became available for constructing new types of indexes.[1] But unless heavily subsidized (as in India and the United States), these schemes rarely achieved scale. The overall low demand for these products led to additional research and pilot programs to better understand the barriers holding back IBI.

The farmer demand problem

A key question was understanding low farmer demand: was the price of commercially provided insurance just too high (although heralded as being cheaper than conventional MPCI, IBI premiums were often found to be well above actuarially fair levels), was the quality of IBI products just too poor (as a result of *basis risk*), or were farmers constrained from adopting IBI due to other reasons, such as limited liquidity, low levels of financial literacy and education, or lack of trust in the product and insurance providers (Binswanger-Mkhize 2012; Hazell and Timu 2021)?

Basis risk emerged as a key challenge that could undermine uptake of IBI even when offered at affordable premiums. For some of the early products, the payouts determined by the index were very poorly correlated with farmers' losses (Clarke 2016). Moreover, studies in Bangladesh (Clarke et al. 2016; Hill et al. 2019) and India (Hill et al. 2016) confirmed previous findings in the literature[2] that farmers' demand for insurance was fairly price inelastic (that is, not responsive to price changes) and, in some cases, sufficiently inelastic that premium subsidies had limited ability to raise demand. Together, these studies led to considerable work on how to improve the indexes (Carter 2009) and to an examination of the sensitivity of demand to basis risk. Ultimately, it became clear that further innovation was needed to develop affordable, high-quality insurance products that better fit farmers' needs.

The evidence produced by IFPRI and other researchers also pointed to other factors constraining potential demand. Household wealth was found to be positively associated with the demand for insurance (Clarke and Kumar 2016; Hill et al. 2013), and experimental evidence was consistent with liquidity constraints dampening demand (Casaburi and Willis 2018; Liu et al.

1 Good examples were pilot programs initiated by nongovernmental organizations, researchers, and charitable foundations. For example, Oxfam launched the Horn of Africa Risk Transfer for Adaptation (HARITA) initiative in Ethiopia; the Syngenta Foundation launched the Kilimo Salama project (later ACRE Africa) in Kenya; and ILRI led the launch of the Index Based Livestock Insurance in Kenya.

2 For example, see Cole et al. (2013), Karlan et al. (2014), and Mobarak and Rosenzweig (2013).

2020). The impact of farmers' risk aversion on uptake of agricultural insurance was found to be mixed; in some cases, findings aligned with theoretical predictions (Casaburi and Willis 2018; Liu et al. 2020) that demand would increase with low- to moderate-risk aversion but eventually decrease at higher levels of risk aversion, as basis risk made benefits from the insurance uncertain (Hill et al. 2016). Women farmers were in some cases shown to be less likely to buy insurance than men (Akter et al. 2016; Clarke and Kumar 2016; Delavallade et al. 2015). Demand could be fostered by intensive training and learning through experience (Cai and Song 2017; Hill et al. 2016, 2019). Trust was also found to be an important determinant of demand, with higher uptake from farmers who had prior experience with the insurer (Giné et al. 2008), who had received an endorsement from a trusted local source (Cole et al. 2013) or government agents (Akter et al. 2016), or who had received a payout themselves or witnessed others in their community receive a payout (Cole et al. 2014; Hill et al. 2016; Stein 2018).

Importantly, although many insurance pilots struggled with low levels of demand, several studies demonstrated the considerable potential of insurance to increase productivity and improve welfare: when farmers trusted they were insured, they undertook higher-risk but also higher-return activities, with up to a 30 percent increase in farm investments.[3] While quantifying the returns to increased input investment is not trivial, since these typically vary across settings and years (depending on weather conditions and prices), even assuming a relatively low average return to input use, the reported additional investments can lead to an average increase in income growth of 1 to 9 percent.

New types of insurance products

Motivated by the challenges of low demand, combined with the finding that transferring risk away from farmers through insurance could result in increased investments in agriculture, from about 2010 onward IFPRI and other researchers focused on designing and testing new types of financially viable insurance products with potential to be more attractive to farmers.

3 See Mobarak and Rosenzweig (2013) for rainfall-index insurance in India; Elabed and Carter (2015) for area-yield insurance in Mali; Karlan et al. (2014) for rainfall-index insurance in Ghana; Cai et al. (2015) for swine insurance in China; Cai (2016) for area-yield insurance in China; Fuchs and Wolff (2016) for rainfall-index insurance in Mexico; Jensen, Barrett, and Mude (2017) for livestock insurance in Kenya; Hill et al. (2019) for rainfall-index and area-yield insurance in Bangladesh; Bulte et al. (2020) for multi-peril crop insurance in Kenya; and Stoeffler et al. (2022) for area-yield insurance in Burkina Faso.

Remote-sensing products using higher spatial resolution satellite imagery—such as the European Space Agency's Sentinel suite—were developed to improve localized damage assessment for weather-based IBI (Black et al. 2016; Enenkel et al. 2018). In fact, some of the most successful index insurance products offered at scale use remote-sensing technology, such as vegetation index–based livestock insurance (Chantarat et al. 2013; Jensen et al. 2024) and rainfall index–based replanting guarantees. For example, ACRE Africa, a major agricultural insurance service provider in Africa, provides farmers with cash to buy new seed if rains fail after planting. Nevertheless, the capacity of these products to estimate localized damage remains limited—particularly when it comes to very small and fragmented fields in LMICs. They also face difficulties in reliably predicting actual plant or animal health from remotely sensed data, especially when identifying crop damage in intercropped fields or due to visibility issues from cloud cover. Thus, while useful in some contexts, satellite products were not able to fully resolve the issues with earlier products.

Area-yield indexes underpin the largest public crop insurance schemes in places where crop yield measurement for insured areas can be done cost-effectively, namely India (Clarke et al. 2016) and China (Cai et al. 2020). If agroecological zones are well defined and a sufficiently large number of fields can be sampled for crop yield measurement within an insured area, the basis risk associated with area-yield contracts can be much lower than in weather-based IBI. However, the costs and logistics of measuring crop yields at a large scale and on a timely basis make it challenging to provide area-yield insurance in many contexts. An innovation known as gap insurance or conditional audits attempts to retain the benefits of area-yield IBI while reducing costs by combining it with weather IBIs. Specifically, these insurance products add a layer of area-yield coverage to redress failures of a primary weather-based index product, reducing the scale at which crop yields need to be measured in a given season by focusing only on areas where the underlying weather index fell short and failed to trigger payouts for farmers who claim to have experienced crop losses. When the product was first tested in a small pilot in Ethiopia (Berhane et al. 2015; Flatnes et al. 2018) and later at a larger scale in Bangladesh by IFPRI researchers (Hill et al. 2019), demand proved to be quite high, although it was not clear whether this resulted from the product's design or from other contextual factors.

Geographic variability in growing conditions, however, is not the only obstacle that index products face. Data limitations and resource availability at the design phase typically result in indexes that cater to a hypothetical representative farmer but are too rigid to meet the needs of highly heterogeneous farmer populations. IFPRI researchers attempted to mitigate this issue by offering farmers a menu of smaller insurance products, or "insurance coupons," each optimized to cover a narrow region of risk (for example, a certain level of excess rainfall in a given calendar month), allowing farmers to purchase a combination or bundle of coupons best suited to their needs. Although trials in Ethiopia and Uruguay showed that farmers valued this innovation, demand remained relatively low (Ceballos and Robles 2020; Hill and Robles 2011).

An additional approach was to market index products to existing informal insurance networks, such as funeral groups or other social networks in which members help each other in times of financial hardship. These products rely on social transfers within the group to provide insurance for idiosyncratic risks that are likely experienced by only a few individual members at a time. These risks are expensive to insure through formal insurance because individual losses are difficult to monitor, but group members tend to have more information about one another than insurers do. Formal insurance, in turn, can focus on covering only covariate risks that are more likely to affect multiple members at once, and are therefore more difficult to cover through informal insurance arrangements (Dercon et al. 2014). Although this innovation did increase demand in an IFPRI pilot in Ethiopia, its large-scale applicability remains limited to settings with similar preexisting institutional arrangements. More recent work, which takes informal risk-sharing into account when designing and marketing formal insurance, caters to rural-urban migrants as a way for them to insure against poor weather conditions at home and has shown relatively high demand, indicating promise for further work in this area (Kazianga and Wahhaj 2020).

To better address basis risk at low cost, IFPRI researchers recently tested a picture-based insurance (PBI) model that relies on smartphone pictures of crop fields to assess damage remotely and at a low cost, thereby combining features of indemnity (field-level damage estimation) and index products (lower cost). Pictures of farmers' fields, taken by farmers themselves, were sufficient for identifying moderate and severe damage remotely and at a low cost across a number of crops, without widespread fraud (Ceballos et al. 2019). This approach was also shown to increase demand across studies in India, Kenya, and Ethiopia (Kramer, Pattnaik et al., 2024; Kramer, Waweru et al., 2024; Kramer, Porter, and Bizuayehu, 2023), albeit not always sufficiently to bring willingness to pay close to commercial premium levels (Ceballos and Kramer 2019). Impacts on productivity differed across research sites. In India, the innovation had no impacts on input use, agricultural practices, or technology adoption, whereas in Kenya, PBI improved fertilizer use; but mainly among male farmers, despite increasing insurance take-up mainly among female farmers. Importantly, the innovation seems to be valued by farmers and social enterprises that partnered in these impact evaluations, including ACRE Africa and Dvara E-Registry, which plan to add PBI to their portfolios of financial services. Moreover, PBI can be used either to assess crop damage directly or in combination with other approaches, such as to operationalize the conditional audits or crop yield measurement for farmers who claim to have experienced losses in the gap insurance approach described above.

Beyond basis risk, various studies have looked at other avenues for fostering farmer demand for agricultural insurance products. Recognizing the finding that agricultural risk imposes a higher burden on women (Dercon and Krishnan 2000; Hoddinott and Kinsey 2000; Quisumbing et al. 2018), innovations that consider gender roles, such as by reformulating insurance products to directly address the risks that women face (Arteaga et al. 2023; Hobbs 2022), also hold promise to generate demand (Timu and Kramer 2023). Relaxing farmers' liquidity constraints can boost uptake as well. For example, allowing farmers to defer premium payments to the end of the agricultural season translates into substantial increases in demand (Belissa et al. 2019). This approach works well in settings where insurance is purchased every year (as in the case of swine insurance in China, examined in Liu et al. 2020), which allows insurers to collect premiums for the previous season when the new coverage period begins, or in contract farming settings where premiums can be deducted from farmers' revenues at harvest (Casaburi and Willis 2018). While much of the agricultural insurance sold in the United States is paid for at the end of the insured period, the possibility of default may partly explain why such a model is still rare in LMICs.

Finally, insurance has been effectively bundled with a broad array of products and services, such as credit,[4] which helps address liquidity constraints, and with agricultural inputs, such as seeds of drought-tolerant varieties, which has been associated with larger impacts on demand (Boucher et al. 2024; Kramer, Waweru, and Malacarne 2023). These studies point to a general finding that tying insurance to other products or services and thereby insuring aspects of the production process (whether that is borrowing, input or irrigation costs, or livestock) appears more commercially promising than insuring crop production more generally.

4 Examples include Giné and Yang (2009); Kramer, Pattnaik et al. (2024); Meyer et al. (2017); Ndegwa et al. (2020); and Shee et al. (2019).

While many of the interventions described above have been successful in raising demand, providing agricultural insurance at scale is likely to require a combination of these. Unfortunately, there do not appear to be any studies of such multipronged approaches. Moreover, while some of the models above—such as bundling insurance with products and services—appear to boost demand considerably, they still leave a large fraction of agricultural production uninsured and often exclude vulnerable farmers who may not normally use the products and services insured.

Future directions

Significant advances have been made in understanding why farmers' demand for insurance remains low and in designing new products that address demand-side obstacles as well as some supply-side barriers. Despite these advances, agricultural insurance has not taken off at scale without substantial premium subsidies.

There are several promising lines of inquiry that may yet help to shift the paradigm. These include: (1) exploring new digital innovations beyond those discussed above (including artificial intelligence), (2) multidimensional studies of combinations of some of the individual interventions analyzed so far to capture possible synergies between them, (3) linking insurance with new risk-reducing technologies and weather forecasting, and (4) linking market-based insurance approaches with the informal risk-sharing networks on which many poor households rely, including remittances (Jack and Suri 2014). The diversity of farmers' insurance needs also deserves greater recognition, as publicly supported programs often take a one-size-fits-all approach in their quest for scale.

It is important to recognize, however, that insurance can only address certain types of production risks, and that these are often not the most important from the perspective of shortfalls in total farm or household income. It is also important to recognize that farmers have other ways of managing risk, including diversifying into nonfarm sources of income.[5] Other public interventions like social safety nets and disaster relief programs may also provide households with some protection against agricultural losses. Agricultural insurance is, at best, only one of the tools households can use for managing risk, and the

5 Nonfarm income now accounts for over half of total income for rural households in LMICs, and some studies have shown it can be a useful buffer against agricultural losses (Hazell and Timu 2024).

BOX 10.2 Role of risk management strategies and insurance by type of farm household and risk

Type of risk (severity of losses increases from top to bottom)	Type of household (capacity to manage risk increases from left to right)			
	Chronically poor	Nonpoor, subsistence-oriented small farms	Commercially oriented small farms	Medium and large commercial farms
	Risk management stragies and insurance			
Idiosyncratic	Income diversification and regular social protection transfers	Self-insurance (e.g., on-farm and off-farm diversification, credit and savings, membership in a local risk-pooling group), plus asset and personal insurance.		
Intermediate	As above, plus some forms of heavily subsidized insurance	As above, plus climate-smart farming; agricultural and credit insurance might help but may have to be subsidized from high to low across groups.		
Catastrophic	As above, plus disaster assistance	As above, plus disaster assistance and/or catastrophe insurance; catastrophe insurance may have to be subsidized from high to low across groups.		

Idiosyncratic risks comprise frequent but typically low-impact losses, such as modest yield and production losses at farm or plot level. **Intermediate risks**, such as frost, hail, or flood damage, involve less frequent but larger losses that may be correlated within groups of households. **Catastrophic risks** are infrequent events, such as a severe floods or drought, that have high and systemic impacts affecting many households within a region.

Chronically poor households are generally the most vulnerable to risk, with limited options for avoiding or coping with losses. Their primary challenge is protecting household consumption of food and meeting other basic needs, and they are a major target group for social protection programs (see Chapter 11). If insurance has a role for this group, it is probably to protect specific assets like livestock or savings and would need to be heavily subsidized. **Nonpoor, subsistence-oriented farms** have sufficient resources to avoid chronic poverty, but repeated losses can push them into poverty. Subsidized insurance against intermediate risks may be helpful for protecting their incomes and assets and may enable them to invest more in their farms and transition to commercial farming. **Commercially oriented small farms** might also benefit from insurance against intermediate risks, though more to protect against loss of specific agricultural investments than to stabilize household income or consumption. Because of their links to value chains, these farms may have access to insurance through intermediaries that is bundled with services such as credit and modern inputs. **Medium and large commercial farms** face the highest levels of production risk but also have the greatest capacity to self-manage risk. This group is the most viable clientele for private insurers either directly or through value chain intermediaries and may need little if any premium subsidy. In addition to the subsidies indicated above, subsidies may also be relevant for small-scale meso-insurance schemes that target institutions rather than individuals, for example counties, banks, cooperatives, or savings groups, and underwrite some of the covariate production and market risks. Subsidies may also be justified if insurance is to tackle some of the more catastrophic risks at affordable premium rates and reduce the need for disaster assistance payments.

Source: Adapted from Kramer et al. (2022).

design and development of insurance products need to be better integrated within a broader and more holistic risk-management perspective that considers all these alternatives.

As we seek to design better insurance products, it seems crucial to pay more attention to market and risk segmentation (see Box 10.2) (Kramer et al. 2022). For many smallholders, insurance is best seen as part of a holistic income-support approach, while commercial farmers have more options for managing risk and may benefit more from specific types of indemnity or IBI products to protect agricultural investments. The rationale for premium subsidies should distinguish between commercial and smallholder farmers. Subsidies for smallholders, especially those below or near the poverty line, might best be formally linked to safety net and public relief programs (see Chapter 11),[6] whereas for commercial farmers, subsidies should be directed at overcoming market development problems such as insufficient weather stations, helping to educate farmers about the value of insurance, and underwriting more extreme catastrophic events. Subsidies might also be targeted according to the type of risk being insured. For example, if subsidies were targeted to insurance against catastrophic risks (too expensive for most farmers to insure on their own), there might be more scope for private insurers to offer more affordable insurance for less covariate risks. To support government decision-making, research needs to provide empirical cost-benefit analyses of different types of insurance subsidies and of the cost-effectiveness of insurance

6 A good example is the Hunger Safety Net Program (HSNP) in Kenya, which supports poor households on a regular basis but is also empowered to scale up its support to a much wider range of households in the event of major droughts.

compared to alternative instruments for achieving the same goals of protecting incomes and promoting productivity.

Even within a well-balanced scheme of agricultural risk management with appropriate involvement of the public sector and access to financial tools to manage frequent risks, developing private insurance markets requires fostering several interventions at scale and a strong commitment from the private sector, something that has not yet been seen. Most pilot programs have been initiated and developed by researchers, donors, governments, or nongovernmental organizations, and only rarely by the private sector (Hazell et al. 2024). Successful agricultural insurance also requires strong public support and coordination, considering that programs that have successfully scaled up to date have involved heavy premium subsidies.

References

Akter, S., T.J. Krupnik, F. Rossi, and F. Khanam. 2016. "The Influence of Gender and Product Design on Farmers' Preferences for Weather-Indexed Crop Insurance." *Global Environmental Change* 38: 217–229.

Arteaga, J., M. Carter, and A. Hobbs. 2023. "Insuring Those Who Bear the Risk: The Impact of Gender-Inclusive Insurance in Kenya." NBER Working Paper 31639. National Bureau of Economic Research, Cambridge, MA, 2023.

Belissa, T., E. Bulte, F. Cecchi, S. Gangopadhyay, and R. Lensink. 2019. "Liquidity Constraints, Informal Institutions, and the Adoption of Weather Insurance: A Randomized Controlled Trial in Ethiopia." *Journal of Development Economics* 140: 269–278.

Berhane, G., S. Dercon, R. Hill, W. Bank, and A. Taffesse. 2015. "Formal and Informal Insurance: Experimental Evidence from Ethiopia." Paper presented at the International Association of Agricultural Economists Conference, Milan, Italy, August 8–14.

Binswanger-Mkhize, H.P. 2012. "Is There Too Much Hype about Index-Based Agricultural Insurance?" *Journal of Development Studies* 48 (2): 187–200.

Black, E., H. Greatrex, M. Young, and R. Maidment. 2016. "Incorporating Satellite Data into Weather Index Insurance." *Bulletin of the American Meteorological Society* 97 (10): 203–206.

Boucher, S.R., M.R. Carter, J.E. Flatnes, T.J. Lybbert, J.G. Malacarne, P.P. Mareyna, and L.A. Paul. 2024. "Bundling Genetic and Financial Technologies for More Resilient and Productive Small-Scale Farmers in Africa." *The Economic Journal* 134 (662): 2321–2350.

Bulte, E., F. Cecchi, R. Lensink, A. Marr, and M. Van Asseldonk. 2020. "Does Bundling Crop Insurance with Certified Seeds Crowd-In Investments? Experimental Evidence from Kenya." *Journal of Economic Behavior & Organization* 180: 744–757.

Cai, J. 2016. "The Impact of Insurance Provision on Household Production and Financial Decisions." *American Economic Journal: Economic Policy* 8 (2): 44–88.

Cai, J., Y. Chen, H. Fang, and L.A. Zhou. 2015. "The Effect of Microinsurance on Economic Activities: Evidence from a Randomized Field Experiment." *Review of Economics and Statistics* 97 (2): 287–300.

Cai, J., A. de Janvry, and E. Sadoulet. 2020. "Subsidy Policies and Insurance Demand." *American Economic Review* 110 (8): 2422–2453

Cai, J., and C. Song. 2017. "Do Disaster Experience and Knowledge Affect Insurance Take-Up Decisions?" *Journal of Development Economics* 124: 83–94.

Carter, M.R. 2009. *Intelligent Design of Index Insurance for Smallholder Farmers and Pastoralists.* 2020 Vision Focus Brief 17-06. Washington, DC: IFPRI.

Carter, M.R., and C.B. Barrett. 2006. "The Economics of Poverty Traps and Persistent Poverty: An Asset-Based Approach." *Journal of Development Studies* 42 (2): 178–199.

Casaburi, L., and J. Willis. 2018. "Time Versus State in Insurance: Experimental Evidence from Contract Farming in Kenya." *American Economic Review* 108 (12): 3778–3813.

Ceballos, F., and B. Kramer. 2019. "From Index to Indemnity Insurance Using Digital Technology: Demand for Picture-Based Crop Insurance." IFPRI Discussion Paper 01890. IFPRI, Washington, DC.

Ceballos, F., B. Kramer, and M. Robles. 2019. "The Feasibility of Picture-Based Insurance (PBI): Smartphone Pictures for Affordable Crop Insurance." *Development Engineering* 4: 100042.

Ceballos, F., and M. Robles. 2020. "Demand Heterogeneity for Index-Based Insurance: The Case for Flexible Products." *Journal of Development Economics* 146: 102515.

Chantarat, S., A.G. Mude, C.B. Barrett, and M.R. Carter. 2013. "Designing Index-Based Livestock Insurance for Managing Asset Risk in Northern Kenya." *Journal of Risk and Insurance* 80 (1): 205–237.

Clarke, D.J. 2016. "A Theory of Rational Demand for Index Insurance." *American Economic Journal: Microeconomics* 8 (1): 283–306.

Clarke, D.J., and N. Kumar. 2016. "Microinsurance Decisions: Gendered Evidence from Rural Bangladesh." *Gender, Technology and Development* 20 (2): 218–241.

Clarke, D.J., O. Mahul, K. Rao, and N. Verma. 2016. "Weather Based Crop Insurance in India." World Bank Policy Research Working Paper 5985. World Bank, Washington, DC.

Cole, S., X. Giné, J. Tobacman, P. Topalova, R. Townsend, and J. Vickery. 2013. "Barriers to Household Risk Management: Evidence from India." *American Economic Journal: Applied Economics* 5 (1): 104–135.

Cole, S., D. Stein, and J. Tobacman. 2014. "Dynamics of Demand for Index Insurance: Evidence from a Long-Run Field Experiment." *American Economic Review* 104 (5): 284–290.

Dandekar, V.M. 1976. "Crop Insurance in India." *Economic and Political Weekly* 11 (26): A61–A80.

Delavallade, C., F. Dizon, R.V. Hill, and J.P. Petraud. 2015. "Managing Risk with Insurance and Savings: Experimental Evidence for Male and Female Farm Managers in the Sahel." Policy Research Working Paper Series 7176. World Bank, Washington, DC.

Dercon, S., R.V. Hill, D. Clarke, I. Outes-Leon, and A.S. Taffesse. 2014. "Offering Rainfall Insurance to Informal Insurance Groups: Evidence from a Field Experiment in Ethiopia." *Journal of Development Economics* 106: 132–143.

Dercon, S., and J. Hoddinott. 2004. "Health, Shocks and Poverty Persistence." In *Insurance Against Poverty*, ed. S. Dercon, 124–136. Oxford, UK: Oxford University Press.

Dercon, S., and P. Krishnan. 2000. "In Sickness and in Health: Risk Sharing Within Households in Rural Ethiopia." *Journal of Political Economy* 108 (4): 688–727.

Elabed, G., and M. Carter. 2015. "Ex-Ante Impacts of Agricultural Insurance: Evidence from a Field Experiment in Mali." Unpublished paper.

Enenkel, M., C. Farah, C. Hain, et al. 2018. "What Rainfall Does Not Tell Us—Enhancing Financial Instruments with Satellite-Derived Soil Moisture and Evaporative Stress." *Remote Sensing* 10 (11): 1819. Flatnes, J.E., M.R. Carter, and R. Mercovich. 2018. "Improving the Quality of Index Insurance with a Satellite-Based Conditional Audit Contract." Feed the Future Innovation Lab for Markets, Risk and Resilience Working Paper. University of California, Davis.

Fuchs, A., and H. Wolff. 2016. "Drought and Retribution: Evidence from a Large-Scale Rainfall-Indexed Insurance Program in Mexico." World Bank Policy Research Working Paper 7565. World Bank, Washington, DC.

Giné, X., R. Townsend, and J. Vickery. 2008. "Patterns of Rainfall Insurance Participation in Rural India." *World Bank Economic Review* 22 (3): 539–566.

Giné, X., and D. Yang. 2009. "Insurance, Credit, and Technology Adoption: Field Experimental Evidence from Malawi." *Journal of Development Economics* 89 (1): 1–11.

GIZ (German International Cooperation Society). 2021. *Innovations and Emerging Trends in Agricultural Insurance for Smallholder Farmers: An Update*. Eschborn, Germany.

Gollier, C. 2024. "Enhanced Financial Services for Improved Climate Resilience: Some Insights from Economic Theory." *Financial Protection Forum Blog*. Financial Protection Forum, April 22.

Hazell, P. 1992. "The Appropriate Role of Agricultural Insurance in Developing Countries." *Journal of International Development* 4 (6): 567–581.

Hazell, P., C. Pomareda, and A. Valdés, eds. 1986. *Crop Insurance for Agricultural Development: Issues and Experience*. Baltimore: Johns Hopkins University Press.

Hazell, P., and A.G. Timu. 2024. "What's Holding Back Private Sector Agricultural Insurance?" IFPRI Discussion Paper 2316. IFPRI, Washington, DC.

Hazell, P., and A.G. Timu. 2021. "Review and Synthesis of IFPRI's PIM Funded Program of Work on Agricultural Insurance, 2012-2020." IFPRI Discussion Paper 2079. IFPRI, Washington, DC.

Hazell, P., and A. Valdés. 1984. "Choosing the Right Role for Crop Insurance." *Ceres* 17: 17–20.

Hill, R.V., N. Gorgulu, B. Brunckhorst, M. Nguyen, S. Hallegatti, and E. Naikal. 2024. "One-Fifth of the World's Population at High Risk from Climate-Related Hazards." *Nature Portfolio*, preprint.

Hill, R.V., J. Hoddinott, and N. Kumar. 2013. "Adoption of Weather-Index Insurance: Learning from Willingness to Pay Among a Panel of Households in Rural Ethiopia." *Agricultural Economics* 44: 385–398.

Hill, R.V., N. Kumar, N. Magnan, et al. 2019. "Ex Ante and Ex Post Effects of Hybrid Index Insurance in Bangladesh." *Journal of Development Economics* 136 (8): 1–17.

Hill, R.V., and M. Robles. 2011. "Flexible Insurance for Heterogeneous Farmers: Results from a Small-Scale Pilot in Ethiopia." IFPRI Discussion Paper 1092. IFPRI, Washington, DC.

Hill, R.V., M. Robles, and F. Ceballos. 2016. "Demand for a Simple Weather Insurance Product in India: Theory and Evidence." *American Journal of Agricultural Economics* 98 (4): 1250–1270.

Hobbs, A. 2022. "Gender Roles and the Decision to Insure." Working Paper. San Francisco, University of San Francisco.

Hoddinott, J., and B. Kinsey. 2000. "Adult Health in the Time of Drought." FCND Discussion Paper 79. IFPRI, Washington, DC.

Jack, W., and T. Suri. 2014. "Risk Sharing and Transaction Costs: Evidence from Kenya's Mobile Money Revolution." *American Economic Review* 104 (1): 183–223.

Jensen, N.D., C.B. Barrett, and A.G. Mude. 2017. "Cash Transfers and Index Insurance: A Comparative Impact Analysis from Northern Kenya." *Journal of Development Economics* 129: 14–28.

Jensen, N.D., F.P. Fava, A.G. Mude, et al. 2024. *Escaping Poverty Traps and Unlocking Prosperity in the Face of Climate Risk: Lessons from Index-Based Livestock Insurance.* Cambridge, UK: Cambridge University Press.

Karlan, D., R. Osei, I. Osei-akoto, and C. Udry. 2014. "Agricultural Decisions After Relaxing Credit and Risk Constraints." *Quarterly Jounrnal of Economics* 129 (2): 597–652.

Kazianga, H., and Z. Wahhaj. 2020. "Will Urban Migrants Formally Insure Their Rural Relatives? Family Networks and Rainfall Index Insurance in Burkina Faso." *World Development* 128: 104764.

Kramer, B., P. Hazell, H. Alderman, F. Ceballos, N. Kumar, and A.G. Timu. 2022. "Is Agricultural Insurance Fulfilling Its Promise for the Developing World? A Review of Recent Evidence." *Annual Review of Resource Economics* 14 (1): 291–311.

Kramer, B., S. Pattnaik, P.S. Ward, and Y. Xu. 2024. "Impacts of an Innovative Credit + Insurance Bundle for Marginalized Farmers: Evidence from a Cluster Randomized Trial in Odisha, India." IFPRI Discussion Paper 2288. IFPRI, Washington, DC.

Kramer, B., M. Porter, and S. Wassie Bizuayehu. 2023. "Basis Risk, Social Comparison, Perceptions of Fairness and Demand for Insurance: A Field Experiment in Ethiopia." Presented at the 2024 AAEA/ASSA Annual Meeting, January 5–7, 2024, San Antonio, Texas.

Kramer, B., C. Waweru, F. Cecchi, B. Kivuva, and L. Waithaka. 2024. "See It Grow: A Randomized Evaluation of Digital Innovations in Crop Insurance to Increase Insurance and Fertilizer Demand in Kenya." Paper presented at the 2024 AAEA Annual Meeting, July 28–30, New Orleans, LA.

Kramer, B., C. Waweru, and J.G. Malacarne. 2023. "Summing the Parts: How Does 'Bundling' Affect Willingness-to-Pay for Seeds and Insurance in a Sample of Kenyan Farmers?" Paper presented at the 2023 Agricultural and Applied Economics Association (AAEA) Annual Meeting, Washington, DC, July 23–25.

Liu, Y., K. Chen, and R.V. Hill. 2020. "Delayed Premium Payment, Insurance Adoption, and Household Investment in Rural China." *American Journal of Agricultural Economics* 102 (4): 1177–1197.

Lowder, S.K., J. Skoet, and T. Raney, 2016. "The Number, Size, and Distribution of Farms, Smallholder Farms, and Family Farms Worldwide." *World Development* (87): 16–29.

Meyer, R., P. Hazell, and P. Varangis. 2017. "Unlocking Smallholder Credit: Does Credit-Linked Agricultural Insurance Work?" Working Paper. International Labor Organization, Washington, DC; International Finance Corporation, Geneva.

Miranda, M.J. 1991. "Area-Yield Crop Insurance Reconsidered." *American Journal of Agricultural Economics* 73 (2): 233–242.

Mobarak, A.M., and M.R. Rosenzweig. 2013. "Informal Risk Sharing, Index Insurance, and Risk Taking in Developing Countries." *American Economic Review* 103 (3): 375–380.

Ndegwa, M.K., A. Shee, C.G. Turvey, and L. You. 2020. "Uptake of Insurance-Embedded Credit in Presence of Credit Rationing: Evidence from a Randomized Controlled Trial in Kenya." *Agricultural Finance Review* 80 (5): 745–766.

Quisumbing, A.R., N. Kumar, and J.A. Behrman. 2018. "Do Shocks Affect Men's and Women's Assets Differently? Evidence from Bangladesh and Uganda." *Development Policy Review* 36 (1): 3–34.

Shee, A., C.G. Turvey, and L. You. 2019. "Design and Rating of Risk-Contingent Credit for Balancing Business and Financial Risks for Kenyan Farmers." *Applied Economics* 51 (50): 5447–5465.

Skees, J., P. Hazell, and M. Miranda. 1999. "New Approaches to Crop Yield Insurance in Developing Countries." EPTD Discussion Paper 55. IFPRI, Washington, DC.

Stein, S. 2018. "Dynamics of Demand for Rainfall Index Insurance: Evidence from a Commercial Product in India." *World Bank Economic Review* 32 (3): 692–708.

Stoeffler, Q., M. Carter, C. Guirkinger, and W. Gelade. 2022. "The Spillover Impact of Index Insurance on Agricultural Investment by Cotton Farmers in Burkina Faso." *World Bank Economic Review* 36 (1): 114–140.

Swiss Re. 2019. *Agricultural Insurance: A Growth Story*. Zurich.

Timu, A.G., and B. Kramer. 2023. "Gender-Inclusive, -Responsive, and -Transformative Agricultural Insurance: A Literature Review." *Global Food Security* 36: 100672.

SECTION IV

Transforming Lives and Livelihoods

Chapter 11

SOCIAL PROTECTION PROGRAMS
Building the Evidence and Defining Priorities

Daniel O. Gilligan, Akhter Ahmed, Harold Alderman, Alan de Brauw, Melissa Hidrobo, Kalle Hirvonen, John Hoddinott, Jessica Leight, Shalini Roy, and Alemayehu Seyoum Taffesse

Daniel O. Gilligan is director, Poverty, Gender, and Inclusion Unit (PGI), IFPRI; **Akhter Ahmed** is country representative, Bangladesh, PGI, IFPRI-Bangladesh; **Harold Alderman** is a research fellow emeritus, Director General's Office, IFPRI; **Alan de Brauw** is a senior research fellow, Markets, Trade, and Institutions Unit, IFPRI; **Melissa Hidrobo** is a senior research fellow, PGI, IFPRI; **Kalle Hirvonen**, **Jessica Leight**, and **Shalini Roy** are senior research fellows, PGI, IFPRI; **John Hoddinott** is a nonresident fellow, PGI, IFPRI, and H.E. Babcock Professor of Food & Nutrition Economics and Policy, Cornell University; and **Alemayehu Seyoum Taffesse** is a senior research fellow and program leader, Ethiopia Strategy Support Program, Development Strategies and Governance Unit, IFPRI-Ethiopia.

Key messages

- Over the past 50 years, social protection programs to reduce poverty, food insecurity, and vulnerability and improve economic opportunity have become increasingly prominent in low- and middle-income countries.
- Social protection policy, and in particular social assistance, has evolved over this period from a focus on food-based subsidies, rations, and distribution systems to include cash transfers and graduation and economic inclusion programs that aim to strengthen the poverty-reducing effects of social assistance.
- Research on social protection has contributed to major changes in evidence-based policymaking for poverty reduction. The use of randomized controlled trials and other rigorous empirical methods has facilitated the evaluation of large-scale programs and the generation of new evidence around the impact of social protection on a wide range of outcomes as well as the effect of complementary programs that seek to expand those positive impacts.

- Modern social protection programs have greatly increased in number, scale, and complexity, with program designs tailored to different contexts and increasingly informed by evidence. These programs have also been adapted to changes in the needs of at-risk populations, evolving policy objectives, and the increasing prevalence of crises driven by conflict, extreme weather, and health shocks.

Moving forward, research on social protection programs should focus on:

- **Building understanding** on how context impacts the effectiveness of social protection, including through better evidence on how to adapt these programs for an increasingly crisis-prone world and design them for fragile settings.
- **Strengthening the evidence base** on how to refine the design and delivery of social protection to expand the benefits within households (for example, to further benefit women and youth) and across households (to expand their effects on local economies and thereby further benefit poor households).
- **Identifying scalable designs** for social protection programs to foster greater poverty alleviation. Evidence is needed that compares the impact, sustainability, and cost-effectiveness of different models, and explores how to better integrate them.

Government social protection programs to reduce poverty, food insecurity, and vulnerability in low- and middle-income countries (LMICs) have become increasingly prominent over the last 50 years. Social protection programs have increased in number, scale, and complexity as program designs have been tailored to different contexts and increasingly informed by evidence of their effectiveness. New technologies have facilitated less costly, better-targeted delivery strategies, as well as adaptation to changes in the cost of in-kind and cash modalities.[1] More recently, social protection programs

1 Social protection programs can be grouped into three categories: (1) social assistance including noncontributory transfers such as food and in-kind transfers, food subsidies or rations, unconditional cash transfers, conditional cash transfers, school meals, and public works; (2) social insurance, including contributory programs such as old-age and disability pensions, sick leave, and health insurance; and (3) labor market programs that can be contributory or noncontributory, including job training, employment services, wage subsidies, and unemployment insurance (see World Bank 2018). In this chapter, we focus primarily on social assistance programs, which are more prevalent in LMICs.

have also adapted to evolving policy objectives and changes in the needs of at-risk populations, including those driven by the increasing prevalence of crises due to conflict, climate change, and health shocks.

The inception of modern social assistance programs was broadly catalyzed by world food shortages in 1973–1974. In the aftermath of these shortages, governments in LMICs expanded programs to stabilize food access, including food subsidies, grain reserves, and food-based humanitarian assistance to respond to food crises, with these programs growing in prevalence throughout the 1980s and early 1990s. As improvements in communications and transportation infrastructure fostered greater market integration in the 1990s, governments introduced targeted cash-based assistance programs, which were often facilitated by newly available electronic card payment systems. The commercial supply response through markets was generally sufficient to meet this new source of demand without triggering inflation. Over time, governments introduced a variety of cash-based transfer modalities, including conditional cash transfers, unconditional cash transfers, and eventually, "cash-plus" models, often implemented at a large scale. More recently, multisectoral graduation and economic inclusion programs have sought to expand the poverty-reducing effects of these programs (see Box 11.1 for a description of the major types of social assistance programs).

The emergence of these trends in social protection policy coincided with a period of rapid expansion in the application of randomized controlled trials (RCTs). These evaluation approaches, together with complementary expansions in the use of qualitative assessments and cost-effectiveness analyses for impact assessment, led to major changes in evidence-based policymaking for poverty reduction. IFPRI's research played a prominent role in this shift by evaluating large-scale social protection programs across a wide swathe of low- and lower-middle-income countries,[2] testing the effectiveness of delivery strategies, and working with partners in government and nongovernmental organizations to examine complementary programs that would expand the impact of these investments (Quisumbing et al. 2020). IFPRI's research has provided evidence on the effectiveness of numerous social protection program modalities and on their impacts on a broad set of outcomes, while often examining the underlying mechanisms that contribute to these results. In particular, IFPRI has played a leading role in research on the synergistic effect of

2 These countries have included Bangladesh, Brazil, Ecuador, Egypt, El Salvador, Ethiopia, Honduras, India, Kenya, Malawi, Mali, Mexico, Mozambique, Nicaragua, Niger, Pakistan, Somalia, South Africa, Sri Lanka, Turkey, Uganda, and Yemen.

BOX 11.1 Major types of social assistance programs

FOOD SUBSIDIES OR FOOD RATIONS. Food subsidies provide consumers access to food at below-market prices and can be targeted or provide general price support. Food rations provide food directly to consumers for free or at reduced prices. Egypt has longstanding food subsidies for bread, pasta, and sugar. India's Public Distribution System is a prominent example of a food ration program.

SCHOOL MEALS. Food provided to children at school, often as breakfast or lunch, or take-home rations for children enrolled in school are examples of school meal programs. School meals are usually subsidized or free. School meals are provided for a significant share of students in most countries in the world.

CONDITIONAL CASH TRANSFERS (CCTs). Cash transfers are provided routinely (often monthly or quarterly) to poor households who meet certain conditions, including that school-age children must be enrolled and attending school, preschool children must receive required vaccinations, and women should attend nutrition behavior change communication sessions (if they have children under two years of age) or women's empowerment interventions. CCT programs gained popularity in Latin America in the 2000s; a prominent example is Mexico's PROGRESA program.

UNCONDITIONAL CASH TRANSFERS (UCTs). Cash transfers are provided routinely (often monthly or quarterly) to poor households without placing any requirements on the actions of the household. UCT programs can be implemented by governments or nongovernmental organizations and are implemented globally.

CASH-PLUS PROGRAMS. Cash transfers targeted to the poor are delivered along with additional components, such as nutrition education, livelihood training, financial literacy, and mental health services, which are intended to expand the program's benefits.

PUBLIC WORKS. Cash or food transfers are provided to the poor in exchange for work, often physical labor to build community assets. Prominent examples include India's Mahatma Gandhi National Rural Employment Guarantee Act program and Ethiopia's Productive Safety Net Program.

GRADUATION MODEL OR ECONOMIC INCLUSION PROGRAMS. These programs provide a package of interventions intended to provide a "big push" that increases household welfare and reduces poverty. Programs such as BRAC's Targeting the Ultrapoor Program typically provide an asset transfer, consumption support (routine cash transfers), financial services (savings and credit), business training (on the use of the asset), and mentoring or coaching.

social protection programs on gender inequality and women's empowerment, and on strategies to expand their impact on maternal and child nutrition and health.

In this chapter, we summarize trends in the development of social protection programs across three epochs, namely (1) food-based subsidies, rations, and distribution systems, (2) the proliferation of cash transfer programs, and (3) the emergence of graduation and economic inclusion programs. Throughout, we discuss the contribution of research to changing program approaches and social protection policies, highlighting selected seminal papers and showcasing IFPRI's contribution to evidence and research.

Food subsidies, rations, and food distribution systems

Throughout the 1950s and 1960s, concerns grew over the ability of the world to feed its growing population (Ehrlich 1968). These concerns were accelerated by the events of 1972–1973, when severe weather caused a steep decline in global grain production, followed by a spike in global food prices that was exacerbated by protective agricultural and trade policies in Russia and the United States (Hopkins and Puchala 1978). To address the resulting global food crisis, governments voiced a need for policy options to stabilize food prices and bolster food availability for vulnerable households (Timmer 2010). Popular policies in this period centered on a variety of price stabilization interventions, including broad-based staple food subsidies. Influential research examined the optimal design of food subsidies to minimize poverty, such as locating ration shops in poor communities, subsidizing only inframarginal quantities (those typically consumed at prevailing prices), and subsidizing foods primarily consumed by poor households (Besley and Kanbur 1988).

Empirical research by IFPRI researchers and others showed that food subsidies reduced food costs and stabilized prices, but noted that broad food subsidies in Egypt and elsewhere were poorly targeted and less effective than other options, such as ration cards and supplementary feeding programs, for improving food security and nutrition among individuals in poor households (Ahmed et al. 2001; Alderman et al. 1986; Kennedy and Alderman 1987; Pinstrup-Andersen 1988). A substantial body of literature on commodity price stabilization policies, such as domestic buffer stocks and international commodity agreements, similarly concluded that such policies benefited better-off households more than the poor (Bellemare et al. 2013; Newbery and

Stiglitz 1981). Ration shops, often a legacy of wartime rationing, were repurposed to distribute subsidized flour at stable prices for consumers in partially targeted schemes in Pakistan and elsewhere. In Egypt and Pakistan, food in ration shops was sourced by statutory procurement policies, which were also widely used in sub-Saharan Africa, but this system depressed prices for farmers (Alderman 1988; de Haen and von Braun 1983).

Growing awareness of global hunger, particularly in the aftermath of famines resulting from the 1974 floods in Bangladesh and the 1984 drought in Ethiopia, led to the development of emergency food assistance to respond to food shortages triggered by weather or trade shocks. This often entailed creating costly ad hoc food distribution systems and, with regard to US food assistance, inefficient approaches to monetizing food aid (Barrett and Maxwell 2005). Work by IFPRI and others showed that these responses had positive impacts, but were not nearly as effective as they could have been (Webb et al. 1992). For example, research on Ethiopia showed that food aid during the 1990s largely offset the negative effect of harvest failure on child stunting (Yamano et al. 2005), and emergency food assistance following the 2002 drought effectively increased household food consumption, with larger impacts from food-for-work for better-off households and from direct support for the poorest households (Gilligan and Hoddinott 2007). This emphasis on the impact of delivering food assistance and other safety net programs increased attention on effective strategies for targeting transfers to the poor (Coady et al. 2004b). Other research examined the effect of transfers on poverty dynamics over time, recognizing that transfers may play an insurance role in the face of impending shocks (Baulch and Hoddinott 2000).

From 2000 to 2020, governments vastly expanded investments in school meal programs as a means to transfer food directly to children and improve schooling outcomes (Bundy et al. 2024). IFPRI contributed to a growing body of research showing that school meal programs improve school attendance, often strengthen student learning (Alderman et al. 2012; Aurino et al. 2023), and can address unmet nutritional needs, such as with anemia in adolescent girls, including in a crisis (Adelman et al. 2019). Distributional issues around the benefits of school meals have received considerable attention: early evidence showed that food resources (calories and other nutrients) transferred to a child through school meals led to an equivalent increase in the child's total intake of those calories and nutrients and did not displace at-home consumption (Afridi 2011; Jacoby 2002). Later studies examined the impact of school-based meal programs on the nutritional status of other vulnerable household members including preschool children, finding that a take-home ration modality improved weight-for-age for preschool-age siblings of school-age children in Burkina Faso. In contrast, a take-home ration had no effect on anthropometry of the targeted child's preschool siblings in Uganda, but meals provided at school improved height-for-age for the preschool siblings of targeted children (Adelman et al. 2019; Kazianga et al. 2014). IFPRI's more recent work on school meals investigates effects of school meal programs on local agriculture and the need for more evidence on cost-effectiveness (Alderman et al. 2024).

Cash transfers

Cash transfers provided by the state have a long history (Gentilini 2024), but starting in the 1990s, governments in LMICs, particularly in Latin America, began to formalize targeted cash transfers to poor households that required these households to fulfil conditions related to child schooling, health, and nutrition. These "conditional cash transfer" (CCT) programs were more effectively targeted to the poor than food subsidies, were potentially less distortionary, and provided beneficiaries with greater control over use of the resources. Still, many were concerned that direct cash grants to the poor would not be used toward government priorities such as food security, would create dependency by reducing adult work, or would be misused through spending on alcohol.

IFPRI contributed pioneering research on the impacts of these CCT programs. From 1998 to 2000, the government of Mexico collaborated with IFPRI and academic partners on a groundbreaking randomized impact evaluation of its PROGRESA CCT program. The PROGRESA evaluation was the first RCT of a modern large-scale social protection program, documenting its impacts on a wide range of outcomes including food security, nutrition, and women's status.[3] This collaboration marked a turning point both in government policies for poverty reduction and in research to evaluate these policies. The evidence of PROGRESA's large benefits and cost-effectiveness, alongside evidence of limited work disincentives (Behrman and Skoufias 2006), contributed to a shift in how governments and donors regarded CCT programs, driving the rising popularity of these models throughout Latin America and beyond. Moreover, the PROGRESA evaluation fostered a culture of rigorous collaborative evaluations of social protection programs

3 On evaluation of PROGRESA, see Adato et al. (2003), Behrman (2010), Schultz (2004), and Skoufias (2005).

(Behrman 2010) and promoted the broader use of RCTs throughout development economics to assess the effectiveness of large-scale government programs. After the PROGRESA evaluation, IFPRI conducted impact evaluations of national CCT programs in Brazil, Honduras, Nicaragua, and El Salvador.

In the 2000s, cash transfer programs expanded to sub-Saharan Africa, the Middle East and North Africa, and South Asia. Compared to those in Latin America, programs in these regions were less likely to include health and education conditions due to fewer resources for enforcement, limited value of unenforced conditions (Baird et al. 2014), and a desire to avoid excluding vulnerable populations with limited access to health clinics or schools. Instead, programs often included additional complementary activities (the "plus" components in the "cash-plus" model) such as trainings, linkages to other services and sectors, such as health insurance, or transfers linked to participation in public works projects. IFPRI continued to work with governments to evaluate their large-scale safety net programs; examples include collaborating with the government of Ethiopia to evaluate several phases of its public works program, the Productive Safety Net Program (Berhane et al. 2014), with the government of Mali on its unconditional cash-plus program, Programme de Filets Sociaux Jigisémèjiri (Sessou et al. 2024), and with the government of Egypt on its conditional cash transfer program, Takaful (El-Enbaby et al. 2024).

As the popularity of cash transfer programs grew, debates arose over which program features would best promote key policy objectives. IFPRI's research has been at the frontier of guiding these design decisions. Attention to gender has been a hallmark of this work; for example, the decision to make women the primary recipients in many cash transfer programs was informed by the Institute's research that intrahousehold resource allocations mattered (Quisumbing and Maluccio 1999), and that income earned by women, as compared to men, was more likely to be used for children's needs (Haddad et al. 1997; Hoddinott et al. 1995). IFPRI was also among the first to examine whether making transfers conditional on schooling mattered for educational impacts, showing that conditionality increased primary-to-secondary-school transitions (de Brauw and Hoddinott 2011). The Institute both contributed to and consolidated the knowledge base on which forms of targeting are most effective (Coady et al. 2004a). To address a persistent policy debate on the merits of cash transfers relative to food rations or vouchers, IFPRI built on its earlier work in Bangladesh (Ahmed et al. 2010) by collaborating with the World Food Programme on a multicountry study that compared the impact of cash, food, and vouchers in Bangladesh, Ecuador, Niger, Uganda, and Yemen.[4] The Institute's influential portfolio of studies demonstrated that the optimal modality depended on policymakers' objectives and contextual factors, but cash transfers were generally more cost-effective than food transfers at improving household food security and dietary diversity (Margolies et al. 2015).

More recently, governments have sought to design cash transfer programs to achieve a range of objectives beyond food security—notably, to improve nutrition. IFPRI's research has been instrumental in informing the effectiveness and promising designs of nutrition-sensitive cash transfer programs.[5] An influential collaboration with the World Food Programme in Bangladesh showed that providing mothers with cash transfers alone did not significantly improve child nutritional status, while providing them with a combination of cash and high-quality nutrition education delivered through behavior change communication sessions significantly reduced child malnutrition (Ahmed, Hoddinott, and Roy 2024). This evidence on the importance of well-designed complementary "plus" components, meant to inform and empower caregivers (often mothers) to use transfers toward better diets and health, influenced the design of large-scale nutrition-sensitive programs in Bangladesh and several other countries.

As cash transfer programs have become more multisectoral and have scaled up to reach more than 1 billion people globally, the evidence has grown regarding their long-term benefits and the broader impacts that extend beyond policymakers' primary objectives. IFPRI has played a leading role in demonstrating the multifaceted benefits of cash transfer programs. The Institute's influential body of research shows how cash transfers can reduce women's experience of intimate partner violence (IPV) (Buller et al. 2018), findings which have informed the World Health Organization's recommendations for preventing violence against women in LMICs. The Institute has also demonstrated the impacts of cash transfers on women's decision-making power and empowerment (Hidrobo et al., forthcoming; Peterman et al. 2021), child development (de Walque et al. 2017; Fernald and Hidrobo 2011), mental health (Hidrobo et al. 2023), social capital and trust (Evans et al. 2019), migration (Hidrobo et al. 2022), and agriculture (Hoddinott et al. 2012),

4 This work includes Ahmed, Hoddinott et al. (2024), Gilligan and Roy (2014), Hidrobo et al. (2014), Hoddinott et al. (2019), and Schwab (2019).

5 This literature includes Behrman and Hoddinott (2005), Leroy et al. (2008), Manley et al. (2022), and Olney et al. (2022).

among other areas. Amid fiscal constraints and mounting pressure to invest in programs with long-term benefits, IFPRI has also provided rigorous evidence on which program designs can promote sustained post-program impacts. Studies show that adding complementary "plus" components to cash transfers can increase the sustainability of benefits across outcomes including household poverty, livelihoods, and experience of IPV, thereby enhancing the cost-effectiveness of the initial investments (Ahmed et al. 2025; Roy et al. 2024). This broader range of benefits from programs operating at scale has helped to bolster ongoing support from donors and implementers.

As global crises escalate, there is growing interest in how cash transfer programs can contribute to the policy response. IFPRI's recent studies show significant promise for these programs to help poor rural households respond to crises by protecting food security, livelihoods, and nutrition, including during pandemics (Abay et al. 2023; Ahmed et al. 2024), humanitarian and conflict situations (Kurdi 2021), and climate hazards (Ahmed, Bakhtiar, Hidrobo et al. 2024; Hidrobo et al. 2024). Research on optimal designs to respond to crises has shown that providing anticipatory cash transfers, given in advance of shocks, can improve upon emergency transfers given after shocks, by improving coping, preemptive actions, and resilience capacities (de Brauw et al. 2025). IFPRI's research has also demonstrated that community assets produced by public works programs can have environmental benefits, showing, for example, that the reforestation activities implemented through Ethiopia's Productive Safety Net Programme increased tree cover, an important strategy to mitigate climate change (Hirvonen et al. 2022).

Economic inclusion programs

During the last 10 years, governments, donors, and other stakeholders have increasingly focused on designing multifaceted interventions to facilitate a sustainable exit from poverty by targeted households. This development in policy innovation and the associated academic literature builds on the widely documented benefits of cash transfer programs, but it represents a shift toward multifaceted interventions (generally inclusive of training, coaching, and savings groups, as described in more detail below) with a more explicit focus on building new livelihood activities and associated streams of income that could lead to longer-term exits from poverty. This approach sometimes also incorporates an explicit goal of reduced reliance on cash transfers. The shift in focus is partly motivated by evidence from some high-profile trials in Uganda and Kenya demonstrating that the effects of time-limited cash transfers did not persist over the longer term (Blattman et al. 2020; Haushofer and Shapiro 2018). A recent systematic review found some evidence of decline in the effects of cash and cash-plus interventions over time, but the available evidence is generally drawn from evaluations conducted less than two years post-program, suggesting that relatively little is known about the longer-term effects of transfers (Leight, Hirvonen, and Zafar 2024).

While the precise elements of these multifaceted interventions can differ across contexts, they typically center around a larger lump-sum transfer (often described as a livelihoods transfer) in conjunction with some short-term (time-limited) consumption support, financial inclusion, and training and/or coaching.[6] These models are often described as graduation programs or economic inclusion programs; we preferentially use the latter term here.[7]

The first major contribution to this literature was a six-country randomized trial by Banerjee and colleagues (2015) demonstrating that a multifaceted intervention modeled on programming delivered by BRAC, a major nongovernmental organization in Bangladesh, had significant effects on reducing poverty and enhancing welfare along multiple domains (including consumption, income, assets, and health) (Banerjee et al. 2015). A similarly large trial of BRAC's intervention in Bangladesh, which centered on transfers of livestock and skills to extremely poor women, found equally positive effects at four and seven years post-intervention (Bandiera et al. 2017). Long-term follow-up surveys showed that the positive effects of these interventions persisted and even grew in India (Banerjee et al. 2021) and Bangladesh (Balboni et al. 2022), but attenuated over time in Ethiopia (Barker et al. 2024). Another recent trial suggested that incorporating a psychosocial intervention targeting women could enhance cost-effectiveness (Bossuroy et al. 2022). However, the literature on psychosocial interventions is highly variable, with some showing limited effects; the specific structure and objectives of what are considered psychosocial interventions also vary widely.

While many economic inclusion programs target women, impacts on measures of female empowerment are mixed. The multicountry BRAC

6 There is also the broader category of "cash-plus" interventions, which includes programs that disburse cash in conjunction with other interventions that can vary greatly in their goals and level of intensity. While "cash plus" often primarily describes programs in which the "plus" component is relatively low cost or low intensity, there is some continuity between more intensive cash-plus interventions and economic inclusion programs.

7 One risk of the term "graduation model" is that it can lead to confusion as to what "graduation" refers to, such as graduation from a particular program (that is often time limited, by definition), or from poverty or a state of vulnerability. For this reason, we opt to use the term "economic inclusion programs."

study showed no effects on women's empowerment (Banerjee et al. 2015), while studies in Niger and the Democratic Republic of Congo found positive impacts (Bossuroy et al. 2022). In efforts to make programs more gender transformative, some programs have tried to layer on gender norms interventions. An IFPRI study on male engagement groups found that adding this programming to other livelihood and nutrition interventions led to sustained impacts on men's gender-equitable attitudes and behaviors (Alderman, Gilligan, Hidrobo et al. 2025), but this did not necessarily lead to improvements in women's empowerment (Alderman et al. 2021). Moreover, these efforts are often intensive, requiring a deep understanding of context-specific social norms and specialized expertise outside the social protection sector (Hidrobo et al. 2025).

As the literature on economic inclusion interventions has grown rapidly, policymaker interest in experimenting with multiple types of models has also increased. In particular, the original interventions evaluated by Banerjee and colleagues (2015) were costly (between US$1,500 and $6,000 per household in total); although estimates suggested that benefits exceeded costs, there might nonetheless be reasonable doubts about the feasibility of scale-up (see Box 11.2 for a more detailed overview of the evidence on cost and cost-effectiveness of these models). IFPRI researchers have made major contributions to exploring the effectiveness of lower-cost economic inclusion models, including those embedded within existing social protection systems. In Ethiopia, a model centered around lower-cost livestock (poultry) transfers, savings groups, and training generally did not lead to transformative effects on poverty (Leight, Alderman et al. 2024) though some evidence showed improvements in nutritional knowledge and practices and household resilience to shocks (Alderman, Gilligan, Leight et al. 2025; Hirvonen et al. 2024). More recent work suggests that a nutrition-sensitive model, including short-term maternal cash grants, can be effective in shifting early childhood feeding practices and enhancing child development (Gilligan et al. 2024).

The Institute's researchers have also generated novel evidence around the effectiveness of graduation models in fragile contexts, such as urban Somalia, where a graduation model intervention significantly reduced poverty within the first three years (Leight, Hirvonen et al. 2024). These findings point to the importance of further experimentation, in conjunction with policymakers, to explore the effectiveness and cost-effectiveness of economic inclusion programs implemented at scale and over the longer term. Overall, cost-effectiveness of food transfers, cash transfers, and economic inclusion programs remains an area where more evidence is needed (Box 11.2).

BOX 11.2 Cost and cost-effectiveness of food transfers, cash transfers, and economic inclusion programs

One growing area of research interest involves examining evidence about cost and how the total costs of food and cash transfer programs and economic inclusion programs, including costs for delivery and administration, compare to the estimated benefits. Such comparisons make it possible to draw conclusions about the cost-effectiveness of these programs or the impact on specific outcomes per dollar of investment. Doing so, however, is not always straightforward.

For example, IFPRI's comparative work on the costs associated cash and food transfers benefited from the fact that those studies included the collection of detailed cost data; this showed that the cost of cash delivery is between a third and half the cost of delivering comparable transfers in other forms, such as food or vouchers (Hidrobo et al. 2014; Margolies and Hoddinott 2015).

However, it is much more challenging to assess the cost and cost-effectiveness of economic inclusion programs, as well as the whole range of cash-plus interventions that lie on the spectrum between simple cash transfers and more complex interventions. In part, this is because the cost structure of these programs varies greatly, and detailed information is not always provided. It also reflects the fact that the benefits of such interventions take multiple forms, not all of which are readily monetized. For example, the graduation models in the Banerjee and colleagues (2015) study have costs per beneficiary of between US$1,000 and $2,000, including a total transfer value of $500 to $1,000. A more recent, psychosocial-oriented graduation model implemented in Niger and analyzed reports costs per beneficiary household of between $530 and $650, including $320 of transfer value (all in PPP terms) (Bossuroy et al. 2022). In another study, a relatively lighter cash-plus intervention encompassing savings groups and training (similar to a graduation model) shows an estimated cost of $215 per beneficiary household in Uganda (of which around $80 is cash transfers) (Sedlmayr et al. 2020).

In these three examples, the total cost of the programming, inclusive of cash-plus interventions, is around twice the value of the transfer itself, implying that for the program to be cost-effective, the positive effects of the more complex intervention should be at least double any potential positive effects of cash itself (Leight et al. 2024). While these patterns may not generalize to other economic inclusion interventions, understanding more about the effectiveness of these interventions relative to their cost is crucial to informing evidence-based decisions around scaling.

The forward-looking agenda

The next 25 years will present new challenges for the design and delivery of social protection programs to accelerate poverty reduction, reduce food insecurity and malnutrition, and improve opportunities and well-being for poor populations. Based on our understanding of emerging trends and existing evidence, we have identified three issues that will be important areas for research on social protection leading up to 2050.

First, more evidence is needed on how social protection programs can better respond to a world increasingly prone to crises from increasing extreme weather events, conflict, and health shocks. Evidence on adaptive social protection systems provides guidance on how to leverage existing safety nets designed to address chronic poverty and ensure their capacity to respond to new shocks (Gilligan et al. 2022; Hirvonen 2023). More recently, governments and international humanitarian actors such as the World Food Programme have been piloting anticipatory action models of social assistance that rely on early warning systems for droughts, floods, pests, and health shocks, which, when triggered, allow transfers to reach targeted households and communities in the weeks and months before the shock occurs to strengthen the response (de Brauw 2025; de Brauw et al. 2025; Pople et al. 2024). These strategies, while promising, are attempts to substitute for missing insurance markets. The last 50 years have produced remarkably few examples of insurance schemes available at scale to the poor in LMICs to protect their assets, livelihoods, and health (see Chapter 10). Recent evidence on index-based livestock insurance is encouraging (Jensen et al. 2024), and promising efforts to introduce insurance as part of social protection programs are underway in Ethiopia and elsewhere, but progress remains limited; further evidence on social insurance is needed.

Relatedly, we need to build more evidence on designing effective social assistance in fragile settings. In the period leading up to the COVID-19 pandemic, extreme poverty and malnutrition were increasingly concentrated in fragile states such as Somalia, Sudan, and the Democratic Republic of Congo, as well as in fragile areas within states. In such contexts, formal state capacity to target and reliably deliver cash transfers and other services is weak, requiring simple, safe approaches to deliver transfers and support services (see Chapter 13).

Second, more evidence is needed on how to refine social protection program designs and delivery modalities to expand the benefits within and across households (Neihaus and Suri, forthcoming). Much has been learned about how the design of social protection programs shapes benefits for women, not only by targeting cash transfers to women but also by including design features and complementary components that account for women's time burdens, safety, and restricted access to jobs, markets, and land (Hidrobo et al. 2025). Still, we must continue to refine these gender-sensitive design features to be applicable in more settings. Also, less is known about how to design social protection programs to better benefit youth, including to improve educational attainment and early labor market outcomes. Moreover, a better understanding of the contributions of social protection, including gender-responsive models, to local economies, alongside the costs, is needed to convince policymakers. Similarly, although previous research provided important lessons on targeting social assistance in the 2000s (see Box 11.3 for an overview of IFPRI's research on targeting social assistance programs), attention to this topic lagged until recently, with new work informing the strengths and weaknesses of different targeting approaches (such as community-based targeting, proxy means scores, and categorical targeting) (Abay et al. 2024; Brown et al. 2018). Recent evidence on the general equilibrium effects of cash transfer interventions suggests that program targeting could be designed to optimize these positive spillover effects, particularly to the extent that they further benefit poor households (Egger et al. 2022). Greater evidence on the heterogeneity of impact from economic inclusion programs could offer suggestions on how to better target these programs to those most likely to benefit, while other eligible households continue to receive consumption support.

Third, and relatedly, research on social protection should continue to focus on cost-effectiveness and identifying scalable social protection designs that foster greater poverty alleviation. Cash-plus designs have shown great promise for expanding the impact of cash transfers to outcomes such as nutrition, women's empowerment, and mental health, potentially improving their cost-effectiveness. As noted earlier, multifaceted economic inclusion programs have a positive internal rate of return but are expensive, raising questions about their scalability. Evidence is needed that compares the impact, sustainability, and cost-effectiveness of these models, as well as considering how the models could be better integrated (Allen et al. 2024). A key part of this research agenda on scalable, cost-effective social protection designs in LMICs involves finding better ways to integrate targeted transfers with strategies to promote better livelihoods and job outcomes. The literature on asset transfers shows promise to address this need, while the evidence on jobs and training programs is far more mixed and likely depends on context-specific labor market conditions.

BOX 11.3 Targeting social assistance programs

Targeting social assistance programs involves identifying needy households, often those who are poor or most vulnerable to economic shocks. Effective targeting ensures that limited resources reach those in genuine need, increasing program impact and reducing costs. Common methods include categorical targeting, geographical targeting, community-based targeting, self-selection, and means or proxy means testing (Hoddinott 2001).

These approaches are often combined for greater effectiveness. Categorical targeting directs benefits to people based on identifiable characteristics such as age, gender, disability, or access to land. It is administratively simple and may be effective when resources are scarce. Geographic targeting prioritizes benefits for areas with the largest population of poor people. In community-based targeting, local leaders select beneficiaries based on their knowledge of households' well-being and circumstances, often using criteria unavailable under more administrative approaches. Under self-selection, program benefits are available to all, but program design incentivizes participation by the poor only. Public works programs exemplify self-targeting: due to the work requirement and low wages, only the poor tend to participate. This method is cost-effective and often politically acceptable. Proxy means testing (PMT) relies on regression models to predict consumption levels or poverty status using proxies such as household characteristics, assets, and housing conditions to identify poor households.

IFPRI has made substantial contributions to evidence on the effectiveness of targeting social assistance programs and on the relative performance of different targeting approaches. A review of targeting approaches of 122 social assistance programs across 48 countries in 2004 found that the median program transferred 25 percent more to poor individuals (the poorest two quintiles) than a universal allocation (Coady et al. 2004a). Targeting was better in richer countries, where governments were held more accountable, and where inequality was higher. Targeting also improved with means testing, geographic targeting, and work requirements. Studies by current and former IFPRI researchers addressed the effectiveness of targeting of programs in Albania (Alderman 2002), Mexico (Coady and Parker 2009), and Peru (Stifel and Alderman 2005); the effectiveness of self-targeting for food subsidies (Alderman 2001); and welfare losses due to targeting using alternative welfare indicators (Skoufias and Coady 2007).

A recent IFPRI and World Bank study of community-based targeting in Ethiopia experimented with the fiscal constraint and discretion over targeting criteria (Abay et al. 2024). Results show that community leaders avoided exclusion errors more than inclusion errors and reduced transfer size rather than coverage under tighter budgets. Greater discretion improved access for poorer or conflict-affected households that could be missed under conventional targeting criteria. Leaders are more vulnerable to favoritism when budgets are larger and when they lack discretion.

PMT methods, which are widely used in developing countries, have received mixed support in the literature on targeting. A core critique is that PMTs are prone to exclusion errors (Kidd et al. 2017; Veras et al. 2008). Instead, targeting using a simple demographic scorecard often does as well or better at reaching the poor (Brown et al. 2018). Also, adding a small cost of applying for a program targeted using a PMT can improve targeting performance through self-selection (Alatas et al. 2016).

In response to the limitations of PMTs, IFPRI researchers applied an alternative approach for Bangladesh, using nationally representative survey data to identify indicators for categorical targeting that are strongly correlated with poverty, easily observable, verifiable, and straightforward to collect. At the request of Bangladesh's Ministry of Women and Children Affairs (MoWCA) and the World Food Programme, IFPRI applied the categorical targeting approach to improve the accuracy of targeting for Bangladesh's Vulnerable Group Development program and the Mother and Child Benefit program. MoWCA later adopted this approach to enhance the targeting efficiency of both programs (Ahmed 2018a, 2018b).

The current political climate highlights that social protection programs will need to become more financially stable and prioritize competing demands and objectives. For example, in cash-plus or multisectoral programs, households are more likely to invest the transfer received to boost outcomes relevant to the complementary programming provided. This behavior is beneficial and improves efficiency as households leverage the transfer to increase the benefits of the "plus" component, but it sharpens the trade-offs in program design, such as between promoting investment in productive assets or improving child nutrition.

IFPRI's ongoing research and strong partnerships with governments, international humanitarian actors, academics, and donors position the Institute to be at the forefront of informing social protection strategies to improve well-being for poor and vulnerable households over the next 25 years. IFPRI will work closely with these partners to identify cost-effective solutions that are context specific, scalable, and inclusive.

References

Abay, K.A., G. Berhane, D.O. Gilligan, K. Tafere, and A. Seyoum Taffesse. 2024. "Targeting Social Assistance in Fragile Settings: An Experiment on Community-Based Targeting." IFPRI Discussion Paper 2289. IFPRI, Washington, DC.

Abay, K., G. Berhane, J. Hoddinott, and K. Tafere. 2023. "COVID-19 and Food Security in Ethiopia: Do Social Protection Programs Protect?" *Economic Development and Cultural Change* 71 (2): 373–402.

Adelman, S., D.O. Gilligan, J. Konde-Lule, and H. Alderman. 2019. "School Feeding Reduces Anemia Prevalence in Adolescent Girls and Other Vulnerable Household Members in a Cluster Randomized Controlled Trial in Uganda." *Journal of Nutrition* 149 (4): 659–666.

Adato, M.; B. de la Brière, D. Mindek, A.R. Quisumbing. 2003. "The Impact of PROGRESA on Women's Status and Intrahousehold Relations." In *Household Decisions, Gender, and Development: A Synthesis of Recent Research*, ed. A.R. Quisumbing, 213–218. Washington, DC: IFPRI.

Afridi, F. 2011. "The Impact of School Meals on School Participation: Evidence from Rural India." *Journal of Development Studies* 47 (11): 1636–1656.

Ahmed, A. 2018a. *Assessing Indicators for Selection of Participants for the Vulnerable Group Development (VGD) Program in Bangladesh*. Dhaka: IFPRI.

Ahmed, A. 2018b. "Assessing Indicators for Targeting the Needy under the Improved Maternity and Lactating Mother Allowance Program in Urban Bangladesh." Dhaka: IFPRI.

Ahmed, A., M. Bakhtiar, M. Hidrobo, et al. 2024. *Social Assistance and Adaptation to Flooding in Bangladesh*. CGIAR Research Initiative on Gender Equality Brief. Washington, DC: IFPRI.

Ahmed, A., M. Bakhtiar, J. Hoddinott, and S. Roy. 2024. "Does Nutrition-Sensitive Social Protection Build Longer-Term Resilience? Experimental Evidence from Bangladesh." IFPRI Discussion Paper 2282. IFPRI, Washington, DC.

Ahmed, A., H. Bouis, T. Gutner, and H. Lofgren. 2001. *The Egyptian Food Subsidy System*. IFPRI Research Report 119. Washington, DC.

Ahmed, A., M. Hidrobo, J. Hoddinott, B. Kolt, S. Roy, and S. Tauseef. 2025. "Sustainable Poverty Reduction through Social Assistance: Modality, Context, and Complementary Programming in Bangladesh." *American Economics Journal: Applied Economics* 17 (2): 102–126.

Ahmed, A., J. Hoddinott, and S. Roy. 2024. "Food Transfers, Cash Transfers, Behavior Change Communication and Child Nutrition: Evidence from Bangladesh." *World Bank Economic Review*. 39 (2): 439–472.

Ahmed, A., J. Hoddinott, S. Roy, and E. Sraboni. 2024. "Transfers, Nutrition Programming, and Economic Well-Being: Experimental Evidence from Bangladesh." *World Development* 173: 106414.

Ahmed, A., A. Quisumbing, M. Nasreen, J. Hoddinott, and E. Bryan. 2010. *Comparing Food and Cash Transfers to the Ultra-Poor in Bangladesh*. IFPRI Research Monograph 163. Washington, DC: IFPRI.

Alatas, V., A. Banerjee, R. Hanna, B.A. Olken, R. Purnamasari, and M. Wai-Poi. 2016. "Self-Targeting: Evidence from a Field Experiment in Indonesia." *Journal of Political Economy* 124 (2): 371–427.

Alderman, H. 1988. "The Twilight of Flour Rationing in Pakistan." *Food Policy* 13 (3): 245–256.

Alderman, H. 2001. "Multi-Tier Targeting of Social Assistance: Role of Inter-governmental Transfers." *World Bank Economic Review* 15 (1): 33–53.

Alderman, H. 2002. "Do Local Officials Know Something We Don't? Decentralization of Targeted Transfers in Albania." *Journal of Public Economics* 83 (3): 375–404.

Alderman, H., D. Bunday, and A. Gelli. 2024. "School Meals Are Evolving: Has the Evidence Kept Up?" *World Bank Research Observer* 39 (2): 159–176.

Alderman, H., D.O. Gilligan, M. Hidrobo, J. Leight, M. Mulford, and H. Tambet. 2025. "Men Can Cook: Effectiveness of a Light-Touch Men's Engagement Intervention to Change Attitudes and Behaviors in Rural Ethiopia." *World Development* 185: 106781.

Alderman, H., D.O. Gilligan, and K. Lehrer. 2012. "The Impact of Food for Education Programs on School Participation and Education Attainment in Northern Uganda." *Economic Development and Cultural Change* 61 (1): 187–218.

Alderman, H., D.O. Gilligan, J. Leight, T. Mamo, M. Mulford, and H. Tambet. 2025. "Scalable Nutrition Interventions in a Graduation Model Program: Experimental Evidence from Ethiopia." *Economic Development and Cultural Change* 73 (3): 1559–1604.

Alderman, H., and J. Von Braun. 1986. "Egypt's Food Subsidy Policy." *Food Policy* 11 (3): 223–237.

Allen IV, J., D.O. Gilligan, S. Kurdi, and B. Yassa. 2024. "Would You Rather? Household Choice between Cash Transfers or an Economic Inclusion Program." IFPRI MENA Working Paper 44. IFPRI, Washington, DC.

Aurino, E., A. Gelli, C. Adamba, I. Osei-Akoto, and H. Alderman. 2023. "Food for Thought?: Experimental Evidence on the Learning Impacts of a Large-Scale School Feeding Program." *Journal of Human Resources* 58 (1): 74–111.

Baird, S., F.H.G. Ferreira, B. Özler, and M. Woolcock. 2014. "Conditional, Unconditional and Everything in between: A Systematic Review of the Effects of Cash Transfer Programmes on Schooling Outcomes." *Journal of Development Effectiveness* 6 (1): 1–43.

Balboni, C., O. Bandiera, R. Burgess, M. Ghatak, and A. Heil. 2022. "Why Do People Stay Poor?" *The Quarterly Journal of Economics* 137 (2): 785–844.

Bandiera, O., R. Burgess, N. Das, S. Gulesci, I. Rasul, and M. Sulaiman. 2017. "Labor Markets and Poverty in Village Economies." *The Quarterly Journal of Economics* 132 (2): 811–870.

Banerjee, A., E. Duflo, N. Goldberg, et al. 2015. "A Multifaceted Program Causes Lasting Progress for the Very Poor: Evidence from Six Countries." *Science* 348 (6236): 1260799.

Banerjee, A., E. Duflo, and G. Sharma. 2021. "Long-Term Effects of the Targeting the Ultra Poor Program." *American Economic Review: Insights* 3 (4): 471–486.

Barker, N., D. Karlan, C. Udry, and K. Wright. 2024. "The Fading Treatment Effects of a Multifaceted Asset-Transfer Program in Ethiopia." *American Economic Review: Insights* 6 (2): 277–294.

Barrett, C.B., and D.G. Maxwell. 2005. *Food Aid after Fifty Years: Recasting Its Role*. New York: Routledge.

Baulch, R., and J. Hoddinott, eds. 2000. *Economic Mobility and Poverty Dynamics in Developing Countries*. London: Frank Cass.

Behrman, J.R. 2010. "The International Food Policy Research Institute (IFPRI) and the Mexican PROGRESA Anti-poverty and Human Resource Investment Conditional Cash." *World Development* 38 (10): 1473–1485.

Behrman, J.R., and J. Hoddinott. 2005. "Program Evaluation with Unobserved Heterogeneity and Selective Implementation: the Mexican Progresa Impact on Child Nutrition." *Oxford Bulletin of Economics and Statistics* 67: 547–569.

Behrman, J.R., and E. Skoufias. 2006. "Mitigating Myths About Policy Effectiveness: Evaluation of Mexico's Antipoverty and Human Resource Investment Program." *Annals of the American Academy of Political and Social Science* 606 (1): 244–275.

Bellemare, M.F., C.B. Barrett, and D.R. Just. 2013. "The Welfare Impacts of Commodity Price Volatility: Evidence from Rural Ethiopia." *American Journal of Agricultural Economics* 95 (4): 877–899.

Berhane, G., D.O. Gilligan, J. Hoddinott, N. Kumar, and A. Seyoum Taffesse. 2014. "Can Social Protection Work in Africa? The Impact of Ethiopia's Productive Safety Net Programme." *Economic Development and Cultural Change* 63 (1): 1–26.

Besley, T., and R. Kanbur. 1988. "Food Subsidies and Poverty Alleviation." *The Economic Journal* 98 (392): 701–719.

Blattman, C., N. Fiala, and S. Martinez. 2020. "The Long-Term Impacts of Grants on Poverty: Nine-Year Evidence from Uganda's Youth Opportunities Program." *American Economic Review: Insights* 2 (3): 287–304.

Bossuroy, T., M. Goldstein, B. Karimou, et al. 2022. "Tackling Psychosocial and Capital Constraints to Alleviate Poverty." *Nature* 605 (7909): 291–297.

Brown, C., M. Ravallion, and D. van de Walle. 2018. "A Poor Means Test? Econometric Targeting in Africa." *Journal of Development Economics* 134: 109–124.

Buller, A.M., A. Peterman, M. Ranganathan, A. Bleile, M. Hidrobo, and L. Heise. 2018. "A Mixed-Method Review of Cash Transfers and Intimate Partner Violence in Low- and Middle-Income Countries." *World Bank Research Observer* 33 (2): 218–258.

Bundy, D.A.P., U. Gentilini, L. Schultz, et al. 2024. "School Meals, Social Protection and Human Development: Revisiting Trends, Evidence, and Practices in South Asia and Beyond." Social Protection & Jobs Discussion Paper 2401. World Bank, Washington, DC.

Coady, D., M. Grosh, and J. Hoddinott. 2004a. "Targeting Outcomes Redux." *World Bank Research Observer* 19 (1): 61–85.

Coady, D., M. Grosh, and J. Hoddinott. 2004b. *The Targeting of Transfers in Developing Countries: Review of Experience and Lessons.* Washington, DC: World Bank and IFPRI.

Coady, D., and S. Parker. 2009. "Targeting Social Transfers to the Poor in Mexico." IMF Working Paper 060. International Monetary Fund, Washington, DC.

de Brauw, A. 2025. *Cost Effectiveness of Anticipatory Action: Lesotho, Madagascar, and Mozambique.* CGIAR Initiative on Fragility, Conflict, and Migration Technical Report. Montpellier, France: CGIAR System Organization.

de Brauw, A., B. Balana, K. Andam, et al. 2025. *Anticipatory Cash Transfers for Climate Resilience: Evidence from an Experimental Study in Northern Nigeria.* Washington, DC: IFPRI.

de Brauw, A., and J. Hoddinott. 2011. "Must Conditional Cash Transfer Programs Be Conditioned to Be Effective? The Impact of Conditioning Transfers on School Enrollment in Mexico." *Journal of Development Economics* 96 (2): 359–370.

de Haen, H., and J. von Braun. 1983. *The Effects of Food Price and Subsidy Policies on Egyptian Agriculture.* IFPRI Research Report 42. Washington, DC: IFPRI.

de Walque, D., L. Fernald, P. Gertler, and M. Hidrobo. 2017. "Cash Transfers and Child and Adolescent Development." In *Child and Adolescent Health and Development, Disease Control Priorities*, eds. G.P.D. Bundy, N. de Silva, S. Horton, and D. T. Jamison, 3rd ed., vol. 8, 325–342. Washington, DC: World Bank.

Egger, D., J. Haushofer, E. Miguel, P. Niehaus, and M. Walker. 2022. "General Equilibrium Effects of Cash Transfers: Experimental Evidence from Kenya." *Econometrica* 90 (6): 2603–2643.

Ehrlich, P. 1968. *The Population Bomb.* Ballantine Books: New York.

El-Enbaby, H., D.O. Gilligan, N. Karachiwalla, Y. Kassim, and S. Kurdi. 2024. "Cash Transfers, Gender Norms and Women's Control over Decision-Making in Egypt." IFPRI MENA Working Paper 25. IFPRI, Washington, DC.

Evans, D.K., B. Holtemeyer, and K. Kosec. 2019. "Cash Transfers Increase Trust in Local Government." *World Development* 114: 138–155.

Evans, D.K., and K. Kosec. 2023. "Cash Transfers, Trust, and Inter-household Transfers: Experimental Evidence from Tanzania." *World Bank Economic Review* 37 (2): 221–234.

Fernald, L.C.H., and M. Hidrobo. 2011. "Ecuador's Cash Transfer Program (Bono de Desarrollo Humano) and Child Development in Infants and Toddlers: A Randomized Effectiveness Trial." *Social Science and Medicine* 72 (9): 1437–1446.

Fiszbein, A., and N.R. Schady. 2009. *Conditional Cash Transfers: Reducing Present and Future Poverty*. Washington, DC: World Bank Publications.

Gentilini, U. 2024. *Timely Cash: Lessons from 2,500 Years of Giving People Money*. Oxford, UK: Oxford University Press.

Gilligan, D.O., S. Devereux, and J. Tenzing. 2022. *Social Protection: Designing Adaptive Systems to Build Resilience to Climate Change*. In *2022 Global Food Policy Report: Climate Change and Food Systems*, 58–63. Washington, DC: IFPRI.

Gilligan, D.O., K. Hirvonen, J. Leight, and H. Tesfaye. 2024. *Implementer-Led Evaluation and Learning (IMPEL) Evaluation of SPIR II RFSA-Midline Survey Report*. Washington DC: IFPRI.

Gilligan, D.O., and J. Hoddinott. 2007. "Is There Persistence in the Impact of Emergency Food Aid? Evidence on Consumption, Food Security and Assets in Rural Ethiopia." *American Journal of Agricultural Economics* 89 (2): 225–242.

Gilligan, D.O., and S. Roy. 2014. *Resources, Stimulation, and Cognition: How Transfer Programs and Preschool Shape Cognitive Development in Uganda*. Washington, DC: IFPRI.

Haddad, L.J., J.F. Hoddinott, and H. Alderman. 1997. *Intrahousehold Resource Allocation in Developing Countries: Models, Methods, and Policies*. Washington, DC: IFPRI.

Haushofer, J., and J. Shapiro. 2018. *The Long-Term Impact of Unconditional Cash Transfers: Experimental Evidence from Kenya*. Busara Center for Behavioral Economics, Nairobi.

Heath, R., M. Hidrobo, and S. Roy. 2020. "Cash Transfers, Polygamy, and Intimate Partner Violence: Experimental Evidence from Mali." *Journal of Development Economics* 143: 102410.

Hidrobo, M., and L. Fernald. 2013. "Cash Transfers and Domestic Violence." *Journal of Health Economics* 32 (1): 304–319.

Hidrobo, M., J. Hoddinott, A. Peterman, A. Margolies, and V. Moreira. 2014. "Cash, Food, or Vouchers? Evidence from a Randomized Experiment in Northern Ecuador." *Journal of Development Economics* 107: 144–156.

Hidrobo, M., N. Karachiwalla, and S. Roy. 2023. "The Impacts of Cash Transfers on Mental Health and Investments: Experimental Evidence from Mali." *Journal of Economic Behavior & Organization* 216: 608–630.

Hidrobo, M., V. Mueller, and S. Roy. 2022. "Cash Transfers, Migration, and Gender Norms." *American Journal of Agricultural Economics* 104 (2): 550–568.

Hidrobo, M., V. Mueller, S. Roy, C. Noreyni Fall, C. Lavaysse, and A. Belli. 2024. "Rainy Day Funds? How Men and Women Adapt to Heavy Rainfall Shocks and the Role of Cash Transfers in Mali." IFPRI Discussion Paper 2301. IFPRI, Washington, DC.

Hidrobo, M., A. Peterman, and L. Heise. 2016. "The Effect of Cash, Vouchers, and Food Transfers on Intimate Partner Violence: Evidence from a Randomized Experiment in Northern Ecuador." *American Economic Journal: Applied Economics* 8 (3): 284–303.

Hidrobo, M., A. Peterman, N. Kumar, M. Lambon-Quayefio, S. Roy, D.O. Gilligan and F. Paz. "Social Protection and Gender: Policy, Practice and Research." In *The Handbook of Social Protection: Evidence to Inform Policy in Low- and Middle-Income Countries*, eds. Rema Hanna and Ben Olken. Cambridge, MA: MIT Press, forthcoming.

Hirvonen, K. 2023. "Social Protection: Adaptive Safety Nets for Crisis Recovery." In *2023 Global Food Policy Report: Rethinking Food Crisis Responses*, 52–61. Washington, DC: IFPRI.

Hirvonen, K., D.O. Gilligan, J. Leight, H. Tambet and V. Villa. 2023. "Do Ultra-Poor Graduation Programs Build Resilience against Droughts? Evidence from Rural Ethiopia." IFPRI Discussion Paper 2206. IFPRI, Washington, DC.

Hirvonen, K., E.A. Machado, A.M. Simons, and V. Taraz. 2022. "More Than a Safety Net: Ethiopia's Flagship Public Works Program Increases Tree Cover." *Global Environmental Change* 75: 102549.

Hoddinott, J. 2001. "Targeting: Principles and Practice." In *Food Security in Practice: Methods for Rural Development Projects*, 89–97. Washington, DC: IFPRI.

Hoddinott, J., G. Berhane, D.O. Gilligan, N. Kumar, and A. Seyoum Taffesse. 2012. "The Impact of Ethiopia's Productive Safety Net Programme and Related Transfers on Agricultural Productivity." *Journal of African Economies* 21 (5): 761–786.

Hoddinott, J., and L. Haddad. 1995. "Does Female Income Share Influence Household Expenditures? Evidence from Côte D'Ivoire." *Oxford Bulletin of Economics and Statistics* 57 (1): 77–96.

Hoddinott, J., S. Sandstrom, and J. Upton. 2019. "The Impact of Cash and Food Transfers: Evidence from a Randomized Intervention in Niger." *American Journal of Agricultural Economics* 100 (4): 1032–1049.

Hopkins, R.F., and D.J. Puchala, eds. 1978. *The Global Political Economy of Food*. St. Paul, MN: University of Minnesota Press.

Jacoby, H.G. 2002. "Is There an Intrahousehold 'Flypaper Effect'? Evidence from a School Feeding Programme." *The Economic Journal* 112 (476): 196–221.

Jensen, N.D., F.P. Fava, A.G. Mude, et al. 2024. *Escaping Poverty Traps and Unlocking Prosperity in the Face of Climate Risk: Lessons from Index-Based Livestock Insurance*. Cambridge, UK: Cambridge University Press.

Kazianga, H., D. de Walque, and H. Alderman. 2014. "School Feeding Programs, Intrahousehold Allocation and the Nutrition of Siblings: Evidence from a Randomized Trial in Rural Burkina Faso." *Journal of Development Economics* 106: 15–34.

Kennedy, E.T., and H. Alderman. 1987. *Comparative Analyses of Nutritional Effectiveness of Food Subsidies and Other Food-Related Interventions*. Washington, DC: IFPRI and World Health Organization-UNICEF.

Kidd, S., B. Gelders, and D. Bailey-Athias. 2017. "Exclusion by Design: An Assessment of the Effectiveness of the Proxy Means Test Poverty Targeting Mechanism." Extension of Social Security (ESS) Working Paper 56. International Labour Organization, Geneva.

Kurdi, S. 2021. "The Nutritional Benefits of Cash Transfers in Humanitarian Crises: Evidence from Yemen." *World Development* 148: 105664.

Leight, J., H. Alderman, D.O. Gilligan, M. Hidrobo, and M. Mulford. 2024. "Can a Light-Touch Graduation Model Enhance Livelihood Outcomes? Evidence from Ethiopia." IFPRI Discussion Paper 2203. IFPRI, Washington, DC.

Leight, J., K. Hirvonen, N. Karachiwalla, and D. Rakshit. 2024. *Building Pathways out of Poverty for Ultra-Poor Internally Displaced People in Baidoa*. Washington, DC: IFPRI.

Leight, J., K. Hirvonen, and S. Zafar. 2024. "The Effectiveness of Cash and Cash Plus Interventions on Livelihoods Outcomes: Evidence from a Systematic Review and Meta-Analysis." IFPRI Discussion Paper 2262. IFPRI, Washington, DC.

Leroy, J.L., A. García-Guerra, R. García, C. Dominguez, J. Rivera, and L.M. Neufeld. 2008. "The Oportunidades Program Increases the Linear Growth of Children Enrolled at Young Ages in Urban Mexico." *Journal of Nutrition* 138 (4): 793–798.

Manley, J., H. Alderman, and U. Gentilini. 2022. "More Evidence on Cash Transfers and Child Nutritional Outcomes: A Systematic Review and Meta-Analysis." *BMJ Global Health* 7 (4): e008233.

Margolies, A., and J. Hoddinott. 2015. "Costing Alternative Transfer Modalities." *Journal of Development Effectiveness* 7 (1): 1–17.

Neihaus, P., and T. Suri. "Cash Transfers." In *The Handbook of Social Protection: Evidence to Inform Policy in Low- and Middle-Income Countries*, eds. R. Hanna and B. Olken. Cambridge, MA: MIT Press, forthcoming.

Newbery, D.M.G., and J.E. Stiglitz. 1981. *The Theory of Commodity Price Stabilization*. Oxford, UK: Oxford University Press.

Olney, D.K., A. Gelli, N. Kumar, H. Alderman, A. Go, and A. Raza. 2022. "Social Assistance Programme Impacts on Women's and Children's Diets and Nutritional Status." *Maternal & Child Nutrition* 18 (4): e13378.

Peterman, A., B. Schwab, S. Roy, M. Hidrobo, and D.O. Gilligan. 2021. "Measuring Women's Decisionmaking: Indicator Choice and Survey Design Experiments from Cash and Food Transfer Evaluations in Ecuador, Uganda and Yemen." *World Development* 141: 105387.

Pinstrup-Andersen, P., ed. 1988. *Food Subsidies in Developing Countries: Costs, Benefits, and Policy Options*. Baltimore: Johns Hopkins University Press for IFPRI.

Pople, A., R. Hill, S. Dercon, and B. Brunckhorst. 2024. *The Importance of Being Early: Anticipatory Cash Transfers for Flood-Affected Households*. Oxford, UK: University of Oxford.

Quisumbing, A., D.O. Ahmed, J. Gilligan, et al. 2020. "Randomized Controlled Trials of Multi-sectoral Programs: Lessons from Development Research." *World Development* 127: 104822.

Quisumbing, A.R. and J.A. Maluccio. 1999. *Intrahousehold Allocation and Gender Relations: New Empirical Evidence (English)*. Policy Research Report on Gender and Development Working Paper Series 2. Washington, DC: World Bank.

Roy, S., M. Hidrobo, J. Hoddinott, and A. Ahmed. 2019. "Transfers, Behavior Change Communication, and Intimate Partner Violence: Post-Program Evidence from Rural Bangladesh." *Review of Economics and Statistics* 101 (5): 865–877.

Roy, S., M. Hidrobo, J. Hoddinott, B. Kolt, and A. Ahmed. 2024. "Can Transfers and Complementary Nutrition Programming Reduce Intimate Partner Violence Four Years Post-Program?" *Journal of Human Resources* 59 (6): 1714.

Sedlmayr R., A. Shah, and M. Sulaiman. 2020. "Cash-Plus: Poverty Impacts of Alternative Transfer-Based Approaches." *Journal of Development Economics* 144: 102418.

Sessou, F.E., M. Hidrobo, S. Roy, and L. Huybregts. 2024. "Educational Impacts of an Unconditional Cash Transfer Program in Mali." *Economics of Education Review* 101: 102547.

Schultz, T.P. 2004. "School Subsidies for the Poor: Evaluating the Mexican Progresa Poverty Program." *Journal of Development Economics* 74 (1): 199–250.

Schwab, B. 2019. "In the Form of Bread? A Randomized Comparison of Cash and Food Transfers in Yemen." *American Journal of Agricultural Economics* 102 (1): 91–113.

Skoufias, E. 2005. *PROGRESA and Its Impacts on the Welfare of Rural Households in Mexico*. IFPRI Research Report 139. Washington, DC: IFPRI.

Skoufias, E., and D. Coady. 2007. "Are the Welfare Losses from Imperfect Targeting Important?" *Economica* 74 (296): 756–776.

Stifel, D., and H. Alderman. 2005. "Targeting at the Margin: The 'Glass of Milk' Subsidy Program in Peru." *Journal of Development Studies* 41 (5): 839–864.

Timmer, C.P. 2010. "Reflections on Food Crises Past." *Food Policy* 35: 1–11.

Valli, E., A. Peterman, and M. Hidrobo. 2019. "Economic Transfers and Social Cohesion in a Refugee-Hosting Setting." *Journal of Development Studies* 55 (Supp. 1): 128–146.

Veras, F., R. Peres, and R. Guerreiro. 2008. "Evaluating the Impact of Brazil's Bolsa Família: Cash Transfer Programmes in Comparative Perspective." *Latin American Research Review* 45 (2): 173–190.

Webb, P., J. von Braun, and Y. Yohannes. 1992. *Famine in Ethiopia: Policy Implications of Coping Failure at National and Household Levels*. IFPRI Research Report 92. Washington, DC: IFPRI.

World Bank. 2018. *The State of Social Safety Nets 2018*. Washington, DC.

Yamano, T., H. Alderman, and L. Christiaensen. 2005. "Child Growth, Shocks, and Food Aid in Rural Ethiopia." *American Journal of Agricultural Economics* 87 (2): 273–288.

Chapter 12

NUTRITION AND DIETS
Research and Action, Looking Back to Move Forward

Stuart Gillespie, Marie T. Ruel, Jef L. Leroy, Deanna Olney, and Anusara Singhkumarwong

Stuart Gillespie is a non-resident senior fellow, IFPRI; **Marie T. Ruel** is a senior research fellow, Nutrition, Diets, and Health Unit (NDH), IFPRI; **Jef L. Leroy** is a senior research fellow, NDH, IFPRI; **Deanna Olney** is director of NDH, IFPRI; and **Anusara Singhkumarwong** is a nutritionist, World Food Programme.

Key messages

- Over the past 50 years, the nature of malnutrition has shifted, with overweight and obesity growing steadily, while problems of undernutrition have declined, albeit slowly. Every country is now facing multiple burdens of malnutrition, often within the same communities, households, and even individuals.
- Research has evolved in its contribution and response to changing trends in nutrition policy and development; the focus has shifted from multisectoral planning to the relative isolation of nutrition from other sectors, and then back to a greater emphasis on nutrition as a key driver and key outcome of development.
- Research has played a critical role in providing evidence of what works, how, and at what cost for improving nutrition through agriculture, social protection, education, and women's empowerment programs and policies.
- Countries are now increasingly demanding knowledge and insights on *how* to drive progress in nutrition policy and programming, and especially on how to leverage food systems to achieve sustainable healthy diets, with a sharper focus on political economy and equity.

Looking to the future, research on nutrition policy and programming should aim to:

- **Continue building the evidence base** needed to facilitate food systems transformation, including developing evaluation tools and methods to capture systemwide effects of interventions and refining modeling tools and techniques to capture diet and nutrition outcomes more accurately.
- **Strengthen the understanding of how food environment interventions affect diet and nutrition outcomes**; this includes research on how best to describe and analyze complex food environments and identify which interventions work best, in which contexts, and at what cost.
- **Investigate the potential role and benefit of public–private partnerships** with companies whose products and practices support healthy diets and nutrition.
- **Examine how interventions implemented by sectors such as health,** water and sanitation, social protection, education, and women's empowerment can complement and enhance the positive impacts of food systems transformation on sustainable healthy diets, nutrition, and health.
- **Invest in capacity strengthening and policy research** that generates evidence of what works and how within complex food systems, while also identifying how this evidence can be best used to drive change.

The last half-century has seen major changes in the nature of malnutrition around the world, as well as in our understanding of its manifestations and key drivers, the people most affected, and the policies and programs developed to address it. In 1975, the primary concern was undernourishment and hunger, or insufficient dietary energy (UN 1974). At the beginning of the 21st century, the focus started to shift to the "double burden of malnutrition," with overweight and obesity joining undernutrition and micronutrient deficiencies to become a major public health challenge around the world (Gillespie and Haddad 2001). Over the decades, evidence has accumulated on the links between malnutrition and a range of noncommunicable diseases (NCDs) and on the pivotal importance of healthy diets and nutritional well-being in driving human and economic development.

In this chapter, we review the evolution of nutrition in both policy and programming before looking to the future. Aligning with IFPRI's mandate, we pay particular attention to agriculture, food systems, and multisectoral approaches to addressing malnutrition.

The chapter comprises three sections. In the first, we examine the last 50 years of policy relevant to nutrition, highlighting key events and priorities. In the second, we focus on how nutrition has been incorporated into agricultural programming, and in the third, we discuss future directions for nutrition policy, programming, and research as the field continues to evolve. Throughout we highlight IFPRI's contributions since its founding, and potential future research contributions.

Evolution of nutrition-relevant policy and research

The global food crisis of the early 1970s led the first-ever World Food Conference to focus on access to food in addition to aggregate food production. Hosted by the Food and Agriculture Organization of the United Nations in 1974, the conference concluded with a proclamation that has been reiterated thousands of times since: "Every man, woman, and child has the inalienable right to be free from hunger and malnutrition."[1] That same year, Donald McLaren's *The Great Protein Fiasco* refuted the long-standing belief that protein deficiency was the main cause of malnutrition (McLaren 1974). This publication followed work in India showing that if people's diets were adequate in calories, their protein intake would also likely be adequate (Sukhatme 1970). Solutions to malnutrition began to be sought in the wider social and economic arena, well beyond technological nutrient fixes.

In 1976, the World Bank published a groundbreaking study showing that economic growth alone was not the solution to malnutrition (Reutlinger and Selowsky 1976), a finding later amplified by Amartya Sen's seminal work on entitlements that emphasized the links between inequality, hunger, and famine: "Starvation is the characteristic of some people not having enough food to eat, not the characteristic of there being not enough food to eat" (Sen 1983).

A growing appreciation for the wider drivers of undernutrition paved the way for nutrition planning bodies to be established in 26 countries around the world. By the 1980s, however, most had ceased operations; although addressing malnutrition requires a multisectoral response, such actions cannot be elaborately choreographed by any singular entity, especially when its funding and political clout is inadequate (Field 1987).

1 The World Food Conference was the key event that precipitated the creation of IFPRI in 1975.

The false promise of nutrition planning led to a period of "nutritional isolationism" in the 1990s, during which nutritionists focused largely on two interventions that needed little involvement from other sectors: micronutrient supplementation and breastfeeding (Gillespie et al. 2003). This relative isolationism occurred despite UNICEF's publication of a groundbreaking conceptual framework on the determinants of malnutrition. Showing the importance of underlying and structural drivers, this framework remains influential to this day (UNICEF 1990). In 1996, the World Food Summit in Rome concluded with a commitment to "implement policies aimed at eradicating poverty and inequality and improving physical and economic access by all, at all times, to sufficient, *nutritionally adequate* and safe food and its effective utilization" (FAO 1996). Quality of diets—not just total calories—was being incorporated into food security definitions and goals.

As the new millennium dawned, global nutrition governance was considered fragmented and dysfunctional (Morris et al. 2008), and other challenges captured donor funding and political attention. The accelerating AIDS epidemic in Africa was one such challenge that had major implications for food and nutrition security. IFPRI responded in 2001 by co-founding the Regional Network on AIDS, Livelihoods, and Food Security (RENEWAL) with the International Service for National Agricultural Research. RENEWAL combined locally prioritized research with capacity strengthening and policy communications in a pioneering "network of networks" approach in eastern and southern Africa. Evidence on the bidirectional links between HIV and food and nutrition security was used to adapt agricultural strategies and actions, as well as to strengthen HIV policies aimed at prevention, treatment, care, and mitigation.[2]

Propelled by a confluence of factors, including the food price spikes of 2007/08, nutrition started to garner attention within the development agenda. In 2008, the first *Lancet* Maternal and Child Nutrition Series highlighted the critical importance of nutrition in the first 1,000 days of a child's life—from conception to the second birthday—and recommended a package of nutrition-specific interventions to support lasting benefits throughout life. The same year, the influential Copenhagen Consensus concluded that nutrition interventions were among the most cost-effective in development (Copenhagen Consensus Center 2008). Global Burden of Disease assessments

2 RENEWAL's influence was acknowledged in the 2006 UN Declaration on AIDS that "all people at all times [to] have access to sufficient, safe, and nutritious food...as part of a comprehensive response to HIV/AIDS."

at this time also elevated the importance placed by the public health and biomedical communities on diets and nutrition (IHME 2025).

Harnessing this growing momentum, the Scaling Up Nutrition (SUN) Movement was launched in 2010. This multicountry network of governments, supported by global actors, generated political momentum and increased discourse on nutrition, albeit sidestepping challenges related to commercial drivers of malnutrition.

Research on governance and the political economy of nutrition picked up in the 2010s, with studies drawing on insights and frameworks from the broader literature on development, political science, and health policy. In 2012, IFPRI launched the multipartner Transform Nutrition consortium, partly to address the question of how enabling environments for nutrition could be cultivated to use existing political and economic resources more effectively. The 2013 *Lancet* Maternal and Child Nutrition series featured IFPRI-led research on nutrition-sensitive programs and the politics of reducing malnutrition (Gillespie et al. 2013). The latter article highlighted the pivotal importance of well-communicated evidence, strong governance, capacity, and financing for generating political momentum and transforming it into on-the-ground impact.

The 2010s also saw the launch of CGIAR's Agriculture for Nutrition and Health (A4NH) program, led by IFPRI. Working with partners in high-burden countries, A4NH investigated the connections between gender-sensitive agriculture, nutrition, and health and their program and policy implications (A4NH 2021). The Leveraging Agriculture for Nutrition in South Asia (LANSA) consortium (2012–2018) worked to change public and political discourse on nutrition-sensitive agriculture in the region (A4NH 2018). In India, the Partnerships and Opportunities to Strengthen and Harmonize Actions for Nutrition in India (POSHAN) project was launched in 2011 to generate, synthesize, and mobilize diverse types of nutrition data and evidence to support national and state-level policy decisions (IFPRI, n.d.-a).

By the mid-2010s, many countries were expressing a demand for knowledge and insights on *how* to drive progress in nutrition policy and programming; evidence of *what* to do was not enough. This demand provided the impetus for IFPRI's Stories of Change in Nutrition (SoC) initiative, launched in 2015 (IFPRI, n.d.-b). Working directly with local researchers and policymakers, SoC focused on countries that had exceeded expectations in reducing child undernutrition, and by 2021, it had published success stories from 23 countries and 4 Indian states (Gillespie et al. 2016). SoC identified seven crucial components for progress—commitment, coherence, accountability, data, capacity, finance, and leadership—and showed how stories can *inspire* as well as inform action. Building on emerging lessons from these initiatives, Transform Nutrition West Africa (TNWA) was launched in 2017 to support decisions and actions to improve maternal and child nutrition through an inclusive process of knowledge generation and mobilization (IFPRI, n.d.-c).

These initiatives—Transform Nutrition, A4NH, LANSA, POSHAN, SoC, and TNWA—were all grounded in a collaborative ethic that bridged international and transdisciplinary boundaries to maximize the relevance and use of global and local knowledge. Going beyond research on policy, they actively engaged partners in the policy process. The initiatives also incorporated capacity and leadership development activities: in partnership with the Institute of Development Studies (IDS), a series of summer schools for policymakers and practitioners engaged several hundred students from across the world (Transform Nutrition 2017). Regional adaptations were later launched in South Asia and West Africa.

As the 2010s progressed, recognition grew of the coexistence of multiple forms of malnutrition—including undernutrition, micronutrient deficiencies, and overweight and obesity—that occur within the same countries, communities, households, and even individuals. The global magnitude of this challenge was documented by the 2020 *Lancet* series on the Double Burden of Malnutrition (Popkin et al. 2020), which also introduced the concept of double-duty actions, defined as interventions, programs, and policies that simultaneously prevent or reduce the risk of undernutrition and of overweight, obesity, or diet-related NCDs (Ruel and Hawkes 2019). The need for double-duty actions was driven by evidence showing that some social protection programs that provided food or cash to poor households in highly food insecure areas had done harm by inadvertently exacerbating obesity in women (Leroy et al. 2013, 2019).

The coexistence of multiple forms of malnutrition also demands research and action to address the commercial determinants of health. As food systems have become increasingly imbalanced and misaligned with the needs of people and planet, a select few transnational corporations have dominated global markets with highly profitable but unhealthy, ultra-processed food (UPF) and beverage products. Between 1962 and 2021, the market capitalization of the UPF sector increased from around US$80 billion to more than $1.6 trillion in real terms (Wood et al. 2023). These products—and the commercial and political practices of the companies that sell them—are increasingly associated with a range of harms, including undernutrition, obesity and diet-related NCDs, and environmental degradation (Lane et al. 2024; Monteiro 2009; Van Tulleken 2023).

As the 2019 *Lancet* Commission on the Global Syndemic of Obesity, Undernutrition, and Climate Change highlighted, concerns were growing around not only the hyper-concentration of market power in the food system, but also the way in which companies and their associations sought to influence policy to prevent regulations promoting nutrition and health (Swinburn et al. 2019). This tension was heightened in the lead-up to the 2021 United Nations Food Systems Summit, where opaque governance led many organizations to boycott the Summit (Nisbett et al. 2021). Conflicts of interest emerged as a major research focus (Freudenberg et al. 2021; Gillespie 2023).

Amid the disproportionate effects of the COVID-19 pandemic and unequal access to vaccines, the early 2020s have seen a sharper focus on equity (Alonso et al. 2023; Gillespie 2020). The *2020 Global Nutrition Report*'s core theme of nutrition equity was informed by A4NH's conceptual work, developed from wider connections with health professionals (Development Initiatives 2020; Harris et al. 2021). This work also aligned with a resurgent agenda to decolonize research and development, fueled in part by the Black Lives Matter movement. IFPRI and IDS researchers reflected on what could be learned from a reassessment of the history of international nutrition (Nelson et al. 2021). A greater appreciation of historically configured power asymmetries—including the perspectives of those with limited power in the global arena—is key to avoiding repeated cycles of top-down "quick fixes" and to driving sustainable progress.

In Figure 12.1, we provide a timeline showing key reports, publications, world events, movements and conferences—as well as shifts in program and policy research focus.[3]

Nutrition-sensitive agriculture: Refocusing the goal

Since the 1980s, research has focused extensively on the linkages between agriculture and nutrition, with significant contributions from IFPRI (Pinstrup-Andersen et al. 1984). Early work demonstrated that agriculture affected nutrition through a variety of pathways, such as by increasing employment and farmer's incomes and by lowering food prices. Agriculture could also negatively affect nutrition by increasing workers' nutrient and energy

3 We recognize that perceptions of key events and trends will differ by organization and individual. Our shared perspective derives from our roles as policy and program researchers within IFPRI during much of this period.

FIGURE 12.1 Timeline

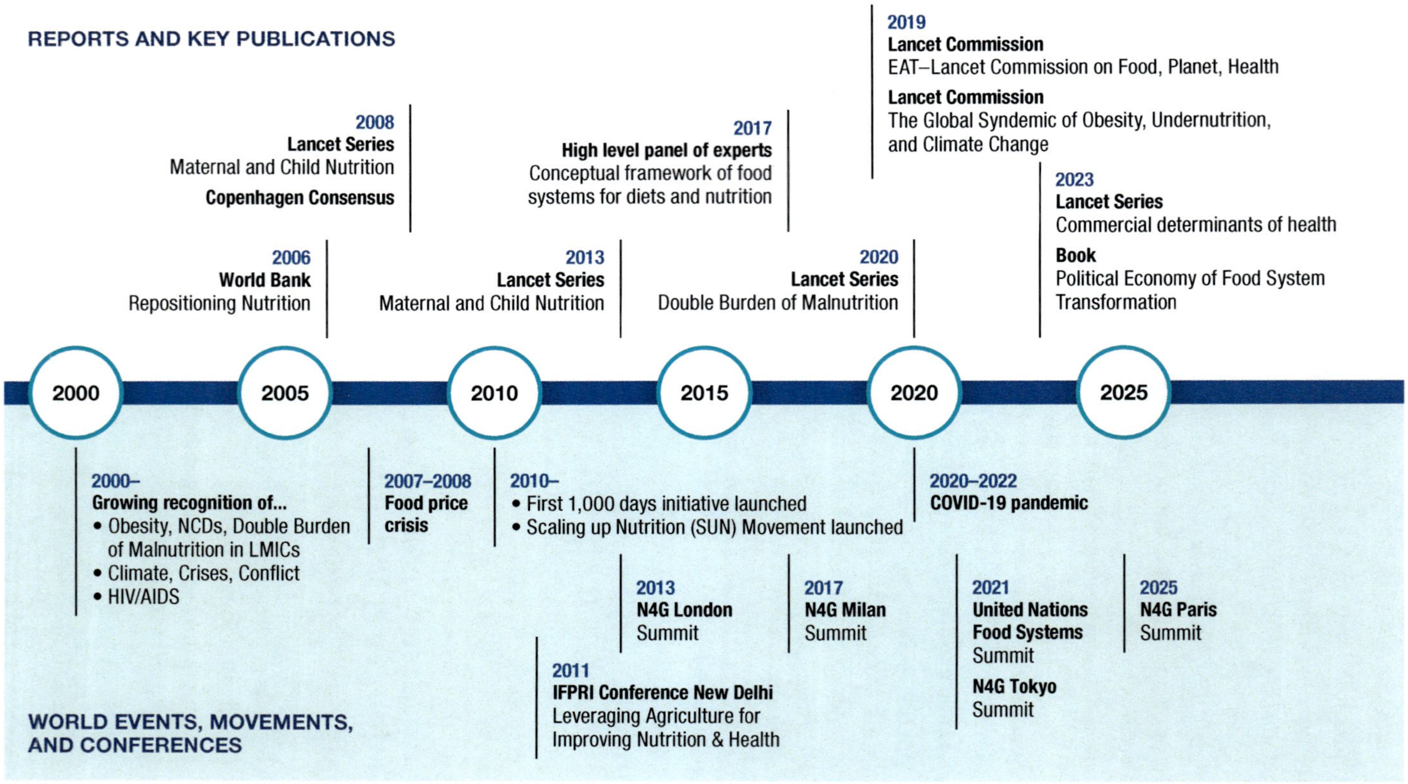

Figure 12.1 continued

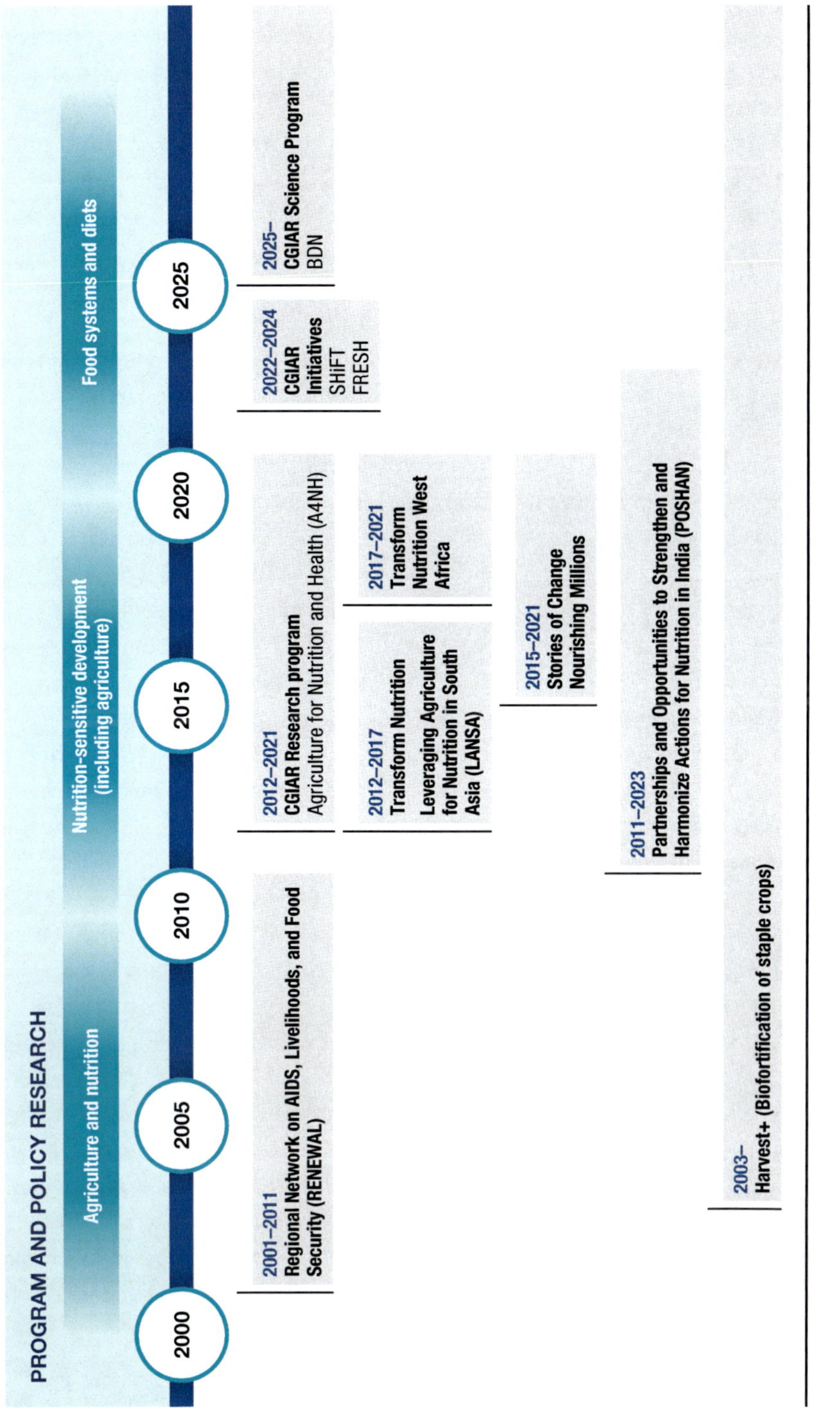

requirements or by reducing women's time available for childcare (Pinstrup-Andersen et al. 1984). In the 1990s, IFPRI's research on the impacts of agricultural commercialization found that income growth was insufficient to improve child nutritional status. Researchers hypothesized that this result was due to the pernicious effects of poor living and water and sanitation conditions and low empowerment of women (von Braun and Kennedy 1994). The pioneering Breeding for Nutrition initiative, later named HarvestPlus, was launched in the late 1990s under the leadership of Howdy Bouis (Bouis and Welch 2010). This innovation successfully delivered staple crops with higher concentrations of bioavailable micronutrients, thus improving micronutrient intake and status among nutritionally vulnerable women and young children (CAST 2020).

The early 21st century saw renewed efforts to better understand the role of agriculture in improving nutrition. Several literature reviews on the linkages between agriculture interventions and nutrition showed beneficial impacts on poverty, production of targeted commodities, and household food security, but no meaningful impacts on child nutrition outcomes or complementary feeding practices.[4] In the 2013 *Lancet* Maternal and Child Nutrition series, IFPRI-led research on nutrition-sensitive programming suggested ways to reshape agriculture programs to enhance their nutrition impact (Ruel and Alderman 2013). Introducing the term "nutrition-sensitive agriculture" (NSA), the researchers highlighted the need for agriculture programs to incorporate specific nutrition goals and interventions, target mother and child dyads during the first 1,000 days of the child's life, and address key underlying determinants of malnutrition. The latter include poverty; food insecurity; inadequate diets; poor caregiving knowledge, practices, and empowerment; and limited access to basic social, health, water and sanitation, and education services.

Increased clarity on the objectives of NSA led to new efforts to develop detailed guidelines and recommendations to strengthen the design, targeting, implementation, and evaluation of NSA programs. Donors, program implementers, and researchers collaborated to carefully design, implement, and rigorously evaluate a new wave of NSA programs (see Box 12.1 for an example of collaboration between IFPRI and the World Food Programme). This evidence showed the new programs had more consistently positive impacts than earlier programs on household food security, consumption of targeted nutritious crops and animal products, and household, maternal, and child dietary diversity

4 Studies include Berti (2004); Leroy and Frongillo (2007); Leroy et al. (2008); Masset et al. (2012); Ruel (2001); Ruel and Alderman (2013); Webb-Girard et al.(2012); and World Bank (2007).

BOX 12.1 Linking nutrition research and implementation: An example of IFPRI-WFP partnership

In 2017, IFPRI and the World Food Programme (WFP) collaborated on the development of guidelines to enhance nutrition-sensitive programming, marking a significant milestone in WFP's global efforts to end malnutrition (WFP 2016). IFPRI and WFP worked together to support country offices in translating these strategies into actionable plans, and selected Sri Lanka as a focal country for this effort.

From 2017 to 2022, WFP's Food for Assets (FFA) program in Sri Lanka worked to address climate-related challenges and improve the resilience of rural agricultural households. Key interventions included the rehabilitation of community water reservoirs and livelihood diversification. Nutrition objectives aimed to diversify the diets of targeted households (including food-insecure households, those with children 6–23 months of age, and those with pregnant or breastfeeding women) and prioritize nutrition-relevant assets, such as backyard and commercial gardening and poultry-rearing along with the integration of nutrition-related social behavior change communication. In 2018, program impact pathways (PIPs), developed through multiple rounds of consultation, helped WFP teams identify opportunities to integrate nutrition objectives and actions into wider program design. The PIPs also provided a basis for subsequent evaluation of program contributions to nutrition outcomes.

Early findings highlighted the importance of integrating nutrition into community asset creation programs to improve diet- and nutrition-related behaviors. Through close collaboration with national research institutes, the study also sought to build local capacity to assess and monitor the impact of nutrition-sensitive programming.

Building on lessons from this work, WFP and IFPRI expanded their collaboration to address urban nutrition challenges in Bangladesh, the Philippines, and Sri Lanka in 2023 and 2024. Opportunities to deepen these efforts include joint advocacy with national governments to support investments in nutrition-relevant implementation research. Lessons from Sri Lanka could bolster advocacy for greater government investment in generating evidence on nutrition-sensitive social protection, agriculture, and school nutrition, as well as other sectoral action.

(Margolies et al. 2022; Ruel et al. 2018; Sharma et al. 2021). Some studies also reported impacts on child health outcomes (such as reductions in common morbidity symptoms) and on maternal health and well-being, nutrition knowledge, and empowerment. Larger benefits were observed in programs that incorporated high-quality health and nutrition social behavior change communication and well-designed women's empowerment sensitization activities. However, impacts on child anthropometric indicators (such as stunting, wasting, and underweight) and micronutrient status were observed only in programs that also included water, sanitation, and hygiene (WASH) interventions to prevent or reduce infections, and/or distributed micronutrient supplements to complement low-quality diets commonly found in areas with limited availability or access to nutrient-rich foods (Ruel et al. 2018; Sharma et al. 2021).

This stronger body of evidence provided key insights into the strengths and limitations of NSA to improve nutrition. It showed that NSA programs, as generally designed, are well suited to addressing some of the underlying determinants of malnutrition (such as household food security, diversity in food consumption, diet quality, and women's empowerment) but are not sufficient to achieve child growth impacts. Based on this evidence, researchers recommended focusing on diets, rather than child anthropometry, as the main achievable goal of NSA programs (Leroy et al. 2020).

The well-documented operational challenges of implementing high-quality multisectoral and multifaceted NSA programs present an additional consideration for shifting to diets as a key outcome of NSA (Leroy et al. 2008; Leroy and Frongillo 2007; Ruel and Alderman 2013; Webb-Girard et al. 2012). Box 12.2 highlights IFPRI's engagement with different sectors in research and action to improve all nutrition outcomes. Achieving impacts on child anthropometry and micronutrient status by adding WASH and micronutrient supplement distribution to all NSA programs would be unrealistic, as it could render programs overly complex and challenging to implement. A related operational question is whether program components provided by different sectors (such as health, water and sanitation, agriculture, and women's empowerment) should be integrated or co-located (Di Prima et al. 2022; Ruel et al. 2018; Sharma et al. 2021). The World Bank's proposed approach—to plan multisectorally and implement sectorally—has not been tested (World Bank 2013), though it holds potential for reducing the complexity, implementation challenges, and cost of NSA programs while also ensuring synergies in impacts. Shifting the focus of NSA programs to improve household and individual food security and diet outcomes, however, is a more realistic and operationally achievable goal.

BOX 12.2 Evaluating nutrition-sensitive multisectoral programs: What works, how, and at what cost

IFPRI's mission is to provide research-based policy solutions that sustainably reduce poverty and end hunger and malnutrition. Ending malnutrition will require not only achieving improvements in food security and diets but also improving—and ultimately preventing—suboptimal nutrition outcomes such as micronutrient deficiencies, wasting, stunting, and overweight and obesity. Multisectoral solutions are critical, but these are challenging to evaluate. IFPRI collaborates with partners to co-design and evaluate multisectoral programs, recognizing the need for rigorous evidence on which programs work, how they function, and their relative cost-effectiveness. The Institute's approach to comprehensive evaluation helps identify not only the impact of these programs, but also barriers, catalysts, and in some cases, cost-effectiveness.

IFPRI's pioneering work in evaluating social protection programs is highlighted in Chapter 11. The Institute's comprehensive evaluations of large-scale programs such as the United States Agency for International Development's Preventing Malnutrition in Children Under 2 Approach (PM2A) and the Alive & Thrive (A&T) Initiative have provided critical evidence about how food assistance programs and behavior change communication (BCC) approaches can be designed to improve diets, nutrition, and well-being (Alive & Thrive 2025; FANTA 2025).

Findings from the PM2A studies in Burundi and Guatemala, for example, provided critical evidence on how the prevention of stunting, wasting, and anemia and improvement of child development outcomes are affected by the timing, duration, and composition of food assistance packages, coupled with health systems strengthening and BCC (Leroy et al. 2016b, 2018, 2021; Olney et al. 2018, 2019). Accompanying studies on cost-effectiveness also highlighted how program timing, duration, and transfer composition are critical to achieving beneficial nutrition outcomes (Heckert et al. 2020).

The A&T Initiative's impact evaluations highlighted the importance of multichannel interventions for increasing exclusive breastfeeding in Bangladesh and Viet Nam, as well as improving child development outcomes in Bangladesh (Frongillo et al. 2017; Menon et al. 2016; Nguyen et al. 2016). Subsequent studies led by IFPRI through the A&T Initiative include research on the impact of strengthening nutrition services during antenatal care to improve antenatal visit attendance and maternal nutrition practices in Burkina Faso (Kim et al. 2023) and the use of school-based programs to improve adolescent nutrition in Ethiopia. IFPRI has also contributed to understanding the impact of school meal programs on nutritional status and other outcomes (Adelman et al. 2019; Chakrabarti et al. 2021; Kazianga et al. 2014).

Future directions

Food systems

In recent years, there has been growing attention to the need to transform food systems to improve diets, planetary health, and livelihoods (UN 2023). Such a transformation must address barriers to improving the healthiness of diets by increasing the desirability, affordability, accessibility, and availability of nutritious foods—a challenge that will also require tackling food environments replete with unhealthy foods (IFPRI 2024). While the potential of transformation is promising, the process is complex and unpredictable, beset by competing demands and the need to carefully consider trade-offs around nutrition and health, the environment, gender equity, and economic growth (IFPRI 2024). Eating a more planet-friendly diet that includes significantly less animal-source food (ASF), for instance, can lead to substantial deficits in micronutrient adequacy (Kazianaga et al. 2014; Kim et al. 2023), especially for the most nutritionally vulnerable populations, such as women of reproductive age, adolescents, and young children in low- and middle-income countries (LMICs) (Beal 2024; Leonard et al. 2024).

Despite the growing focus on food systems transformation, evidence is needed to demonstrate that systemwide interventions are feasible and effective means of improving diets and nutrition. The work of designing, implementing, and evaluating food systems interventions presents several methodological challenges (Neufeld et al. 2024), including the complexity of designing approaches to address multiple forms of malnutrition and simultaneously changing various food system components in a given geographical space. The interconnectivity of different food system components presents another challenge, as changes at any level could have both positive and negative effects. These competing effects can have wide-ranging implications for diets, income, and other relevant outcomes, as well as the broader interests of public and private sector stakeholders.

Through the new CGIAR Science Programs, including the Better Diets and Nutrition (BDN) Science Program, IFPRI will continue building the evidence needed to improve diets through food systems transformation. For example, the CGIAR Research Initiative on Fruit and Vegetables for Sustainable Healthy Diets has co-designed end-to-end approaches to increase the intake of fruits and vegetables in Sri Lanka and Tanzania; the effectiveness of these approaches for increasing dietary intake and vegetable production is currently being evaluated (Bliznashka et al. 2023). The BDN Science Program will continue these efforts, expanding them to include aquatic foods and ASFs. These approaches aim to simultaneously improve the desirability, affordability, accessibility, and availability of healthy foods.

A significant challenge, however, involves the need to isolate the impact of intervention outcomes from external factors and document "general equilibrium" effects that extend throughout the system. Traditional methods, in which participants are randomly assigned to a treatment or a control group, can offer insights into what is possible across a set of interventions, but these are unlikely to capture systemwide effects. Alternative evaluation methods thus need further development (Neufeld et al. 2024). In addition, modeling techniques to predict the systemwide impacts of policy changes need to be refined to more accurately capture nutrition outcomes, such as the healthiness of diets and population-level prevalence of micronutrient deficiencies and overweight and obesity.

Food environments

The last 15 years have seen growing attention to food environments—the places where people interact with and access food as part of the wider food system (Turner et al. 2018). Food environments are undergoing dramatic changes in LMICs, which are giving rise to several challenges, including the ubiquity and extensive marketing of UPFs; the limited accessibility and affordability of nutritious fresh foods such as fruits, vegetables, and ASF, and the need to ensure the safety of these foods; and the growing importance of the digital food environment. Within this digital environment, primarily unhealthy foods are being marketed and advertised through social media, a new avenue that is difficult to monitor or regulate (Fretes et al. 2024).

Evidence on the effectiveness of food environment interventions is limited. Mandatory front-of-package labeling and taxes on select unhealthy foods and beverages have been shown to reduce consumption of unhealthy products, but more evidence is needed on the impacts on total diet or nutrition outcomes.[5] The United Kingdom's 5 A Day campaign, which promotes the daily consumption of five portions of fruits and vegetables, was shown to increase consumption of these food groups by around half a portion per day (Castiglione and Mazzocchi 2019). In collaboration with other CGIAR Centers and in-country partners, IFPRI is engaged in research on food environments in LMICs. Future work will focus on how best to describe, qualify, and analyze

5 On front-of package labeling, see Beal (2024); Bliznashka et al. (2023); Leonard et al. (2024); and Neufeld et al. (2024). On taxes, see Andreyeva et al. (2022); Caro et al. (2018); Cawley et al. (2019); Colchero et al. (2017); and Taillie et al. (2017).

complex food environments and to identify which food environment interventions work best, in which contexts, and at what cost. Interventions being considered for evaluation include policies and actions to improve the availability, desirability, and affordability of healthy foods, including perishable fruits and vegetables and ASFs; information and behavior change campaigns to promote sustainable healthy foods and diets; retail-level interventions such as product placement; restrictions on the marketing of unhealthy foods to vulnerable groups, such as children; and policies for regulation of the digital food environment, which is currently limited by the absence of valid metrics for monitoring (Fretes et al. 2024).

Private sector

The private sector and public–private partnerships (PPPs) are essential to achieve the scalable and sustainable changes needed in food systems to address barriers to consumption of healthy diets. However, guardrails must be put in place to ensure that contributions from the private sector are positive (Kraak et al. 2012). Doing so must include a closer examination of the role of multistakeholder initiatives and platforms intended to support food systems transformation. Multinational corporations with a focus on food, agriculture, and health have tremendous market, financial, and, increasingly, political power in global food systems (van den Akker et al. 2024). The well-established negative influence of many products and practices—including the marketing of breastmilk substitutes (Anttila-Hughes et al. 2018; Rollins et al. 2023) and the saturation of food environments with heavily marketed, hyperpalatable, and unhealthy UPFs—calls for strong regulation, enhanced systems of accountability, and careful consideration for any future engagement.

Many private sector actors, including small and medium enterprises (SMEs), have the capacity, knowledge, and motivation to drive positive change, as shown by the example of large-scale food fortification (Garrett 2018). As resources continue to be constrained, PPPs that adhere to shared pro-nutrition principles are critical to achieving scalable and sustainable solutions. To garner the sustained involvement of the private sector, actions are needed to make safe, sustainable, and healthy diets profitable for actors across the food system, as well as making them affordable and desirable for consumers. This could be achieved by increasing the availability, accessibility, affordability, and desirability of perishable nutrient-rich foods: promising solutions include innovations in crop breeding, production, and postharvest processes, as well as leveraging marketing expertise from the private sector to increase the desirability of foods such as fruits and vegetables and to sustain support for fortification of staple foods.

Alternative proteins

Alternative proteins derived from plants, microorganisms, insects, and cultivated proteins are drawing greater attention due to the environmental footprint of these foods, which is considerably lower than that of ASF (Bryant 2022). Reducing production costs, increasing consumer acceptance, and understanding the nutritional effects of adding alternative proteins to diets in LMICs require additional research (Mukherji et al. n.d.). Given that micronutrient deficiencies in these regions are highly prevalent, further research is also needed to understand whether these products provide similar amounts of bioavailable micronutrients as ASF.

Beyond the focus on diets and the food system

The recent focus on food systems and food environments risks shifting attention away from sectors such as health, social protection, and education that play a critical role in improving nutrition outcomes. Improvements in food systems are essential but insufficient to lead to better nutrition outcomes, as they cannot address underlying determinants such as limited purchasing power and poor health. In addition, food systems may cease to operate in situations of conflict, natural disasters, or weather shocks. These breakdowns can lead to humanitarian crises and problems of acute malnutrition, which need to be addressed through other sectors. IFPRI will continue its work on identifying effective and affordable interventions through comprehensive evaluations of programs implemented by different sectors (Leroy et al. 2016a). These studies will pay particular attention to how to address poor-quality implementation and limited coverage, which constrain the impact of these interventions (Quisumbing et al. 2020; Warren et al. 2020). Implementation research will include a focus on equity and the empowerment of women (Njuki et al. 2023), youth, and Indigenous populations as critical actors in evolving food systems.

Overall, we need to generate evidence of what works and how, while also identifying how this evidence can be best used for positive program and policy changes. This will require new approaches to capacity sharing and policy research, as well as engagement, to ensure that the evidence being generated is demand driven and can be used to inform the global evidence base and address context-specific needs.

References

Adelman, S., Daniel O Gilligan, Joseph Konde-Lule, and Harold Alderman. 2019. "School Feeding Reduces Anemia Prevalence in Adolescent Girls and Other Vulnerable Household Members in a Cluster Randomized Controlled Trial in Uganda." *Journal of Nutrition* 149 (4): 659–666.

A4NH (Agriculture for Nutrition and Health). 2018. "FLAGSHIP 4: Supporting Policies, Programs, and Enabling Action through Research (SPEAR), Highlight Three: Culmination of Leveraging Agriculture for Nutrition in South Asia (LANSA)." In *A4NH 2018 Annual Report*. Washington, DC.

A4NH. 2021. "Pathways to Progress: The Story of Agriculture for Nutrition and Health and What Comes Next." Webinar, November 4.

Alive & Thrive. 2025. "Better Nutrition for Women and Children: Better Outcomes for All." Accessed January 10. www.aliveandthrive.org/en

Alonso, S., M.D. Angel, E. Muunda, E. Kilonzi, G. Palloni, D. Grace, and J.L. Leroy. 2023. "Consumer Demand for Milk and the Informal Dairy Sector Amidst COVID-19 in Nairobi, Kenya." *Current Developments in Nutrition* 7 (4): 100058.

Andreyeva, T., K. Marple, S. Marinello, T.E. Moore, and L.M. Powell. 2022. "Outcomes Following Taxation of Sugar-Sweetened Beverages." *JAMA Network Open* 5 (6): e2215276.

Anttila-Hughes, J., L.C.H. Fernald, P. Gertler, P. Krause, E. Tsai, and B. Wydick. 2018. "Mortality from Nestlé's Marketing of Infant Formula in Low and Middle-Income Countries." NBER Working Paper 24452. National Bureau of Economic Research, Cambridge, MA.

Beal, T. 2024. "Environmentally Protective Diets May Come with Trade-Offs for Micronutrient Adequacy." *American Journal of Clinical Nutrition* 119 (4): 872–873.

Berti, P.R., J. Krasevec, and S. FitzGerald. 2004. "A Review of the Effectiveness of Agriculture Interventions in Improving Nutrition Outcomes." *Public Health Nutrition* 7 (5): 599–609.

Bliznashka, L., N. Kumar, J. Kinabo, et al. 2023. *Improving Fruit and Vegetable Intake and Production in Tanzania: An Evaluation of the FRESH End-to-End Approach*. FRESH Research Brief 1. Washington, DC: IFPRI.

Bouis, H.E., and R.M. Welch. 2010. "Biofortification—A Sustainable Agricultural Strategy for Reducing Micronutrient Malnutrition in the Global South." *Crop Science* 50 (Suppl 1): S-20–S-32.

Bryant, C.J. 2022. "Plant-Based Animal Product Alternatives Are Healthier and More Environmentally Sustainable than Animal Products." *Future Foods* 6 (December):100174.

Caro, J.C., C. Corvalan, M. Reyes, A. Silva, B. Popkin, and L.S. Taillie. 2018. "Chile's 2014 Sugar Sweetened Beverage Tax and Changes in Prices and Purchases of Sugar-Sweetened Beverages: An Observational Study in an Urban Environment." *PLoS Medicine* 15 (7): e1002597.

CAST (Council for Agricultural Science and Technology). 2020. "Food Biofortification: Reaping the Benefits of Science to Overcome Hidden Hunger: A Paper in the Series on the Need for Agricultural Innovation to Sustainably Feed the World by 2050." Issue Paper 69. Ames, IA.

Castiglione, C., and M. Mazzocchi. 2019. "Ten Years of Five-a-Day Policy in the UK: Nutritional Outcomes and Environmental Effects." *Ecological Economics* 157 (March): 185–194.

Cawley, J., A.M. Thow, K. Wen, and D. Frisvold. 2019. "The Economics of Taxes on Sugar-Sweetened Beverages: A Review of the Effects on Prices, Sales, Cross-Border Shopping, and Consumption." *Annual Review of Nutrition* 39: 317–338.

Chakrabarti, S., S.P. Scott, H. Alderman, P. Menon, and D.O. Gilligan. 2021. "Intergenerational Nutrition Benefits of India's National School Feeding Program." *Nature Communications* 12: 4248.

Colchero, M.A., J. Rivera-Dommarco, B.M. Popkin, and S.W. Ng. 2017. "In Mexico, Evidence of Sustained Consumer Response Two Years after Implementing a Sugar-Sweetened Beverage Tax." *Health Affairs* 36 (3): 564–571.

Copenhagen Consensus Center. 2008. *Copenhagen Consensus 2008: Results*. Copenhagen.

Development Initiatives. 2020. *Development Initiatives 2020 Global Nutrition Report: Action on Equity to End Malnutrition*. Bristol, UK.

Di Prima, S., E.P. Wright, I.K. Sharma, E. Syurina, and J.E.W. Broerse. 2022. "Implementation and Scale-Up of Nutrition-Sensitive Agriculture in Low- and Middle-Income Countries: A Systematic Review of What Works, What Doesn't Work and Why." *Global Food Security* 32: 100595.

FANTA (Food and Nutrition Technical Assistance). 2025. "PM2A: Preventing Malnutrition in Children Under 2 Approach." Accessed January 10. www.fantaproject.org/focus-areas/maternal-and-child-health-and-nutrition/pm2a

FAO (Food and Agriculture Organization of the United Nations). 1996. "Rome Declaration on World Food Security and World Food Summit Plan of Action." World Food Summit, November 13–17, 1996. Rome.

Field, J.O. 1987. "Multisectoral Nutrition Planning: A Post-Mortem." *Food Policy* 12 (1): 15–28.

Fretes, G., M. Quinn, and J.L. Leroy. 2024. "Food Environments: Improving Their Healthfulness." In *Global Food Policy Report 2024: Food Systems for Healthy Diets and Nutrition*, 46–52. Washington, DC: IFPRI.

Freudenberg, N., K. Lee, K. Buse, et al. 2021. "Defining Priorities for Action and Research on the Commercial Determinants of Health: A Conceptual Review." *American Journal of Public Health* 111 (12): 2202–2211.

Frongillo, E.A., P.H. Nguyen, K.K. Saha, et al. 2017. "Large-Scale Behavior-Change Initiative for Infant and Young Child Feeding Advanced Language and Motor Development in a Cluster-Randomized Program Evaluation in Bangladesh." *Journal of Nutrition* 147 (2): 256–263.

Garrett, G.S., C. Manus, and A. Bleuthner. 2018. "The Importance of Public–Private Collaboration in Food Fortification Programs." In *Food Fortification in a Globalized World*, eds. M.G. Venkatesh Mannar and R. Hurrell, 113–120. Amsterdam: Elsevier.

Gillespie, S. 2020. "Epidemics and Food Systems: What Gets Framed, Gets Done." *Food Security* 12 (4): 895–898.

Gillespie, S. 2023. "Beyond Reasonable Doubt: Avoiding Conflict of Interest in Nutrition Research." ANH Academy [blog], September 28.

Gillespie, S., and L. Haddad. 2001. *Attacking the Double Burden of Malnutrition in Asia and the Pacific*. Manila: Asian Development Bank.

Gillespie, S., L. Haddad, V. Mannar, P. Menon, and N. Nisbett. 2013. "The Politics of Reducing Malnutrition: Building Commitment and Accelerating Progress." *Lancet* 382 (9891): 552–569.

Gillespie, S., J. Hodge, S. Yosef, and R. Pandy-Lorch R, eds. 2016. *Nourishing Millions: Stories of Change in Nutrition*. IFPRI: Washington, DC.

Gillespie, S., M. McLachlan, and R. Shrimpton. 2003. *Combating Malnutrition: Time to Act*. Human Development Network Health, Nutrition, and Population Series. Washington, DC: World Bank.

Harris, J., W. Tan, B. Mitchell, and D. Zayed. 2021. "Equity in Agriculture–Nutrition–Health Research: A Scoping Review." *Nutrition Review* 80 (1): 78–90.

Heckert, J., J.L. Leroy, D.K. Olney, S. Richter, E. Iruhiriye, and M.T. Ruel. 2020. "The Cost of Improving Nutritional Outcomes through Food-Assisted Maternal and Child Health and Nutrition Programmes in Burundi and Guatemala." *Maternal & Child Nutrition* 16 (1): e12863.

IFPRI (International Food Policy Research Institute). n.d.-a. "Partnerships and Opportunities to Strengthen and Harmonize Actions for Nutrition in India (POSHAN)." Accessed January 10, 2025. https://poshan.ifpri.info/

IFPRI. n.d.-b. "Stories of Change in Nutrition." Accessed January 10, 2025. www.ifpri.org/project/stories-change-nutrition/

IFPRI. n.d.-c. "Transform Nutrition West Africa." Accessed January 10, 2025. https://view.ceros.com/ifpri/project-highlights/p/1

IFPRI. 2024. *Global Food Policy Report 2024: Food Systems for Healthy Diets and Nutrition*. Washington, DC.

IHME (Institute for Health Metrics and Evaluation). 2025. "Global Burden of Disease (GBD) History." www.healthdata.org/research-analysis/about-gbd/history

Kazianga, H., D. de Walque, and H. Alderman. 2014. "School Feeding Programs, Intrahousehold Allocation and the Nutrition of Siblings: Evidence from a Randomized Trial in Rural Burkina Faso." *Journal of Development Economics* 106: 15–34.

Kim, S.S., R.R. Zagré, C.T. Ouédraogo, et al. 2023. "Intensified Nutrition Interventions in Antenatal Care Services Increased Consumption of Iron and Folic Acid Supplements and Early Breastfeeding Practices in Burkina Faso: Results of a Cluster-Randomized Program Evaluation." *Journal of Nutrition* 153 (10): 3058–3067.

Kraak, V.I., P.B. Harrigan, M. Lawrence, P.J. Harrison, M.A. Jackson, and B. Swinburn. 2012. "Balancing the Benefits and Risks of Public–Private Partnerships to Address the Global Double Burden of Malnutrition." *Public Health Nutrition* 15 (3): 503–517.

Lane, M.M., E. Gamage, S. Du, et al. 2024. "Ultra-Processed Food Exposure and Adverse Health Outcomes: Umbrella Review of Epidemiological Meta-Analyses." *BMJ* (February): e077310.

Leonard, U.M., C.L. Leydon, E. Arranz, and M.E. Kiely. 2024. "Impact of Consuming an Environmentally Protective Diet on Micronutrients: A Systematic Literature Review." *American Journal of Clinical Nutrition* 119 (4): 927–948.

Leroy, J., and E. Frongillo. 2007. "Can Interventions to Promote Animal Production Ameliorate Undernutrition?" *Journal of Nutrition* 137 (10): 2311–2316.

Leroy, J.L., P. Gadsden, T. González de Cossío, and P. Gertler. 2013. "Cash and In-Kind Transfers Lead to Excess Weight Gain in a Population of Women with a High Prevalence of Overweight in Rural Mexico." *Journal of Nutrition* 143 (3): 378–383.

Leroy, J.L., D.K. Olney, N. Nduwabike, and M.T. Ruel. 2021. "Tubaramure, a Food-Assisted Integrated Health and Nutrition Program, Reduces Child Wasting in Burundi: A Cluster-Randomized Controlled Intervention Trial." *Journal of Nutrition* 151 (1): 197–205.

Leroy, J.L., D.K. Olney, and M.T. Ruel. 2016a. "Evaluating Nutrition-Sensitive Programs: Challenges, Methods, and Opportunities." In *Achieving a Nutrition Revolution for Africa: The Road to Healthier Diets and Optimal Nutrition*, 130–146. Washington, DC: IFPRI.

Leroy, J.L., D. Olney, and M. Ruel. 2016b. "Tubaramure, a Food-Assisted Integrated Health and Nutrition Program in Burundi, Increases Maternal and Child Hemoglobin Concentrations and Reduces Anemia: A Theory-Based Cluster-Randomized Controlled Intervention Trial." *Journal of Nutrition* 146 (8): 1601–1608.

Leroy, J.L., D. Olney, and M. Ruel. 2018. "Tubaramure, a Food-Assisted Integrated Health and Nutrition Program, Reduces Child Stunting in Burundi: A Cluster-Randomized Controlled Intervention Trial." *Journal of Nutrition* 148 (3): 445–452.

Leroy, J.L., D.K. Olney, and M.T. Ruel. 2019. "PROCOMIDA, a Food-Assisted Maternal and Child Health and Nutrition Program, Contributes to Postpartum Weight Retention in Guatemala: A Cluster-Randomized Controlled Intervention Trial." *Journal of Nutrition* 149 (12): 2219–2227.

Leroy, J.L., M.T. Ruel, and D.K. Olney. 2020. *Measuring the Impact of Agriculture Programs on Diets and Nutrition*. Washington, DC: IFPRI.

Leroy, J.L., M. Ruel, E. Verhofstadt, and D.K. Olney. 2008. *The Micronutrient Impact of Multisectoral Programs Focusing on Nutrition: Examples from Conditional Cash Transfer, Microcredit with Education, and Agricultural Programs*. Washington, DC: Micronutrient Forum.

Margolies, A., C.G. Kemp, E.M. Choo, et al. 2022. "Nutrition-Sensitive Agriculture Programs Increase Dietary Diversity in Children under 5 Years: A Review and Meta-Analysis." *Journal of Global Health* 12: 08001.

Masset, E., L. Haddad, A. Cornelius, and J. Isaza-Castro. 2012. "Effectiveness of Agricultural Interventions That Aim to Improve Nutritional Status of Children: Systematic Review." *BMJ* (344): d8222.

McLaren, DS. 1974. "The Great Protein Fiasco." *Lancet* 304 (7872): 93–96.

Menon, P., P.H. Nguyen, K.K. Saha, et al. 2016. "Impacts on Breastfeeding Practices of At-Scale Strategies That Combine Intensive Interpersonal Counseling, Mass Media, and Community Mobilization: Results of Cluster-Randomized Program Evaluations in Bangladesh and Viet Nam." *PLoS Med* 13 (10): e1002159.

Monteiro, C.A. 2009. "Nutrition and Health. The Issue Is Not Food, nor Nutrients, So Much as Processing." *Public Health Nutrition* 12 (5): 729–731.

Morris, S.S., B. Cogill, and R. Uauy. 2008. "Effective International Action Against Undernutrition: Why Has It Proven So Difficult and What Can Be Done to Accelerate Progress?" *Lancet* 371 (9612): 608–621.

Mukherji, A., C. Arndt, J. Arango, et al. n.d. *Achieving Agricultural Breakthrough: A Deep Dive into Seven Technological Areas*. Montpellier, France: CGIAR System Organization.

Nelson, E.M., N. Nisbett, and S. Gillespie. 2021. "Historicising Global Nutrition: Critical Reflections on Contested Pasts and Reimagined Futures." *BMJ Global Health* 6 (11): e006337.

Neufeld, L.M., S. Nordhagen, J.L. Leroy, et al. 2024. "Food Systems Interventions for Nutrition: Lessons from 6 Program Evaluations in Africa and South Asia." *Journal of Nutrition* 154 (6): 1727–38.

Nguyen, P.H., S.S. Kim, T.T. Nguyen, et al. 2016. "Exposure to Mass Media and Interpersonal Counseling Has Additive Effects on Exclusive Breastfeeding and Its Psychosocial Determinants among Vietnamese Mothers." *Maternal & Child Nutrition* 12 (4): 713–725.

Nisbett, N., K. Buse, J. Collin, L. Hoey, and L. Westerman. 2021. "UN Food Systems Summit Leaders Must Not Remain Silent on Its Inadequate Rules of Engagement with Commercial Actors." BMG Opinion [blog], September 1.

Njuki, J., S. Eissler, H. Malapit, R. Meinzen-Dick, E. Bryan, and A. Quisumbing. 2023. "A Review of Evidence on Gender Equality, Women's Empowerment, and Food Systems." In *Science and Innovations for Food Systems Transformation*, eds. J. von Braun, K. Afsana, L.O. Fresco, and M.H. Ali Hassan, 165–189. Cham, Switzerland: Springer International Publishing.

Olney, D.K., J. Leroy, L. Bliznashka, and M.T. Ruel. 2018. "PROCOMIDA, a Food-Assisted Maternal and Child Health and Nutrition Program, Reduces Child Stunting in Guatemala: A Cluster-Randomized Controlled Intervention Trial." *Journal of Nutrition* 148 (9): 1493–1505.

Olney, D.K., J.L. Leroy, L. Bliznashka, and M.T. Ruel. 2019. "A Multisectoral Food-Assisted Maternal and Child Health and Nutrition Program Targeted to Women and Children in the First 1000 Days Increases Attainment of Language and Motor Milestones among Young Burundian Children." *Journal of Nutrition* 149 (10): 1833–1842.

Pinstrup-Andersen, P., A. Berg, and M. Forman. 1984. *International Agricultural Research and Human Nutrition*. Occasional Papers 3. Washington, DC: IFPRI.

Popkin, B.M., C. Corvalan, and L.M. Grummer-Strawn. 2020. "Dynamics of the Double Burden of Malnutrition and the Changing Nutrition Reality." *Lancet* 395 (10217): 65–74.

Quisumbing, A.R., A. Ahmed, D.O. Gilligan, et al. 2020. "Randomized Controlled Trials of Multi-Sectoral Programs: Lessons from Development Research." *World Development* 127: 104822.

Reutlinger, S., and M. Selowsky. 1976. "Malnutrition and Poverty: Magnitude and Policy Options." World Bank Staff Occasional Paper 23. Johns Hopkins University Press, Baltimore and London.

Rollins, N., E. Piwoz, P. Baker, et al. 2023. "Marketing of Commercial Milk Formula: A System to Capture Parents, Communities, Science, and Policy." *Lancet* 401 (10375): 486–502.

Ruel, M.T. 2001. *Can Food-Based Strategies Help Reduce Vitamin A and Iron Deficiencies? A Review of Recent Evidence*. Washington, DC: IFPRI.

Ruel, M.T., and H. Alderman. 2013. "Nutrition-Sensitive Interventions and Programmes: How Can They Help to Accelerate Progress in Improving Maternal and Child Nutrition?" *Lancet* 382 (9891): 536–551.

Ruel, M.T., and C. Hawkes. 2019. "Double Duty Actions to Tackle All Forms of Malnutrition (P10-053-19)." *Current Developments in Nutrition* 3 (Suppl 1): 814.

Ruel, M.T., A.R. Quisumbing, and M. Balagamwala. 2018. "Nutrition-Sensitive Agriculture: What Have We Learned So Far?" *Global Food Security* 17: 128–153.

Sen, A. 1983. *Poverty and Famines*. Oxford, UK: Oxford University Press.

Sharma, I.K., S. Di Prima, D. Essink, and J.E.W. Broerse. 2021. "Nutrition-Sensitive Agriculture: A Systematic Review of Impact Pathways to Nutrition Outcomes." *Advances in Nutrition* 12 (1): 251–275.

Sukhatme, P.V. 1970. "Size and Nature of the Protein Gap." *Nutrition Reviews* 28 (9): 223–226.

Swinburn, B.A., V.I. Kraak, S. Allender, et al. 2019. "The Global Syndemic of Obesity, Undernutrition, and Climate Change: The Lancet Commission Report." *Lancet* 393 (10173): 791–846.

Taillie, L.S., J.A. Rivera, B.M. Popkin, and C. Batis. 2017. "Do High vs. Low Purchasers Respond Differently to a Nonessential Energy-Dense Food Tax? Two-year Evaluation of Mexico's 8% Nonessential Food Tax." *Preventive Medicine (Baltimore).* 105: S37–S42.

Transform Nutrition. 2017. "Transforming Leadership in Nutrition." Medium.

Turner, C., A. Aggarwal, H. Walls, et al. 2018. "Concepts and Critical Perspectives for Food Environment Research: A Global Framework with Implications for Action in Low- and Middle-Income Countries." *Global Food Security* 18: 93–101.

UN (United Nations). 1974. *Report on the World Food Conference.* Rome.

UN. 2023. *Making Food Systems Work for People and Planet: UN Food Systems Summit +2: Report of the Secretary-General.* New York.

UNICEF. 1990. *Strategy for Improved Nutrition of Children and Women in Developing Countries.* UNICEF Policy Review. New York.

Van den Akker, A., A. Fabbri, S. Slater, A.B. Gilmore, C. Knai, and H. Rutter. 2024. "Mapping Actor Networks in Global Multi-Stakeholder Initiatives for Food System Transformation." *Food Security* 16: 1223–1234.

Van Tulleken, C. 2023. *Ultra-Processed People: Why Do We All Eat Stuff That Isn't Food … and Why Can't We Stop?* Stephens Point, WI: Cornerstone Press.

von Braun, J., and E. Kennedy. 1994. *Agriculture Commercialization, Economic Development and Nutrition.* Baltimore: Johns Hopkins University Press.

Warren, A.M., E.A. Frongillo, and R. Rawat. 2020. "Building Implementation Science in Nutrition." *Advances in Nutrition* 11 (5): 1392–1398.

Webb-Girard, A., J.S. Self, C. McAuliffe, and O. Olude. 2012. "The Effects of Household Food Production Strategies on the Health and Nutrition Outcomes of Women and Young Children: A Systematic Review." *Paediatric and Perinatal Epidemiology* 26 (Suppl 1): 205–222.

Wood, B., E. Robinson, P. Baker, G. Paraje, M. Mialon, C. van Tulleken, and G. Sacks. 2023. "What Is the Purpose of Ultra-Processed Food? An Exploratory Analysis of the Financialisation of Ultra-Processed Food Corporations and Implications for Public Health." *Globalization and Health* 19 (1): 85.

World Bank. 2007. *From Agriculture to Nutrition. Pathways, Synergies and Outcomes.* Washington, DC.

World Bank. 2013. *Improving Nutrition through Multisectoral Approaches.* Washington, DC.

WFP (World Food Programme. 2016. *Nutrition Policy (2017–2021). Draft: Informal Consultation. 25 November 2016.* Rome.

Chapter 13

FRAGILITY AND CONFLICT
Addressing Crises and Building Resilient Food Systems

Kibrom Abay, Kate Ambler, Jeffrey R. Bloem, Katrina Kosec, and Khalid Siddig

Kibrom Abay is a senior research fellow, Development Strategies and Governance Unit (DSG), IFPRI; **Kate Ambler** is a senior research fellow, Markets, Trade, and Institutions Unit (MTI), IFPRI; **Jeffrey R. Bloem** is a research fellow, MTI, IFPRI; **Katrina Kosec** is a senior research fellow, Poverty, Gender, and Inclusion Unit, IFPRI; and **Khalid Siddig** is a senior research fellow, DSG, IFPRI-Kenya.

Key messages

- Poverty, hunger, and undernourishment are increasingly concentrated in fragile and conflict-affected settings. As the incidence of conflict grows and climate change compounds fragility, projections show two-thirds of the world's extreme poor will live in these places by 2030.
- Early research used cross-country and macro-level analysis to understand links between poverty, economic growth, and conflict. A shift to household-level surveys highlighted the vicious circle of conflict and hunger and the need to consider conflict prevention within hunger interventions.
- Research on conflict impacts has spurred development and aid organizations to integrate short-term relief with long-term resilience- and peace-building efforts.
- Evaluations of specific programs and policies—before, during, and after crises—inform adaptations and improvements. This includes substantial work on mechanisms for effective food aid, conducted in partnership with providers. Much of this work points to the benefits of food aid, but it also highlights possible challenges in delivering food aid in protracted conflicts.
- Research is difficult in fragile and conflict-affected settings, but new tools such as satellite data and cellphone surveys have opened new possibilities.

Looking to the future, research and policy should:

- **Pursue a food-system-wide perspective.** More research on how conflict affects the entire food system—from local food systems to governance at all levels—is needed to inform investments for facilitating recovery and stabilizing food systems.
- **Support adaptation and contextualization of interventions.** Building resilient food systems in fragile and conflict-affected settings requires a comprehensive approach that addresses both short-term local humanitarian needs and long-term development and peace-building objectives.
- **Build local governance capacity.** Improving local governance in fragile and conflict-affected settings requires both top-down and bottom-up approaches to strengthen institutions, improve service delivery, and foster trust between governments and citizens.
- **Document differences in how vulnerable groups fare and adapt to conflicts.** Further empirical work is needed to understand what types of interventions may best serve these groups.
- **Address ethical and practical challenges of research in fragile settings.** As technology rapidly evolves, using it to support actionable research while continuing to protect individuals and partners on the ground is critical.

Hunger and malnutrition are, more than ever before, concentrated in fragile and conflict-affected areas around the world. This chapter summarizes and reflects on the evolution of food policy research conducted in these areas over the past 50 years and looks ahead at challenges and opportunities anticipated for the next 25 years. Looking back, we identify major themes and topics in the literature—including links between food insecurity and conflict, the effectiveness of food aid, the role of climate change, the consequences of price shocks, and resilience-building among displaced people—highlighting important contributions from IFPRI researchers and other scholars. Looking ahead, we consider how policy solutions will need to evolve to address the critical challenges that fragility and conflict present for building resilient food systems.

We define a "fragile setting" as a place characterized by a low level of institutional and governance capacity that significantly impedes the state's ability to function effectively, maintain peace, or foster economic and social development (World Bank 2024). A "conflict-affected setting" is a place characterized by acute

FIGURE 13.1 Poverty and fragility

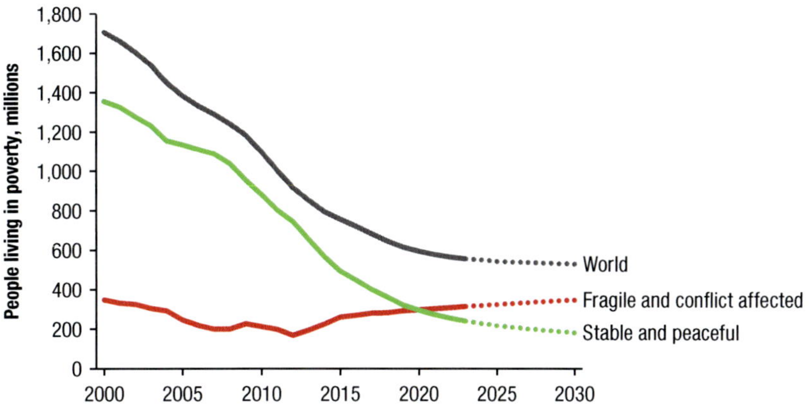

Source: Mahler et al. (2020b).
Note: Projections indicated by dotted lines.

physical insecurity and the use of violence by a group—including state forces, organized non-state groups, or other irregular entities—with a political purpose or motivation. Such force can be two-sided, involving armed conflict between groups, or one-sided, targeting civilians (World Bank 2024). Poverty and hunger are increasingly concentrated in fragile and conflict-affected settings. Although the number of people living in poverty has declined globally since 2000, this decline has occurred mostly in stable and peaceful settings (Figure 13.1). In fragile and conflict-affected settings, in contrast, poverty declined only slightly between 2000 and 2012, and since 2012, the number of people estimated to be living in poverty in these settings has increased. Most strikingly, as of 2020, more extremely poor people were living in fragile and conflict-affected settings than in stable and peaceful ones. Projections suggest that by 2030, two-thirds of the world's extreme poor will live in fragile and conflict-affected settings.[1]

Undernourishment and stunting, two important indicators of hunger, are also concentrated in fragile and conflict-affected settings. The global rate of undernourishment declined from roughly 13 percent in 2001 to about

1 These dynamics are driven by persistent rates of poverty in countries categorized as fragile or conflict-affected in all years between 2000 and 2020, increasing rates of poverty in countries that became fragile or conflict-affected between 2000 and 2020, and declining rates of poverty in countries that became stable and peaceful or were always stable and peaceful between 2000 and 2020 (Mahler et al. 2020a).

FIGURE 13.2 Prevalence of stunting and undernourishment in fragile and conflict-affected settings

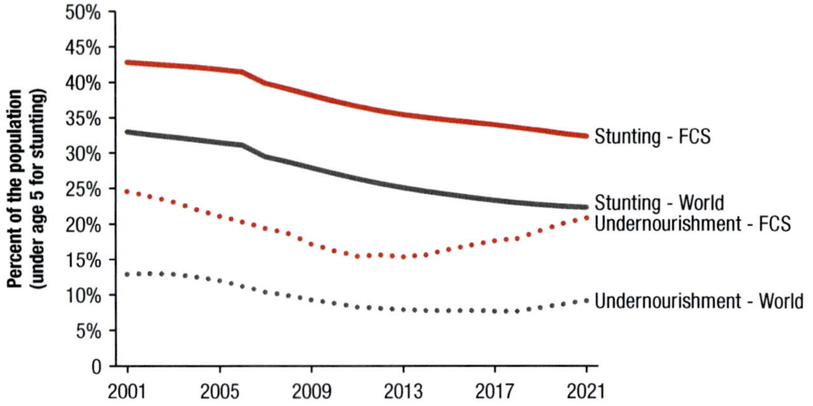

Source: Stunting data are from UNICEF, the World Health Organization, and World Bank joint child malnutrition estimates. Undernourishment data are from the Food and Agriculture Organization of the United Nations. Both datasets are published in the World Bank's World Development Indicators. https://databank.worldbank.org/source/world-development-indicators

Note: Stunting rate indicates the percentage of children under five years old who are stunted. Undernourishment indicates the share of the population with insufficient caloric intake to meet the minimum energy requirements necessary for a healthy individual. FCA = fragile and conflict-affected areas.

9 percent in 2021, but in fragile and conflict-affected settings, the rate of undernourishment is on average roughly 10 percentage points higher, and this gap has been growing in recent years (Figure 13.2). Similarly, rates of stunting (among children under five years old) are consistently 10 percentage points higher in fragile and conflict-affected settings than the global rate.

The coalescence of poverty and hunger with fragility and conflict suggests the importance of analyzing conflict as part of development research. Moreover, the incidence of conflict in the world today appears to be growing—with the past 3 years being the most violent of the last 30 (Rustad 2024). While conflict has not always been a central subject within development research (Blattman and Miguel 2010; Collier and Hoeffler 1998), it is increasingly identified as a key impediment to development (Collier et al. 2003; Corral et al. 2020; Gates et al. 2012; Ray and Esteban 2017; Verwimp et al. 2019). In response, international development and humanitarian organizations have increased their efforts to prevent or reduce conflict and its adverse effects. They are also working to connect short-term humanitarian relief with long-term development and peace-building efforts to create more sustainable solutions, described as working along the "humanitarian-development-peace nexus." All of this is happening amid the ongoing climate crisis, which also fuels—at times leading to and other times exacerbating—conflict, poverty, and hunger within fragile settings.

Early work on conflict

Early research on food systems in fragile and conflict-affected settings, conducted in the 1980s, applied economic models and theories to explain the causes of civil wars. This work highlighted economic drivers of conflict, such as the availability of natural resources, and emphasized how poverty and weak governance increase vulnerability to conflict (Collier and Hoeffler 1998; Collier et al. 2003; Collier 2007). Cross-country empirical work triggered major debates on the direction of causality between conflict on the one hand and poverty and economic growth on the other (for example, Blattman and Miguel 2010).

In the 1990s and 2000s, researchers began studying how fragility and conflict influence hunger, health, and nutrition. The cross-country and macro-level analyses that dominated the previous decade were tested using household survey data (Messer et al. 1998; Smith and Haddad 2000; von Braun et al. 1998). Other studies examined how policy responses affected conflict and hunger (for example, de Waal 1997). An IFPRI book described conflict as a policy failure that was driving famine in Africa (von Braun et al. 1998), and others highlighted the vicious cycle in low- and middle-income countries driving conflict and various indicators of hunger, including child malnutrition (Messer et al. 1998; Smith and Haddad 2000). Taken together, this stream of work stressed that emergency and development interventions, including food aid, should consider conflict prevention in the targeting and delivery of interventions.[2]

Although these macro-level contributions helped shape the debate on the causes and consequences of conflict, this research faced major methodological limitations and data shortcomings that challenged the identification of causal relationships between conflict and key development and welfare outcomes at the sectoral, community, and household levels (Blattman and Miguel 2010). Thus, in the 2010s, research began to focus more directly on the micro-level causes and consequences of armed conflict. Using household and other micro-level data, several studies offered alternative perspectives on fragility and conflict, including

2 Additional work includes that of Pender and Hazell (2000) assessing how sustainable development can be promoted in "less-favored" areas, which included settings facing multiple sources of fragility and conflict. Alderman et al. (2006) examined the effect of conflict on educational attainment among children in Zimbabwe. Gilligan and Hoddinott (2007) studied the effect of emergency food aid in conflict-affected rural Ethiopia.

the complex relationship between conflict and poverty (for example, Justino et al. 2013; World Bank 2011). As part of this pivot, IFPRI produced several flagship reports drawing attention to fragility and conflict (Fan 2011; von Grebmer et al. 2010, 2015). Notably, IFPRI's *2014–2015 Global Food Policy Report* included a chapter highlighting how conflict contributed to humanitarian tragedies characterized by widespread food and nutrition insecurity in places such as Gaza, Syria, and Yemen (Breisinger et al. 2015). The direct parallels with current conflicts around the world illustrate the persistence of these crises.

Generally, this body of research acknowledges the bidirectional relationship between fragility and conflict on the one hand and food insecurity and famines on the other. This work demonstrates that, while there are large incentives for peace (Blattman 2022), the presence of conflict in fragile settings makes it difficult to rebuild a stable and peaceful environment. Incentives for peace stem from the enormous costs associated with conflict in fragile settings, in terms of both human life and livelihoods. Once actors turn to conflict, however, the perceived costs and benefits become skewed by uncertainty, intangible incentives, and misperceptions, among other factors, making conflict seem like the only option (Blattman 2022). These political, social, and psychological forces push against prevailing economic forces to generate an environment that challenges the re-establishment of stability and peace.

Impact evaluations of interventions in fragile and conflict-affected settings

Early scholarship on the impact of conflict on development outcomes spurred international aid and development agencies to respond to conflict-related crises through new types of relief and emergency measures. Initial efforts focused on promoting diets with sufficient caloric intake via short-term humanitarian relief interventions. However, concerns soon emerged about the sustainability and potential unintended consequences of these programs, prompting a shift toward integrating short-term humanitarian relief with long-term development and peace-building efforts. Beginning in the 2010s, IFPRI and other researchers deepened this exploration, analyzing the effects of specific policies and programming designed to combat food insecurity and build resilient food systems in fragile and conflict-affected settings. This body of research spans programming implemented before, during, and after (that is, during recovery) crises, whether caused by or exacerbating fragility, and provides evidence to inform strategies that balance immediate humanitarian needs with sustainable, systemic solutions in fragile contexts.

Examining mechanisms for effective food aid

Food aid, typically provided by humanitarian agencies, aims to reduce hunger and food insecurity in fragile and conflict-affected areas. Much of the work by IFPRI researchers on food aid has been conducted in close partnership with governments and development organizations. In rural Ethiopia, for example, IFPRI has evaluated the impact of emergency aid and the country's main social protection program, the Productive Safety Net Program, for many years (see Chapter 11).[3] The challenges associated with delivering and evaluating food aid in fragile and conflict-affected settings have required close collaboration among implementing partners and research organizations. For example, to inform the World Food Programme's (WFP) food assistance in fragile and conflict-affected settings, IFPRI conducted a series of studies looking at the impact of different delivery modalities, including studies in Ecuador, Mali, Niger, and Yemen.[4] Each of these studies generally found that food aid helps reduce hunger in these settings; they also provided critical guidance on the relative effectiveness of specific distribution mechanisms (for example, in-kind versus cash) and complementary policies. This work contributed to IFPRI becoming an important knowledge partner for WFP, a partnership that continues today.[5]

The literature testing these food systems interventions in fragile and conflict-affected settings was elevated and expanded by a 2017 conference on Social Protection in Contexts of Fragility and Forced Displacement, organized by UNICEF. Contributions from IFPRI researchers included a study testing the effects of an emergency school feeding program on children's educational outcomes during a period of conflict in Mali (Aurino et al. 2019), a study

---

3 This extensive work includes Berhane et al. (2014), Gilligan and Hoddinott (2007), Gazeaud and Stephane (2023), Gilligan et al. (2009), Hoddinott et al. (2012), Knippenberg and Hoddinott (2019), and Sabates-Wheeler et al. (2022).

4 An experiment implemented in northern Ecuador tested the relative effects of cash transfers, in-kind food transfers, or food vouchers (Hidrobo et al. 2014). The study found that all modalities improved the quantity and quality of food consumed among recipients, but in-kind transfers led to the largest increase in calories consumed, while food vouchers led to the largest increases in dietary diversity. In Niger, a study of the relative effects of distributing cash or in-kind food transfers found that households receiving cash made more bulk purchases, and households receiving in-kind transfers experienced improved dietary diversity (Hoddinott et al. 2018). In Yemen, a similar study found that cash transfers led to greater dietary diversity and the in-kind transfers led to higher calorie intake (Schwab 2019). Finally, in Mali, a study of the effects of WFP's food assistance program in conflict-affected areas found that aid boosted household food expenditures, food consumption, and children's height (Tranchant et al. 2019).

5 IFPRI researchers continue close collaboration with WFP, including most recently through a set of projects under the CGIAR Research Initiative on Fragility, Conflict, and Migration in 2023–2024, which were designed to provide specific guidance to WFP offices. www.cgiar.org/initiative/fragility-conflict-and-migration/; www.ifpri.org/interactive/fcm-wp5/

assessing the effect of providing food aid on child nutritional outcomes during active conflict in Niger (Brück et al. 2019), and another estimating the effect of cash transfers, in-kind food transfers, and food vouchers on social cohesion among Colombian refugees and poor Ecuadorians (Valli et al. 2019).

While many studies find that the distribution of food aid leads to benefits for the receiving population, some research demonstrates that food aid may also generate some negative effects.[6] A recent literature review concluded that existing studies lack a convincing way of identifying causal effects, and that more rigorous research and more analysis are needed of the channels through which food aid may influence conflict and its intensity (Koppenberg et al. 2023). The possible politicization of food aid distribution is another important consideration in assessing its effectiveness. For example, a study of the political economy of food aid in Madagascar following a devastating cyclone found that aid provided by the government was more prevalent in areas with better radio coverage and stronger political support for the existing government, while aid provided by international agencies was less affected by media or political factors (Francken et al. 2012). Additionally, a study of the distribution of cash and food transfers in Malawi showed that these humanitarian aid transfers were disproportionately targeted to areas where the government perceived the need to garner votes to support its candidates in the next elections (Duchoslav et al. 2023).

Climate change and conflict

As scientists began to emphasize how the ongoing climate crisis is leading to irreversible climate change, researchers started to document how extreme weather events associated with climate change can contribute to and exacerbate conflict in fragile settings (Burke et al. 2009; Hsiang et al. 2013). IFPRI and other researchers have analyzed the effects of extreme weather events on a variety of outcomes, including livelihoods, aspirations, and migration. For example, a study in Somalia examining the link between extreme weather events and conflict highlighted how, in combination, they are an obstacle to policies and interventions to promote peace, stability, and food security (Maystadt and Ecker 2014). In rural Pakistan, exposure to flooding destroyed individuals' aspirations for the future, but government disaster

6 For example, an influential study of US food aid suggests that food aid may, in part, fuel conflict and prolong fragility (Nunn and Qian 2014). This controversial finding indicated that alternative humanitarian programs may be needed to support food security instead of, or in addition to, food aid. However, concerns have been raised about the study's methodology, calling into question the validity of these findings (Christian and Barrett 2024).

relief was shown to ameliorate this reduction in aspirations (Kosec and Mo 2017). Also in Pakistan, extreme heat stress was found to motivate individual migration from rural to urban areas (Mueller et al. 2014). Further study of this trend found that, although climate-induced migration allows migrants to increase their food consumption, they also experience worse physical and mental health outcomes (Chen et al. 2019). These findings suggest that policies and programming should support beneficial migration while also mitigating potential negative physical and psychological consequences.

Economic shocks, prices, and conflict

Amid a world characterized by recurring economic and international food crises, another topic that continues to attract debate and scholarship is the role of economic and price shocks in conflict, including civil war and violent conflicts (Andersen et al. 2022; Bazzi and Blattman 2014; Blair et al. 2021; Dube and Vargas 2013; Ubilava et al. 2023). This debate continues to generate more nuanced understanding. However, a meta-analysis showed that, while the relationship between commodity price changes and the likelihood of conflict is weak on average, an increase in the price of agricultural commodities tends to reduce conflict, while an increase in the price of capital-intensive commodities, such as oil and minerals, tends to increase conflict (Blair et al. 2021).[7] These findings offer indirect evidence on the potential of price stabilization policies to contribute to reducing food insecurity and conflict in fragile settings. For example, in Ethiopia, an evaluation of several strategies for stabilizing cereal prices during a drought found that strategic grain reserves and interventions specifically targeted to vulnerable households can help mitigate the adverse effects of droughts on food security (Dorosh et al. 2018). A study of the country's grain market and disaster management system confirmed the important role of public grain stock programs in stabilizing prices and supporting food security during periods of crisis and fragility (Rashid et al. 2018).

Building resilience among displaced people

More recently, the CGIAR Science Program on Food Frontiers and Security (and its predecessor, the CGIAR Research Initiative on Fragility, Conflict, and Migration) has contributed to impact evaluations of policies and programs aiming to build resilient food systems in fragile and conflict-affected settings (Kosec, Läderach, and Ruckstuhl 2023). This includes several

7 Dube and Vargas (2013) and Blair et al. (2021) offer comprehensive theoretical explanations and mechanisms to justify these patterns and empirical relationships.

ongoing projects. For example, a randomized controlled trial in Gombe, Nigeria, studied the effects of promoting a bundle of biofortified seeds and other agricultural inputs among both internally displaced people and residents of their host community (Amare et al. 2024). In Baidoa City, Somalia, a study implemented in partnership with World Vision examined the effect of an ultra-poor graduation program (that is, a multifaceted program aiming to "graduate" households from poverty) among internally displaced people (Leight et al. 2024). Work supported by this initiative increasingly considers the role of local governments and how citizen trust of these governments is formed. For example, recipients of Pakistan's national unconditional cash transfer program were found to become more satisfied with local government compared with those who did not receive the cash transfer—suggesting that cash transfers can contribute to building social cohesion and stability (Kosec and Mo 2024). Additionally, in rural Mali, an effective and trusted local government proved important in effectively deploying emergency aid (Bleck et al. 2024) (see also Chapter 15).

New research tools

In fragile and conflict-affected settings, reliable data are often scarce, and research is difficult (see Box 13.1). Armed conflict disrupts service delivery, security, and markets—and with them, data collection efforts. This presents a critical challenge for designing and testing policies and programming. In recent years, however, new data collection methods—including remotely sensed satellite data and computer-assisted telephone surveys—have enabled rapid monitoring of economic activity, food prices, poverty, and food security during conflict events. For example, in the absence of reliable data on economic outcomes in conflict-affected contexts, researchers have used nighttime light emissions to measure economic activity (Shortland et al. 2013; Weidmann and Schutte 2017). Similarly, combined satellite data on nitrogen dioxide concentration in the air and nighttime light intensity were used to track shifting patterns of economic activity amid escalated armed conflict in Sudan in early 2023 (Guo et al. 2023).

Conflict event datasets

Among the most important conflict research tools are conflict event datasets, including Armed Conflict Location and Event Data (ACLED) (Raleigh et al. 2010) and the geo-referenced event dataset from the Uppsala Conflict Data Program (UCDP) (Sundberg and Melander 2013). These datasets use

BOX 13.1 IFPRI's Country Programs in fragile and conflict-affected settings

Conducting research in fragile and conflict-affected settings is particularly complicated due to security issues, logistical barriers, and disruptions to traditional data collection mechanisms. Through its Country Programs, IFPRI is overcoming some of these challenges by leveraging innovative methodologies and flexible partnerships that enable the generation of timely, context-specific evidence to inform emergency responses, resilience-building, and recovery strategies.

To overcome the limitations of traditional data collection methods amid active conflicts, IFPRI employs satellite imagery and remote sensing. Similarly, high-frequency computer-assisted telephone interviews (CATI) have enabled researchers to gather real-time data on household food security and livelihoods facilitating real-time monitoring of economic conditions and disruptions in economic activities amid active conflict. These CATI surveys have been used in Sudan for both rural and urban household surveys (Kirui et al. 2024a; IFPRI and UNDP 2024a, 2024b). These were implemented in collaboration with humanitarian (World Food Programme) and development (United Nations Development Programme) partners operating in Sudan and have generated important insights for their work (Abushama et al. 2023). In Myanmar, telephone surveys conducted as part of the Myanmar Household Welfare Survey provided critical insights into food security and nutrition amid political and economic turmoil (MAPSA 2022). In addition, the rising costs of diets and their impact on food systems in Myanmar have been monitored through frequent surveys, such as the March 2024 round of agrifood systems monitoring (MAPSA 2024).

These innovative methods help generate critically needed information in conflict areas while also ensuring research continuity during crises. The remote data collection methods and data sources can be combined with conventional surveys, administrative databases, and model-based forecasts, which can ultimately enhance the robustness of empirical findings. By combining cutting-edge tools with conventional data sources, IFPRI has produced actionable insights that address immediate humanitarian needs while laying the groundwork for sustainable development. This dual approach exemplifies our commitment to navigating the complexities of fragile contexts and transforming evidence into effective policies that foster resilience and recovery.

local media reports to catalog conflict events, which are geo-coded, dated, and include information about the actors involved. IFPRI researchers and others have used these datasets for a variety of studies. For example, ACLED data were combined with high-frequency phone survey data from Ethiopia to provide near real-time assessments of changes in household welfare amid active conflict (Abay et al. 2023). In Myanmar, ACLED data were linked with data from phone surveys of actors within the rice value chain to document how conflict can slow food system changes that are commonly associated with structural transformation (Goeb et al. 2024). The UCDP data were used to study how exposure to conflict influences child nutrition in Yemen (Ecker et al. 2024). Finally, ACLED data were used to analyze how conflict influences civic engagement by women and men in 13 African countries, as well as how having women in office mitigates conflict's negative impacts on women (Kosec et al. 2023).

Despite the increased availability and use of conflict event data, these data suffer from limitations and are likely to include measurement error. Two broad issues can affect the reliability of conflict event data: selection and veracity problems. First, it is inevitable that local or international media will not cover all conflict events (Weidmann 2015). Selection of a conflict event for inclusion in the database is determined by supply-side factors (such as remoteness of the event) and demand-side factors (for example, small-scale events or those that do not involve actors of interest to domestic or international audiences may be systematically excluded). Second, the accuracy of information in the database is highly variable. For example, in Afghanistan, a comparison of UCDP data with a military database revealed that events in more remote areas are subject to more measurement error in both the location and severity of the conflict event (Weidmann 2015); this finding led the study's author to recommend that analyses mitigate biases by avoiding the use of geographic identifiers at any scale below the district level. A study on the role of cellphone coverage found that conflict events are more likely to be reported in areas with better coverage (Weidmann 2016). Such biases are serious and must be considered in any research using conflict event data—and findings must be interpreted with these caveats in mind.[8]

---

8 Aggregation bias is another source of measurement error in conflict data (Rockmore 2017; Rockmore and Barrett 2022; Sharma and Gibson 2020). Because of data limitations, most existing studies use aggregate measures and exposure to conflict, sometimes aggregated at first or second subnational administrative levels—a process that can generate classical and sometimes nonclassical measurement errors that can ultimately distort statistical inferences.

Data on food systems in fragile and conflict-affected areas

Other tools that enable rapid analysis of the functioning of food systems in fragile and conflict-affected settings include the Data in Emergencies (DIEM) database managed by the Food and Agriculture Organization of the United Nations (FAO) as well as IFPRI's Food Security Portal and the Excessive Food Price Variability Early Warning System.[9] The DIEM database offers high-frequency data on shocks and livelihoods in countries prone to multiple shocks, with the objective of informing humanitarian and related responses. IFPRI's tracking tools, which provide daily information on food prices, played an important role in providing timely analysis during the first year of the Russia–Ukraine war, in 2022 and 2023 (Glauber and Laborde 2023). Additionally, tools including a UN dashboard that tracks the distribution and supply of humanitarian supplies and the FAO's remote-sensing data collection efforts enabled rapid assessments of the severity of famine in Gaza in late 2023 and 2024 (Vos et al. 2024).[10]

Looking ahead

Amid the increasing concentration of poverty and hunger in fragile and conflict-affected settings, the need for targeted and context-sensitive interventions has never been more critical. Looking ahead, novel interventions, programming, and policies are needed to break free from the vicious cycles in which conflict fuels poverty and hunger, which in turn exacerbate conflict. Beyond addressing immediate poverty impacts and food needs, however, interventions should focus on building the resilience of food systems. While we have highlighted many important studies and lessons, critical knowledge gaps remain regarding how to build food system resilience amid compounding crises and how strategies should be differentiated across contexts. Five policy lessons and areas for future work emerge.

Pursue a food-system-wide perspective. More research is needed on *how* conflict and crises disrupt the functioning of key sectors in local food systems, agrifood value chains, input and output markets, and governance at all levels. This knowledge can inform government investments to support more stable food systems. While existing work on how smallholder farmers respond and adapt to exposure to different types of conflict (for example, Adelaja and George 2019; Bloem et al. 2025; George et al. 2021) provides useful insights,

9 https://data-in-emergencies.fao.org/ ; www.foodsecurityportal.org/

10 UNRWA dashboard, www.unrwa.org/what-we-do/gaza-supplies-and-dispatch-tracking

additional work is needed for a more comprehensive understanding of effects throughout the food system.

Support adaptation and contextualization of interventions. Adaptation and contextualization are key for ensuring impact and sustainability of interventions. Although social protection programs, technological innovations, and other policy interventions effectively support poverty reduction efforts in stable settings, understanding what works for fragile and conflict-affected settings requires further experimentation and learning. Building resilient food systems in these settings requires a comprehensive approach that address local realities—such as mobility limitations, nonfunctional markets, dysfunctional governance, and insecurity and theft—while also considering ways to incorporate long-term development and peace-building objectives. Tailored social protection—including cash, in-kind transfers of food or assets, and vouchers—should be deployed based on the specific needs and institutional conditions of each context, should be designed mindfully to avoid unintended consequences (such as exacerbating conflict), and should empower the most vulnerable, including women and marginalized groups. For example, to ensure sustainability, food aid might be paired with complementary programs such as livelihood support and community-building efforts, which can reinforce stability and promote lasting development. To address the complex mental health challenges of living in fragile settings (Alloush and Bloem 2022), cash transfers can be combined with group therapy services (Alderman et al. 2024). Overall, it is essential to strengthen the focus on combining short-term humanitarian efforts with initiatives that build resilience and foster social cohesion.

Build local governance capacity. Resilient and inclusive governance is critical to tackling fragility and conflict with policy solutions and programming. Building local governance capacity in fragile and conflict-affected settings requires both top-down and bottom-up approaches to strengthen institutions, improve service delivery, and foster trust between governments and citizens. From the top down, training in budgeting, fiscal planning, resource allocation, and leadership development can enhance public sector capacity—ensuring that local officials are better equipped to manage aid and deliver essential services. From the bottom up, engaging citizens through participatory planning, feedback mechanisms, and civic education fosters accountability and strengthens the social contract between communities and their leaders. For example, in Nigeria, training women to advocate for their policy needs increases local leaders' responsiveness to them (Adida et al. 2024). In Brazil, participatory budgeting initiatives, which allow citizens to directly influence municipal spending, yield more equitable resource allocation and increased trust in government (Avritzer 2009). And in Mali, involving local leaders and key community members—including elected officials, traditional authorities, and civil society representatives—in aid distribution can ensure that knowledge of local conditions informs aid delivery (Bleck et al. 2024). Additionally, investments that broadly strengthen citizens' trust in government—such as contextually appropriate social protection programs—can demonstrate the state's commitment to supporting citizens, thus promoting trust (Kosec and Mo 2024). Together, these efforts can both foster stability and create a foundation for broader service improvements and long-term development in fragile and conflict-affected areas.

Document differences among vulnerable groups. Empirical work on how different groups, including youth and women, fare and adapt to recurring conflicts, and what types of intervention may best serve these groups, remains an important research gap (see Box 13.2). For example, while inadequate access to land and jobs combined with a growing youth population, especially in Africa, has fueled violent extremism and conflict, youth are also central to peace-building, innovation, and development. Further, while women are often excluded from influencing policy decisions about how to improve service delivery, reduce poverty, and support individual livelihoods, efforts that include their voices have shown substantial promise for bringing about more inclusive policies (see Chapter 14). Additional research is needed to understand how both the growing youth population and recent innovations to reduce structural barriers for women in food systems can be harnessed for development, peace, and prosperity.

Address ethical and practical challenges of research in fragile settings. An immediate challenge and frontier for researchers relates to how we can safely and ethically conduct research in fragile and conflict-affected areas. Data collection in these settings is fraught with logistical and ethical challenges, such as ensuring participants' safety and minimizing harm in volatile environments. Innovative tools like satellite imagery, remote sensing, and conflict event datasets offer new opportunities for timely analysis but must be employed carefully to address biases and measurement errors. As technology rapidly evolves, including artificial intelligence, using it to support actionable research while continuing to protect individuals and partners on the ground is critical.

Through these key priorities, we will be able to inform and support the development of resilient food systems in fragile and conflict-affected settings around the world. This work informs and contributes to efforts that

BOX 13.2 Research on women in fragile and conflict-affected settings

Women face unique vulnerabilities in fragile and conflict-affected settings, where existing structural barriers limit their ability to respond and adapt effectively to crises (Bendavid et al. 2021). Lack of access to social and financial services, limitations on mobility, and low levels of political agency all disadvantage women in these settings (Swinnen and Kosec 2023). Fragility and conflict can erode the social networks and other institutions on which women rely and can separate them from their husbands and families. More data and research are needed to fully understand the conditions of women in these settings and the policy solutions that can support them (Malapit and Brown 2023). IFPRI researchers are at the forefront of new research to address these gaps.

For example, IFPRI's Sudan office has been collecting data on women in the agrifood sector during the ongoing conflict. Interviews with agrifood entrepreneurs in Kassala and Gedaref States highlight gendered experiences in navigating supply chain disruptions and rising input costs. Women, particularly those working in crop processing, have adapted by diversifying income sources and relocating businesses closer to their homes. However, many also scaled back operations, compromising both their economic security and community food supplies. Women's aspirations often include expanding businesses and achieving financial independence, but these ambitions are constrained by the war's economic toll (Kirui et al. 2024b, 2024c).

In another example, an IFPRI research team and its partners focus on anticipatory action programming in Nigeria that is designed to take preemptive action between the forecast of a disaster, such as a flood, and the event itself. They are studying how these programs can be designed to reach, benefit, and empower women—including by helping them guide programming decisions that affect their own resilience to disasters. The work extends our understanding of the challenges of reaching women with anticipatory action programming and the prerequisites for implementing programs that can go further to empower women (Gonzales et al. 2024).

Increasing women's political agency can play a role in promoting community resilience in fragile settings. In a study involving 5,800 women across three states in Nigeria where women's political participation is low, IFPRI researchers evaluated the impact of (1) a six-month training program aimed at increasing women's collective efficacy and (2) the combination of that program with a training for men aimed at increasing their allyship for women's empowerment. Results showed that trained women were more politically engaged across several measures, had increased measures of self-efficacy, and showed improvements across several measures of economic empowerment, but no additional impact was attributable to the addition of the men's training (Adida et al. 2024). This study is evidence of one successful approach for increasing women's political agency—a key component for promoting stability in fragile settings.

help support local livelihoods and also promotes global peace, security, and stability. Among other efforts, IFPRI is contributing to achieving these five key priorities through the new CGIAR Science Program on Food Frontiers and Security (2025–2030), which will continue tackling these research needs over the next five years. This program fills a critical gap in CGIAR's ambitious effort to achieve food and nutrition security by focusing on support to neglected, "frontier" food systems: fragile and conflict-affected, urban and peri-urban, and island food systems. All three food systems are characterized by heavy constraints to food production, natural resource management concerns, vulnerable populations, high risk from climate change and global and regional trade shocks, and challenges posed by displacement and migration. The program has strong potential for advancing research on these five recommendations.

References

Abay, K.A., K. Tafere, G. Berhane, J. Chamberlin, and M.H. Abay. 2023. "Near-Real-Time Welfare and Livelihood Impacts of an Active War: Evidence from Ethiopia." *Food Policy* 119: 102526.

Abushama, H., D. Resnick, K. Siddig, and O.K. Kirui. 2023. *Political and Economic Drivers of Sudan's Armed Conflict: Implications for the Agri-Food System*. SSSP Working Paper 15. IFPRI, Washington, DC.

Adelaja, A., and J. George. 2019. "Effects of Conflict on Agriculture: Evidence from the Boko Haram Insurgency." *World Development* 117: 184–195.

Adida, C., K. Kosec, J. Kyle, C.H. Mo, L. Arriola, D. Adeyanju, and R. Fisher. 2024. *Strengthening Women's Voice and Agency in Nigeria: Evidence from a Randomized Control Trial on Women's Advocacy and Men's Allyship Trainings*. IFPRI Policy Note. Washington, DC: IFPRI.

Alderman, H., D.O. Gilligan, M. Hidrobo, J. Leight, M. Mulford, H. Tambet, and T. Mamo. 2024. "Deepening Social Protection Systems: Enhancing Livelihoods and Health in Ethiopia." Policy Seminar. IFPRI, Washington, DC, April 17.

Alderman, H., J. Hoddinott, and B. Kinsey. 2006. "Long-Term Consequences of Early Childhood Malnutrition." *Oxford Economic Papers* 58 (3): 450–474.

Alloush, M., and J.R. Bloem. 2022. "Neighborhood Violence, Poverty, and Psychological Well-Being." *Journal of Development Economics* 154: 102756.

Amare, M., K. Ambler, T. Bamiwuye, J.R. Bloem, R. Misra, and J. Wagner. 2024. *Take-Up and Demand for a Bundle of Biofortified Seeds and Agricultural Inputs in Northeastern Nigeria*. Fragility, Conflict, and Migration Initiative Policy Brief. Washington, DC: IFPRI.

Andersen, J.J., F.M. Nordvik, and A. Tesei. 2022. "Oil Price Shocks and Conflict Escalation: Onshore versus Offshore." *Journal of Conflict Resolution* 66 (2): 327–356.

Aurino, E., J-P. Tranchant, A.S. Diallo, and A. Gelli. 2019. "School Feeding or General Food Distribution? Quasi-Experimental Evidence on the Educational Impacts of Emergency Food Assistance during Conflict in Mali." *Journal of Development Studies* 55 (Suppl 1): 7–28.

Avritzer, L. 2009. *Participatory Institutions in Democratic Brazil.* Baltimore: Johns Hopkins University Press.

Bazzi, S., and C. Blattman. 2014. "Economic Shocks and Conflict: Evidence from Commodity Prices." *American Economic Journal: Macroeconomics* 6 (4): 1–38.

Bendavid, E., T. Boerma, N. Akseer, et al. 2021. "The Effects of Armed Conflict on the Health of Women and Children." *Lancet* 397 (10273): 522–532.

Berhane, G., D.O. Gilligan, J. Hoddinott, N. Kumar, and A.S. Taffesse. 2014. "Can Social Protection Work in Africa? The Impact of Ethiopia's Productive Safety Net Programme." *Economic Development and Cultural Change* 63 (1): 1–26.

Blair, G., D. Christensen, and A. Rudkin. 2021. "Do Commodity Price Shocks Cause Armed Conflict? A Meta-Analysis of Natural Experiments." *American Political Science Review* 115 (2): 709–716.

Blattman, C. 2022. *Why We Fight: The Roots of War and the Paths to Peace.* New York: Viking Press.

Blattman, C., and E. Miguel. 2010. "Civil War." *Journal of Economic Literature* 48 (1): 3–57.

Bleck, J., L. Carrillo, K. Kosec, J. Kyle, and M. Somano. 2024. "Unlocking Locally Led Resilience amid Conflict and Climate Stress: Views from Community Leaders in Mali on Development Priorities, Aid Distribution, and Anticipatory Action." IFPRI Discussion Paper 2272. IFPRI, Washington, DC.

Bloem, J.R., A. Damon, D.C. Francis, and H. Mitchell. 2025. "Herder-Related Violence, Labor Allocation, and the Gendered Response of Agricultural Households." *Journal of Development Economics* 176: 103512.

Breisinger, C., O. Ecker, and J.-F. Trinh Tan. 2015. "Conflict and Food Insecurity: How Do We Break the Links?" In *2014–2015 Global Food Policy Report*, 51–59. Washington, DC: IFPRI.

Brück, T., O.M. Diaz Botia, N.T.N. Ferguson, J. Ouedraogo, and Z. Ziegelhofer. 2019. "Assets for Alimentation? The Nutritional Impact of Assets-Based Programming in Niger." *Journal of Development Studies* 55 (SI): 55–74.

Burke, M.B., E. Miguel, S. Satyanath, J.A. Dykema, and D.B. Lobell. 2009. "Warming Increases the Risk of Civil War in Africa." *PNAS* 106 (49): 20670–20674.

Chen, J., K. Kosec, and V. Mueller. 2019. "Moving to Despair? Migration and Well-Being in Pakistan." *World Development* 113: 186–203.

Christian, P., and C.B. Barrett. 2024. "Spurious Regressions and Panel IV Estimation: Revisiting the Causes of Conflict." *The Economic Journal* 34 (659): 1069–1099.

Collier, P. 2007. *The Bottom Billion: Why the Poorest Countries Are Failing and What Can Be Done About It.* Oxford, UK: Oxford University Press.

Collier, P., V.L. Elliott, H. Hegre, A. Hoeffler, M. Reynal-Querol, and N. Sambanis. 2003. *Breaking the Conflict Trap: Civil War and Development Policy.* World Bank Policy Research Report. Washington, DC: World Bank; Oxford, UK: Oxford University Press.

Collier, P., and A. Hoeffler. 1998. "On Economic Causes of Civil War." *Oxford Economic Papers* 50 (4): 563–573.

Corral, P., A. Irwin, N. Krishnan, and D.G. Mahler. 2020. *Fragility and Conflict: On the Front Lines of the Fight Against Poverty.* Washington, DC: World Bank Publications.

de Waal, A. 1997. *Famine Crimes: Politics and the Disaster Relief Industry in Africa.* London: African Rights and The International African Institute.

Dorosh, P., J. Smart, B. Minten, and D. Stifel. 2018. "Droughts, Cereal Prices, and Price Stabilization Options in Ethiopia." Strategy Support Program Working Paper 126. IFPRI, Washington, DC.

Dube, O., and J.F. Vargas. 2013. "Commodity Price Shocks and Civil Conflict: Evidence from Colombia." *Review of Economic Studies* 80 (4): 1384–1421.

Duchoslav, J., E. Kenamu, and J. Thunde. 2023. "Targeting Hunger or Votes? The Political Economy of Humanitarian Transfers in Malawi." *World Development* 165: 106179.

Ecker, O., A. Al-Malk, and J.-F. Maystadt. 2024. "Civil Conflict, Cash Transfers, and Child Nutrition in Yemen." *Economic Development and Cultural Change* 72 (4): 2069–2100.

Fan, S. 2011. "Overview: Major Food Policy Developments in 2011." In *2011 Global Food Policy Report*, 1–10. Washington, DC: IFPRI.

Francken, N., B. Minten, and J.F.M. Swinnen. 2012. "The Political Economy of Relief Aid Allocation: Evidence from Madagascar." *World Development* 30 (3): 486–500.

Gates, S., H. Hegre, H.M. Nygard, and H. Strand. 2012. "Development Consequences of Armed Conflict." *World Development* 40 (9): 1713–1722.

Gazeaud, J., and V. Stephane. 2023. "Productive Workfare? Evidence from Ethiopia's Productive Safety Net Program." *American Journal of Agricultural Economics* 105 (1): 265–290.

Gilligan, D.O., and J. Hoddinott. 2007. "Is There Persistence in the Impact of Emergency Food Aid? Evidence on Consumption, Food Security, and Assets in Rural Ethiopia." *American Journal of Agricultural Economics* 89 (2): 225–242.

Gilligan, D.O., J. Hoddinott, and A.S. Taffesse. 2009. "The Impact of Ethiopia's Productive Safety Net Programme and Its Linkages." *Journal of Development Studies* 45 (10): 1684–1706.

Glauber, J., and D. Laborde. 2023. *The Russia-Ukraine Conflict and Global Food Security*. Washington, DC: IFPRI.

Goeb, J., B. Minten, N. Aung, Z.W. Aung, A.M. Zu, and M.T. Htar. 2024. "Conflict, Inclusivity, and Transformation of the Rice Value Chain in Myanmar." Paper presented at the 32nd International Conference of Agricultural Economists, August 2–7, New Delhi.

Gonzales, T., K. Kosec, J. Kyle, et al. 2024. "How Can Anticipatory Action Programming Support Women? Application of the Reach-Benefit-Empower-Transform Framework in Nepal and Nigeria." IFPRI Discussion Paper 2298. IFPRI, Washington, DC.

Guo, Z., H. Abushama, K. Siddig, O.K. Kirui, K. Abay, and L. You. 2023. "Monitoring Indicators of Economic Activities in Sudan amidst Ongoing Conflict Using Satellite Data." *Defence and Peace Economics* 35 (8): 1–17.

Hidrobo, M., J. Hoddinott, A. Peterman, A. Margolies, and V. Moreira. 2014. "Cash, Food, or Vouchers? Evidence from a Randomized Experiment in Northern Ecuador." *Journal of Development Economics* 107: 144–156.

Hoddinott, J., G. Berhane, D.O. Gilligan, N. Kumar, and A. Seyoum Taffesse. 2012. "The Impact of Ethiopia's Productive Safety Net Programme and Related Transfers on Agricultural Productivity." *Journal of African Economies* 21 (5): 761–786.

Hoddinott, J., S. Sandstrom, and J. Upton. 2018. "The Impact of Cash and Food Transfers: Evidence from a Randomized Intervention in Niger." *American Journal of Agricultural Economics* 100 (4): 1032–1049.

Hsiang, S.M., M. Burke, and E. Miguel. 2013. "Quantifying the Influence of Climate on Human Conflict." *Science* 341 (6151): 1235367.

IFPRI (International Food Policy Research Institute) and UNDP (United Nations Development Programme). 2024a. *Livelihoods in Sudan amid Armed Conflict: Evidence from a National Rural Household Survey*. Washington, DC: IFPRI.

IFPRI and UNDP. 2024b. *The Socioeconomic Impact of Armed Conflict on Sudanese Urban Households*. Washington, DC: IFPRI.

Justino, P., T. Brück, and P. Verwimp, eds. 2013. *A Micro-Level Perspective on the Dynamics of Conflict, Violence, and Development*. New York: Oxford University Press.

Kirui, O.K., M. Ahmed, A.S. Taffesse, H. Abushama, and K. Siddig. 2024a. "Sudan Rural Household Survey 2023: Sampling and Implementation Procedures for the First Round." Sudan Strategy Support Program Working Paper 18. IFPRI, Washington, DC.

Kirui, O.K., K. Siddig, M. Fisher, H. Abushama, M. Ahmed, M. Raouf, and A.S. Taffesse. 2024b. *Evolution of Food Insecurity in Sudan during the Ongoing Conflict*. Sudan Strategy Support Program Policy Note 8. Washington, DC: IFPRI.

Kirui, O.K., K. Siddig, M. Fisher, M. Cavicchioli, and J. Chamberlin. 2024c. *Supporting Sudan's Entrepreneurs in Crisis: Policy Insights from Medium, Small and Micro Enterprises*. Sudan Strategy Support Program Policy Note 9. Washington, DC: IFPRI.

Knippenberg, E., and J. Hoddinott. 2019. "Shocks, Social Protection, and Resilience: Evidence from Ethiopia." Ethiopia Strategy Support Program Working Paper 109. IFPRI, Washington, DC.

Koppenberg, M., A.K. Mishra, and S. Hirsch. 2023. "Food Aid and Violent Conflict: A Review and Empiricist's Companion." *Food Policy* 121: 102542.

Kosec, K., J. Kyle, and H. Takeshima. 2023. "When Women Hold Local Office: Implications for Individual Women's Political Participation amid Conflict and Climate Shocks in Africa." IFPRI Discussion Paper 2230. IFPRI, Washington, DC.

Kosec, K., P. Läderach, and S. Ruckstuhl. 2023. "Fragility, Conflict, and Migration: A New CGIAR Initiative for a Fragile World." *CGIAR Blog*, May 22.

Kosec, K., and C.H. Mo. 2017. "Aspirations and the Role of Social Protection: Evidence from a Natural Disaster in Rural Pakistan." *World Development* 97: 49–66.

Kosec, K., and C.H. Mo. 2024. "Does Relative Deprivation Condition the Effects of Social Protection Programs on Political Support? Experimental Evidence from Pakistan." *American Journal of Political Science* 68 (2): 832–849.

Leight, J., K. Hirvonen, N. Karachiwalla, and D. Rakshit. 2024. *Building Pathways out of Poverty in Baidoa: A Randomized Controlled Trial: Evidence from the Midline Survey*. Learning Brief March 2024. Washington, DC: IFPRI.

Mahler, D.G., T. Vishwanath, and A. Irwin. 2020a. "Living in a Two-Speed World: Extreme Poverty Is Declining in Most Countries but Stagnating in Economies in Fragility or Conflict." *World Bank Data Blog*, March 2.

Mahler, D.G., T. Vishwanath, and A. Irwin. 2020b. "What Could Stop the World from Ending Poverty? Think Fragility and Conflict." *World Bank Data Blog*, February 27.

Malapit, H., and L. Brown. 2023. "Gender: Promoting Equality in Fragile and Conflict-Affected Settings." In *2023 Global Food Policy Report*, 62–71. Washington, DC: IFPRI.

MAPSA (Myanmar Agriculture Policy Support Activity). 2022. *The State of Food Security and Nutrition in Myanmar: Findings from the Myanmar Household Welfare Survey 2021–2022*. Myanmar Strategy Support Program Research Note 88. Washington, DC: IFPRI.

MAPSA. 2024. *Monitoring the Agri-Food System in Myanmar: The Rising Costs of Diets–March 2024 Survey Round*. Myanmar Strategy Support Program Research Note 107. Washington, DC: IFPRI.

Maystadt, J.-F., and O. Ecker. 2014. "Extreme Weather and Civil War: Does Drought Fuel Conflict in Somalia Through Livestock Price Shocks?" *American Journal of Agricultural Economics* 96 (4): 1157–1182.

Messer, E., M.J. Cohen, and J. d'Costa. 1998. "Food from Peace: Breaking the Links between Conflict and Hunger." Food, Agriculture, and the Environment Discussion Paper 24. IFPRI, Washington, DC.

Mueller, V., C. Gray, and K. Kosec. 2014. "Heat Stress Increases Long-Term Human Migration in Rural Pakistan." *Nature Climate Change* 4: 182–185.

Nunn, N., and N. Qian. 2014. "US Food Aid and Civil Conflict." *American Economic Review* 104 (6): 1630–1666.

Pender, J., and P. Hazell. 2000. "Promoting Sustainable Development in Less-Favored Areas." In *A 2020 Vision for Food, Agriculture, and the Environment*. Washington, DC: IFPRI.

Raleigh, C., A. Linke, H. Hegre, and J. Karlsen. 2010. "Introducing ACLED: An Armed Conflict Location and Event Dataset." *Journal of Peace Research* 47 (5): 651–660.

Rashid, S., P. Dorosh, and D. Alemu. 2018. "Grain Markets, Disaster Management, and Public Stocks: Lessons from Ethiopia." *Global Food Security* 19: 31–39.

Ray, D., and J. Esteban. 2017. "Conflict and Development." *Annual Review of Economics* 9: 263–293.

Rockmore, M. 2017. "The Cost of Fear: The Welfare Effect of the Risk of Violence in Northern Uganda." *World Bank Economic Review* 31 (3): 650–669.

Rockmore, M., and C.B. Barrett. 2022. "The Implications of Aggregated Measures of Exposure to Violence for the Estimated Impacts on Individual Risk Preferences." *World Development* 157: 105925.

Rustad, S.A. 2024. *Conflict Trends: A Global Overview, 1946-2023*. PRIO Paper. Oslo: Peace Research Institute.

Sabates-Wheeler, R., K. Hirvonen, J. Lind, and J. Hoddinott. 2022. "Expanding Social Protection Coverage with Humanitarian Aid: Lessons on Targeting and Transfer Values from Ethiopia." *Journal of Development Studies* 58 (10): 1981–2000.

Schwab, B. 2022. "Comparing the Productive Effects of Cash and Food Transfers in a Crisis Setting: Evidence from a Randomized Experiment in Yemen." *Journal of Development Studies* 55: 29–54.

Sharma, H., and J. Gibson. 2020. "Mean-Reverting Errors and Spatially Aggregated Conflict Data: Impacts of Civil War on International Migration from Nepal." In *Civil Conflict and International Migration in Nepal*, Chapter 3. Hamilton, New Zealand: University of Waikoto.

Shortland, A., K. Christopoulou, and C. Makatsoris. 2013. "War and Famine, Peace and Light? The Economic Dynamics of Conflict in Somalia 1993–2009." *Journal of Peace Research* 50 (5): 545–561.

Smith, L.C., and L.J. Haddad. 2000. *Explaining Child Malnutrition in Developing Countries: A Cross-Country Analysis*. Research Report 111. Washington, DC: IFPRI.

Sundberg, R., and E. Melander. 2013. "Introducing the UCDP Georeferenced Event Dataset." *Journal of Peace Research* 50 (4): 523–532.

Swinnen, J., and K. Kosec. 2023. "The Road to Resilience: Rethinking Responses to Food Crises." In *Global Food Policy Report 2023: Rethinking Food Crisis Responses*, 6–19. Washington, DC: IFPRI.

Tranchant, J.-P., A. Gelli, L. Bliznashka, et al. 2019. "The Impact of Food Assistance on Food Insecure Populations during Conflict: Evidence from a Quasi-Experiment in Mali." *World Development* 119: 185–202.

Ubilava, D., J.V. Hastings, and K. Atalay. 2023. "Agricultural Windfalls and the Seasonality of Political Violence in Africa." *American Journal of Agricultural Economics* 105 (5): 1309–1332.

Valli, E., A. Peterman, and M. Hidrobo. 2019. "Economic Transfers and Social Cohesion in a Refugee-Hosting Setting." *Journal of Development Studies* 55 (SI): 128–146.

Verwimp, P., P. Justino, and T. Brück. 2019. "The Microeconomics of Violent Conflict." *Journal of Development Economics* 141: 102297.

von Braun, J., T. Teklu, and P. Webb. 1998. *Famine in Africa: Causes, Responses, and Prevention.* Bonn, Germany: Centre for Development Research, University of Bonn.

von Grebmer, K., J. Bernstein, A. de Waal, N. Prasai, S. Yin, and Y. Yohannes. 2015. *2015 Global Hunger Index: Armed Conflict and the Challenge of Hunger.* Bonn, Germany: Welthungerhilfe; Washington, DC: IFPRI; Dublin: Concern Worldwide.

von Grebmer, K., M.T. Ruel, P. Menon, et al. 2010. *2010 Global Hunger Index: The Challenge of Hunger.* Washington, DC: IFPRI; Bonn, Germany: Welthungerhilfe; Dublin: Concern Worldwide.

Vos, R., I. Elouafi, and J. Swinnen. 2024. "Famine in Gaza, Questions for Research and Preventative Action." *Nature Food* 5: 346–348.

Weidmann, N.B. 2015. "On the Accuracy of Media-Based Conflict Event Data." *Journal of Conflict Resolution* 59 (6): 1129–1149.

Weidmann, N.B. 2016. "A Closer Look at Reporting Bias in Conflict Event Data." *American Journal of Political Science* 60 (1): 206–218.

Weidmann, N.B., and S. Schutte. 2017. "Using Night Light Emissions for the Prediction of Local Wealth." *Journal of Peace Research* 54 (2): 125–140.

World Bank. 2024. *Classification of Fragile and Conflict-Affected Situations.* World Bank Brief. Washington, DC.

Chapter 14

GENDER RESEARCH
Metrics and Policies for Greater Equity and Inclusion

Agnes Quisumbing, Ruth Meinzen-Dick, Hazel Malapit, Jessica Heckert, Berber Kramer, and Bjorn Van Campenhout

Agnes Quisumbing is a senior research fellow, Poverty, Gender, and Inclusion Unit (PGI), IFPRI; **Ruth Meinzen-Dick** is a research fellow emeritus, IFPRI; **Hazel Malapit** is a senior research coordinator, PGI, IFPRI; **Jessica Heckert** is a senior research fellow, PGI, IFPRI; **Berber Kramer** is a senior research fellow, Markets, Trade, and Institutions Unit, IFPRI; and **Bjorn Van Campenhout** is a research fellow, Innovation Policy and Scaling Unit, IFPRI.

Key messages

- Over the past 50 years, research on gender in development has evolved in parallel with the growing awareness of women's role in economic development, the importance of gender relations both within and outside the household, and the recognition that women's empowerment and gender equality are important goals in themselves.
- Lessons learned from early research on women and gender showed that challenging the assumptions underlying household models can build understanding of household behavior and help identify policy instruments to influence that behavior.
- Evidence that increasing women's control of resources led to better outcomes for children inspired the design and implementation of conditional cash transfer programs worldwide, with impact evaluations of these programs subsequently generating new evidence on household behavior.
- Recognition of women's empowerment and gender equality as intrinsically valuable created impetus to improve their measurement, motivate policy action, monitor progress, and ascertain what works, ultimately leading to the development of metrics such as the Women's Empowerment in Agriculture Index.
- Research on policies, programs, and innovations that take gender into account highlights that intentional gender programming and

transformative approaches—not simply having empowerment objectives—are needed to improve women's empowerment and achieve gender equality in the longer run.

Looking ahead, gender research should aim to address:

- **Intersectionality.** Engage with aspects of disadvantage that may compound gender inequalities, such as wealth, class, caste, and ethnicity, and recognize that neither women nor men are homogeneous groups.
- **Life cycle and demographic factors.** Consider how gendered vulnerabilities and opportunities differ and evolve throughout the life cycle, with particular attention to both youth and the elderly; identify critical investments to ensure the success and well-being of young women and men throughout their working years; and assess the gender and generational implications of migration for rural areas and agricultural labor.
- **Structural and institutional change.** Examine constraints to individual agency posed by structural conditions and social norms, including through applying tools to measure and address gender inequalities in rural institutions and governance, exploring the role of aspirations and locus of control in overcoming poverty, and involving men in interventions.

Research on gender in development has evolved in parallel with the growing awareness of women's role in economic development, the importance of gender relations both within and outside the household, and the recognition that women's empowerment and gender equality are important goals in themselves. This evolution has led to meaningful insights for the design and implementation of development policies and programs. In this chapter, we trace the evolution of gender research in the context of the development discourse on gender, focusing on agrifood systems, and identify three challenges that future research should address: intersectionality, life cycle and demographic changes, and structural and institutional barriers to gender equality.

From women in development to gender and development: Parallel evolutions of research and discourse

Although women's role as mothers, wives, and caregivers has long been recognized, research in the 1970s, including the ground-breaking *Woman's Role in*

Economic Development (Boserup 1970) which first called attention to women's role as agricultural producers, inspired much of the work on women in development through the early 1980s (Beneria et al. 2016). This work highlighted how the socially constructed gender division of labor varies by place and changes over time, and how the process of modernization harms women, who are either disadvantaged relative to men and/or experience an absolute loss in status over time. Discussions of the so-called female farming system in sub-Saharan Africa (in which women farm plots separately from men and specialize in different types of crops, with women traditionally cultivating food crops and men focusing on cash crops) as contrasted with the farming systems of South Asia (where farming is more intensive in male labor and family members work together on the family farm) shaped the way agricultural economists and development researchers viewed women in agriculture. This paved the way toward reducing, though not completely eliminating, the undercounting and "invisibility" of women in agriculture (Dixon 1982).

Research on women in agriculture did not always consider their involvement within the context of their relationships with men or other family and community members, however. Titles of research programs, such as the International Rice Research Institute's Women in Rice Farming Systems program, were emblematic of this singular focus on women, as were important United Nations conventions, such as the 1995 United Nations Conference on Women in Beijing. IFPRI's policy report for the conference framed women as representing the three pillars of food security: producers, income earners, and caregivers (Quisumbing et al. 1995). But the exclusive focus on women masked the importance of broader gender relations, an oversight that was impossible to ignore. For example, IFPRI research on agricultural commercialization and nutrition highlighted both the importance of women's control of income for child nutrition and the tendency for men to control income from women's crops once those crops became profitable (von Braun and Kennedy 1994).

The women's movement has been called a force that "propelled the institutionalization of the field of gender and development in the 1970s and beyond," and it has continued to influence the work of governments, their foreign assistance programs, and nongovernmental organizations, as manifested by the Women in Development approach, which explicitly recognizes women's economic and productive roles (Beneria et al. 2016, 8). The increased attention to women in development was accompanied by a focus on including women in official statistics, as well as growing discontent with widely accepted models of household behavior based on the unitary model of the household. This model assumed that either all household members have the same preferences, pool all resources, and agree on all decisions, or that one household member makes the decisions for everyone (Becker 1981). The unitary model has been challenged, both theoretically and empirically, by collective models of household behavior that allow decision-makers to have different preferences and do not assume that households have a common goal (Chiappori 1992). These challenges arose from the growing empirical literature on household decision-making in low- and middle-income countries (LMICs), which was stimulated by theoretical developments in modeling household behavior and intrahousehold allocation and by the increased availability of sex-disaggregated data on labor, asset ownership, and decision-making with which to test these household models (Strauss and Thomas 1995).

Early work that juxtaposed the unitary model of the household against the collective model—often using a bargaining framework in which household decision-makers, usually the husband and wife, negotiated when making household decisions—frequently rejected the assumptions of the unitary model. Although tests of the unitary versus collective models can be conducted using household-level outcomes such as budget shares, comparisons using individual-level outcomes (such as individual health and nutrition indicators) are more powerful. These comparisons can test whether individuals with more bargaining power favor specific household members owing to their age, gender, position in the household, or other characteristics. IFPRI's research program on agricultural commercialization and nutrition gathered individual-level data on nutritional status (often for women and children under five years of age), as well as data on variables that could be broadly interpreted as proxies of men's and women's bargaining power, thus providing a wealth of data with which to test these models in different country contexts (von Braun and Kennedy 1994). The early 1990s were a fertile time for testing alternative household models and measures of bargaining power—and for the findings of this research to influence policy, such as the targeting of extension services and the design of conditional cash transfer programs, among others (see Chapters 8 and 11).

In the conclusion to their volume of studies on intrahousehold resource allocation by economists, demographers, sociologists, and anthropologists, Haddad and colleagues (1997) argued that using the unitary model of the household as a guideline for policy prescriptions may lead to policy failures because: (1) public transfers may have different effects, depending on who within the household receives the transfer, (2) nonrecipients of the transfer may act to counter the intent of the program or policy, (3) information may not reach the persons in the household who need it most, since the person receiving the information does not necessarily share it, and (4) policymakers may be limited to a smaller set of instruments with which to affect development outcomes. The unitary model predicts that only changes in prices and household incomes can shift household behavior and that such changes will benefit all members of a household. In contrast, the collective model recognizes that there can be significant differences among household members and posits that household allocation can be affected by a range of policies, such as changes in property rights and access to credit, public works schemes, and legal and institutional rights (see Chapter 6).

Much of the empirical evidence supporting the collective model of the household comes from data on rural households in LMICs. This evidence includes studies showing differential propensities to spend income controlled by men or women (Hoddinott and Haddad 1995), differential effects of men's and women's assets on consumption (Doss 2006) and household expenditures and schooling (Quisumbing and Maluccio 2003), incomplete sharing of risk within the household (Dercon and Krishnan 2000; Doss 2006; Goldstein and Udry 2008), and inefficiency of resource allocation between plots managed by men and women (Udry 1996). Overall, these findings have influenced the design of programs that targeted women, such as conditional cash transfer programs and other social protection programs more broadly. In turn, these programs have generated a large body of evidence that resources controlled by women are associated with investments in child schooling, health, and nutrition, as shown by both observational and experimental studies.[1] The focus of collective models on intrahousehold bargaining has emphasized rivalry and competition among household members, ignoring areas of shared resources and interests. But the activities that individuals do together—form households, share ownership and control over some resources, work together on family farms, produce some output jointly, have and raise children together, and share in some consumption—indicate that there are gains to jointness within families and households. Therefore, more recent studies call for greater attention to joint decision-making and resource control to better understand household dynamics (Doss and Meinzen-Dick 2015; Doss and Quisumbing 2020; Fafchamps and Quisumbing 2008). Expanding the focus from women to gender creates opportunities to understand the challenges that men face, both in collaboration and competition with women.

---

1 Observational studies are reviewed by Quisumbing et al. (2003), among others. Yoong, Rabinovich, and Diepeveen (2012) review experimental studies of resource transfers to women versus men. Results from Yoong, Rabinovich, and Diepeveen (2012) are discussed later in this chapter.

As attention to gender issues within the household has risen, so too has awareness of the need to address gender inequalities in assets within communities and society at large. An early conference on gender and property rights identified commonalities in the constraints to women's rights to agricultural land, water, and forests, and the consequences of this for women's productivity, equity, and empowerment, and the natural environment.[2] This work has contributed to research and policy efforts to strengthen women's property rights, as discussed in Chapter 6 on tenure.

Linkages between gender and development outcomes

The evidence against the unitary model of the household was foundational to the notion that increasing resources controlled by women—whether human, physical, social, or political capital—was important to realizing positive development outcomes. This research, to which IFPRI has contributed, has primarily influenced policy and practice in the areas of social protection, nutrition, and agriculture (Haddad et al. 1997).

The positive relationship between women's human capital, which is usually measured through educational attainment, and child nutrition and education outcomes has long been established by observational studies (Quisumbing 2003; Smith et al. 2003). A systematic review of programs transferring economic resources to women found that: (1) targeting transfers to women can improve children's well-being, especially through investments in children's health and education, and (2) the gender of the transfer recipient affects the outcomes of some programs, which is consistent with the predictions of the collective model. However, (3) increasing women's control of transfers does not guarantee positive outcomes, and (4) outcomes may depend on the type of program offered (Yoong et al. 2012). For instance, evidence from Sri Lanka finds that men entrepreneurs secure permanent increases in income after receiving cash transfers, though women entrepreneurs do not (De Mel et al. 2009).

The targeting of conditional transfers to women was a key element of Mexico's PROGRESA (Programa Nacional de Educación, Salud, y Alimentación) and its successor program, Oportunidades. PROGRESA was one of the first large-scale programs to target transfers to women, a decision based on evidence from then-emerging literature on intrahousehold allocation showing

2 This e-conference included several key papers: Lastarria-Cornheil (1997), Meinzen-Dick et al. (1997), Rocheleau and Edmunds (1997), and Zwarteveen (1997).

that resources controlled by women were associated with better child health and nutrition outcomes (Skoufias 2005). In a follow-up study with PROGRESA beneficiaries 20 years later, researchers found that children who were exposed to the program *in utero*, as infants, or as children transitioning from primary to secondary school had better educational outcomes, higher income expectations, greater geographic mobility (including international migration), and delayed marriage and childbearing (Araujo and Macours 2021). A recent review of nutritional impacts of social assistance programs found that when women or children were not specifically targeted, there was little evidence of positive impact across the assessed diet and nutrition-related outcomes of women and children. The review recommended that programs designed to improve women's and children's diets and nutrition-related outcomes should continue to target women and/or children with the transfers (Olney et al. 2022).

The *2010–2011 State of Food and Agriculture Report*, published by the Food and Agriculture Organization of the United Nations (FAO), highlighted how increasing women's control of resources could lead to gains in agricultural productivity. These findings were based on simulations that provided women farmers with the same resources as men (FAO 2011). In the FAO's 2023 report on the status of women in agrifood systems, simulations suggest that achieving gender equality at scale in agriculture and agrifood systems could increase global gross domestic product and reduce food insecurity (FAO 2023).

The important connections between gender and agriculture are also illustrated by an agriculture–nutrition linkages framework that identifies six pathways through which agricultural interventions can impact nutrition (Ruel and Alderman 2013). Of these, three focus explicitly on women and gender: women's social status and empowerment through increased access to and control over resources; women's time through participation in agriculture, which can be either positive or negative for their own nutrition and that of their children; and women's health and nutrition through engagement in agriculture.

Measuring women's empowerment and gender equality

The recognition of women's empowerment and gender equality as Sustainable Development Goal 5 (SDG5) provided an impetus for gender researchers to develop better metrics for women's empowerment. Perspectives differ on the measurement of women's empowerment, but the field has generally coalesced around a conceptual definition of empowerment that defines it as the process by which people expand their ability to make strategic life choices, particularly in contexts in which this ability had been denied to them (Kabeer 1999). The need for empowerment metrics reflects the rationale for measuring poverty: measurement matters if it can motivate political action, monitor progress, and help guide the design of effective policy actions, so that researchers, program designers, and policymakers can discern what works and what does not (Atkinson 2019).

In the past, many development practitioners were concerned with women's empowerment and gender equality only insofar as they could be used to achieve objectives on health and nutrition (Galiè et al. 2019; Heckert et al. 2019; Sraboni 2014), productivity (Diiro 2018), and resource management (Sodhi et al. 2010). Now, however, spurred by the adoption of women's empowerment and gender equality as an SDG, these are increasingly regarded as goals in themselves, within both agricultural research and development and food systems interventions (Elias et al. 2021; Pyburn and van Eerdewijk 2021). The Women's Empowerment in Agriculture Index (WEAI) is one of the most widely used empowerment metrics, used by 279 organizations in 69 countries as of April 2025. Unlike other indexes that rely on aggregate-level indicators, the WEAI is a survey-based index that draws from interviews of the primary adult woman and man in the same household, and, as with multidimensional poverty indexes, is additive and decomposable (Alkire et al. 2013). These properties enable policymakers to easily understand how indicators are combined and identify nuances underlying each component indicators and the constraints to empowerment for different subgroups.

The original WEAI consists of five domains (production, resources, income, leadership, and time) and 10 indicators, and the Abbreviated WEAI, developed later, is a shorter, 6-indicator version (Malapit et al. 2017). This information summarizes women's empowerment achievements in the five domains, and enables computation of the gender parity index, which compares a woman's empowerment to that of the man in her household. The WEAI was developed in consultation with stakeholders and researchers working on gender issues and piloted in three countries (Bangladesh, Guatemala, and Uganda) with IFPRI's long-term data collection partners, and was launched in 19 countries through the Feed the Future Initiative's baseline monitoring survey. Baseline findings show that lack of access to and decisions over credit, excessive workloads, and lack of group membership are the top empowerment constraints across countries (Malapit et al. 2014). Moreover, focusing on the component indicators has led to a rich understanding of trade-offs between different domains of empowerment.

Agricultural development projects subsequently adopted and modified the WEAI metrics in an attempt to monitor the impact of their projects on women's empowerment, but this proliferation of WEAI versions did not permit comparisons across projects and portfolios. The Gender, Agriculture, and Assets Project, Phase 2 (GAAP2) was a portfolio of 13 agricultural development projects in nine countries in South Asia and sub-Saharan Africa that used qualitative and survey tools to develop a project-level WEAI (pro-WEAI) (Malapit et al. 2019). The pro-WEAI can be used to measure the impacts of interventions on aspects of empowerment considered essential to a project's mission and to make comparisons across a portfolio of different projects. Each project applied to join the portfolio and was represented by a project implementor and an evaluation partner, ensuring that both research rigor and on-the-ground experience were reflected.

The process of developing the pro-WEAI led to the creation of the Reach, Benefit, Empower, and Transform (RBET) Framework as IFPRI researchers and partners examined projects' strategies in pursuit of empowerment objectives (Johnson et al. 2018; Kleiber et al. 2019).[3] In this framework (Table 14.1), projects that *reach* women include them in program activities, while those that *benefit* women improve their well-being outcomes, including income, health, and nutrition. Indicators to assess reach include the number of women and men attending training or extension programs; indicators of benefits include income earned by women or women's nutritional status. But neither "reach" nor "benefit" explicitly addresses empowerment, so projects aiming to *empower* women need to go beyond these objectives to facilitate women's ability to make and implement strategic life choices. Gender-transformative approaches emphasize interventions that aim to *transform* formal and informal systems, institutions, and markets that limit women's opportunities into those that support gender equality (Kleiber et al. 2019; Morgan et al. 2023). These approaches often aim to holistically change gender norms at the community and societal levels, address structural and institutional barriers, and mobilize the power of the collective. Most projects conceptualize "empower" as occurring at the individual level, whereas transformation occurs at a higher level and involves changes in norms and structures that may lie beyond the purview of individual agricultural development projects (Larson et al. 2024). Applying the RBET framework can facilitate discussion about the desired

3 The RBET framework was developed by IFPRI, and the CGIAR Research Program on Fish Agri-Food Systems further developed it to include "transform." See Johnson et al. (2018) and Kleiber et al. (2019).

TABLE 14.1 The Reach, Benefit, Empower, and Transform (RBET) framework

	Reach	Benefit	Empower	Transform
Definition	Include women in program activities	Increase women's well-being	Strengthen the ability of women to make and implement life choices	Go beyond the woman and her household to change systems, gender norms, and power relations on a larger scale
Objective	Ensure that women have the same opportunity to access program activities as men: • Address barriers to participation, such as program information, timing, or location of meetings and trainings	Require more than reaching women: • Women value the intervention • Direct benefits accrue to women • Women's needs, preferences, and constraints are considered in the intervention design	Go beyond reaching and benefiting women: • Increase women's agency • Shift gender norms and attitudes among participants	Go beyond empowering individual women: • Involve men • Change gender norms at community and societal levels • Address structural and institutional barriers • Mobilize the power of the collective

Source: Quisumbing et al. (2023), adapted from Johnson et al. (2018) and Morgan et al. (2023).

level of impact and identify actions and indicators related to empowerment and transformation, rather than merely reach or benefit. Better understanding of the strategies undertaken, and the theory underlying which aspects of agency can change as a result, can point to specific indicators of agency that should be measured to assess impacts.

Using the pro-WEAI and RBET framework to evaluate the GAAP2 portfolio and the UN Joint Program for Rural Women's Economic Empowerment (JP RWEE) illustrates the value of having a standardized metric to evaluate empowerment impacts of development projects. A synthesis of mixed methods impact evaluations of 11 GAAP2 projects highlights the need for projects to design their strategies specifically for empowerment, rather than assume that projects aiming to reach and benefit women automatically empower them (Quisumbing et al. 2024). In the GAAP2 portfolio, the projects that succeeded in empowering women were intentional about their project strategies, designed activities specific to cultures and contexts, and paid attention to unintended consequences (such as backlash from men or increased workload for women).

For example, a homestead food production project in Bangladesh, which provided women with gardening materials and trainings, found that the women often sold surpluses of the vegetables they produced and maintained control over the income generated, either saving it or spending it on personal needs, family needs, additional food, or their children's education (Dupuis et al. 2022). Beneficiaries of this project reported greater confidence and motivation when they saw their gardens becoming productive. In a project that organized rural women in Ethiopia into savings and credit associations, the opportunity for women to earn money helped them develop a positive self-image and be proactive about their rights and responsibilities (Hillesland et al. 2022). Women participating in a poultry value chain project in Burkina Faso said raising poultry increased their self-confidence in their skills and capacities. For these women beneficiaries, gaining financial independence was critical, as it helped them to rely less on their husbands' permission or direction for how to spend money (Eissler et al. 2020).

For the JP RWEE project, IFPRI was brought in to synthesize results across four country programs that used WEAI metrics in their impact evaluations. The synthesis across these four studies highlights the potential of group-based approaches, the importance of involving men, and the need to be mindful of increased workload resulting from women's involvement in livelihood programs (Quisumbing et al. 2023). In Ethiopia, Kyrgyzstan, and Nepal, women talked about their workload increasing owing to the time required to participate in groups and livelihood interventions. However, while trainings increased women's time burden in Kyrgyzstan, participation in the Gender Action Learning System (GALS) led to a rebalancing of household duties among family members, allowing women to direct their time to income-generating activities (University of Central Asia 2021). In Ethiopia, women expressed that their actual time use was less important to them than control over how they spent their time, or what has been called time-use agency (Eissler et al. 2022; Mulema et al. 2021). Even if they increased time spent on productive work, women themselves may perceive this as positive, if they benefit from the returns.

Comparing the distribution of impacts from GAAP2 and JP RWEE on composite indicators of empowerment and gender equality shows that a larger proportion of JP RWEE projects had positive impacts on women's empowerment indicators than did the GAAP2 portfolio, reflecting JP RWEE's more women-focused programming. Some of the JP RWEE projects also implemented GALS, an approach that involves the whole household in setting goals and identifying activities to achieve those goals. But both programs reported far less impact on men's empowerment, whether positive or negative, and the achievement of gender parity within the household, suggesting that shifting norms to achieve gender parity is more difficult.

The availability of empowerment metrics and development outcomes from the same datasets enables a more direct analysis of links between women's empowerment and those outcomes, as compared to using proxy indicators.

TABLE 14.2 Women's empowerment in relation to food systems outcomes

Extent of agreement among studies	Number of studies		
	Low (1–3 studies)	Medium (4–6 studies)	High (7–13 studies)
Low			
Medium		• Women's nutrition and diets • Household-level food security and diets	• Agricultural production
High	• Life satisfaction • Children's educational outcomes • Water, sanitation, and health • Poverty transitions		• Child nutrition and diets

Source: Myers et al. (2023).

A study reviewing empirical research on the association between women's empowerment, measured using the WEAI, and primary food systems outcomes across a range of different contexts confirms that fostering women's empowerment may have a positive effect on many food systems–related outcomes (Myers et al. 2023). Table 14.2 presents the amount of evidence on these outcomes and the extent of agreement among the studies. Women's empowerment seems to be closely linked to improved child nutrition and is also associated with household-level food security and, to a lesser degree, with women's own diets and nutrition. Although there is evidence that women's empowerment and intrahousehold gender equity are also positively associated with agricultural productivity, the context and type of crop matter. As often occurs with studies involving empowerment, household, societal, and institutional factors such as wealth, class, and norms frequently play critical moderating roles. Although women's empowerment is an important goal in itself, demonstrating its positive association with other development outcomes helps to build a constituency and policy support for programs and policy to enhance women's empowerment and gender equality.

Lessons learned from metrics for measuring empowerment across multiple domains and aspects of women's lives, as well as growing demand for such metrics, highlight the need for similar empowerment metrics that can be used at the national level and in large multi-topic surveys. Such metrics need to be collected with shorter survey modules and capable of measuring women's empowerment across livelihood strategy, life cycle stage, or period of structural transformation. To meet this demand, IFPRI, along with the World Bank Living Standards Measurement Studies unit, Emory University, and multiple partners in the pilot countries of Bangladesh, Guatemala, Malawi, and Nepal led the development of the recently launched Women's Empowerment Metric for National Statistical Systems (WEMNS) (Seymour et al. 2024). These partners included the National Statistics Office in Malawi, independent research institutes in Nepal, and long-term data collection partners in Bangladesh and Guatemala that had been involved in piloting the WEAI.

WEMNS consists of 12 indicators of instrumental, intrinsic, and collective agency, as well as agency-enabling resources. Findings from the pilot studies suggest that programming and policy investments to address women's lack of confidence with engaging their communities, leading in organizations, and influencing spending decisions can contribute meaningful gains to women's empowerment globally. Currently, WEMNS data are being collected at scale by the FAO in Liberia, Sierra Leone, and Tanzania. The availability and analysis of nationally representative data on women's empowerment, along with data on agriculture, diet, and food security outcomes, from these countries and elsewhere has the potential to inform priority investments to support women's empowerment, lead to a clearer understanding of the linkages between women's empowerment and other key development outcomes, and support the monitoring of progress toward women's empowerment and gender equality.

Challenges for future work

Intersectionality

Gender research has not fully engaged with other aspects of disadvantage that may compound gender inequalities, such as wealth, class, caste, and ethnicity. In Viet Nam, for example, IFPRI's ongoing work with the Alliance of Bioversity International and CIAT and Wageningen University & Research shows that in certain ethnic minority communities, men are less empowered than women (Lan Nguyen et al. 2024). Intersectional approaches also recognize that women (and men) are not homogeneous groups—a high-caste or wealthy woman in South Asia may be more disempowered than a low-caste or poorer woman due to the influence of patriarchy (Doss et al. 2022). Similarly, a woman's role in the household can also make a great difference in her empowerment. While past gender research often compared outcomes for male and female household heads, women in male-headed households typically achieve lower empowerment scores than do female household heads, and mothers-in-law are often more empowered than their daughters-in-law.

Life cycle and demographic factors

Intersectional approaches can also be used to consider how gendered vulnerabilities and opportunities differ and evolve throughout the life cycle, with particular attention to both youth and the elderly. First, considering the rural livelihoods of youth (often defined as those ages 15 to either 24 or 34), the focus on the paid work of youth frequently emphasizes the roles of young men. Research that considers the productive and reproductive roles of young women and men is important for understanding their unique experiences and needs to support the development of their livelihoods alongside complementary roles in their families and communities (Heckert et al. 2021). In addition to addressing immediate concerns about young people's vulnerabilities and lack of land, capital, or other resources, there is also a need to identify critical investments to prepare young women and men for continued success and well-being as they advance through their prime working years.

While the current rhetoric has largely focused on youth, there has been little recognition of either the vulnerabilities or increased knowledge and experience associated with aging, the increase in caregiving work for the "sandwich" generation who are simultaneously caring for young children and aging parents, or the inequalities related to an individual's position in the household (for example, daughters-in-law relative to mothers-in-law).

Migration has particularly important implications for the gender and generational composition of rural areas and agricultural labor. Both qualitative and quantitative studies in areas with high levels of male outmigration reveal important changes in the scope of women's agency within households and community institutions (Doss et al. 2022; Meinzen-Dick et al. 2022). The movement of men out of agricultural labor affects women's productive and reproductive workloads, as well as gender dynamics and shifts in rural gender relations (Doss et al. 2021). In Nepal, a country with high rates of male outmigration, migration may increase women's mobility and decision-making roles after their husbands migrate, especially among Dalit and Janajati households, but wives with migrant husbands face stress from added responsibilities and workload, both domestic and care work and especially productive nondomestic work, which can lead to disempowerment (Kaspar 2005; Sijapati et al. 2017).

Structural and institutional change

While exercising agency is at the core of women's empowerment, individual agency is constrained by structural conditions and gendered social norms, which are the informal institutions that dictate appropriate behavior for women and men. Thus, addressing structural and institutional barriers to gender equality may be the most important step to transform gender relations. Women's voice and agency beyond the household level are critical for access to essential resources and decision-making power, which can contribute to resilience to climate change (Hidrobo et al. 2024; Njuki et al. 2022; Takeshima et al. 2022). The *2024 Global Gender Gap Report* highlights slow progress in women's empowerment in various spheres, including political empowerment; at the current rate of progress, it will take 134 years to close the gender gap. An interdisciplinary approach is key to addressing structural and institutional barriers, with important roles for social psychology, sociology, and political sciences (WEF 2024). Governments, international organizations, development organizations, the private sector, and civil society, especially women's organizations, can all play an important role in addressing these barriers. There are a growing number of tools to measure and address gender inequalities in rural institutions and governance, including the Social Institutions and Gender Index (SIGI) and the Women's Empowerment in Agrifood Systems (WEAGov) assessment tool (ElDidi et al. 2021; OECD 2023). Understanding the interplay between the mutually reinforcing areas of empowerment and structural/institutional change will be an important area of future research to highlight potential entry points for meaningful change across the system.

Recent research highlights the crucial role of aspirations and locus of control—the degree to which people believe that they, rather than outside forces, can control outcomes of events in their own lives—in overcoming poverty, showing that a positive mindset can empower individuals to strive for better outcomes (Jayachandran 2018). A study in Kyrgyzstan finds that women with higher aspirations are more likely to espouse egalitarian gender attitudes, as are their husbands, who themselves have higher aspirations. These women play a greater role in household decision-making, and their spouses are more likely to agree with them about women's roles in decision-making (Kosec et al. 2022).

Women often perceive that they have limited agency over their lives, which can reduce their motivation to pursue progress and change. Role models offer a powerful mechanism to boost aspirations for the future. Successful women entrepreneurs, government officials, or private-sector leaders offer examples of the possibilities that can be achieved. Exposure to a female role model in popular media can significantly improve the educational outcomes of female students (Riley 2024). The influence of role models extends beyond media and can be mediated by various actors, including teachers, politicians, family members, and siblings.

To address systemic and institutional barriers effectively, involving men in interventions is critical. In some contexts, women's decision-making is perceived as threatening to masculinity, even if women report their spouses consulting them before taking a decision (Meinzen-Dick et al. 2019). Spousal approval may dictate women's access to credit, either in terms of women needing to seek permission to participate in credit groups or men exerting undue influence on how borrowed funds should be spent. However, projects that involved men have been able to slowly change behavior and task-sharing. For example, discussions with program staff in the Nepal JP RWEE project indicate that men willingly took on additional reproductive work and shared women's workload. The positive impact of GALS-related interventions on men and women alike in Kyrgyzstan is consistent with emerging evidence from other countries (for example, a gender and nutrition agricultural intervention in Bangladesh that trained husbands and wives together) that involving both men and women in gender-transformative approaches, rather than focusing on women alone, may be key to effective and sustainable programs (Quisumbing et al. 2021, 2023). By not involving men, program designers may be making incorrect assumptions about men's ability and willingness to change. For example, a study in Saudi Arabia reveals that men often misjudge their peers' support for women working outside the home (Bursztyn et al. 2020). Correcting these misconceptions has been shown to increase the likelihood of women securing higher-paying jobs. Similarly, in Uganda, exposing men to videos that challenge gender stereotypes in agriculture has been shown to enhance women's empowerment within the sector (Lecoutere et al. 2023).

Concluding remarks

This retrospective and prospective view of gender research shows that challenging the assumptions underlying household models, whether the unitary model or the reliance of many collective models on the bargaining model, may be key to understanding household behavior and discovering policy instruments to influence that behavior. The finding that increasing women's control of resources improved children's education, health, and nutrition inspired the design and rigorous testing of Mexico's PROGRESA, which led to the design and implementation of conditional cash transfer programs worldwide. In turn, impact evaluations of these new programs—and their expansion into different countries, contexts, and areas of intervention—generated new evidence that expanded our understanding of household behavior.

Many of the programs targeting women viewed women's empowerment as instrumental to achieving development objectives. However, the recognition of women's empowerment and gender equality as intrinsically valuable created the impetus to measure them better, motivate policy action, monitor progress, and ascertain what works. This led to the development of empowerment metrics, such as the WEAI and its variants and the recently developed WEMNS, in collaboration with country and institutional partners.

IFPRI and partners have long engaged in research and development of innovative measurements and tools on gender norms and on women's voice and leadership in policy processes and governance. At the project level, collaborating with partners on the co-design and rigorous testing of gender-responsive and gender-transformative policies, programs, and innovations has shown that merely having empowerment objectives is not enough. Intentional gender programming and transformative approaches are needed to increase women's empowerment and achieve gender equality in the longer run. Generating evidence and taking action are two important parts of a process for policy research to influence programs and policies, and for those policies themselves to expand the evidence base for our understanding of gender dynamics and household behavior. Acknowledging the complementarity of research and action will enable us to better address the research and policy gaps in the areas of intersectionality, life cycle and demographic factors, and structural and institutional change.

References

Alkire, S., R. Meinzen-Dick, A. Peterman, A.R. Quisumbing, G. Seymour, and A. Vaz. 2013. "The Women's Empowerment in Agriculture Index." *World Development* 52: 71–91.

Araujo, M.C., and K. Macours. 2021. "Education, Income and Mobility: Experimental Impacts of Childhood Exposure to Progresa After 20 Years." IDB Working Paper 01288. Inter-American Development Bank, Washington, DC.

Atkinson, A.B. 2019. *Measuring Poverty Around the World.* Princeton: Princeton University Press.

Becker, G.S. 1981. *A Treatise on the Family.* Cambridge, MA: Harvard University Press.

Beneria, L., G. Berik, and M.S. Floro. 2016. *Gender, Development, and Globalization: Economics As If All People Mattered.* New York and London: Routledge.

Boserup, E. 1970. *Woman's Role in Economic Development.* New York: St. Martin's Press.

Bursztyn, L., A.L. González, and D. Yanagizawa-Drott. 2020. "Misperceived Social Norms: Women Working Outside the Home in Saudi Arabia." *American Economic Review* 110 (10): 2997–3029.

Chiappori, P. 1992. "Collective Labor Supply and Welfare." *Journal of Political Economy* 100 (3): 437–467.

De Mel, S., D. McKenzie, and C. Woodruff. 2009. "Are Women More Credit Constrained? Experimental Evidence on Gender and Microenterprise Returns." *American Economic Journal: Applied Economics* 1 (3): 1–32.

Dercon, S., and P. Krishnan. 2000. "In Sickness and in Health: Risk Sharing Within Households in Rural Ethiopia." *Journal of Political Economy* 108 (4): 688–727.

Diiro, G.M., G. Seymour, M. Kassie, G. Muricho, and B.W. Muriithi. 2018. "Women's Empowerment in Agriculture and Agricultural Productivity: Evidence from Rural Maize Farmer Households in Western Kenya." *PLoS ONE* 13 (5): e0197995.

Dixon, R.B. 1982. "Women in Agriculture: Counting the Labor Force in Developing Countries." *Population and Development Review* 8 (3): 539–566.

Doss, C. 2006. "The Effects of Intrahousehold Property Ownership on Expenditure Patterns in Ghana." *Journal of African Economies* 15 (1): 149–180.

Doss, C.R., and R. Meinzen-Dick. 2015. "Collective Action Within the Household: Insights from Natural Resource Management." *World Development* 74 (10): 171–183.

Doss, C.R., R. Meinzen-Dick, A. Pereira, and R. Pradhan. 2022. "Women's Empowerment, Extended Families and Male Migration in Nepal: Insights from Mixed Methods Analysis." *Journal of Rural Studies* 90: 13–25.

Doss, C.R., A. Qaisrani, K. Kosec, V. Slavchevska, A. Galiè, and N. Kawarazuka. 2021. "From the 'Feminization of Agriculture' to Gender Equality." In *Advancing Gender Equality Through Agricultural and Environmental Research: Past, Present, and Future*, eds. R. Pyburn and A. van Eerdewijk, 297–327. Washington, DC: IFPRI.

Doss, C.R., and A.R. Quisumbing. 2020. "Understanding Rural Household Behavior: Beyond Boserup and Becker." *Agricultural Economics* 51 (1): 47–58.

Dupuis, S., M. Hennink, A.S. Wendt, J.L. Waid, M.A. Kalam, S. Gabrysch, and S. Sinharoy. 2022. "Women's Empowerment Through Homestead Food Production in Rural Bangladesh." *BMC Public Health* 22 (1): 1–11.

ElDidi, H., K. Kosec, and R. Meinzen-Dick. 2021. "Gender in Rural Institutions and Governance: A Review of Existing Tools." CGIAR GENDER Platform Working Paper 002. Nairobi: CGIAR GENDER Platform.

Elias, M., S.M. Cole, A.R. Quisumbing, A.M.P. Valencia, R.S. Meinzen-Dick, and J. Twyman. 2021. "Assessing Women's Empowerment in Agricultural Research." In *Advancing Gender Equality Through Agricultural and Environmental Research: Past, Present, and Future*, eds. R. Pyburn and A. van Eerdewijk, 329–362. Washington, DC: IFPRI.

Eissler, S., J. Heckert, E. Myers, G. Seymour, S. Sinharoy, and K. Yount. 2022. "Measuring Women's Empowerment: Gender and Time-Use Agency in Benin, Malawi and Nigeria." *Development and Change* 53: 1010–1034.

Eissler, S., A. Sanou, J. Heckert, et al. 2020. "Gender Dynamics, Women's Empowerment, and Diets: Qualitative Findings from an Impact Evaluation of a Nutrition-Sensitive Poultry Value Chain Intervention in Burkina Faso." IFPRI Discussion Paper 1913. IFPRI, Washington, DC.

Fafchamps, M., and A.R. Quisumbing. 2008. "Household Formation and Marriage Markets in Rural Areas." In *Handbook of Development Economics*, vol. 4, eds. T.P. Schultz and J.A. Strauss, 3187–3247. Amsterdam: North-Holland.

FAO (Food and Agriculture Organization of the United Nations). 2011. *The State of Food and Agriculture 2010-2011: Women in Agriculture: Closing the Gender Gap for Development.* Rome.

FAO. 2023. *The Status of Women in Agrifood Systems.* Rome.

Galiè, A., N. Teufel, L. Korir, I. Baltenweck, A. Webb Girard, P. Dominguez-Salas, and K.M. Yount. 2019. "The Women's Empowerment in Livestock Index." *Social Indicators Research* 142 (2): 799–825.

Goldstein, M., and C. Udry. 2008. "The Profits of Power: Land Rights and Agricultural Investment in Ghana." *Journal of Political Economy* 116 (6): 981–1022.

Haddad, L., J. Hoddinott, and H. Alderman. 1997. *Intrahousehold Resource Allocation in Developing Countries: Methods, Models, and Policy.* Baltimore: Johns Hopkins University Press.

Heckert, J., D.K. Olney, and M.T. Ruel. 2019. "Is Women's Empowerment a Pathway to Improving Child Nutrition Outcomes in a Nutrition-Sensitive Agriculture Program?: Evidence from a Randomized Controlled Trial in Burkina Faso." *Social Science & Medicine* 233: 93–102.

Heckert, J., A. Pereira, C. Doss, E. C. Myers, and A. Quisumbing. 2021. "Structural Transformation and Gendered Transitions to Adulthood Among Rural Youth: Cross-National Evidence from Low- and Middle-Income Countries." *Journal of Development Studies* 57 (4): 614–634.

Hidrobo, M., K. Kosec, H.N. Gartaula, B. Van Campenhout, and L. Carrillo. 2024. "Making Complementary Agricultural Resources, Technologies, and Services More Gender-Responsive." *Global Food Security* 42: 100778.

Hillesland, M., S. Kaaria, E. Mane, M. Alemu, and V. Slavchevska. 2022. "Does a Joint United Nations Microfinance 'plus' Program Empower Female Farmers in Rural Ethiopia? Evidence Using the Pro-WEAI." *World Development* 156: 105909.

Hoddinott, J., and L. Haddad. 1995. "Does Female Income Share Influence Household Expenditures? Evidence from Cote D'Ivoire." *Oxford Bulleting of Economics and Statistics* 57 (1): 77–96.

Jayachandran, S. 2018. "Think Positive, Climb Out of Poverty? It Just Might Work." *New York Times*. July 13.

Johnson, N., M. Balagamwala, C. Pinkstaff, S. Theis, R. Meinzen-Dick, and A. Quisumbing. 2018. "How Do Agricultural Development Projects Empower Women? Linking Strategies with Expected Outcomes." *Journal of Gender, Agriculture and Food Security* 3 (2): 1–19.

Kabeer, N. 1999. "Resources, Agency, Achievements: Reflections on the Measurement of Women's Empowerment." *Development and Change* 30 (3): 435–464.

Kaspar, H. 2005. *I Am the Household Head Now! Gender Aspects of Out-Migration for Labor in Nepal*. Kathmandu: Nepal Institute of Development Studies.

Kleiber, D., P. Cohen, C. Gomese, and C. McDougall, et al. 2019. *Gender Integrated Research for Development in Pacific Coastal Fisheries*. Program Brief: FISH-2019-02. Penang, Malaysia: CGIAR Research Program on Fish Agri-Food Systems.

Kosec, K., K. Akramov, B. Mirkasimov, J. Song, and H. Zhao. 2022. "Aspirations and Women's Empowerment: Evidence from Kyrgyzstan." *Economics of Transition and Institutional Change* 30 (1): 101–134.

Lan Nguyen, T.T., M. van den Bergh, T. Stomph, D. Nabuuma, and B. Kramer. 2024. "Empowerment Unveiled: Gender Dynamics and the Impact of Nutrition-Sensitive Agricultural Interventions Among Ethnic Minority Groups in Northern Vietnam." Unpublished manuscript.

Larson, A., R. Meinzen-Dick, S. Trautman, et al. 2024. *From Reach to Transformation: Leveraging the RBET Framework to Secure Women's Land and Resource Rights. Global Initiative for Gender Transformative Approaches for Securing Women's Resource Rights (WRR)*. Bogor, Indonesia: Center for International Forestry Research; Nairobi: World Agroforestry.

Lastarria-Cornhiel, S. 1997. "Impact of Privatization on Gender and Property Rights in Africa." *World Development* 25 (8): 1317–1333.

Lecoutere, E., D.J. Spielman, and B. Van Campenhout. 2023. "Empowering Women through Targeting Information or Role Models: Evidence from an Experiment in Agricultural Extension in Uganda." *World Development* 167: 106240.

Malapit, H. J., C. Pinkstaff, K.Sproule, et al. 2017. "The Abbreviated Women's Empowerment in Agriculture Index (A-WEAI)." IFPRI Discussion Paper 1647. IFPRI, Washington, DC.

Malapit, H.J., A. Quisumbing, R.S. Meinzen-Dick, et al. 2019. "Development of the Project-Level Women's Empowerment in Agriculture Index (pro-WEAI)." *World Development* 136: 675–692.

Malapit, H.J., K. Sproule, C. Kovarik, et al. 2014. *Measuring Progress Toward Empowerment: Women's Empowerment in Agriculture Index: Baseline Report*. Washington, DC: IFPRI.

Meinzen-Dick, R.S., L.R. Brown, H. Feldstein, and A.R. Quisumbing. 1997. "Gender and Property Rights: Overview." *World Development* 25 (8): 1299–1302.

Meinzen-Dick, R.S., P. Pradhan, and W. Zhang. 2022. "Migration and Gender Dynamics of Irrigation Governance in Nepal." *International Journal of the Commons* 16 (1): 1–18.

Meinzen-Dick, R.S., D. Rubin, M. Elias, A.A. Mulema, and E. Myers. 2019. "Women's Empowerment in Agriculture: Lessons from Qualitative Research." IFPRI Discussion Paper 1797. IFPRI, Washington, DC.

Morgan, M., A. Larson, S. Trautman, E. Garner, M. Elias, and R. Meinzen-Dick. 2023. *Gender Transformative Approaches to Strengthen Women's Land and Resource Rights*. Bogor, Indonesia: Centre for International Forestry Research; Nairobi: World Agroforestry; Rome: International Fund for Agricultural Development.

Mulema, A.A., B. Boonabaana, L. Debevec, L. Nigussie, M. Alemu, and S. Kaaria. 2021. "Spiraling Up and Down: Mapping Women's Empowerment Through Agricultural Interventions Using the Community Capitals Framework in Rural Ethiopia." *Community Development* 52 (1): 113–130.

Myers, E., J. Heckert, S. Faas, H.J. Malapit, R.S. Meinzen-Dick, K. Raghunathan, and A.R. Quisumbing. 2023. "Is Women's Empowerment Bearing Fruit? Mapping Women's Empowerment in Agriculture Index (WEAI) Results Using the Gender and Food Systems Framework." IFPRI Discussion Paper 2190. IFPRI, Washington, DC.

Njuki, J., S. Eissler, H. Malapit, R. Meinzen-Dick, E. Bryan, and A. Quisumbing. 2022. "A Review of Evidence on Gender Equality, Women's Empowerment, and Food Systems." *Global Food Security* 33: 100622.

OECD (Organisation for Economic Co-Operation and Development). 2023. *SIGI 2023 Global Report: Gender Equality in Times of Crisis*. Social Institutions and Gender Index. Paris.

Olney, D.K., A. Gelli, N. Kumar, H. Alderman, A. Go, and A. Raza. 2022. "Social Assistance Programme Impacts on Women's and Children's Diets and Nutritional Status." *Maternal & Child Nutrition* 18: e13378.

Pyburn, R., and A. van Eerdewijk, eds. 2021. *Advancing Gender Equality through Agricultural and Environmental Research: Past, Present, and Future*. Washington, DC: IFPRI.

Quisumbing, A.R., ed. 2003. *Household Decisions, Gender, and Development: A Synthesis of Recent Research*. Washington, DC: IFPRI.

Quisumbing, A., A. Ahmed, J. Hoddinott, A. Pereira, and S. Roy. 2021. "Designing for Empowerment Impact in Agricultural Development Projects: Experimental Evidence from the Agriculture, Nutrition, and Gender Linkages (ANGeL) Project in Bangladesh." *World Development* 146: 105622.

Quisumbing, A.R., R. Brown, H.S. Feldstein, L.J. Haddad, and C. Peña. 1995. *Women: The Key to Food Security*. Food Policy Report 5. Washington, DC: IFPRI.

Quisumbing, A., B. Gerli, S. Faas, et al. 2023. "Assessing Multicountry Programs Through a "Reach, Benefit, Empower, Transform' Lens." *Global Food Security* 37: 100685.

Quisumbing, A.R., and J.A. Maluccio. 2003. "Resources at Marriage and Intrahousehold Allocation: Evidence from Bangladesh, Ethiopia, Indonesia, and South Africa." *Oxford Bulletin of Economics and Statistics* 65 (3): 283–327.

Quisumbing, A., R. Meinzen-Dick, H.J. Malapit, et al. 2024. "Enhancing Agency and Empowerment in Agricultural Development Projects: A Synthesis of Mixed Methods Impact Evaluations from the Gender, Agriculture, and Assets Project, Phase 2 (GAAP2)." *Journal of Rural Studies* 108: 103295.

Riley, E. 2024. "Role Models in Movies: The Impact of Queen of Katwe on Students' Educational Attainment." *Review of Economics and Statistics* 106 (2): 334–351.

Rocheleau, D., and D. Edmunds. 1997. "Women, Men and Trees: Gender, Power and Property in Forest and Agrarian Landscapes." *World Development* 25 (8): 1351–1371.

Ruel, M.T., and H. Alderman. 2013. "Nutrition-Sensitive Interventions and Programmes: How Can They Help to Accelerate Progress in Improving Maternal and Child Nutrition?" *Lancet* 382 (9891): 536–551.

Seymour, G., J. Heckert, A. Quisumbing, et al. 2024. "Measuring Women's Empowerment in National Surveys: Development of the Women's Empowerment Metric for National Statistical Systems (WEMNS)." IFPRI Discussion Paper 2254. IFPRI, Washington, DC.

Sijapati, B., A.S. Lama, J. Baniya, J. Rinck, K. Jha, and A. Gurung. 2017. *Labour Migration and the Remittance Economy: The Socio-Political Impact.* Kathmandu: CESLAM/Social Science Baha.

Skoufias, E. 2005. *PROGRESA and Its Impacts on the Welfare of Rural Households in Mexico.* Research Report 139. Washington, DC: IFPRI.

Smith, L.C., U. Ramakrishnan, A. Ndiaye, L.J. Haddad, and R. Martorell. 2003. *The Importance of Women's Status for Child Nutrition in Developing Countries.* Research Report 131. Washington, DC: IFPRI.

Sodhi, N.S., P. Davidar, and M. Rao. 2010. "Empowering Women Facilitates Conservation." *Biological Conservation* 143 (5): 1035–1036.

Sraboni, E., H.J. Malapit, A. Quisumbing, and A. Ahmed. 2014. "Women's Empowerment in Agriculture: What Role for Food Security in Bangladesh?" *World Development* 61: 11–52.

Strauss, J., and D. Thomas. 1995. "Human Resources: Empirical Modeling of Household and Family Decision." In *Handbook of Development Economics*, vol. 3, eds. J. Behrman and T. N. Srinivasan, 1883–2023. Amsterdam: Elsevier.

Takeshima, H., K. Raghunathan, and K. Kosec. 2022. "Climate Change and Women's Voice and Agency beyond the Household: Insights from India." IFPRI Discussion Paper 2148. IFPRI, Washington, DC.

Udry, C. 1996. "Gender, Agricultural Production, and the Theory of the Household." *Journal of Political Economy* 104 (5): 1010–1046.

University of Central Asia. 2021. "Qualitative Study on the GAL and GALS/BALI Implementation in Kyrgyzstan." Unpublished report.

von Braun, J., and E.T. Kennedy. 1994. *Agricultural Commercialization, Economic Development, and Nutrition*. Washington, DC: IFPRI.

WEF (World Economic Forum). 2024. *2024 Global Gender Gap Report*. Geneva.

Yoong, J., L. Rabinovich, and S. Diepeveen. 2012. *The Impact of Economic Resource Transfers to Women Versus Men: A Systematic Review*. Technical Report. London: EPPI-Centre, Social Science Research Unit, Institute of Education, University of London.

Zwarteveen, M.Z. 1997. "Water: From Basic Need to Commodity: A Discussion on Gender and Water Rights in the Context of Irrigation." *World Development* 25 (8): 1335–1349.

SECTION V

Effecting Change

Chapter 15

POLITICAL ECONOMY AND GOVERNANCE
Agriculture and Food Policy from Local to Global

Jordan Kyle and Danielle Resnick, with Jonathan Mockshell

Jordan Kyle is a research fellow, Poverty, Gender, and Inclusion Unit, IFPRI; **Danielle Resnick** is a senior research fellow, Markets, Trade, and Institutions Unit, IFPRI; and **Jonathan Mockshell** is a scientist, Alliance of Bioversity and CIAT.

Key messages

- Policymaking is shaped by political economy and governance factors such as incentives, institutional structures, ideological biases, and power dynamics. Over the past 50 years, these factors have intersected with major trends affecting the international development policy landscape, with important implications for agriculture and food policy.
- Major government reforms have shifted the influence of different actors in national policy processes, given greater leverage to local actors in decision-making and implementation, and elevated attention to power asymmetries in global governance.
- Research on governance and political economy has highlighted limits to overcoming power imbalances and capacity constraints, as well as the importance of incentives, ideas, and coalitions in the politics of agricultural policy reform.
- In-depth case studies, active engagement with government decision-makers, and innovative tools have supported the policymaking process and generated evidence on the types of institutional arrangements and enabling environments that can ensure investments are translated into effective service delivery and development outcomes.

Looking ahead, research should address:

- **Decentralization and local engagement.** Assess the impacts of decentralization on service provision and ways to improve subnational capacity to support the delivery of policy goals, and identify the role that local leaders can play in more integrated food system policy approaches.

- **Political economy of policy reforms and distribution.** Strengthen understanding of who has voice and power in policymaking processes, evaluate whether multistakeholder platforms and food policy councils are equitable and socially inclusive, and assess opportunities to overcome political economy challenges and ensure cooperation.
- **State capacity.** Build better understanding of how states can improve their capacity to manage trade-offs across sectors and tackle implementation of policies that affect multiple ministries, as well as how leadership and electoral turnover affect implementation of national food systems transformation pathways.
- **Global governance.** Identify ways to bolster global governance for food systems, building on the strengths of both multilateral objectives and multistakeholder viewpoints, and support the establishment of accountability platforms that transparently track how institutions are delivering on promises to tackle hunger, food security, and sustainable agriculture.

Policymaking is shaped by evidence as well as by political economy and governance factors such as incentives, institutional structures, ideological biases, and power dynamics. Over the past several decades, these factors have intersected with significant trends affecting the international development policy landscape, with important implications for agriculture and food policy. Informing evidence-based policymaking has been central to IFPRI's mandate since its inception. This chapter examines four key areas to which IFPRI has made significant contributions in research, methodology, and policy: decentralization, agriculture and food policy reform processes, political economy of distribution, and state capacity. Looking forward, challenges such as urbanization, debt distress, climate change, conflict, rapid technological advances, and corporate concentration of supply chains will increase food system complexity and cross-national externalities. Given these challenges, the chapter concludes by addressing a fifth critical area: the need to build effective and legitimate global institutions for food systems governance.

Decentralization and local engagement

Decentralization—which involves the transfer of autonomy, resources, and accountability from upper to lower tiers of government—has been one of the most important governance reforms in recent decades. This reform gained popularity in the early 1990s due to multiple pressures, including dissatisfaction with the results of centralized planning, structural adjustment reforms, demands for regional autonomy, and democratization across the Global South. A broad recognition of the limits of state capacity and the ability of local groups to manage resources has also driven decentralization reforms (see Chapter 6) (Meinzen-Dick et al. 2008). Although decentralization can, in some cases, lead to positive development outcomes, it is not a panacea.

Research on the consequences of decentralization for development outcomes has focused primarily on two pathways. In the first, decentralization brings policy decisions closer to those affected by them, creating the opportunity to leverage local knowledge, to incorporate local preferences, to enhance citizen participation, and to improve accountability and responsiveness of government. This first pathway emphasizes decentralization's potential to improve quality of governance, and studies have tested its impacts on public service delivery, public investment choices, and outcomes for citizens like health, education, and poverty. Second, decentralization expands the number of decision-making actors, necessitating coordination for policy coherence across jurisdictions and attention to potential for elite capture, which occurs when powerful individuals or groups dominate decision-making or divert resources for private gain. This second pathway highlights how decentralization can complicate the achievement of policy goals that span jurisdictions or reinforce existing local power structures.

The global literature on decentralization's impact through these pathways, to which IFPRI has contributed, finds mixed effects of decentralization on public investments and services, recognizing that "the devil is in the details" when it comes to understanding whether services and programs work as intended in rural areas (Raabe et al. 2010). For example, uneven subnational government capacities can result in unequal public service delivery and policy outcomes across localities (Gingerich 2013; Ziblatt 2008). In this vein, IFPRI research found that staffing local governments with bureaucrats with greater knowledge of local agricultural conditions and higher intrinsic motivation results in higher access to agricultural extension services for rural households in Nepal and matters more for policy outcomes than traditional capacity indicators like budget and staff education levels (Kyle and Resnick 2019a).

Additionally, local norms and inequalities often exclude some groups from decision-making. In settings where women face limited mobility and restrictive gender norms, they are less likely to contact local officials or speak up in community meetings (see Chapter 14) (Kosec et al. 2024; Palaniswamy et al. 2019). Variations in the level of democratization can also condition whether decentralization results in more pro-poor policies (Faguet 2005) and the mix of public investment choices made by local governments (Kosec and Mogues 2020). Even when decentralization succeeds in enhancing accountability to local populations, electoral pressures can push subnational governments to prioritize infrastructure projects because of their high visibility to citizens, such as the construction of roads, health clinics, and schools, at the expense of agricultural services (Resnick 2022).

Increasingly, governments are seeking analysis of the impacts of decentralization on service provision and on how to improve subnational capacity so that local governments can deliver policy goals (see Box 15.1). As mayors and other local leaders start to gain greater policy autonomy globally, bolstered by transnational networks such as the Milan Urban Food Policy Pact and efforts to localize the Sustainable Development Goals (SDGs), it is critical that research identify the role that such local leaders can play in more integrated food systems policy approaches—either as complements or substitutes to national policy processes. For instance, IFPRI work in Mali highlights how local leaders, including both mayors and traditional authorities, are pivotal in helping vulnerable populations manage climate crises (Bleck et al. 2024). This research finds that when crises occur, local leaders can leverage their knowledge of communal needs as well as the trust of local populations to help governments and donors to identify beneficiaries of humanitarian support, handle transport and logistics of aid distribution, provide advice on local needs, and manage local disputes around aid distribution. Identifying approaches to enhance the decision-making power of women, youth, and marginalized groups within local governance is also critical to ensure that localization does not reinforce preexisting inequalities (Adida et al. 2024; IFPRI 2010).

Political economy of policy reforms

Decentralization processes often occurred in parallel to structural adjustment reforms during the 1980s and 1990s (see Chapters 3 and 18). As governments were compelled by donor agencies and macroeconomic instability to shift resources and priorities, a rich scholarship on the political economy of agricultural policy reforms emerged to understand when and why the preferences of rural agricultural producers had more influence in policy reform processes (Anderson et al. 2013; Krueger 1992; Swinnen 2010). Some pointed to the primacy of decision-makers' policy beliefs as a reason why agriculture faced distortions in developing countries prior to structural adjustment reforms

BOX 15.1 Government demand for governance research: Cases from Ghana and Nepal

As governments consider public sector reforms, they need to better understand how these reforms may impact agricultural service provision, financing, and incentive structures. Moreover, governments sometimes see a disjunction between policies and outcomes in rural areas, and they want to understand whether public sector reforms could help close those gaps.

IFPRI has provided demand-driven analysis on the impacts of governance reforms, both through its country offices and in partnership with local institutes. For example, in 2012, Ghana began to devolve responsibility for its agriculture sector to more than 200 metropolitan, municipal, and district assemblies (MMDAs), a process that involved transferring the responsibilities of agricultural staff and budgets from national ministries to the MMDAs. Several years later, IFPRI partnered with Ghana's Center for Democratic Development to assess the impacts of devolution on agricultural service provision by using district-level budget data, along with surveys of district agricultural bureaucrats and almost 1,000 households. The findings showed that devolution had a negative impact on expenditures on agricultural extension because local citizens preferred their MMDAs to prioritize investments in physical infrastructure rather than low-visibility services such as extension. The research was used to inform Ghana's Ministry of Food and Agriculture that the underinvestment in extension could be reversed by addressing time lags in the distribution of resources to the MMDAs and halting the continuous subdivision of MMDAs, which reduced their internally generated revenue (Resnick 2022).

Similarly, in 2015, Nepal's constitutional referendum and resulting transition to a federal system entailed major shifts for the agriculture sector. First, authority and autonomy for many agricultural and livestock activities was shifted to seven newly created provincial governments and more than 700 local governments. Second, elected executives, rather than appointed ones, would oversee these local governments, which would be able to determine their own policy priorities. An analysis of lessons learned from countries that experienced similar reforms—as well as primary data collection with national policymakers, local bureaucrats, and rural households, conducted in collaboration with Nepal's Integrated Institute for Development Studies—provided critical guidance to the Ministry of Agricultural and Livestock Development about how to implement reforms with consideration for clear lines of institutional authority, accountability, autonomy, incentives, and coordination across the new tiers of government when the reform process proceeded in 2016 (Kyle and Resnick 2019b). Overall, this type of granular research on how decentralization reforms interact with local context can support policymakers in ensuring that policy reforms translate into the desired policy outcomes on the ground and avoid some of its pitfalls.

(Grindle and Thomas 1989). One study (Krueger 1992), for instance, highlighted that governments historically levied distortionary agricultural export taxes as a way to finance rapid industrialization, while another (Van de Walle 1989) claimed that aspirations for food self-sufficiency compelled distortionary measures during the 1970s and 1980s.

Others emphasized that the nature of post-structural-adjustment reforms for agriculture reflected regime changes. A study in Africa showed that distortions that were negatively affecting agricultural producers decreased when much of the region transitioned to multiparty democracy in the 1990s because rural voters had more influence over politicians (Bates and Block 2013). Elsewhere, researchers focused on the importance of the ideological leanings of the political party in power (Olper 2007; Swinnen 2010), the role of the media (Olper and Swinnen 2013), and whether elites had linkages to agriculture businesses (Tyce 2020; Widner 1994) as reasons for agricultural policy decisions. More recently, the role of coalitions in agricultural policy reforms has garnered more attention, with researchers noting that traditional distinctions between rural producers and urban consumers are no longer realistic, especially given the rise of new interest groups within food systems (such as processors, banks, insurance companies), and the importance of global agricultural value chains (see Chapter 7) (Swinnen 2015; Swinnen and Resnick 2023).

Some analysts further combined these insights with theoretical advancements in the public policy field, such as the Advocacy Coalition Framework (ACF) (Sabatier and Jenkins-Smith 1993). The ACF highlights how individuals leverage their beliefs about the causes of and solutions to a policy problem within a particular policy subsystem (such as agriculture, health, education, or environment) to form coalitions with likeminded partners, sometimes with the aid of "policy brokers" who foster compromise and minimize conflict.

As part of a robust agenda on participatory policymaking analyses, IFPRI researchers and partners adapted the ACF and utilized narrative approaches to study the role of beliefs in agricultural subsidy policies and programs, including in India (Birner et al. 2011), Ghana and Uganda (Mockshell and Birner 2015), and Senegal (Mockshell and Birner 2020), as well as in advisory services programs in Uganda (Rwamigisa et al. 2018). These studies showed that relevant policy reforms were hindered by contrasting belief systems and causal assumptions and corresponding competing advocacy (and veto) coalitions. For instance, the study in Senegal uncovered two distinct coalitions that diverged in their policy beliefs about the contributions of subsidies to agriculture;

namely, government policymakers saw subsidies as indispensable to address poor soil fertility and low agricultural productivity while donors viewed them as costly and inefficient (Mockshell and Birner 2020).

To ensure that many of the insights above could be translated into applied political economy analysis, IFPRI and collaborating institutions developed replicable tools and frameworks to both understand and anticipate policy decisions. The NetMap tool, originally applied to northern Ghana and since adopted by many organizations, provides a low-cost, interview-based mapping approach to uncover sources of conflict and cooperation, examine informal and formal networks, and highlight power relations (Schiffer and Hauck 2010). NetMap enables communities to offer their unique perspective on the policy process, ensuring that researchers and government officials properly interpret it. The Kaleidoscope Model, an applied framework for understanding the drivers of policy change, focuses on five stages of the policy process—agenda setting, design, adoption, implementation, and evaluation and reform—and identifies 16 key variables and conditions that facilitate change (Resnick et al. 2018). By integrating power dynamics and political conflict into the analysis of policymaking and recognizing the diverse range of actors, including donors, civil society, the private sector, and government, who can influence the policy cycle, the model provides valuable insights for identifying entry points into the policy process (see Figure 15.1). The Kaleidoscope Model has been applied to many areas of policy reform, including fertilizer subsidies, large-scale food fortification, and environmental sustainability (Baca et al. 2024; Dhehibi et al. 2023; Resnick et al. 2018).

The Women's Empowerment in Agrifood Governance (WEAGov) assessment framework builds on the Kaleidoscope Model to examine the role that women's perspectives play in the policymaking process (Ragasa et al. 2022). In partnership with Krea University in India and the Agricultural Policy Research Network in Nigeria, IFPRI has piloted the framework with over 400 agrifood policy experts to evaluate the extent to which women's voices, needs, and preferences are included in the policy cycle (Ragasa et al. 2023, 2024). This tool has yielded actionable insights for policymakers. For example, agrifood policy experts involved in the WEAGov assessment in Nigeria highlighted lack of funding, implementation resources, and monitoring efforts toward gender targets already set within national agrifood policies as key constraints for improving policy outcomes for women. They identified that greater coordination with the Ministry of Budget and Economic Planning before budgets are set for each line ministry could improve responsiveness to women's preferences in agrifood policies (CGIAR 2023).

FIGURE 15.1 The Kaleidoscope Model of policy change

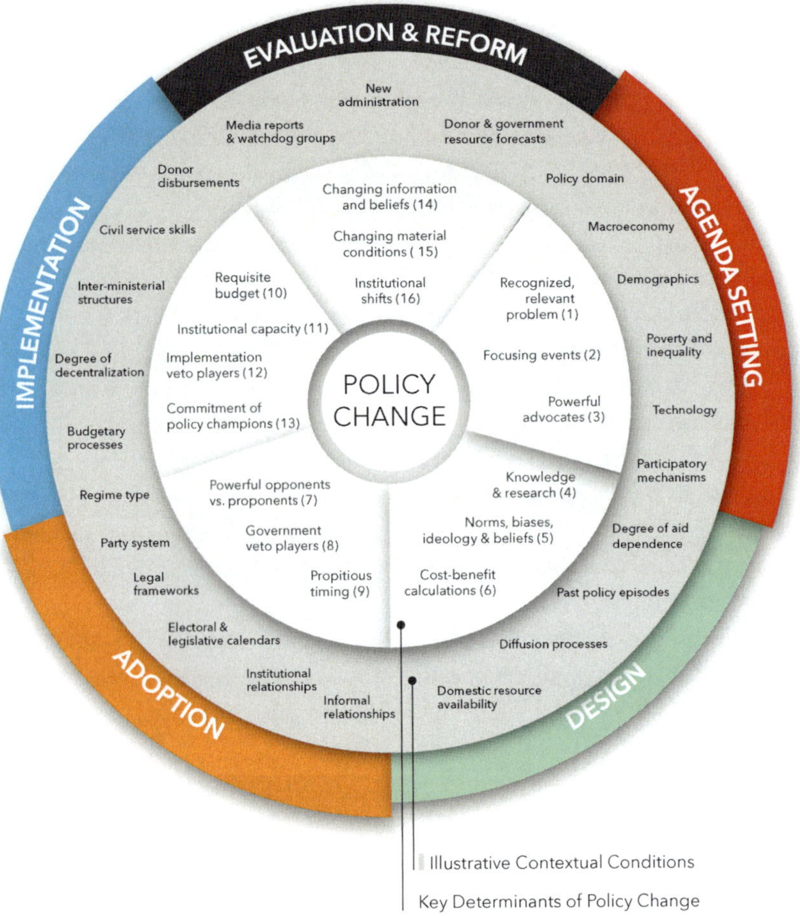

Source: Resnick et al. (2018).

Understanding who has voice and power in policymaking processes continues to be a critical issue, especially given that inclusive participation remains central to food policy discussions, most notably with the dialogues approach that emerged from the 2021 United Nations Food Systems Summit (UNFSS). The growth in multistakeholder platforms and food policy councils across all regions of the world has spurred further interest in understanding the extent to which participation leads to genuinely transformative outcomes (see Chapter 13) (Moragues-Faus 2021), leading IFPRI and its partners to assess whether these platforms are equitable and socially inclusive (Ratner et al. 2022).

Political economy of distribution

Following structural adjustment reforms, which were seen as overly donor-driven, the 2005 Paris Declaration on Aid Effectiveness became a major impetus in the power dynamics that dominated the subsequent decade of development, including in the agrifood sector. The Paris Declaration established five key principles—ownership, harmonization, alignment, results-based outcomes, and mutual accountability between donors and partner countries for development results—to which heads of state committed to accelerate the Millennium Development Goals (OECD 2005). These principles were intended to rectify longstanding imbalances between high-income northern donors and donor recipients in the Global South. To enhance ownership and donor coordination, the Declaration advocated for two-thirds of aid, by 2010, to be distributed through budget support or sectorwide programming—both of which would enable partners to decide how to allocate funding according to their own national development and sectoral strategies (Lundsgaarde and Engberg-Pedersen 2019).

In Africa, the confluence of budget support and expenditure targets under the Maputo Declaration of the African Union's Comprehensive Africa Agriculture Development Program (CAADP) resulted in the revival of direct government support to agriculture, most notably through input subsidy programs (Jayne and Rashid 2013). As evidence emerged that such programs might not be appropriately targeted or the best use of resources, scholars at IFPRI and various universities began to focus on the political economy of distribution. By combining data on subsidy distribution with local election data, researchers uncovered distinct political economy reasons why subsidy programs might be retained despite questionable impacts on productivity, cost effectiveness, and poverty reduction. For instance, in Ghana, fertilizer subsidies were disproportionately targeted to anti-incumbent districts in an attempt at vote mobilization (Banful 2011). By contrast, in countries such as Zambia, fertilizer subsidies more frequently went to ruling party districts (Mason et al. 2017). In Malawi, subsidies were not targeted along partisan lines but did benefit the ruling party in subsequent elections (Dionne and Horowitz 2016).[1] Extensive research on fertilizer and electricity subsidies in India also showed that concern over losing farmer votes proved a key barrier to removing subsidies (Birner et al. 2011).

Many governments disproportionately favored input subsidies due to their high visibility and easy attributability, even as they displaced funding for longer-term but higher-return investments in agricultural research and development (Mogues 2015; Mogues and Do Rosario 2016). Qualitative discourse analysis revealed the emergence of two opposing coalitions. Government partners, on one side, emphasized the narrative of poor soil fertility and claimed that input subsidies could increase incomes, food security, and jobs for youth. Donors, on the other side, marshalled narratives about poor implementation of subsidies, displacement of the private sector, poor targeting, and corruption (Mockshell and Birner 2015, 2020). Such narratives reveal that some divisions over policy direction, which the Paris Declaration intended to minimize, instead remained quite sizable between donors and recipient countries.[2]

The intersection between narratives, electoral incentives, and agricultural productivity is also relevant to the broader agenda on repurposing global agricultural support: of the approximately US$800 billion in total support, more than $86 billion goes to input subsidies (see Chapters 3 and 18) (Gautam et al. 2022). Incentivizing communities to accept a reduction in subsidies for their reinvestment into ecosystem service payments, agricultural research and development, or other domains requires grappling with political economy dynamics (Vos et al. 2023). In many cases, this has been achieved by compensating potential losers from reform, earmarking subsidies for these new investments, and taking advantage of windows of opportunity, including shifts in political administrations. Overcoming political economy challenges will be critical

1 These dynamics may be very different in other policy areas. For instance, research in Malawi has found that partisan political economy targeting may be more prominent for humanitarian aid transfers (Duchoslav et al. 2023).

2 Twenty years on, the Paris Aid principles have been further battered by the rise in many more aid actors—foundations, philanthropists, and nontraditional donors—as well as sovereign wealth creditors, and the growing use of local NGOs as delivery mechanisms (Dietrich 2013). As a result, national governments are again losing ownership over aid resources, with one study highlighting that four out of five aid projects are managed by nongovernmental agencies (World Bank 2024).

to ensure global cooperation on repurposing subsidies: even though all countries may benefit in the long term from smarter investments in sustainable food systems, governments may fear alienating powerful constituencies in the short term. For example, analyses of fuel subsidies in Indonesia highlight that when corruption is common, citizens may strongly oppose reforms because they lack trust in their government to replace subsidies with something better (Kyle 2018).

Unpacking state capacity

The return of input subsidy programs in the 2000s presaged the resurgence of state-led development, which remains pronounced to this day. Inspired by the example of the so-called Asian miracle economies, some development economists began to emphasize the importance of industrial upgrading for the growth of low- and middle-income countries (LMICs) and highlighted need for governments to play a central role in facilitating that process (see Chapter 3) (Lin 2013; Page and Tarp 2017; Rodrik 2007). IFPRI contributed to this discussion with a robust research program on structural transformation processes in Africa, delving into multiple case studies and tracing patterns of productivity growth, both within agriculture and between agriculture and other sectors (McMillan et al. 2014, 2017). A key finding is that neither structural transformation nor fundamentals like high human capital and strong institutions alone can drive rapid and sustained growth; both are needed, and policy sequencing affects growth trajectories. On the ground in Africa, agro-industrial policy became increasingly prominent, including creation of growth corridors, agro-industrial zones, and agricultural growth poles. The goal of these policy actions was to add value through integrated investments in infrastructure and logistics, as well as trade, tax, and regulatory incentives (Picard et al. 2017).

Yet, for agro-industrial efforts to succeed, they need to be guided by well-capacitated states that can work with the private sector to provide missing inputs such as research and development, identify new products that can be profitably produced, and facilitate upstream and downstream coordination (Hausmann et al. 2008). Consequently, scholars have examined why and when governments are incentivized to take on this role and for which agricultural value chains, noting the importance of the "political settlement"—that is, the distribution of power and influence between political and business elites (Behuria 2020; Whitfield et al. 2015). Research in Ghana highlighted how a state that was strongly democratic but weakly capacitated could be a barrier to agriculture-led industrialization, causing the state to be driven more by electoral considerations than transformation goals (Resnick 2019). In parallel, in-depth case studies on the political economy of agricultural commercialization showed how the state subjected key agricultural value chains to ad hoc policy interventions, such as import and export bans, which aimed to extract electoral gains rather than generate transformative benefits (Chinsinga and Naess 2022).

With the shift from structural transformation to food systems transformation, there is a growing need to consider implications for policy coherence (Dewi et al. 2024). This will require better understanding of how states can improve their capacity to manage trade-offs across sectors and tackle implementation of policies that affect multiple ministries (Resnick and Swinnen 2023). In doing so, much can be learned from IFPRI's prior work on governments' improvements to multisectoral coordination for nutrition policies (Gillespie and van den Bold 2017), on-the-ground experience supporting Ethiopia's Agricultural Transformation Agency (Zandstra et al. 2024), and cross-ministerial surveys of bureaucratic staff about incentives and political interference in service provision (Birner et al. 2012). Following the UNFSS, burgeoning research work also considers how ministry leadership and electoral turnover affect implementation of national food systems transformation pathways (Resnick 2024). All of these cases have emphasized the need for much more research on the black box of food systems bureaucracies to better understand ministerial motivations and performance criteria to implement complex, multisectoral policies.

Global governance

The UNFSS represents one example of the broader shift toward multi-stakeholderism in the global governance of food systems, occuring in tandem with a growing skepticism of the multilateral system. While multilateralism commits national governments to implement decisions, multi-stakeholderism brings together diverse participants—from civil society, business, and government—to collaboratively decide on solutions (Gleckman 2018). From the end of World War II to the mid-2000s, multilateralism predominated in global food and agricultural policy through institutions such as the United Nations' Food and Agriculture Organization, International Fund for Agricultural Development, World Food Programme, and World Health Organization as well as the World Trade Organization and the World Bank. Multilateral policy processes, such as negotiations through the UN's Congress of Party (CoP) summits, also facilitated inter-governmental cooperation.

Several concurrent factors began to gradually weaken the multilateral system. These included the rapid expansion of nongovernmental organizations and private enterprises through the political and economic liberalization of the 1980s and 1990s, a lack of agreement on agricultural trade disciplines since the WTO's Doha Round started in 2001 (see Chapter 16), and the growth of new global bodies, such as the Group of 20 (G20), in the wake of the 2008 global food and financial crisis (Clapp and Murphy 2013; Gleckman 2018; Lele 2021; Paarlberg 2002). The 2008 financial crisis also stimulated the World Economic Forum to launch a Global Redesign Initiative to revamp global governance through multi-stakeholderism (WEF 2010), and an emphasis on multistakeholder partnerships is now embedded within SDG 17. More recently, growing populism, geopolitical tensions, and a sense of unequal power on the part of LMICs within the post-World War II institutional framework have further eroded the legitimacy of multilateralism (Resnick and Swinnen 2023). Convening processes that sometimes include UN entities and sometimes exclude them, including the G20's Global Agriculture and Food Security Program in 2010 and its Global Alliance Against Hunger and Poverty in 2024 as well as the G7's Global Alliance for Food Security in 2022, have further fragmented multilateralism.

Yet, while disappointment with multilateral cooperation is not new, there are few viable alternatives. Using the UNFSS as an example, many scholars and activists have been just as disappointed with the multistakeholder approach, arguing that it creates entry points for corporate influence and allows governments to outsource responsibility for decision-making and implementation (Anderl and Hißen 2023; Canfield et al. 2021). At the same time, intergovernmental coordination remains essential, both for responding to crises, such as rising food prices from trade protectionism (Martin et al. 2024), and addressing longer-term collective action issues, such as escalating sovereign debt, rapid technological change, and climate change.

In the coming decades, IFPRI and its partners can play a key role in identifying ways to bolster global governance for the food system, building on the strengths of both multilateral objectives and multistakeholder viewpoints. Such an agenda can consider the viability of different global governance mechanisms proposed in recent years, such as the Food Systems Stability Board comprised of national representatives from key ministries and which would serve as a food systems risk body to help countries anticipate and mitigate food crises through early warning systems and monitoring of potential shocks to trade and price stability (Dixson-Decleve et al. 2021). Building on recent IFPRI work on global commitments to SDG 2 (Zorbas et al. 2024), there is also a need to establish accountability platforms that transparently track how institutions are delivering on their promises to tackle hunger, food security, and sustainable agriculture. Such efforts could be enhanced by bolstering the institutional efforts of the UN's Committee on Food Security, established in 1974 and reformed in 2009, and whose primary mandate is to promote policy convergence toward improved food security. As part of CGIAR's Policy Innovations Program (2025–2030), greater attention will be given to additional options for addressing the frailties in global governance of the food system.

Conclusions

Over the last 50 years, development trends have shifted the influence of different actors in national policy processes, given greater leverage to local actors in decision-making and implementation, and elevated attention to power asymmetries in global governance. These trends have intersected with important policy issues in the food and agricultural spheres, including input subsidies and agro-industrialization. Although space constraints prevent this chapter from providing an exhaustive review of all key political economy and governance areas relevant to food and agricultural policy, it does highlight the ways IFPRI and its partners have contributed to the governance and political economy scholarship around major policy trends. This has been achieved not only through in-depth case studies and active engagement with government decision-makers but also by creating innovative tools to navigate the policymaking process and generating evidence on the types of institutional arrangements and enabling environments that can ensure investments are translated into effective service delivery and development outcomes. Notably, these advancements have coincided with a broader recognition in the international development community of the centrality of political economy analysis for the success of what otherwise seem to be technical policy reforms (Whaites et al. 2023).

This research has shown that decentralization and local engagement are not panaceas for overcoming power imbalances and capacity constraints. It has also highlighted the importance of incentives, ideas, and coalitions in the politics of agricultural policy reform. Looking ahead, global challenges such as climate change, environmental degradation, and rising inequality raise new questions about the governance and political economy of food systems. While past research emphasized governance arrangements believed to be most effective for rural services and agrifood systems, we do not yet know enough about governance arrangements that are most adaptive to rapid change or that best foster policy coherence for achieving the SDGs. Even as the development landscape becomes more unpredictable and complex in the coming years, IFPRI and its CGIAR partners will remain committed to ensuring research recommendations are politically grounded and to advancing knowledge on the institutional prerequisites for, and political risks to, sustainable and healthy food systems.

References

Adida, C., K. Kosec, J. Kyle, C.H. Mo, L. Arriola, D. Adenyanju, and R. Fisher. 2024. *Strengthening Women's Voice and Agency in Nigeria: Evidence from a Randomized Control Trial on Women's Advocacy and Men's Allyship Training*. IFPRI Policy Note. Washington, DC: IFPRI.

Anderl, F., and M. Hißen. 2023. "How Trust Is Lost: The Food Systems Summit 2021 and the Delegitimation of UN Food Governance." *European Journal of International Relations* 30 (1): 151–175.

Anderson, K., G. Rausser, and J. Swinnen. 2013. "Political Economy of Public Policies: Insights from Distortions to Agricultural and Food Markets." *Journal of Economic Literature* 51 (2): 423–477.

Baca, M.F.D., L. Moreno Lerma, and S. Burkart. 2024. "How Do Sustainability Policies Emerge in the Colombian Political System? A Kaleidoscope Model Analysis of the Policy for Sustainable Cattle 2022–2050." *Cleaner and Circular Bioeconomy* 7: 100075.

Banful, A.B. 2011. "Old Problems in the New Solutions? Politically Motivated Allocation of Program Benefits and the 'New' Fertilizer Subsidies." *World Development* 39 (7): 1166–1176.

Bates, R.H., and S.A. Block. 2013. "Revisiting African Agriculture: Institutional Change and Productivity Growth." *Journal of Politics* 75 (2): 372–384.

Behuria, P. 2020. "The Domestic Political Economy of Upgrading in Global Value Chains: How Politics Shapes Pathways for Upgrading in Rwanda's Coffee Sector." *Review of International Political Economy* 27 (2): 348–376.

Birner, R., S. Gupta, and N. Sharma. 2011. *The Political Economy of Agricultural Policy Reform in India Fertilizers and Electricity for Irrigation*. Washington, DC: IFPRI.

Birner, R., M. Sekher, and K. Raabe. 2012. *Reforming the Public Administration for Food Security and Agricultural Development*. Washington, DC: IFPRI.

Bleck, J., L. Carrillo, J. Gottlieb, K. Kosec, J. Kyle, and M. Soumano. 2024. "Unlocking Locally-Led Resilience Amid Conflict and Climate Stress." IFPRI Discussion Paper 2272. IFPRI, Washington, DC.

Canfield, M., M.D. Anderson, and P. McMichael. 2021. "UN Food Systems Summit 2021: Dismantling Democracy and Resetting Corporate Control of Food Systems." *Frontiers in Sustainable Food Systems* 5: 1–15.

CGIAR. 2023. "Tool for Raising Women's Voice Gains Traction in Nigeria." CGIAR Research Initiative on Gender Equality. Montpellier, France.

Chinsinga, B., and L.O. Naess. 2022. *The Political Economy of Agricultural Commercialisation: Insights from Crop Value Chain Studies in Sub-Saharan Africa*. Brighton, UK: Agricultural Policy Research in Africa, Institute of Development Studies.

Clapp, J., and S. Murphy. 2013. "The G20 and Food Security: A Mismatch in Global Governance?" *Global Policy* 4 (2): 129–138.

Dewi, D., D. Aytekin, K.R. Schneider, N. Covic, J. Fanzo, S. Nordhagen, and D. Resnick. 2024. "Defining and Measuring Policy Coherence for Food System Transformation: A Scoping Review." *Global Food Security* 43: 100803.

Dhehibi, B., A.M. Oumer, A. Frija, K.T. Akramov, and M. Baum. 2023. *Revisiting the Kaleidoscope Model for Policy Analysis and Exploring Its Applicability to the Complex Agrifood Systems of Countries in Central and West Asia and North Africa: Methodological Guideline.* Beirut: International Center for Agricultural Research in the Dry Areas.

Dietrich, S. 2013. "Bypass or Engage? Explaining Donor Delivery Tactics in Foreign Aid Allocation." *International Studies Quarterly* 57 (4): 698–712.

Dionne, K.Y., and J. Horowitz. 2016. "The Political Effects of Agricultural Subsidies in Africa: Evidence from Malawi." *World Development* 87: 215–226.

Dixson-Decleve, S., J.A. Ocampo, and F. Salim. 2021. "The Case for a Food Systems Stability Board." *Project Syndicate*, April 5.

Duchoslav, J., E. Kenamu, and J. Thunde. 2023. "Targeting Hunger or Votes? The Political Economy of Humanitarian Transfers in Malawi." *World Development* 165: 106179.

Faguet, J.-P. 2005. "The Effects of Decentralization on Public Investment: Evidence and Four Lessons from Bolivia and Colombia." LSE Development Research Centre Working Paper No. 62. Crisis States Research Centre, London.

Gautam, M., D. Laborde, A. Mamum, W. Martin, V. Pineiro, and R. Vos. 2022. *Repurposing Agricultural Policies and Support*. Washington, DC: World Bank and IFPRI.

Gillespie, S., and M. van den Bold. 2017. "Agriculture, Food Systems, and Nutrition: Meeting the Challenge." *Global Challenges* 1 (3): 1600002.

Gingerich, D. 2013. "Governance Indicators and the Level of Analysis Problem: Empirical Findings from South America." *British Journal of Political Science* 43 (3): 505–540.

Gleckman, H. 2018. *Multistakeholder Governance and Democracy: A Global Challenge*. London and New York: Routledge.

Grindle, M.W., and J.W. Thomas. 1989. "Policy Makers, Policy Choices, and Policy Outcomes: The Political Economy of Reform in Developing Countries." *Policy Sciences* 22 (3/4): 213–248.

Hausmann, R., D. Rodrik, and C. Sabel. 2008. *Reconfiguring Industrial Policy: A Framework with an Application to South Africa*. Cambridge, MA: Center for International Development, Harvard University.

IFPRI. 2010. *Gender and Governance in Rural Services: Insights from India, Ghana, and Ethiopia*. Washington, DC: World Bank Publications.

Jayne, T.S., and S. Rashid. 2013. "Input Subsidy Programs in Sub-Saharan Africa: A Synthesis of Recent Evidence." *Agricultural Economics (United Kingdom)* 44 (6): 547–562.

Kosec, K., J. Kyle, S. Narayanan, K. Raghunathan, and S. Ray. 2024. "Claim-Making Under India's Mahatma Gandhi National Rural Employment Guarantee Act (MGNREGA): Barriers and Opportunities for Women's Voice and Agency Over Asset Selection." IFPRI Discussion Paper 2247. IFPRI, Washington, DC.

Kosec, K., and T. Mogues. 2020. "Decentralization Without Democracy." *World Politics* 72 (2): 165–213.

Krueger, A.O. 1992. *The Political Economy of Agricultural Pricing Policy: A Synthesis of the Political Economy in Developing Countries*. Baltimore: Johns Hopkins University Press.

Kyle, J. 2018. "Local Corruption and Popular Support for Fuel Subsidy Reform in Indonesia." *Comparative Political Studies* 51 (11): 1472–1503.

Kyle, J., and D. Resnick. 2019a. "Delivering More with Less: Service Provision in Low Capacity States." *Studies in Comparative International Development* 54 (1): 133–163.

Kyle, J., and D. Resnick. 2019b. "Nepal's Changing Governance Structure and Implications for Agricultural Development." In *Agricultural Transformation in Nepal: Trends, Prospects, and Policy Options*, eds. G. Thapa, A. Kumar, and P.K. Joshi, 573–600. Washington, DC: IFPRI.

Lele, U. 2021. "Introduction." In *Food for All: International Organizations and the Transformation of Agriculture*, eds. U. Lele, M. Agarwal, B. Baldwin, and S. Goswami, 1–5. Oxford, UK: Oxford University Press.

Lin, J.Y. 2013. "From Flying Geese to Leading Dragons: New Opportunities and Strategies for Structural Transformation in Developing Countries." In *The Industrial Policy Revolution II: Africa in the 21st Century*, eds. J.E. Stiglitz, J.L. Yifu, and E. Patel, 50–70. London: Palgrave Macmillan.

Lundsgaarde, E., and L. Engberg-Pedersen. 2019. *The Aid Effectiveness Agenda: Past Experiences and Future Prospects*. Copenhagen: Danish Institute for International Studies.

Martin, W.J., A. Mamun, and N. Minot. 2024. "Food Trade Policy and Food Price Volatility" preprint, available online June 24. Rochester, NY: SSRN.

Mason, N.M., T.S. Jayne, and N. van de Walle. 2017. "The Political Economy of Fertilizer Subsidy Programs in Africa: Evidence from Zambia." *American Journal of Agricultural Economics* 99 (3): 705–731.

McMillan, M., D. Rodrik, and C. Sepulveda. 2017. *Structural Change, Fundamentals, and Growth: A Framework and Case Studies*. Washington, DC: IFPRI.

McMillan, M., D. Rodrik, and Í. Verduzco-Gallo. 2014. "Globalization, Structural Change, and Productivity Growth, with an Update on Africa." *World Development* 63: 11–32.

Meinzen-Dick, R., M. Di Gregorio, and S. Dohrn. 2008. "Decentralization, Pro-Poor Land Policies, and Democratic Governance." CAPRi Working Paper 80. CGIAR and United Nations Development Programme, IFPRI, Washington, DC.

Mockshell, J., and R. Birner. 2015. "Donors and Domestic Policy Makers: Two Worlds in Agricultural Policy-Making?" *Food Policy* 55: 1–14.

Mockshell, J., and R. Birner. 2020. "Who Has the Better Story? On the Narrative Foundations of Agricultural Development Dichotomies." *World Development* 135: 105043.

Mogues, T. 2015. "Political Economy Determinants of Public Spending Allocations: A Review of Theories, and Implications for Agricultural Public Investment." *European Journal of Development Research* 27 (3): 452–473.

Mogues, T., and D. Do Rosario. 2016. "The Political Economy of Public Expenditures in Agriculture: Applications of Concepts to Mozambique." *South African Journal of Economics* 84 (1): 20–39.

Moragues-Faus, A. 2021. "The Emergence of City Food Networks: Rescaling the Impact of Urban Food Policies." *Food Policy* 103: 102107.

OECD (Organisation for Economic Co-Operation and Development). 2005. *Paris Declaration on Aid Effectiveness*. Paris.

Olper, A. 2007. "Land Inequality, Government Ideology and Agricultural Protection." *Food Policy* 32 (1): 67–83.

Olper, A., and J. Swinnen. 2013. "Mass Media and Public Policy: Global Evidence from Agricultural Policies." *World Bank Economic Review* 27 (3): 413–436.

Paarlberg, R. 2002. "Governance and Food Security in an Age of Globalization." Food, Agriculture, and the Environment Discussion Paper 36. IFPRI, Washington, DC.

Page, J., and F. Tarp. 2017. *The Practice of Industrial Policy: Government–Business Coordination in Africa and East Asia*. Oxford, UK: Oxford University Press.

Palaniswamy, N., R. Parthasarathy, and V. Rao. 2019. "Unheard Voices: The Challenge of Inducing Women's Civic Speech." *World Development* 115: 64–77.

Picard, F., M. Coulibaly, and C. Smaller. 2017. *The Rise of Agricultural Growth Poles in Africa*. Investment in Agriculture Policy Brief 6. Winnipeg, Canada: International Institute for Sustainable Development.

Raabe, K., R. Birner, M. Sekher, K.G. Gayathridevi, A. Shilpi, and E. Schiffler. 2010. "How to Overcome the Governance Challenges of Implementing NREGA." IFPRI Discussion Paper 963. IFPRI, Washington, DC.

Raabe, K., M. Sekhar, and R. Birner. 2009. "The Effects of Political Reservations for Women on Local Governance and Rural Service Provision: Survey Evidence from Karnataka." IFPRI Discussion Paper 878. IFPRI, Washington, DC.

Ragasa, C., J. Kyle, P. Kristjanson, and S. Eissler. 2022. "Conceptualizing Women's Empowerment in Agrifood Systems Governance A New Framework." IFPRI Discussion Paper 2153. IFPRI, Washington, DC.

Ragasa, C., J. Kyle, A. Onoja, et al. 2023. "Women's Empowerment in Agrifood Governance (WEAGov) Assessment Framework: A Pilot Study in Nigeria." IFPRI Discussion Paper 2222. IFPRI, Washington, DC.

Ragasa, C., J. Kyle, S. Yasmin, H. Pande, A. Sharma, S. Basu, and D. Najjar. 2024. "Women's Empowerment in Agrifood Governance (WEAGov) Assessment Framework: Insights from India Pilot Study." IFPRI Discussion Paper 2313. IFPRI, Washington, DC.

Ratner, B.D., A.M. Larson, J.P. Sarmiento Barletti, et al. 2022. "Multistakeholder Platforms for Natural Resource Governance: Lessons from Eight Landscape-Level Cases." *Ecology and Society* 27 (2): 270202.

Resnick, D. 2019. "Strong Democracy, Weak State: The Political Economy of Ghana's Stalled Structural Transformation." In *Ghana's Economic and Agricultural Transformation: Past Performance and Future Prospects*, eds. X. Diao, P. Hazell, S. Kolavalli, and D. Resnick, 49–96. Oxford, UK: Oxford University Press.

Resnick, D. 2022. "Does Accountability Undermine Service Delivery? The Impact of Devolving Agriculture in Ghana." *European Journal of Development Research* 34: 1003–1029.

Resnick, D. 2024. *Political Economy for Food Systems Pathways: A New Decision Toolkit*. Geneva: Global Alliance for Improved Nutrition.

Resnick, D., S. Haggblade, S. Babu, S.L. Hendriks, and D. Mather. 2018. "The Kaleidoscope Model of Policy Change: Applications to Food Security Policy in Zambia." *World Development* 109: 101–120.

Resnick, D., and J. Swinnen. 2023. "Introduction." In *The Political Economy of Food System Transformation: Pathways to Progress in a Polarized World*. Oxford, UK: Oxford University Press.

Rodrik, D. 2007. *One Economics, Many Recipes*. Princeton, NJ: Princeton University Press.

Rwamgisa, P., R. Birner, M. Mangheni, and A. Semana. 2018. "How to Promote Institutional Reforms in the Agricultural Sector? A Case Study of Uganda's National Agricultural Advisory Services (NAADS)." *Development Policy Review* 36 (5): 607–627.

Sabatier, P.A., and H.C. Jenkins-Smith. 1993. *Policy Change and Learning: An Advocacy Coalition Approach*. Boulder, CO: Westview Press.

Schiffer, E., and J. Hauck. 2010. "Net-Map: Collecting Social Network Data and Facilitating Network Learning Through Participatory Influence Network Mapping." *Field Methods* 22 (3): 231–249.

Swinnen, J. 2010. "Political Economy of Agricultural Distortions: The Literature to Date." In *The Political Economy of Agricultural Price Distortions*, ed. Kym Anderson, 81–104. Cambridge, UK, and New York: Cambridge University Press.

Swinnen, J. 2015. "Changing Coalitions in Value Chains and the Political Economy of Agricultural and Food Policy." *Oxford Review of Economic Policy* 31 (1): 90–115.

Swinnen, J., and D. Resnick. 2023. "Policy Coalitions in Food Systems Transformation." In *The Political Economy of Food System Transformation: Pathways to Progress in a Polarized World*, eds. D. Resnick and J. Swinnen, 111–132. Oxford, UK: Oxford University Press; Washington, DC: IFPRI.

Tyce, M. 2020. "A 'Private-Sector Success Story'? Uncovering the Role of Politics and the State in Kenya's Horticultural Export Sector." *Journal of Development Studies* 56 (10): 1877–1893.

Van de Walle, N. 1989. "Rice Politics in Cameroon: State Commitment, Capability, and Urban Bias." *Journal of Modern African Studies* 27 (4): 579–599.

Vos, R., W. Martin, and D. Resnick. 2023. "The Political Economy of Reforming Agricultural Support Policies." In *The Political Economy of Food System Transformation: Pathways to Progress in a Polarized World*, eds. D. Resnick and J. Swinnen, 54–75. Oxford, UK: Oxford University Press; Washington, DC: IFPRI.

WEF (World Economic Forum). 2010. *Everybody's Business: Strengthening International Cooperation in a More Interdependent World*. Geneva.

Whaites, A., L.-H. Piron, A.R. Menocal, and G. Teskey. 2023. *Understanding Political Economy Analysis and Thinking and Working Politically*. London: Foreign, Commonwealth and Development Office and Thinking and Working Politically Community of Practice.

Whitfield, L., O. Therkildsen, L. Buur, and A.M. Kjær. 2015. *The Politics of African Industrial Policy: A Comparative Perspective*. Cambridge, UK: Cambridge University Press.

Widner, J.A. 1994. "Single Party States and Agricultural Policies: The Cases of Ivory Coast and Kenya." *Comparative Politics* 26 (2): 127–147.

World Bank. 2024. *Aid Circumvention: The Elusive Dream of Putting Countries in the Driver's Seat*. Washington, DC: World Bank.

Zandstra, T., G.T. Abate, S. Rashid, and N. Minot. 2024. "The Importance of the Agricultural Transformation Agency in Ethiopia." *IFPRI Blog*, May 24.

Ziblatt, D. 2008. "Why Some Cities Provide More Public Goods than Others: A Subnational Comparison of the Provision of Public Goods in German Cities in 1912." *Studies in Comparative International Development* 43: 273–289.

Zorbas, C., D. Resnick, E. Jones, et al. 2024. "From Promises to Action: Analyzing Global Commitments on Food Security and Diets Since 2015." IFPRI Discussion Paper 02238. IFPRI, Washington, DC.

Chapter 16

AGRIFOOD TRADE
Changing Challenges, Changing Perspectives on Policy and Policy Research

Eugenio Díaz-Bonilla, Madhur Gautam, Joseph Glauber, Will Martin, Valeria Piñeiro, Sherman Robinson, Fousseini Traoré, and Rob Vos

Eugenio Díaz-Bonilla is a special advisor to the Inter-American Institute for Cooperation on Agriculture (IICA) and a visiting senior research fellow, Director General's Office, IFPRI; **Madhur Gautam** is a senior research fellow, Markets, Trade, and Institutions Unit (MTI), IFPRI; **Joseph Glauber** is a research fellow emeritus, IFPRI; **Will Martin** is a senior research fellow, MTI, IFPRI; **Valeria Piñeiro** is regional representative for Latin America and the Caribbean, MTI, IFPRI; **Sherman Robinson** is a research fellow emeritus, IFPRI; **Fousseini Traoré** is a senior research fellow, MTI, IFPRI; and **Rob Vos** is a senior research fellow, MTI, IFPRI.

Key messages

- Global agrifood trade has tripled since 1975, driven in part by a growing trade in processed foods.
- Agricultural policies in developing countries have shifted from taxing agriculture to providing positive support and protection, converging to the lower levels of trade protection of developed economies and helping to spur trade.
- IFPRI has been a leader in measuring trade protection provided to agriculture and food sectors and in assessing its impacts on food supplies, poverty, and food security and nutrition.
- IFPRI's research has highlighted the critical role agrifood trade can play in mitigating the adverse impacts of global market shocks and adapting to climate change.
- Negotiations to achieve a fairer multilateral regime for agricultural trade have stalled, but existing rules provide developing countries with space to use trade policy and domestic support measures to simultaneously promote food security and sustainable agriculture.

Looking ahead, policy research is needed to help achieve a trading system that facilitates sustainable agrifood system transformation, including:

- **Deepen analysis of national and international policy coherence,** as trade policy alone will not drive sustainable food systems transformation. This should include research on the "repurposing" of existing large-scale support to agriculture to increase market actors' incentives to adopt sustainable and productivity-enhancing practices, while also considering the influence of exchange rate, fiscal, and monetary policies.
- **Identify and address the distributional consequences of domestic policy reforms and trade**—both between and within countries—for incomes, livelihoods, natural-resource use, and access to food. These consequences determine the political feasibility of policy reform.
- **Expand analysis of food standards** to determine how to align them for safety, dietary quality, and sustainability and how to embed these regulatory measures in World Trade Organization (WTO) rules. Such measures as well as reform of price incentives are needed to induce timely and pervasive food systems change.

Recent global events have highlighted the crucial connections between trade and food security. During the COVID-19 pandemic, restrictions on mobility and trade strained global and national food systems. Russia's war on Ukraine revealed the vulnerability of many food-importing countries to breadbasket failures, resulting, in this case, from dependence on both cereals and fertilizers produced in the Black Sea region. Climate change is placing greater pressure on food systems as more frequent and intense weather shocks—such as the excessive rainfall that affected rice harvests in India and other parts of South Asia in 2023—cause simultaneous crop failures, thus threatening both local and global food security. In response to these events, some affected countries immediately imposed restrictive trade policies that exacerbated food price surges and volatility, instead of providing insulation against these shocks, as was intended (Martin et al. 2024).

Agrifood trade[1] can thus be a double-edged sword. It supports resilience by diversifying supply and making up for local supply shortages, but it also exposes local food systems to global shocks. Agrifood trade serves a number

1 Agrifood products include all agricultural, livestock, fisheries, forestry and related products that are covered under the WTO Agreement on Agriculture.

of objectives. Beyond the big ships, ports, and other infrastructure that bring food to consumers, trade is also about smallholders and rural traders receiving better prices for their products in bigger and higher-value markets in cities and abroad, improving efficiency in the production and use of natural resources, and increasing consumer access to safer and more diversified and nutritious foods. Trade does not always serve each of these objectives equally well, stirring controversy about which trade policies best contribute to food security, better nutrition, and (sustainable) agricultural development.

Among IFPRI's mandates when it was created was to advise on proper policy responses to the global food price shocks of the early 1970s. Since then, IFPRI has remained a leader in research on agrifood trade policies, with its researchers' analyses both influencing and evolving alongside changing perspectives on how trade can help address global challenges. This chapter reviews the evolution of trade research, with a focus on the contributions of IFPRI and others, as well as priorities for future research. Figure 16.1 provides a simplified timeline of the main events in international markets and trade policies and changes in IFPRI's research priorities since 1975, discussed in detail in the next sections.

Major trends in agrifood trade and policies since 1975

Shift in agrifood trade from raw to processed and specialty products

Driven by rising incomes in emerging economies, population growth, changing consumption patterns, and decreasing trade and transportation costs, global agrifood trade has grown continuously since 1975—at an average annual rate of 5.8 percent—and its value has almost tripled (measured in constant dollars). Over the past two decades, the expansion of global trade accelerated, with its value quadrupling in current prices between 2003 and 2023 to reach US$1.9 trillion (Figure 16.2). In addition, the quality of trade data has improved with computerization of customs, especially in developing countries, while IFPRI researchers have played a key role in improving measurement of informal trade flows in Africa (Bouët et al. 2018).

The composition of agrifood trade is also shifting. The share of primary (raw) agricultural commodities in agrifood exports dropped from 88 percent in the 1970s to 77 percent at present, while trade in processed products increased. In developed countries, this share now represents half

FIGURE 16.1 IFPRI's research on agricultural trade

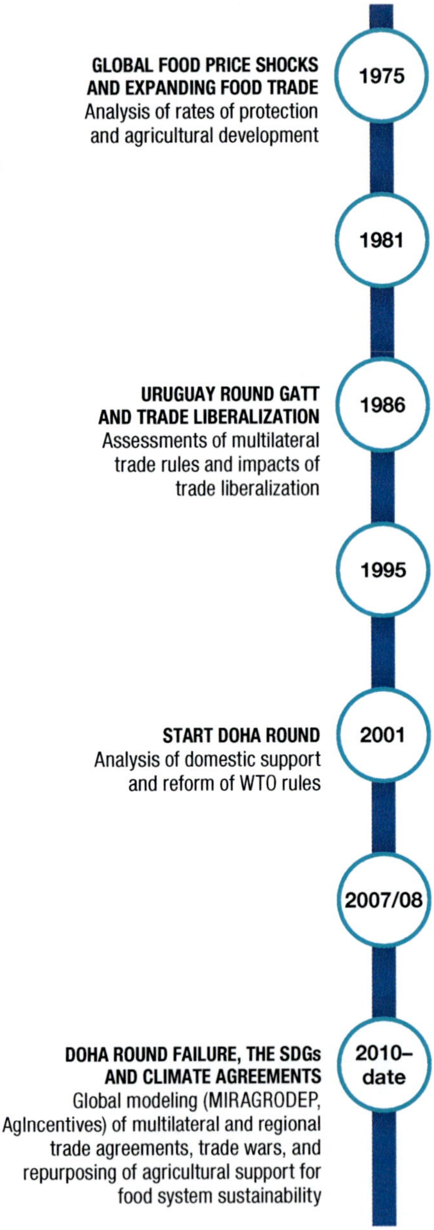

Source: Authors' depiction.

Note: GATT = General Agreement on Tariffs and Trade; LMICs = low- and middle-income countries; SDGs = Sustainable Development Goals; WTO = World Trade Organization.

FIGURE 16.2 The shifting composition of expanding world agrifood exports, 1975–2023

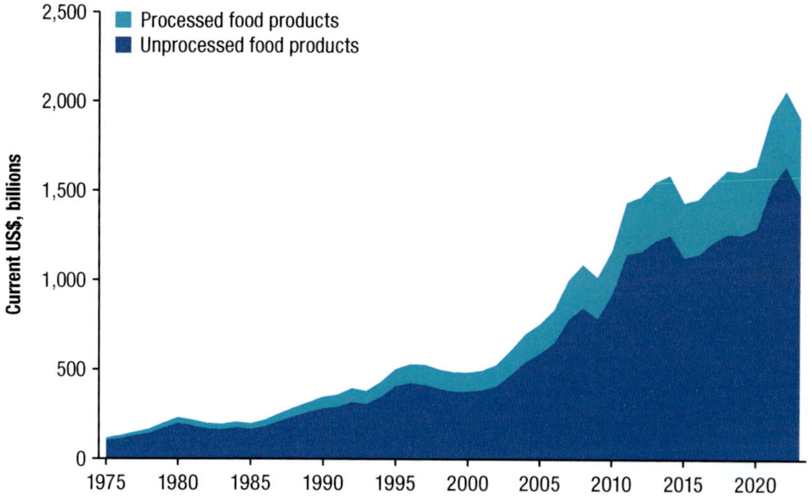

Source: Authors' calculations based on Comtrade data. https://comtradeplus.un.org/

of agrifood imports (UNCTAD 2024). Nonetheless, the trade intensity of agrifood products (that is, the share of traded products in total production) is relatively low—only 13 percent of agrifood products are traded internationally, compared to 30 percent of nonagricultural goods (Gaigné and Gouel 2022).

The share of developing countries in world trade of both processed and unprocessed agrifood products has grown substantially since 1975, as these countries have become major suppliers (exporters) and markets (importers). Their share in global agrifood exports increased from 30 percent in 1975 to 35 percent in 2023, while their share in global imports almost doubled, from 20 percent to more than 35 percent during the same period (Figure 16.3). The trend is largely driven by South–South trade (Aksoy and Ng 2010) and by a few emerging economies, including Argentina, Brazil, Chile, China, India, and, to a lesser extent, South Africa. The Least Developed Countries (LDCs) are lagging far behind (FAO 2018, 2021).

Major changes in agrifood trade policies

These trade trends were accompanied by changing policies for agricultural trade and domestic support (Figure 16.4). In the 1970s, policy approaches differed markedly between rich and poor nations, reflecting differences in

FIGURE 16.3 Developing country share in world agrifood trade, 1975–2023 (percentage)

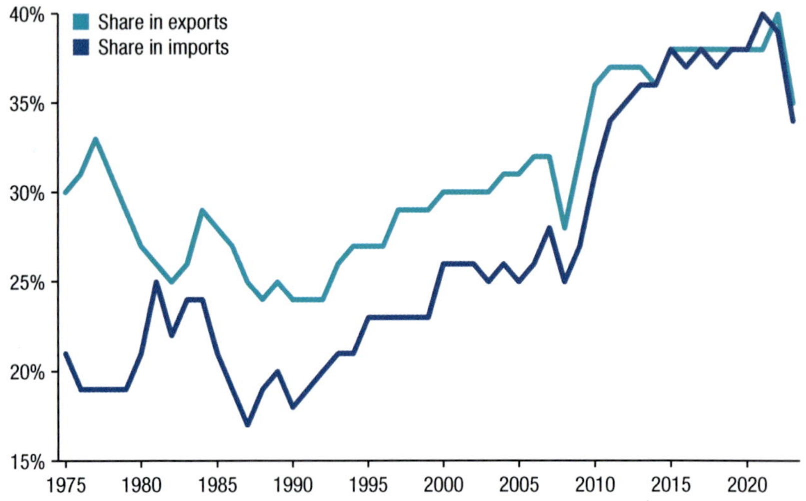

Source: Authors' calculations based on Comtrade data. https://comtradeplus.un.org/

FIGURE 16.4 Protection rates for agriculture in high-income and developing countries, 1957–2022

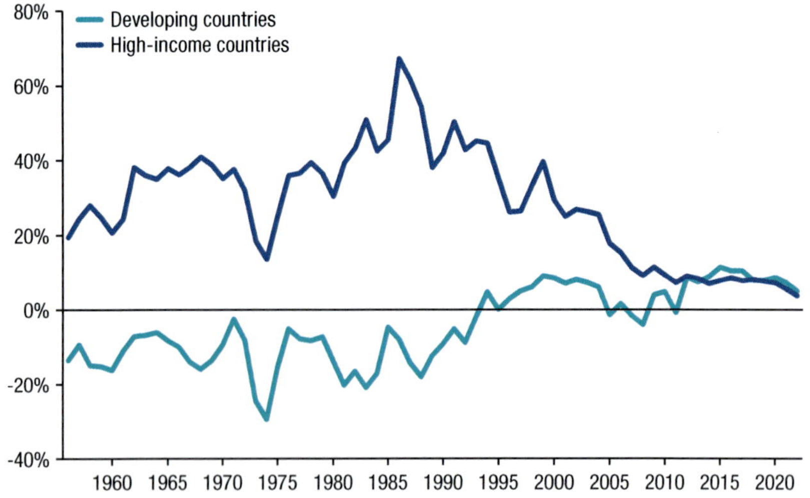

Source: AgIncentives Consortium database. www.agincentives.org

Note: Rates refer to the nominal rate of protection (NRP), as defined in footnote 2 to this chapter.

their level of development. Developed countries provided strong support to their farm sectors, as reflected in an average nominal rate of protection[2] of close to 40 percent, while developing countries taxed their agriculture sectors at about 12 percent on average. Many developing countries prioritized import-substituting industrial development to stimulate growth (see Chapters 3 and 18), and for many low-income countries, import tariffs and taxes on exports were also important sources of fiscal revenues. But, as assessed by studies on the political economy of agricultural protection (for example, Anderson et al. 2013), developing countries' taxation of agriculture decreased as their incomes grew. Furthermore, the Uruguay Round helped to reduce protection in the developed countries (Hathaway and Ingco 1996), as Figure 16.4 shows.

Globally, short-term fluctuations in average protection rates have been strongly influenced by policies enacted to insulate domestic prices and markets from global food price shocks. For instance, both developing and developed food-importing countries lowered protection rates by more than 20 percentage points in attempts to insulate themselves against the food price surge of 1975. Similar responses were observed during the global food price shocks of 2007/08, 2010/11, and 2021/22. As World Bank and IFPRI research has demonstrated, these responses came in tandem with export restrictions imposed by major producer countries, which were also intended to provide insulation from higher prices but which, in practice, exacerbated the rise in world market prices and the volatility of both domestic and world prices (Anderson et al. 2014; Martin et al. 2024).

Protecting or "unprotecting" agriculture? Research on trade policy practice

Two issues have been central to the research agenda on agrifood trade policy over the past 50 years: the measurement of trade and the stance of trade policy—that is, whether agricultural trade was protected or taxed. Guided by measurement of trade and policy impacts, policy reforms have been formulated at country, regional, and global levels. Over time, both the thinking about those reforms and the policy research agenda have evolved, along with perceptions about the key challenges facing agriculture and food systems.

2 The nominal rate of protection (NRP) measures the extent to which a set of agricultural policies affect the market price of a commodity relative to world prices. It is computed as the price difference, expressed as a percentage, between the farmgate price received by producers and an undistorted reference price at the farmgate level. For further explanation, see the AgIncentives Consortium. www.agincentives.org/nominal-rate-of-protection

Early research on agricultural protection and taxation

Much of the initial research relied on simple data and tools, though the level of analytical sophistication has advanced over time. Early work on country policies pointed to the adverse impacts of agricultural taxation in low- and middle-income countries on development prospects and the poor. Pathbreaking work by IFPRI researchers used simple measures of agricultural protection in high-income countries to show that, by subsidizing and protecting their farmers, these countries were depressing farm prices in developing countries, which would stand to benefit from reductions in such protection (Valdés and Zietz 1980). This award-winning research[3] underpinned further work that applied an economywide model to analyze the impact of macroeconomic and trade policies in Argentina and Chile (Cavallo and Mundlak 1982). This line of work reflected a growing interest in how trade and macroeconomic policies were creating biases in price incentives that worked against agricultural development.

IFPRI research indicated that many developing countries were taxing agriculture (Figure 16.4), either explicitly or implicitly, as part of "cheap food" policies intended to keep real wages down to support industrial growth strategies (see Chapter 18).[4] In many instances, these policies held back agricultural productivity growth and rural transformation, which in turn hampered poverty reduction and food security. Development was found to be more successful where policies supporting public investments in rural infrastructure and R&D were implemented alongside trade and domestic food policies that provided incentives for productivity-enhancing agricultural technologies and practices (see Chapter 17) (Bautista and Valdés 1993; Díaz-Bonilla 2017).

As Figure 16.4 shows, high-income countries' protection of agriculture—which included domestic subsidies, import tariffs, and export subsidies—peaked during the mid-1980s, when world prices of agricultural products were extremely low, partly because export subsidies were depressing world prices. Although the political-economy pressures for agricultural protection suggested that increases in protection would continue in high-income countries (Anderson 1995; Anderson et al. 2013), the high cost of export subsidies finally turned the tide against these policies. From the mid-1990s, multilateral trade negotiations

3 The paper received the Quality of Research Discovery Award from the Agriculture & Applied Economics Association in 1983.

4 A compilation of that work using a partial equilibrium framework can be found in Bautista and Valdés (1993). Later IFPRI analyses using a general equilibrium framework re-estimated the presumed bias against agriculture with more varied results (Jensen et al. 2002; Robinson et al. 2001).

under the Uruguay Round made progress in reducing this protection (Hathaway and Ingco 1996). These developed-country reforms also appear to have reduced the problem of protection, though it typically resurfaces in periods of surging food prices because governments tend to use trade policies to insulate consumers against global food price shocks, even though such responses are typically counterproductive (Martin and Anderson 2011; Martin et al. 2024).

Many developing countries grew rapidly during the period of sharp global economic convergence in income levels between 1996 and 2017 (Martin 2019). This rapid economic growth, together with reforms motivated by a recognition of the costs of taxing agriculture, contributed to the shift from taxation to support of agriculture in these countries (Figure 16.4). However, the relatively weak limits ("disciplines," in the language of trade agreements) accepted on developing country agricultural support in the Uruguay Round may have allowed policymakers to repeatedly use trade policies that insulate their markets during periods of high global food prices.

Prominence of research on trade liberalization

The global shift toward trade liberalization and the evolution of the Uruguay Round, which led to the establishment of the World Trade Organization (WTO) in 1995, created a demand for influential agricultural trade policy research at IFPRI that combines broad analysis of global impacts of trade liberalization with detailed assessments of impacts in particular developing countries or regions (Díaz-Bonilla and Robinson 2001; Hewitt 2008).[5] Most of this research favored openness and market orientation—as a means to drive economic efficiency and growth—over interventions and policy-induced trade distortions. But IFPRI's analysts have not shied away from reporting negative consequences of market liberalization for particular countries or undesired collateral outcomes for poverty, food insecurity, and income inequality (for example, Bouët 2006; Vos et al. 2004, 2006) and suggesting tools other than trade policies to address them. The critical issues of links between trade and agrifood safety, product attributes, and quality regulations

5 Trade analysis in the 1990s and early 2000s focused on some of the key issues raised in what was supposed to be the continuation of the negotiations called for by Article 20 of the Agreement on Agriculture. Because food security was invoked as a special consideration in the negotiations, IFPRI's research aimed to clarify and quantify the trade implications with a focus on the special country groups that were considered in the negotiations (such as LDCs and net food-importing developing countries) (Díaz-Bonilla et al. 2000). Another topic was domestic support, including the "development box" or "food security box," with some simulations using a global general equilibrium model showing the potentially negative impacts of some of the proposed instruments (for example, see Díaz-Bonilla et al. 2004; Díaz-Bonilla, Thomas, and Robinson 2002).

have been examined in a number of studies (see Josling et al. 2004; Maertens and Swinnen 2009; Swinnen 2016). This work highlights the double-edged nature of these measures, which, if well designed, can result in improvements in economic welfare but, if not, can lead to interest-group capture and covert protection.

Climate change, food security, and the repurposing of farm support and trade policies

More recent work considers how market distortions that induce inefficiencies in food systems may exacerbate the "hidden costs" associated with the food system (Brooks and Díaz-Bonilla 2025). Recognizing these costs highlights the imperative of transforming food systems to deliver better outcomes for people and the planet, as well as the economy. IFPRI's modeling work, including use of the MIRAGRODEP model (see Box 16.1), captures impacts of policy reform not only on economic efficiency and the terms of trade but also on poverty, hunger, diets, and global greenhouse gas (GHG) emissions (see Chapter 4). This has enabled influential work assessing opportunities and trade-offs of various trade reforms, including (1) regional trade integration such as the African Continental Free Trade Area (AfCFTA) (for example, see Bouët et al. 2022); (2) the repurposing of agricultural support policies (both protective border measures and domestic support) toward fighting climate change in ways aligned with improving food security and diets and reducing poverty (Gautam et al. 2022; Glauber and Laborde 2022; Laborde et al. 2021); (3) identifying the adverse consequences of trade wars (Bouët and Laborde 2018); and (4) assessing the impacts of restrictive food trade policies on global and domestic markets and price stability (for example, Anderson et al. 2014). This modeling work is built upon methodological innovations that have proven useful for both researchers and policymakers. These include the development of better ways to summarize complex trade policy measures (Laborde et al. 2016) and the potentially dangerous consequences of allowing negotiators to exempt sensitive products from reductions in trade negotiations (Jean et al. 2010).

The work on repurposing agricultural support gained prominence at several global forums, including the G20, the G7, and the United Nations (UN) climate negotiations. As an outcome of the UN COP28 in 2023, 159 countries signed the Declaration on Sustainable Agriculture, Resilient Food Systems and Climate Action, committing to "revisit or orient policies and public support related to agriculture and food systems to promote activities which increase incomes, reduce greenhouse gas emissions, and bolster

BOX 16.1 Evolution of the MIRAGRODEP model

Contributed by **Antoine Bouët**, director of the Centre d'Etudes Prospectives et d'Informations Internationales (CEPII), Paris, and non-resident senior fellow, Markets, Trade, and Institutions Unit, IFPRI; and **David Laborde**, director, Agrifood Economics Division, Food and Agriculture Organization of the United Nations (FAO), Rome.

MIRAGRODEP is a dynamic, multisector, multicountry computable general equilibrium model. Calibrated on the Global Trade Analysis Project (GTAP) database, it can incorporate up to 65 sectors and 160 countries or regions into simulations of the world economy. To keep simulations flexible and relatively rapid, scenario analyses generally include just 25 to 30 sectors and 25 to 30 countries or regions. The model has been adapted and elaborated over time in response to changing trade policy environments and to address new food systems challenges. User groups, such as the African Growth and Development Policy (AGRODEP) Modeling Consortium, have been instrumental in ensuring the model's relevance in responding to real-world policy challenges.

MIRAGRODEP has primarily been used to carry out ex ante evaluations of multilateral agreements (the Doha Development Agenda) and regional trade agreements, including various possible trade agreements in Peru, the South Asian Free Trade Area agreement (Bouët et al. 2010), and preferential schemes, such as the European Union's Economic Partnership Agreement with African regional economic communities and the African, Caribbean, and Pacific countries. It has also been used to simulate various trade war scenarios, with and without conclusion of the Doha Development Agenda, to evaluate the impact of a trade war and determine the opportunity cost of not concluding the Agenda (Bouët and Laborde 2017, 2018). The model has also been used to study the overall impact of export restrictions (Laborde et al. 2013) and their impact on food prices (Bouët and Laborde 2010).

Three specific MIRAGRODEP developments stand out. First, the biofuels version of the model has made it possible to detail the use of crops for energy generation, such as manufacturing of fuels from sugarcane, cereals such as maize (ethanol), or vegetable oils (biodiesel) (Laborde 2011; Valin et al. 2010). This research measured the impact of interventions to support the development of the biofuels sector on world agricultural markets, showing adverse impacts in terms of rising food prices, increased food insecurity, and deforestation.

Second, connecting MIRAGRODEP with a household module (POVANA) based on survey data for large numbers of households with different sources of income and consumption structures has facilitated measurement of the impacts of global market shocks and trade reforms on poverty, food security, and nutrition (Laborde et al. 2013). This version was useful for measuring the poverty and food-security impacts of the COVID-19 pandemic (Laborde et al. 2021) and Russia's invasion of Ukraine (Laborde et al. 2023).

Third, a static version of MIRAGRODEP with little sectoral disaggregation is useful for conducting repeated simulations of a trade reform, each over a very short time period, to evaluate all the possible outcomes of a multilateral trade negotiation. It can also be used to estimate costs and benefits and winners and losers of trade reform measures, helping to identify the likely support and opposition to reform proposals on the negotiation table and, hence, the likelihood of successful negotiations (Bouët and Laborde 2010). The same approach has been used to assess options for negotiated plurilateral agreements (Bouët and Laborde 2021).

resilience, productivity, livelihoods, nutrition, water efficiency and human, animal and ecosystem health while reducing food loss and waste, and ecosystem loss and degradation" (COP28 UAE 2023).

This call for action was influenced by the joint work of IFPRI, the World Bank, and the Food and Agriculture Organization of the United Nations (FAO et al. 2022; Laborde et al. 2021; Gautam et al. 2022; Glauber and Laborde 2022). This work modeled approaches to bringing about transformational reductions in agrifood GHG emissions, while also serving objectives of poverty reduction, better diets, and biodiversity conservation through appropriate reorientation of agricultural support (Figure 16.5). The analyses concluded that many popular approaches—such as providing conditional payments to farmers or eliminating all support—would do little to achieve these goals. For instance, conditioning farm support on the adoption of organic farming practices could lead to sizable reductions in emissions where this support is provided, but such gains would be undercut by increased emissions from land-use change, due to the presumed loss of agricultural productivity associated with existing organic farm practices. The more promising approach to repurposing support that generates the necessary transformational reductions in emissions is their use for development of new technologies that both reduce emissions intensities and raise productivity.

When taking account of the range of goals that policymakers must consider, the selection of policy interventions becomes more complex. Important trade-offs need to be considered when trying to align the objective of reducing emissions from agriculture with economic and social goals, including poverty reduction, food security, income growth, and better diets. Global removal

FIGURE 16.5 Global implications of repurposing agricultural support

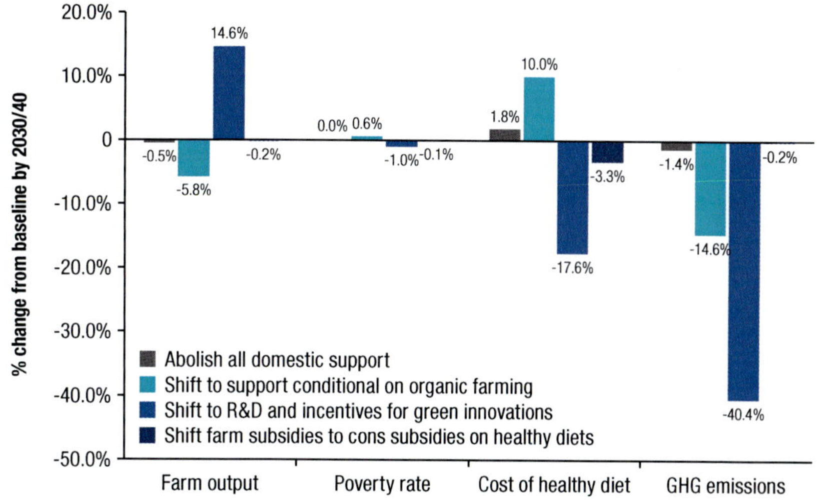

Source: Gautam et al. (2022) and Glauber and Laborde (2022).

Note: Results for the policy scenario are reported as a percentage change from the baseline scenario in 2040 for all indicators, except for the poverty rate, which shows changes in percentage points from the baseline in 2040. The fourth scenario for repurposing existing support toward consumer subsidies shows changes relative to the baseline for 2030. GHG = greenhouse gas.

(that is, removal by all countries) of existing domestic support and trade protection has only a marginal impact on poverty reduction, while global environmental conditionality (requirements linking farm payments to adoption of environment-friendly practices) increases poverty, partly because it raises the cost of healthy diets. Policies that both improve productivity and reduce emissions intensities have the greatest potential for making progress on all these goals (Figure 16.5).

Shaping multilateral and regional trade reforms

Over the past 50 years, trade research by IFPRI and others was largely undertaken to inform regional and multilateral trade negotiations that aimed to facilitate trade in agricultural and food commodities for greater market efficiency and stability. This policy agenda changed substantially over time, with IFPRI's research agenda evolving along with it.

At the beginning of the 1980s, IFPRI researchers developed a global multimarket model to measure the impact of a hypothetical 50 percent across-the-board reduction in tariffs and other trade barriers for 99 commodities in 17 developed countries belonging to the Organisation for Economic Co-operation and Development (OECD) (Valdés and Zietz 1980). Subsequent analyses using general equilibrium models found that (1) trade protection in OECD countries tends to lower world prices and reduce the volume of exports from both developing and developed exporters, while liberalization increases world prices and expands export volumes; (2) tariff escalation (that is, higher tariffs on more processed goods) in developed countries discourages manufacturing growth in developing countries; and (3) food-importing countries are affected by higher world market prices following food trade liberalization in OECD countries, but the final results depend on the response of their own agriculture sectors to improved price incentives (Diao et al. 2005). These findings highlight the food-price dilemma inherent in trade policy: trade protection harms consumers through higher prices while benefiting food producers, whereas trade liberalization benefits consumers through lower prices but harms food producers.

The Uruguay Round (1986–1994), the largest-ever successful international trade negotiation, sought to convert countries' myriad forms of agricultural border protection—such as export bans, licensing, and quotas—into tariffs. But its success was only partial, and these restrictions continue to be imposed in many countries, as do specific tariffs whose protective impact depends on the product price. As a result, assessing the protective impact of complex agricultural trade policies continues to require laborious price comparison approaches (for example, see Josling and Tangermann 2015). To address the need for this information, the AgIncentives Consortium, coordinated by IFPRI, provides an updated and consistent database with measures of agricultural trade protection and domestic support in a comparative format available for researchers and policymakers (Laborde et al. 2024).

Following the Uruguay Round, negotiations on agricultural trade continued under what became known as the Doha Development Agenda (2001–2015). Research supporting this effort resulted in the development of new modeling techniques that could better account for the welfare impacts of the complex policy proposals put forward in these negotiations (Anderson et al. 2005). Using these techniques, it became clear that potential gains could be very large from a Doha Development Agenda that built on the Uruguay Round reductions in protection (Laborde et al. 2017).

Unfortunately, the negotiating modalities developed for Doha failed to recognize the shift in many developing countries' policies—away from taxation and toward protection of agriculture—and collided head-on with the desire of many developing countries to be able to provide this protection.

Despite much careful analysis of the potential adverse impacts of these policies on the welfare of the poor (Anderson and Martin 2006), strong support for allowing developing countries to maintain or increase agricultural protection remained a key source of tension and contributed to the failure of these negotiations (Grant and Mielke 2011).

Fueling some of the debates, IFPRI's trade analysis in the 1990s and early 2000s focused on the implications of special considerations—in particular, food security—that could justify flexible application of trade rules. The research pointed out that, given varying degrees of food insecurity, the country classifications in general use needed to be refined beyond the standard categories of net food importers and LDCs, as many other developing counties also experience significant food insecurity. In addition, the suggested refinements accounted for net food importers that are also large exporters of nonfood products, as in these cases, food import bills tend to be small relative to total export earnings (Díaz-Bonilla 2017).

In addition, IFPRI researchers have assessed the failure of the negotiations for the Doha Round and the WTO Agreement on Agriculture (Brink and Orden 2023; Martin and Mattoo 2011; Orden et al. 2011). Box 16.2 explains some of the controversies and tensions between the WTO multilateral trading system and domestic support for food and agriculture, as well as how IFPRI research helped to outline possible ways out of the impasse.

A specific area of contention relates to public stockholdings (reserves) held for food security reasons. Analyses of the legal and economic aspects, conducted by IFPRI, suggest several ideas that could advance those negotiations (Díaz-Bonilla 2014, 2017; Díaz-Bonilla and Laborde 2015; Glauber and Sinha 2021). This work also found that the existing Agreement on Agriculture already allows developing countries to address most food security issues using public food stockholding, although with some constraints.

The assessments of the failed Doha Round also led to new research on alternative trade arrangements. First, recent IFPRI theoretical research has shown that plurilateral agreements (either exclusive or open variants)[6] can improve welfare if certain conditions are met (Bouët and Laborde 2021). To be approved by WTO members, plurilateral agreements must either include a most-favored nation (MFN) clause or demonstrate that they improve the welfare of non-members. The IFPRI study concludes that inclusive plurilateral

6 The WTO's plurilateral agreements are defined in Annex 4 to the Marrakesh Agreement. These are negotiated and agreed to by subsets of WTO members and bind only those members who have signed them.

BOX 16.2 WTO challenges to facilitating sustainable food systems transformation

Contributed by **David Orden**, professor of Agricultural and Applied Economics, Virginia Polytechnic Institute and State University, Blacksburg, Virginia.

The 1994 World Trade Organization (WTO) Agreement on Agriculture, negotiated when world markets were severely distorted, commits members to limits on tariffs, export subsidies, and trade-distorting domestic support to agriculture. Export subsidies are now prohibited by the Agreement, and many countries have shifted toward less distorting policies. Despite this progress, further negotiations have proved unsuccessful. Limits (disciplines) on domestic support continue to be contested, with China and India along with the European Union, the United States, and Japan providing the highest amounts of support.

The Agreement divides domestic support policies into two broad categories: those with at most minimal or no distorting effects (the "Green Box") and those that are distorting. Green Box policies face no limitations, contributing to the growth in this support globally and marking a step toward repurposing subsidies. Most trade-distorting support is subject to limits. Developing countries generally face constraints on such support when it exceeds 10 percent of production value (*de minimis* 10 percent). The Agreement sets the limit for such support by developed countries at 5 percent of the value of production, but instead of being more restrictive, the rules are more flexible, as for most developed countries, the actual limits are based on their sum of historical support. This allows support levels that can exceed 10 percent of output value. Unsurprisingly, this bifurcation in applying *de minimis* rules has been highly contentious. Developing countries other than China, which joined after the Agreement, have leeway on paper to provide, in addition, unlimited input and investment subsidies to foster rural development, but many lack the resources to do so, while developed countries can continue providing support. Amid this controversy, India has argued for exempting market price support from limits when provided in developing countries for food-security public stockholding programs.

Throughout these complex negotiations, IFPRI research has informed negotiations and public debate. During the Doha Round deliberations in the 2000s, IFPRI studies focused on the price support issues and on options to strengthen support for development (Díaz-Bonilla et al. 2002, 2004). Later work provided shadow projections for key countries' notifications of trade-related measures, which were made available to WTO and the countries involved. These identified policy reforms needed to comply with proposed new commitments and highlighted the severe limits on market price support, as calculated under the WTO, that developing countries could face (Orden et al. 2011).

A decade later, the domestic support issues remain unresolved, and policies in Canada, China, the European Union, India, Korea, and the United States have been challenged in WTO disputes. Recent research has further disentangled the legal and economic aspects of market price support (Díaz-Bonilla 2017; Glauber and Sinha 2021). Other efforts have tracked the implementation, negotiation, and dispute outcomes and have outlined suggestions for reform, including making the rules more consistent across WTO membership, improving how price support is measured, and modifying the rules to address biosecurity, biodiversity, and climate change mitigation (Brink and Orden 2023; Glauber 2022, 2023). These analyses provide an evidence base to draw upon as WTO members seek to move forward toward the still-relevant objective of reducing protection and market-distorting support under the changed market and policy context that has arisen in intervening years.

agreements are the most promising but difficult to achieve, as the addition of an MFN clause reduces the value of the plurilateral agreement for its members, suggesting there is a trade-off between efficiency and acceptability. Second, the rise of bilateral and regional trade agreements has raised important questions about whether "regionalism is a building block or a stumbling block to multilateralism," as Bhagwati (1993) characterized the long-standing debate among economists, policymakers, and trade negotiators. But while multilateralism is stalled, there seems to be considerable potential for continued progress with regional agreements such as the AfCFTA (see Chapter 19 on Africa) (Bouët et al. 2022; Maliszewska and Ruta 2020). The AfCFTA agreement entered into force in May 2019 for the 24 countries that had deposited their instruments of ratification by this date. By the end of December 2022, 54 of the 55 African Union member countries had signed the agreement (only Eritrea has not signed). However, while implementation of the AfCFTA agreement formally began in January 2021, it still faces many operational challenges. Among these are overlapping memberships across the various African regional economic communities, which create inconsistencies in the trade rules. In addition, the substantial informal (agricultural) trade characterized by undocumented cross-border transactions poses a challenge to formalizing intra-Africa trade. Infrastructure deficiencies, inadequate transportation networks, and border facilities further hamper full implementation of the AfCFTA agreement.

Policy challenges and research needs

IFPRI's analytical tools have advanced dramatically over the past 50 years. Key aspects of agricultural trade policy have also improved enormously, most notably the trade opening brought about by the Uruguay Round and the shift of developing country policies away from heavy taxation of agriculture. In terms of products traded, a key change has been a shift away from basic commodities toward processed agricultural products, which typically have lower and less volatile protection rates. However, little progress has been made in reducing the use of trade policies to insulate countries' home markets from changes in world prices—with consequent destabilizing impacts on world prices and, frequently, on countries' domestic prices as well.

Unfortunately, the current political winds are shifting views on trade policy away from the widespread consensus that prevailed during most of the past half-century—a consensus that emphasized the benefits of using markets to raise incomes, buttressed by targeted transfer mechanisms to address income distribution and specific interventions to manage externalities and other market failures. Today, a more interventionist, but still poorly defined, set of policy norms is emerging. Moving forward, priorities in trade policy research should include:

Continuing investigation of the actual impacts and trade-offs of policy interventions, taking advantage of the rapid advances in analytical tools for trade policy analysis, including not only trade and trade-related measures but also macroeconomic policies that affect trade, such as exchange-rate policies (Díaz-Bonilla 2015; Martin 2018). This will require analysis of the effects of policy interventions and clear exposition of their impacts—good and bad—including the consequences of using trade policies to insulate farmers and consumers against global shocks or to advance geopolitical strategies.

Deepening analysis of policy effectiveness and policy coherence at the national and international level. Context-specific assessments of the effectiveness of different trade and domestic policy instruments (and their combinations), which constitutes a vast policy research agenda, will be critical (Vos and Martin 2025). But trade policy alone cannot achieve sustainable food systems transformation. This transformation will require alignment of an array of policies that consider complementarities, potential adverse impacts, and international spillovers. As recent research on repurposing of existing support has concluded, win-win-win gains for people, planet, and prosperity are only possible with strong international policy coordination and domestic policy coherence (Gautam et al. 2022).

Increasing attention to the distributional impacts of trade and domestic policy reforms on incomes, livelihoods, natural resource use, and access to food both between and within countries. For example, climate change in poor countries is much more likely to threaten the livelihoods of very poor people. Such distributional consequences will influence the political feasibility of policy reform. With many pressing questions and new tools available to address them, international agrifood trade must be a very active research area in the coming years.

Ramping up analysis of countries' food standards for safety, dietary quality, and sustainability. These regulatory measures are gaining importance as reliance on price incentive policies alone is unlikely to induce timely and pervasive food system change (Papendick and McNamara 2024). Addressing these standards is an added challenge in WTO implementation and negotiations, and for other key international institutions.

For all these priorities, reaching the policymakers whose decisions would benefit from the evidence and analysis provided by research is critical. The gold standard for researchers is to identify key policy questions and engage with decision-makers before conducting research, and to recognize political-economy constraints and, if necessary, try to address them. This implies both continuous analysis of the political economy of trade policy reform (see Resnick and Swinnen 2023) and deep understanding of decision-makers' needs, which are central not only to refining research questions but also to communicating the answers more effectively for greater impact.

References

Aksoy, M.A., and F. Ng. 2010. "The Evolution of Agricultural Trade Flows." World Bank Policy Research Working Paper 5308. World Bank, Washington, DC.

Anderson, K. 1995. "Lobbying Incentives and the Pattern of Protection in Rich and Poor Countries." *Economic Development and Cultural Change* 43 (2): 401–442.

Anderson, K., M. Ivanic, and W. Martin. 2014. "Food Price Spikes, Price Insulation and Poverty." In *The Economics of Food Price Volatility*, eds. J.-P. Chavas, D. Hummels, and B. Wright, 31–44. Chicago: University of Chicago Press.

Anderson, K., and W. Martin, eds. 2006. *Agricultural Trade Reform and the Doha Development Agenda*. New York: Palgrave Macmillan; Washington, DC: World Bank.

Anderson, K., W. Martin, and D. van der Mensbrugghe. 2005. "Market and Welfare Implications of Doha Reform Scenarios." Conference Paper 331428. Purdue University, Center for Global Trade Analysis, Global Trade Analysis Project, West Lafayette, IN.

Anderson, K., G. Rausser, and J. Swinnen. 2013. "Political Economy of Public Policies: Insights from Distortions to Agricultural and Food Markets." *Journal of Economic Literature* 51 (2): 423–477.

Bautista, R.M., and A. Valdés. 1993. *The Bias against Agriculture: Trade and Macroeconomic Policies in Developing Countries.* Washington, DC: International Center for Economic Growth and IFPRI.

Bhagwati, J. 1993. "New Dimensions in Regional Integration." In *Regionalism and Multilateralism: An Overview*, eds. J. De Melo and A. Panagariya, 22–51. Cambridge, UK: Cambridge University Press.

Bouët, A. 2006. *How Much Will Trade Liberalization Help the Poor?* IFPRI Research Brief 5. Washington, DC: IFPRI.

Bouët, A., and D. Laborde. 2010. "Eight Years of Doha Trade Talks: Where Do We Stand?" *Estey Centre Journal of International Law and Trade Policy* 11 (2): 1–22.

Bouët, A., and D. Laborde. 2017. "Assessing the Potential Cost of a Failed Doha Round." In *Agriculture, Development, and the Global Trading System: 2000–2015*, eds. A. Bouët and D. Laborde Debucquet, 155–192. Washington, DC: IFPRI.

Bouët, A., and D. Laborde. 2018. "US Trade Wars in the Twenty-First Century with Emerging Countries: Make America and Its Partners Lose Again." *The World Economy* 41 (9): 2276–2319.

Bouët, A., and D. Laborde. 2021. "Plurilateral Agreements: A Promising Trade Liberalization Modality?" In *The Road to the WTO Twelfth Ministerial Conference: A Latin American and Caribbean Perspective*, eds. V. Piñeiro, A. Campos, and M. Piñeiro, 131–143. San José, Costa Rica: Inter-American Institute for Cooperation on Agriculture; Washington, DC: IFPRI.

Bouët, A., D. Laborde, and F. Traoré. 2022. "The AfCFTA: The Need for Ambitious Implementation." In *Africa Agriculture Trade Monitor 2022*, eds. A. Bouët, S.P. Odjo, and C. Zaki, 140–171. Kigali, Rwanda: AKADEMIYA2063; Washington, DC: IFPRI.

Bouët, A., S. Mevel, and M. Thomas. 2010. "Is SAFTA Trade Creating or Trade Diverting? A Computable General Equilibrium Assessment with a Focus on Sri Lanka." IFPRI Discussion Paper 950. IFPRI, Washington, DC.

Bouët, A., K. Pace, and J. Glauber. 2018. "Informal Cross-Border Trade in Africa: How Much? Why? And What Impact?" IFPRI Discussion Paper 01783. IFPRI, Washington, DC.

Brink, L., and D. Orden. 2023. *Agricultural Domestic Support under the WTO: Experience and Prospects.* Cambridge, UK: Cambridge University Press.

Brooks, J., and E. Díaz-Bonilla. 2025. "How Helpful Are the 'Hidden Costs of Food Systems' Numbers?" *Food Policy* 131: 102796.

Cavallo, D., and Y. Mundlak. 1982. *Agriculture and Economic Growth in an Open Economy: The Case of Argentina*. IFPRI Research Report 36. Washington, DC: IFPRI.

COP28 UAE (28th United Nations Climate Change Conference–United Arab Emirates). 2023. *COP28 UAE Declaration on Sustainable Agriculture, Resilient Food Systems and Climate Action*. Dubai.

Diao, X., E. Díaz-Bonilla, S. Robinson, and D. Orden. 2005. "Tell Me Where It Hurts, an' I'll Tell You Who to Call: Industrialized Countries' Agricultural Policies and Developing Countries." MTID Discussion Paper 84. IFPRI, Washington, DC.

Díaz-Bonilla, E. 2014. "On Food Security Stocks, Peace Clauses, and Permanent Solutions after Bali." IFPRI Discussion Paper 1388. IFPRI, Washington, DC.

Díaz-Bonilla, E. 2015. *Macroeconomics, Agriculture, and Food Security: A Guide to Policy Analysis in Developing Countries*. Washington, DC: IFPRI.

Díaz-Bonilla, E. 2017. "Food Security Stocks and the WTO Legal Framework." In *Agriculture, Development, and the Global Trading System: 2000–2015*, eds. A. Bouët and D. Laborde, 233–283. Washington, DC: IFPRI.

Díaz-Bonilla, E., X. Diao, and S. Robinson. 2004. "Thinking Inside the Boxes: Protection and Investments in the Development and Food Security Boxes." In *Agricultural Policy Reform and the WTO: Where Are We Heading?*, eds. A. Giavani, M. Bohman, C. Carter, and A. McCalla. Cheltenham, UK: Edward Elgar Publishers.

Díaz-Bonilla, E., and D. Laborde. 2015. "The Bali Agreement: An Assessment from the Perspective of Developing Countries." IFPRI Discussion Paper 1444. IFPRI, Washington, DC.

Díaz-Bonilla, E., and S. Robinson. 2001. "Globalization, Trade Reform and Developing Countries." In *The Unfinished Agenda: Perspectives on Overcoming Hunger, Poverty and Environmental Degradation*, eds. P. Pinstrup-Anderson and R. Pandya-Lorch, 225–232. Washington, DC: IFPRI.

Díaz-Bonilla, E., M. Thomas, and S. Robinson. 2002. "On Boxes, Contents, and Users: Food Security and the WTO Negotiations." IFPRI TMD Discussion Paper 82. IFPRI, Washington, DC.

Díaz-Bonilla, E., M. Thomas, S. Robinson, and A. Cattaneo. 2002. "Food Security and Trade Negotiations in the World Trade Organization: A Cluster Analysis of Country Groups." IFPRI TMD Discussion Paper 59. IFPRI, Washington, DC.

FAO (Food and Agriculture Organization of the United Nations). 2018. *The State of Agricultural Commodity Markets 2018: Agricultural Trade, Climate Change and Food Security*. Rome.

FAO. 2021. *Changing Patterns of Agrifood Trade: The Rising Importance of Developing Countries*. Trade Policy Brief 48. Rome.

FAO, IFAD (International Fund for Agricultural Development), UNICEF, WFP (World Food Programme), and WHO (World Health Organization). 2022. *The State of Food Security and Nutrition in the World 2022: Repurposing Food and Agricultural Policies to Make Healthy Diets More Affordable.* Rome: FAO.

Gaigné, C., and C. Gouel. 2022. "Trade in Agricultural and Food Products." In *Handbook of Agricultural Economics*, vol. 6, eds. C.B. Barrett and D.R. Just, 4845–4931. Amsterdam: Elsevier.

Gautam, M., D. Laborde, A. Mamun, W. Martin, V. Piñeiro, and R. Vos. 2022. *Repurposing Agricultural Policies and Support: Options to Transform Agriculture and Food Systems to Better Serve the Health of People, Economies, and the Planet.* Technical Report. Washington, DC: World Bank and IFPRI.

Glauber, J. 2022. "Trade and Climate Change: The Role of Reforms in Ensuring Food Security and Sustainability." In *2022 Global Food Policy Report: Climate Change and Food Systems*, 28–37. Washington, DC: IFPRI.

Glauber, J. 2023. "Rethinking Trade Rules to Achieve a More Climate Resilient Agriculture." IFPRI Discussion Paper 2164. IFPRI, Washington, DC.

Glauber, J., and D. Laborde. 2022. "Repurposing Food and Agricultural Policies to Deliver Affordable Healthy Diets, Sustainably and Inclusively: What Is at Stake?" Background Paper for *The State of Food Security and Nutrition in the World 2022.* FAO Agricultural Development Economics Working Paper 22-05. FAO, Rome.

Glauber, J., and T. Sinha. 2021. *Procuring Food Stocks under World Trade Organization Farm Subsidy Rules: Finding a Permanent Solution.* IISD Report. Winnipeg: International Institute for Sustainable Development.

Grant, J., and K. Mielke. 2011. "The Special Safeguard Mechanism: Previous Studies and Present Outlook." In *Unfinished Business? The WTO's Doha Agenda,* eds. W. Martin and A. Mattoo, 179–212. Washington, DC: World Bank.

Hathaway, D., and M. Ingco. 1996. "Agricultural Liberalization and the Uruguay Round." In *The Uruguay Round and the Developing Countries*, eds. W. Martin and L.A. Winters, 30–49. Cambridge, UK: Cambridge University Press.

Hewitt, J. 2008. "Impact Evaluation of Research by IFPRI on Agricultural Trade Liberalization, Developing Countries, and WTO's Doha Negotiations." IFPRI Impact Assessment Discussion Paper 28. IFPRI, Washington, DC.

Jean, S., D. Laborde, and W. Martin. 2010. "Formulas and Flexibility in Trade Negotiations: Sensitive Agricultural Products in the World Trade Organization's Doha Agenda." *The World Bank Economic Review* 24 (3): 500–519.

Jensen, H., S. Robinson, and F. Tarp. 2002. "General Equilibrium Measures of Agricultural Policy Bias in Fifteen Developing Countries." IFPRI TMD Discussion Paper 105. IFPRI, Washington, DC.

Josling, T., D. Roberts, and D. Orden. 2004. *Food Regulation and Trade: Toward a Safe and Open Global System.* Washington, DC: Peterson Institute for International Economics.

Josling, T., and S. Tangermann. 2015. "Measuring Levels of Protection in Agriculture: A Survey of Approaches and Results." In *Farm Policies and World Markets: Monitoring and Disciplining the International Trade Impacts of Agricultural Policies*, ed. T. Josling, 17–26. Singapore: World Scientific Publishing.

Laborde, D. 2011. *Assessing the Land Use Change Consequences of European Biofuel Policies: Final Report.* Washington, DC: IFPRI.

Laborde, D., C. Estrades, and A. Bouët. 2013. "A Global Assessment of the Economic Effects of Export Taxes." *The World Economy* 36 (10): 1333–1354.

Laborde, D., T. Lallemant, F. Majeed, A. Mamun, W. Martin, and S. Tokgoz. 2024. *Introducing the Updated AgIncentives Database.* Washington, DC: IFPRI.

Laborde, D., A. Mamun, W. Martin, V. Piñeiro, and R. Vos. 2021. "Agricultural Subsidies and Greenhouse Gas Emissions." *Nature Communications* 12: 2601.

Laborde, D., W. Martin, and D. van der Mensbrugghe. 2017. "Measuring the Impacts of Global Trade Reform with Optimal Aggregators of Distortions." *Review of International Economics* 25 (2): 403–425.

Laborde, D., W. Martin, and R. Vos. 2021. "Impacts of COVID-19 on Global Poverty, Food Security, and Diets." *Agricultural Economics* 52 (3): 375–390.

Laborde, D., G. Matchaya, and F. Traoré. 2023. "Impact of the Russia-Ukraine War on African Agriculture, Trade, Poverty, and Food Systems." In *Africa Agriculture Trade Monitor 2023*, eds. S. Odjo, F. Traoré, and C. Zaki, 146–174. Kigali, Rwanda: AKADEMIYA2063; Washington, DC: IFPRI.

Maertens, M., and J. Swinnen. 2009. "Trade, Standards, and Poverty: Evidence from Senegal." *World Development* 37 (1): 161–178.

Maliszewska, M., and M. Ruta. 2020. *The African Continental Free Trade Area Economic and Distributional Effects.* Washington, DC: World Bank.

Martin, W. 2018. "A Research Agenda for International Agricultural Trade." *Applied Economic Perspectives and Policy* 40 (1): 155–173.

Martin, W. 2019. "Economic Growth, Convergence, and Agricultural Economics." *Agricultural Economics* 50 (S1): 7–27.

Martin, W., and Anderson, K. 2011. "Export Restrictions and Price Insulation during Commodity Price Booms." *American Journal of Agricultural Economics* 94 (2): 422–427.

Martin, W., A. Mamun, and N. Minot. 2024. "Food Trade Policy and Food Price Volatility." IFPRI Discussion Paper 2253. IFPRI, Washington, DC.

Martin, W., and A. Mattoo, eds. 2011. *Unfinished Business? The WTO's Doha Agenda.* Washington, DC: World Bank.

Orden, D., D. Blandford, and T. Josling, eds. 2011. *WTO Disciplines on Agricultural Support: Seeking a Fair Basis for Trade.* Cambridge, UK: Cambridge University Press.

Papendieck, S., and B. McNamara. 2024. "Environmental Concerns and Agricultural Trade: Building a Responsible and Effective Relationship." In *Navigating the Trade Landscape: A Latin American Perspective uilding on the WTO 13th Ministerial Conference,* eds. V. Piñeiro, A. Campos, and M. Piñeiro, 99—120. San Jose: IFPRI and IICA.

Resnick, D., and J. Swinnen, eds. 2023. *The Political Economy of Food System Transformation: Pathways to Progress in a Polarized World.* Oxford, UK: Oxford University Press.

Robinson, S., F. Tarp, and P. Wobst. 2001. "Policy Bias and Agriculture: Partial and General Equilibrium Measures." *Review of Development Economics* 5 (1): 89–104.

Swinnen, J. 2016. "Economics and Politics of Food Standards, Trade, and Development." *Agricultural Economics* 47 (S1): 7–19.

UNCTAD (United Nations Trade and Development). 2024. *Trade in Processed Food.* Geneva.

Valdés, A., and J.A. Zietz. 1980. *Agricultural Protection in OECD Countries: Its Cost to Less-Developed Countries.* Research Report 21. Washington, DC: IFPRI.

Valin, H., B. Dimaranan, and A. Bouët. 2010. "Evaluating the Environmental Cost of Biofuels Policy: An Illustration with Bioethanol." *International Economics* 122 (2): 89–120.

Vos, R., E. Ganuza, S. Morley, and S. Robinson. 2006. *Who Gains from Free Trade? Export-Led Growth, Inequality and Poverty in Latin America.* London: Routledge.

Vos, R., E. Ganuza, S. Morley, S. Robinson, and V. Pineiro. 2004. "Are Export Promotion and Trade Liberalization Good for Latin America's Poor? A Comparative Macro-Micro CGE Analysis." ISS Working Papers - General Series 19158. International Institute of Social Studies of Erasmus University Rotterdam, The Hague.

Vos, R. and W. Martin. 2025. "Options for Reducing Greenhouse Gas Emissions from Agrifood Systems." IFPRI Discussion Paper 2336. IFPRI, Washington, DC. Forthcoming.

Chapter 17

AGRICULTURAL INNOVATION POLICIES
Prioritizing Investments and Promoting Uptake and Impacts at Scale

Catherine Ragasa, David J. Spielman, and John Lynam

Catherine Ragasa is a senior research fellow, Innovation Policy and Scaling Unit (IPS), IFPRI; **David J. Spielman** is director, IPS, IFPRI; and **John Lynam** is an independent scholar.

Key messages

- Innovation in agriculture—including in technological innovation in crop varieties, livestock breeds, and agricultural practices, as well as changes in organizations, policies, and rules—offers a solution to many pressing development problems in low- and middle-income countries, particularly low productivity and poverty.
- Research on the economics of agricultural R&D examined its costs and benefits, demonstrating high returns in terms of productivity and poverty reduction. To support priority-setting and policymaking by governments and funders, researchers developed a set of widely used analytic tool kits for resource allocation and prioritization, including spatial analysis and econometric and computational models, as well as data sets on agricultural investment and capacity.
- Researchers adopted a systems perspective to understand the various actors and social, economic, and political institutions involved in complex innovation processes. The agricultural innovation systems concept pointed to the need for organizational innovation to coordinate diverse actors around efforts to solve complex problems in food systems, and led to the development of various types of multistakeholder innovation platforms and partnerships.
- Recently, innovations research has focused on barriers and enabling policies for scaling promising innovations. Effective scaling is often facilitated by "socio-technical innovation bundles" of policy change, technology options, delivery methods, market links, and coordination mechanisms.

Two successful IFPRI programs—on biofortification and biosafety—offer lessons for a holistic and integrated approach to scaling.

Looking to the future, the review of innovation research points to three key agendas:

- **Develop new methods and integrated frameworks** to better assess multiple policy objectives and inform resource allocation in the context of competing development goals, such as increasing production and addressing climate change. This effort will require systematic integration of spatial analysis, farm models, simulation, and sector and economy-wide models, among other advances.
- **Inform scaling efforts through a better understanding of how to develop best-fit approaches** that consider the local context when scaling innovations for agronomy and natural resource management. Collaborations between research and development organizations on large-scale projects could build evidence on how to create enabling conditions for uptake of innovations.
- **Conduct rigorous impact evaluations**, along with action plans for scalability, of different technological, organizational, and institutional innovations, as well as in-depth analysis of the political economy, policy actions, and enabling environments that support innovations and scale their impacts. It is crucial to focus on which innovative solutions have been shown to be beneficial and how to bring them to diverse farmers' fields.

From the advent of the first farming tools to modern innovations in molecular biology and genetics, technological progress in agriculture has played a critical role in human history and will continue to be essential in tackling the multiplying challenges facing food systems. However, technical change—and accompanying gains in productivity and welfare—has been unevenly distributed across the globe, with some countries and communities benefiting far more than others (Fuglie and Rada 2016; Fuglie et al. 2019; He and Li 2020; Walker et al. 2015).

Given that agricultural science, technology, and innovation are an important solution to the many pressing development challenges facing low- and middle-income countries (LMICs), more attention needs to be directed to this uneven pace of change. Public policy can play a role in shaping social, economic, and individual choices about technical change, and technical change itself is acutely dependent on changes in the social and economic policy environment in which individuals make choices. This chapter reviews the evolution of research on technical change and public policy, from the seminal economic analyses that highlighted the importance of technical change, to innovation systems analyses that broadened our understanding of the technical change process, on to the latest strategies for accelerating change.

To better manage this broad topic, we examine three distinct but overlapping policy agendas on technical change: (1) the agricultural R&D and productivity growth–poverty reduction agenda, (2) the agricultural innovation systems agenda, and (3) the impact-scaling agenda. Within these agendas, we consider three broad types of innovation that reflect long-standing areas of scientific research and innovation. First is biological innovation, or improvements to the genotypic and phenotypic characteristics of cultivated crop varieties and livestock breeds. Second is agronomic innovation, or improvements in the principles and practices applied to the management of agricultural, natural resource, and ecological systems. Third is institutional innovation, or changes in the policies, rules, and norms shared and adhered to by a society, country, or community.[1]

Focusing on the uneven pace of change in LMICs also requires understanding the root causes of the problem: Can it be explained simply by the many competing R&D priorities vying for limited financial and scientific resources? Does the difficulty lie in moving from fundamental research to practical use—that is, from pure basic research to use-inspired basic research to pure applied research? Or is the problem rooted in institutional factors (shared norms, beliefs, and rules in society) or behavioral factors (individual preferences, expectations, and cognitive biases) that shape our willingness to experiment, adopt, and benefit from technical change?

The economics of agricultural R&D and the productivity growth–poverty reduction agenda

Beginning with seminal work in the 1960s and 1970s (Griliches 1964; Hayami and Ruttan 1971), research on agricultural innovation policy has focused on how to produce food, given relative scarcities in the factors of production—land, labor, and capital—and how R&D can increase production and productivity. This focus gave rise to a rich body of research on the benefits and costs

1 We purposely limit this chapter to farm-level innovations; equally important off-farm innovations are discussed in Chapter 7 on food value chains and Chapter 12 on nutrition programs. This chapter focuses on high-level investments, policies, and institutions; equally important discussions on last-mile delivery services and farmers' adoption are found in Chapters 8–10.

of agricultural R&D investments, the rates of return on these investments, and the resulting R&D priorities needed to accelerate agricultural development, food security, and economic growth (Alston et al. 1995, 2000). Early research in this area helped to justify the resources needed for the world's first multilateral program to address a global development challenge—hunger and food insecurity—and thus led to the formation of the Consultative Group on International Agricultural Research, now known as CGIAR (see Box 17.1).

Beginning in the mid-1980s, studies conducted over the next two decades broadly demonstrated high social returns to agricultural R&D in both productivity growth (Alston et al. 2000) and poverty reduction (Fan and Hazell 2001). These findings were based on a wide range of more context-specific studies on the Green Revolution (Hazell and Ramaswamy 1991), natural resource management (Pender 2005), and new varieties, agronomic practices, and other technological advances (Pingali 2022). This work demonstrated the continued importance of public expenditure on the production of public goods for food and agriculture, despite persistently low and volatile public spending by LMICs (Figures 17.1 and 17.2). Unfortunately, this scholarship and evidence did not prevent a precipitous decline in funding for agricultural R&D, which began during the same period due to donor fatigue and declines in national budgets and official development assistance (see Chapter 18) (Byerlee et al. 2002).

FIGURE 17.1 Agricultural research spending by income group, 1981–2016

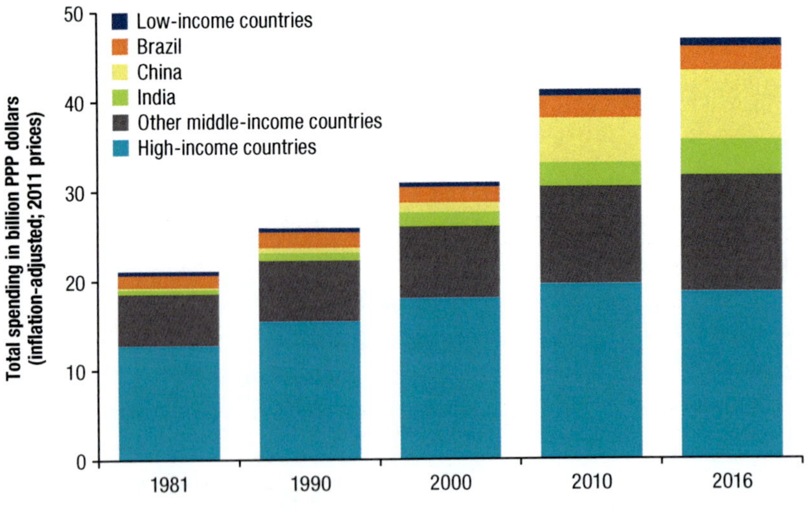

Source: Beintema, Nin-Pratt, and Stads (2020).

BOX 17.1 Mission and structure of CGIAR

CGIAR, known as the Consultative Group on International Agricultural Research when it was created in 1971, is a funding and coordination mechanism for a diverse system of 15 international agricultural research centers (IARCs) focusing largely on the tropics and subtropics. Six IARCs were established before CGIAR's founding, beginning with the International Rice Research Institute in Los Baños, the Philippines, in 1960. Because the first IARCs were closely associated with the Green Revolution that emerged in the late 1960s, these centers were arguably the 20th century's major institutional innovation for foreign assistance to support agricultural development and food security.

In its current structure, CGIAR's mission is to deliver science and innovation that advance the transformation of food, land, and water systems in a climate crisis. The IARC model that emerged in the 1960s, and still largely holds, had four core characteristics. First, IARCs were set up as "centers of excellence" to carry out multidisciplinary research that would generate technology and knowledge spillovers across countries and aimed to create efficiencies through economies of scale and scope. Second, in the medium term, they aimed to compensate for weaknesses in many developing country national research systems while, over the long term, strengthening the capacity of those systems through in-service training and graduate training, often linked to a nearby university. Third, they aimed to reduce political and bureaucratic interference in science by operating as autonomous nongovernmental entities with their own independent but internationally representative boards.

Finally, IARCs were to receive long-term funding from multiple sources, largely from richer countries through their official foreign assistance and philanthropic organizations, complemented by contributions from low- and middle-income and host governments. Agricultural research that would benefit many countries was seen to be a meaningful way for these donor organizations to realize their humanitarian and political objectives. Reflecting the increased understanding of the importance of policies in fostering and scaling innovations, IFPRI was established as one of the IARCs and joined CGIAR in 1979. The IARCs inherited a proven model of crop research but have struggled with a model for natural resources management, as well as stable funding.

Source: Byerlee and Lynam (2020).

FIGURE 17.2 Growth in agricultural research spending by income group, 1981–2016

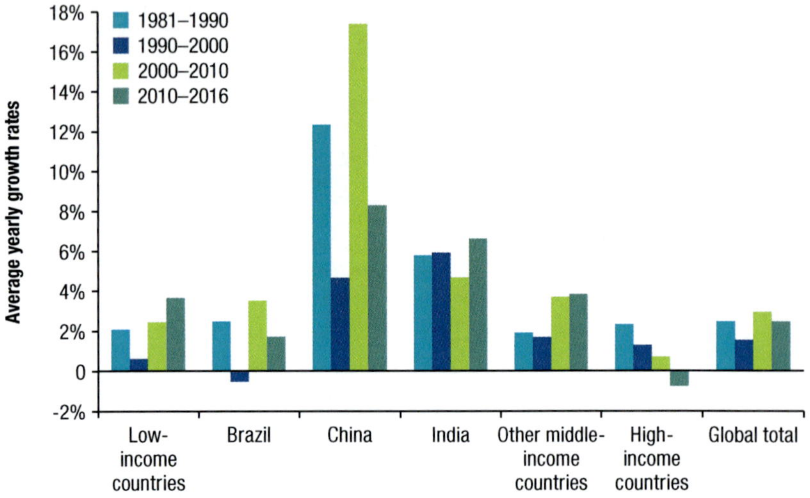

Source: Beintema, Nin-Pratt, and Stads (2020).

Nevertheless, this body of work helped build a set of analytical tool-kits on R&D resource allocation that LMIC researchers still rely on today. The first external program review of IFPRI in 1985 acknowledged that the "allocation of research resources is an elusive problem for which the methodology has not yet been established" (CGIAR TAC 1985, 7). With the publication of *Science under Scarcity* in 1995, IFPRI, together with Julian Alston and George Norton, produced the standard methodological reference on agricultural research evaluation, priority setting, and resource allocation (Alston et al. 1995).

Spatial analysis became a principal tool as the focus of agricultural research shifted from irrigated systems, which were central to the early Green Revolution, to the extensive rainfed farming systems across the developing world. Spatial analysis addressed the need to integrate the diversity and heterogeneity of these farming systems into research priority-setting in response to urgent global sustainability objectives. A good example comes from the International Maize and Wheat Improvement Center (CIMMYT), which in the late 1980s used spatial analysis to defined "mega-environments" (Hartkamp et al. 2000) intended to inform resource allocation in maize research. This approach was linked to the growing capacity within CGIAR's farming systems research, with its focus on smallholder producers

(Collinson 2000). Following on this work, IFPRI's Spatial Analysis Group, formed in 2003, helped in characterizing the research context, differentiating recommendation domains for genetic and natural resource management technologies, and subsequently targeting delivery. Based on identification of similar agroecological zones and farmer systems spanning countries, researchers defined homogeneous production domains. These domains facilitate the development of regional frameworks used to organize research programs that can realize regional spillovers of technology and knowledge (as an example, see Johnson et al. 2011).

IFPRI continues this research today, with the use and improvement of data resources, such as the Spatial Production Allocation Model (SPAM); econometric approaches, such as stochastic frontiers analysis and data envelope analysis; and computational models, such as the Dynamic Research Evaluation for Management (DREAM) model, the International Model for Policy Analysis of Agricultural Commodities and Trade (IMPACT), and the Rural Investment and Policy Analysis (RIAPA) model.[2] These tools are widely used to help policy analysts inform governments and funders about the contribution of agricultural R&D to productivity growth and poverty reduction, and to identify appropriate policy instruments to effect these changes. The tools are also central to the work of the Food and Agriculture Organization of the United Nations (FAO), many US land grant universities, and partners throughout the Global South. Notable examples include IFPRI's partnership with the Association for Strengthening Agricultural Research in Eastern and Central Africa (ASARECA), which analyzed strategic priorities for agricultural development in eastern and central Africa by combining spatial data and analysis with a multimarket simulation model. The model was used to explore options to spur productivity growth, strengthen agricultural markets, and enhance links between economic sectors and countries (Lynam 2016; Omamo et al. 2006). This study provided an empirical and analytical foundation for relatively high-level priority-setting discussions within ASARECA, which then served as a model for the other subregional organizations and a basis for the World Bank's multi-donor trust fund that supports regional R&D frameworks, such as the East Africa Agricultural Productivity Program (Lynam 2016). These resources are, in short, a key asset of IFPRI's research portfolio.[3]

---

2 For more on these models, see www.ifpri.org/modeling-tools/

3 For more on this, see Alston et al. (2000), Fan et al. (2005), Nin-Pratt and Magalhaes (2018), and Pauw and Thurlow (2015).

This work on priority setting was augmented by IFPRI's Agricultural Science and Technology Indicators (ASTI) initiative, a long-term program that collected, analyzed, and shared internationally comparable time-series datasets on agricultural R&D investment and human research capacity in LMICs. Earlier research by IFPRI and partners focused on capacity, essentially in terms of staffing and budget allocation, particularly the primacy of recurrent staff costs over operational funds. Recent research by IFPRI and partners has expanded to link expenditure and staffing capacity to research outputs and outcomes, including agricultural productivity growth (for example, Nin-Pratt et al. 2023). Today, ASTI's work continues through the support of FAO, ensuring that the initiative's data continue to be collected, analyzed, and accessible over the long term, and numerous studies continue to use the ASTI dataset.[4] However, funding remains a problem, despite widespread use and recognition of ASTI's value.

Systems thinking and broader development objectives

Policymakers often need innovative policy strategies and instruments to organize diverse actors around efforts to solve multiple problems. In agriculture, such problems include expanding the market and value chain for a particular crop or commodity or accelerating the widespread uptake of climate-resilient technologies and practices to bolster farmers against droughts, floods, heat, or other adverse weather events and improve their well-being. This task is difficult for policymakers to manage, especially for governments steeped in traditional top-down and centralized approaches to policy design and implementation.

At the center of this work is the recognition that technical change in agriculture is not simply a function of R&D investment. Rather, technical change is part of a broader process of innovation that occurs among a set of heterogeneous actors whose actions and interactions are shaped by social, economic, and political institutions, both formal and informal. This systems perspective continues to address the same productivity growth–poverty reduction challenge as the work of economists and other scientists described in the previous section, but expands to incorporate analysis of organizational and institutional challenges as well as policy options. Novel policy instruments include forming or funding various structures for coordination and

4 Examples include Alston and Pardey (2021), Carden et al. (2019), and Fuglie et al. (2022).

FIGURE 17.3 Innovation process within a changing context in LMICs

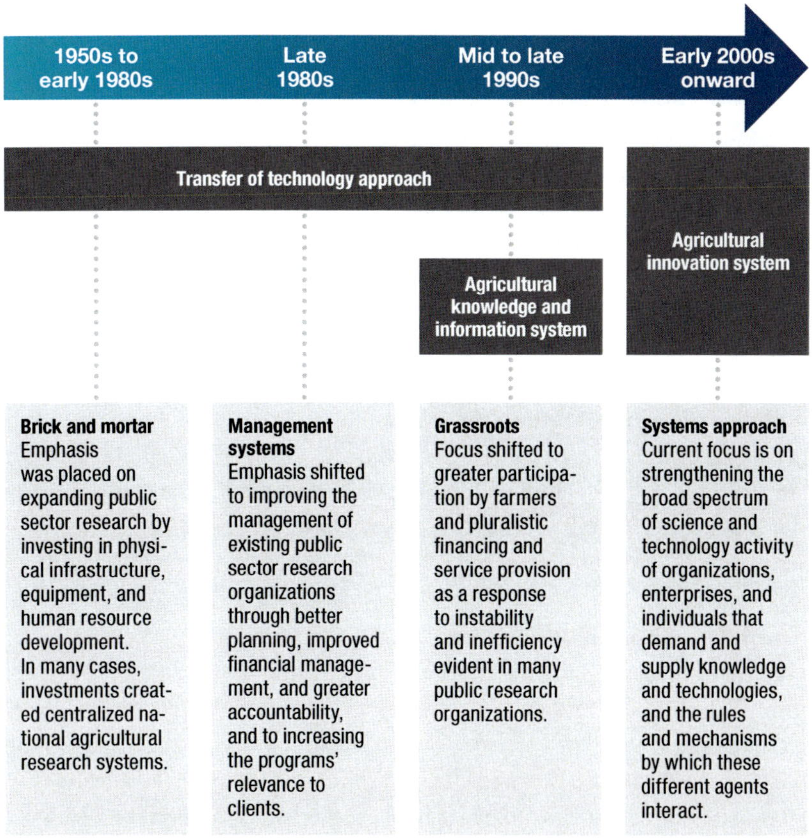

Source: Based on World Bank (2012).

inclusive processes; decentralization, devolution, or de-concentration policies; and reforms of public organizational structures or incentive systems. Policy research in this area includes documenting and assessing the effectiveness, sustainability, and impact of these structures on innovation processes and development outcomes to provide concrete recommendations on how to design, support, enable, and catalyze them.

Understanding of the innovation process has evolved substantially over the last seven decades in LMICs (Figure 17.3). Beginning in the 1950s, the operative model of agricultural innovation was the "transfer of technology" approach (Eicher 1984). In simple terms, this approach described a linear process in which (1) universities trained researchers, (2) the researchers, based primarily, in national agricultural research systems developed technologies, and (3) extension services delivered technologies to farmers, resulting in productivity gains, food security improvements, and other desirable outcomes. The approach relied on scientific contributions from universities in industrialized countries. These contributions were adapted by international agricultural research organizations such as CGIAR, tested by national agricultural research systems in LMICs, and distributed to farmers by their extension agencies (see Chapter 8).

Introduction of the concept of agricultural knowledge and information systems (GFRAS 2012), which provided a more nuanced and nonlinear perspective on the subject, lent greater sophistication to understanding of the innovation process. Gaining popularity in the mid to late 1980s, this concept placed greater emphasis on the roles and coordination of research-extension-education actors and gave greater recognition to farmers' participation in the innovation process. The development of farmer participatory research and demand-led breeding shifted the focus from supply of technology to farmer demand and, in turn, to farmer innovation in adapting technologies within the farming system.

Until the 1970s, public investment largely focused on infrastructure and human resource development in national agricultural research organizations (Figure 17.3). Making meaningful changes to innovation processes, however, also requires some level of scientific, technical, and managerial capacity among national actors to ensure flows of human capital, information, and technology. Consequently, much of the policy research in the 1980s focused on understanding and improving the organizational capacity of public research organizations. Research also focused on the relative investment between R&D and extension and on improving institutional links between the two, especially with the World Bank's funding of training and visit extension. Partly because of the structural adjustment programs of the 1980s, which led to reductions in public spending on government services in the agriculture sectors of many LMICs, attention shifted to the exploration of organizational and institutional alternatives, including decentralization, privatization, contracting, outsourcing, and public–private partnerships (see Chapters 8 and 15). During this period of more than a decade, agricultural innovation was shunted to the margins of economic policy. Both public policy and budget resources were directed toward broader poverty reduction programs and shifted focus from agricultural development to rural development, which encompasses programming in health, education, and other social sector interventions.

It was not until the early 2000s that greater attention was paid to agricultural innovation, this time with a new conceptualization—the agricultural innovation system—inspired by research conducted primarily in the European industrial sector (Hall 2005; Klerkx et al. 2012; World Bank 2012). The agricultural innovation system concept has gained significant popularity, becoming mainstream in science, policy, and practice (Klerkx and Cremaschi 2025). Its central theme—the need to organize and coordinate across multiple actors and processes for greater market participation and social impact—has produced a multifaceted research agenda as well as practical methodologies to evaluate various institutional arrangements and partnerships. Examples include public–private partnerships, farmer collectives, innovation networks, multistakeholder platforms, innovation brokers, and clusters (Maryono et al. 2024; Spielman et al. 2012). These arrangements are designed to coordinate different agents within agricultural markets and reduce transaction costs in market participation. Multistakeholder processes and coordination structures are also developed around sustainability, agroecology, and nutrition (Maryono et al. 2024).

Multistakeholder innovation platforms have been mainstreamed in the largest sustainable intensification program in sub-Saharan Africa (Africa RISING), included in Uganda's national agricultural policy, and used in numerous agricultural projects. Similar approaches have been taken in South Asia through the Cereal Systems Initiative for South Asia (CSISA) and its predecessor, the Rice-Wheat Consortium. Popular market-oriented innovation platforms have been implemented in several programs in Latin America, such as the Papa Andina Project (Devaux et al. 2011). More recently, there has been stronger emphasis on sustainable development (nutrition, gender and social inclusion, climate resilience, agroecology, and sustainability), with stronger voices from social movements globally and with the emergence of new or renamed coordination structures, such as food systems innovation hubs, living labs for people, and transformation labs. Numerous studies show mixed results on the effectiveness, sustainability, and development impact of these approaches, owing to the wide diversity of their structure, degree of formality, objectives, size, and economic and social context.[5] Success stories include the positive large-scale impacts of China's Science and Technology Backyard (An et al. 2024; Zhang et al. 2016), and this empirical evidence can be drawn on to provide comparative and practical lessons on how to invest and support the sustainability and impact of these structures.

5 For example, see Davies et al. (2018), Maryono et al. (2024), and Schut et al. (2018).

IFPRI's work on this new approach dates to the mid-2000s, following its 2004 merger with the International Service for National Agricultural Research (ISNAR). Notable contributions include conceptual work on innovation systems (Spielman 2005), agricultural education and training (Davis et al. 2007), pluralistic extension systems (Birner et al. 2009; Davis et al. 2020; Spielman et al. 2021), private sector innovation (Naseem et al. 2010; Zhou and Babu 2015), organization of national research systems (Babu et al. 2015; Ragasa 2016), and measurement (Spielman and Birner 2008; Spielman and Kelemework 2009). Over the years, these contributions have been complemented by the work of many other organizations—Wageningen University & Research, FAO, the FAO–Agrinatura partnership on Capacity Development for Agricultural Innovation Systems, the Technical Centre for Agricultural and Rural Cooperation, the U.S Agency for International Development's Feed the Future Innovation Labs, the World Bank, the Institute of Development Studies, and other CGIAR Centers—and agriculture innovation systems now constitute a core theme of the CGIAR portfolio (Schut et al. 2024). Emerging assessments of the governance of diverse agricultural innovation systems highlight the need for continued and expanded research on this topic (for example, see Chen et al. 2024; Turner et al. 2020).

Scaling for impact

During the past 10 years, the innovation systems literature has rapidly evolved to address a fundamental question that continues to plague governments and their development partners: why have so many promising innovations failed to be adopted at scale, despite being successfully tested and piloted? These last-mile problems remain an enigma, as does the lack of novel policy options. Randomized, credible impact evaluations of interventions or innovations at small scale could play a critical role in informing scaling (Duflo 2004), yet few such evaluations exist for the agriculture sector. Among the available successful pilots from these randomized evaluations, there is limited evidence that they are scaled up or institutionalized at national or local levels.[6]

Effective scaling is often facilitated by a supportive policy environment that provides incentives for technology adoption and innovation. This bundling of policy change, technology options, delivery methodologies, market

6 This is explored by Abate et al. (2023), Gebreyes et al. (2021), Schut et al. (2020), Sinclair and Coe (2019), Spielman et al. (2021), and Stevenson et al. (2019).

links, and coordination mechanisms has become central to the scaling process, with interventions being piloted and scaled increasingly packaged as "socio-technical innovation bundles" (Barrett et al. 2022) or scaling-ready innovations (Schut et al. 2024). A critical consideration in scaling is the heterogeneity of actors and contexts. Planning within such heterogeneity has marked a return to geospatial analysis, complemented by "large data" and farm typology development (Hassall et al. 2023). Scaling strategies also need to strike a balance between globally scaled practices and local solutions, providing a middle path that avoids the inefficiencies of overly small-scale systems while steering away from the homogenization that global solutions often impose (Barrett et al. 2022). The economics or benefit–cost profiles of innovations often vary across scales (see List 2024), reinforcing the need for iterative impact evaluations at different scales. Further conceptualizations and frameworks on the role of economics in scaling are emerging, and applications in various sectors and programs are ongoing (al-Ubaydli et al. 2021; Halperin et al. 2022), but the economic concept for scaling science in agriculture has not caught up. For several decades, numerous donor agencies and international nongovernmental organizations set up structures and communities of practice and also documented lessons on scaling up development programs. Many scaling frameworks have been proposed (for example, Lam et al. 2020). Emphasis on scaling has become omnipresent within the development community, although many organizations and programs still largely focus on the reach and adoption of technologies and not on sustainability, strengthening of local institutions, or the systems change required for achieving impacts at scale (Woltering et al. 2019). Socio-technical innovation bundling has also been narrowly applied to bundling of inputs and services provided to potential beneficiaries (for example, Banerjee et al. 2024; Boucher et al. 2024; Mishra et al. 2023), rather than to finding solutions through policy, institutional, or systems change.

Within the agrifood system, CGIAR institutionalized the scaling approach with its Scaling for Impact Program. Between 2022 and 2024, CGIAR and partners co-created approximately 100 innovation packages, updated annually, organized in a global Performance and Results Management System, and made publicly available through the CGIAR Results Dashboard. The science of scaling work was initiated under the CGIAR Research Program on Roots, Tubers and Bananas, which published a special issue on the topic in the journal *Agricultural Systems* (Schut et al. 2020). Frameworks and management tools were made available to guide teams in examining their innovations and implementing their scaling strategies (examples include Scaling Scan, Scaling Readiness, and GenderUp, among others; see compilation by Woltering et al. 2019). These tools offer rapid assessment of the scaling process and rely on expert opinion or a few key informants to provide ratings and scores of the different dimensions of scaling. Although they provide generic guidance on scaling pathways, these tools do not provide specific illustrations and guidance on what, where, how, or when to scale, nor do they inform on the ex ante and ex post impact evaluation of innovations, learnings on scaling successes and failures, or comparative cost-effectiveness of alternative scaling approaches. These are major gaps in knowledge and measurement in the scaling science literature. Documentations of lessons, barriers, and accelerators in scaling agricultural innovations are emerging (Barrett 2023; Herrero et al. 2021; Thornton et al. 2024), and these need to be expanded and compared to provide practical guidance for practitioners and decision-makers.

At IFPRI, researchers have contributed to the thinking on scaling pathways and documentation of lessons. In 2011, IFPRI's 2020 Vision Initiative produced a set of policy briefs that contributed to a better understanding of scaling up in agriculture, rural development, and nutrition around the world. These briefs delineate different pathways for scaling up initiatives, identify the key drivers that advance the scaling-up process and the key spaces that enable scaling up, and outline lessons learned (Linn 2012). IFPRI subsequently produced various assessments of agricultural technologies with potential for scaling (for example, Rosegrant et al. 2014); documented lessons on scaling up nutrition programs (for example, Gillespie et al. 2015) (see Chapter 12); illustrated methods to estimate heterogenous marginal effects of technologies and extrapolate effects from plot to national scale to inform scaling strategies or target beneficiaries (for example, Bedi et al. 2022; Komarek et al. 2019); documented lessons for scaling up group-based experimental games as institutional innovations in natural resource management (Meinzen-Dick et al. 2024) (see Chapters 5 and 6); and enhanced capacity and policy dialogues for scaling up natural resource management (for example, the Consortium for Scaling-up Climate Smart Agriculture in South Asia) (see Chapter 5).

HarvestPlus, an IFPRI program with a long history of partnership with CIAT and other CGIAR Centers, emphasized scaling strategies to maximize the global impact of biofortified crops (Box 17.2). Complementary to the collaborative breeding efforts, scaling strategies were developed; studies on the socioeconomic considerations and on nutritional sciences, led by IFPRI researchers with global partners, were foundational to the successful scaling of biofortification in various countries. For other biological innovations, such

BOX 17.2 IFPRI's programs on biofortification and biosafety: Integrated approaches

Biofortification program

The term "biofortification" refers to a plant breeding strategy to increase the micronutrient density of staple food crops and reduce the burden of micronutrient deficiencies in low- and middle-income countries. Initiated in 2003 and housed by IFPRI, the HarvestPlus biofortification program has invested approximately US$500 million in developing, testing, and scaling up biofortified varieties of 12 staple food crops (Bouis et al. 2024). Over 20 years, this relatively modest investment has enriched the diets of millions of people. At a high level of adoption, the benefits of biofortification accrue at little marginal cost, rendering it a highly cost-effective intervention to scale (Lividini et al. 2018).

The rapid scaling of biofortification has been facilitated by a body of evidence proving the efficacy and effectiveness of biofortified crops in improving micronutrient status and health, a rigorous monitoring approach, and a scaling strategy that introduces newly released crops into sustainable and commercial seed, grain, and food systems for broad consumption. HarvestPlus organized its activities under a coordinated, interdisciplinary program spanning advanced genetics, plant breeding, food processing, human nutrition, policy, and more. It works with decision-makers to embed biofortification in national, regional, and multilateral policies and programs; set global standards and targets for seed and grain nutrient concentrations; enable actors along the value chain and promote commercialization to stimulate public and private sector procurement and trade; and develop context-specific partnership approaches for fast diffusion and sustained adoption.

Biosafety program

Biotechnology adoption in developing economies initially progressed slowly in the 1990s due to trade, regulatory, and policy challenges. A significant shift began in 2003 when Brazil joined Argentina as leaders in adoption of genetically modified (GM) crops, with these two countries now accounting for 40 percent of the global GM-planted area. Latin America's GM adoption remains uneven, up from 6 countries in 2003 to 10 today, while Southeast Asia has 9 adopters out of 32 economies. In sub-Saharan Africa, only 6 of 48 economies plant GM crops, but notable progress in the last five years suggests a promising shift.

The policy landscape has evolved quickly, catalyzed by the advent of next-generation biotechnology tools, such as genome editing, that offer the promise of enhanced breeding speed and efficiency. The evolving policy and technology adoption landscape in some African and Asian countries may be traced to several contributing factors, including capacity-strengthening activities offered by different entities and programs, such as IFPRI's Program for Biosafety Systems. These organizations have provided policymakers and stakeholders with sustained technical assistance, advising on rigorous science-based evaluation and approval processes to protect human health and the environment, and capacity-building projects, including tailored, locally led economic research and ex ante and ex post impact assessment of biotech innovations (such as Dzanku et al. 2018; Gouse et al. 2016; Ruhinduka et al. 2020), as well as studies on the cost of regulations (such as Falck-Zepeda et al. 2012; Falck-Zepeda and Gouse 2017), among others. The ultimate goal is to expand the number of safe and available biotech products that meet farmers' needs, making it essential to expand research on biotech innovations, with a particular focus on integrative and coordinated policy, governance, and scaling strategies.

Source: Inputs provided by IFPRI staff members Judy Chambers, Jose Falck-Zepeda, Patricia Zambrano, Jen Foley, and Rewa Misra.

as genetically modified organisms, IFPRI's Program for Biosafety Systems has contributed to the enabling technology and regulatory policy landscapes through sustained country-level technical assistance support, intensive capacity building, and an integrated approach to policy development (combining work on technical, intellectual property, legal, political, and communication issues), supported by a whole-of-government strategy that empowers local leaders and voices across the entire product development cycle. Greater impact of both innovations will directly depend on the further evolution of an enabling policy climate that recognizes and respects local context; generation of credible and broad-based social, political, technical, behavioral, and economic evidence; and communication of this evidence widely and effectively.

Outcomes

IFPRI's contribution to the economics of agricultural R&D has played an important role in sustaining public investment from LMIC governments and from their multilateral, bilateral, and philanthropic development partners and funders (Lynam 2016). Many of the models and tools developed by IFPRI are used today to guide research prioritization and resource allocation to achieve greater efficiency and impact in scaling of R&D investments.

IFPRI's work has also helped to create greater appreciation for the complexity of innovation systems and processes, which challenges the short-term, results-based program management that tends to eclipse other approaches in the development community. There are also signs of renewed emphasis on capacity development and sharing, at both individual and institutional levels, as seen in CGIAR's Capacity Sharing Accelerator program and other initiatives. This trend also demonstrates fundamental changes in strategy at the World Bank, FAO, and CGIAR.[7]

Over the past decade, IFPRI's work has helped advance organizational innovation in the agricultural space. Examples include the introduction of formal entities to better connect actors in research, extension, farming, and industry, including research-extension-farmer-industry linkage systems (REFILS) or district agricultural extension services system (DAESS); more permissive regulatory regimes that encourage public research engagement with the private sector, such as India's National Agricultural Innovation Policy and innovation platforms in Uganda's National Agricultural Policy; and innovative partnerships among actors around a particular innovation opportunity (see examples in Spielman et al. 2012).

Momentum appears to be building around the "decolonization of research" in the agricultural innovation space—that is, a shift in research leadership toward the Global South—and in the rise of national and regional organizations in the Global South, such as AGRA, Akademiya2063, ANAPRI, and a range of policy think tanks and networks in South Asia. The expanding role of IFPRI's regional and country offices is also part of this shift. A parallel shift is underway in the biological sciences—the introduction of participatory plant breeding and participatory varietal selection, which has advanced into citizen science and crowdsourcing approaches that draw on information and communication technology as well as big data to collect and analyze huge amounts of data from farmers. These data-driven, participatory, and transdisciplinary approaches are enabling the testing and evaluation of new varieties to reach more diverse farmers and farming systems faster and make it less costly to accelerate the large-scale adoption of these varieties (Occelli et al. 2024; van Etten et al. 2023).

Still, many scholars and experts have advocated for even greater investment in development of organizational and systemwide capabilities for agricultural innovation (see Hall 2005; Jayne et al. 2023; Klerkx et al. 2012). They have emphasized that although the scientific and technical capacities of individuals in national research organizations have improved over time, organizational

7 For more on this, see World Bank (2012), FAO (2022), TAP (2016), and Schut et al. (2024).

capacities in governance and management have not kept pace (Jayne et al. 2023). The following topics have been highlighted for greater learning and investment: designing and managing transdisciplinary research teams (Forsythe et al. 2024), building sustainable multistakeholder platforms and networks for agricultural innovation (Brouwer et al. 2019), integrating better and more systematic market intelligence into breeding, research, and scaling strategies (Donovan et al. 2022; Rice et al. 2024), and developing and sharing capacity more effectively (see CGIAR Program on Capacity Sharing).

Future research agendas

Our review points to three future research agendas. The first reflects the significant expansion in development objectives for which agricultural research investments are now held accountable (see Chapter 3). More work is needed to evaluate and model this wider set of development outcomes that address the complex realities facing governments today: how to maintain or increase the pace of economic growth and poverty reduction while also addressing the trade-offs associated with climate change adaptation and mitigation, environmental sustainability, the root causes of poor nutrition and health outcomes, and the problems of social inequality, exclusion, and marginalization. New methods and integrated frameworks will be required to better assess multiple policy objectives and better inform resource allocation and prioritization in the context of competing development goals. Driving investment decisions and informing resource allocation will require the systematic integration of spatial analysis, farm models, simulation, sectoral or economywide models using household data, and new approaches to measuring the research cost of innovations into a new framework. More spatially relevant (subnational, local) micro-level data on adoption and productivity impacts of relevant technology innovations and practices are much needed, as they are crucial parameters in macro–micro simulation modeling scenarios that inform the prioritization of resource allocation and assess impacts. Data and analysis on the costs of regulatory and institutional barriers and solutions would also be useful to integrate into investment planning and impact modeling.

The second research agenda focuses on best-fit approaches to scale solutions for agronomy and natural resource management. Although genetic innovations through improved varieties are much easier and more straightforward to deliver and scale widely (primarily through market mechanisms), innovations in systems agronomy or natural resource management in particular are inherently complex, nonlinear, and involve a mix of public and private institutions (see Chapters 5 and 8). Scaling strategies will need to take this reality into account. What may look to the scientist like a context-specific application of well-established principles may represent a difficult set of choices for the farmer. Both genetic innovation and seed choices will need to be evaluated along with agronomic options, within the economic, social, and cultural context of the farmer. Strategic alliances between research organizations and development organizations could help build research on development projects that invest in enabling conditions for uptake of natural resource management practices. Such collaborations often occur in pilot programs, which are typically implemented in artificial contexts. Getting a more representative picture, therefore, will require moving toward collaboration in large-scale programs, especially those implemented by governments. Large-scale impact requires wide adoption of evidence-based innovations (guided by a transdisciplinary approach and iterative evaluations), which necessitates generating innovations suitable for a wide range of contexts and understanding which innovations are suitable for which contexts. A central research focus would explore how relevant institutions differentiate and target their programs and coordinate activities with other institutions.

The third research agenda involves rigorous impact evaluation of technological, social, and institutional innovations at different scales, and in-depth analysis of the political economy and enabling environment that drive or support agricultural innovations and scale their impacts. Many institutional innovations (organization and management practices, incentive systems, formal structures, and platforms) are being funded and piloted; rigorous evidence is needed on how they actually work and how they are sustained so that we can improve their design and make these structures more efficient and effective. Although substantial external support may lead to initial success, regular monitoring and assessment of these innovation platforms and hubs, as well as strategies for their sustainability, will need to be identified and put in place. For the numerous socio-technical bundles also being piloted, rigorous evidence is needed on how best to combine different interventions and elements of a bundle, and how to strike a balance between global or national best practices and locally relevant solutions. Iterative assessments, using participatory approaches as well as innovative and rigorous methods, need to be expanded to evaluate the cost-effectiveness and impacts of agricultural innovations and socio-technical innovation bundles across diverse contexts. Emerging assessments of the governance of agricultural innovation systems need to be expanded to improve understanding of the politics and power dynamics of innovation processes (see Chapter 15), to address power imbalances and improve inclusion (see Chapter 14), and to identify innovative options for budgeting and financing the costs of institutional links and transactions (see Chapter 18). The "how" question will be a necessary focus of future research on agricultural innovation systems and the enabling environment for scaling impacts, with associated requirements for improved analytical methodologies.

Future policy research agendas will need to focus on assessing and designing new strategies for bringing innovative solutions to farmers' fields and achieving sustainable development impacts. Such agendas could include partnership-based scaling models and socio-technical bundles, proactive capacity development for national partners, and a data-driven approach to enhance the enabling environment for public policies and regulations that can facilitate these models and processes—all of which will require research based on transdisciplinary approaches and partnerships with national and local researchers. It is imperative that IFPRI researchers from various disciplines work closely with economists, nutritionists, natural biophysical scientists, sociologists, environmental experts, and legal and political scientists from other organizations to bring to the forefront the best solutions for the many complex challenges facing the world.

References

Abate, G.T., K.A. Abay, J. Chamberlin, Y. Kassim, D.J. Spielman, and Tabe-M.P. Ojong Jr. 2023. "Digital Tools and Agricultural Market Transformation in Africa: Why Are They Not at Scale Yet, and What Will It Take to Get There?" *Food Policy* 116 (2023): 102439.

Alston, J.M., C. Chan-Kang, M.C. Marra, P.G. Pardey, and T.J. Wyatt. 2000. *A Meta-Analysis of the Rates of Return to Agricultural R&D: Ex Pede Herculem?* IFPRI Research Report 113. Washington, DC: IFPRI.

Alston, J.M., G.W. Norton, and P.G. Pardey. 1995. *Science under Scarcity: Principles and Practice for Research Evaluation and Priority Setting.* Ithaca, NY: Cornell University Press.

Alston, J.M., and P.G. Pardey. 2021. "The Economics of Agricultural Innovation." In *Handbook of Agricultural Economics*, vol. 5, eds. C.B. Barrett and D.R. Just, 3895–3980. Amsterdam: Elsevier.

Al-Ubaydli, O. 2021. "How Can Experiments Play a Greater Role in Public Policy? Twelve Proposals from an Economic Model of Scaling." *Behavioural Public Policy* 5 (1): 2–49.

An, Z., Y. Yang, X. Yang, et al. 2024. "Promoting Sustainable Smallholder Farming via Multistakeholder Collaboration." *PNAS* 121 (21): e2319519121.

Babu, S.C., J. Huang, P. Venkatesh, and Y. Zhang. 2015. "A Comparative Analysis of Agricultural Research and Extension Reforms in China and India." *China Agricultural Economic Review* 7 (4): 541–572.

Banerjee, M., P. Mukhopadhy, D. Chadha, K. Goel, H.N. Gartaula, and R. Puskur. 2024. *Socio-Technical Innovation Bundles (STIBs) for Enhancing Women's Empowerment and Resilience: A Synthesis.* New Delhi: CGIAR Initiative on Gender Equality, CGIAR Gender Impact Platform, and International Rice Research Institute.

Barrett, C.B. 2023. "The Political Economy of Bundling Socio-Technical Innovations to Transform Agri-Food Systems." In *The Political Economy of Food System Transformation*, eds. D. Resnick and J. Swinnen, 206–229. Oxford, UK: Oxford University Press.

Barrett, C.B., T. Benton, J. Fanzo, et al., eds. 2022. *Socio-Technical Innovation Bundles for Agri-Food Systems Transformation.* Cham, Switzerland: Springer International Publishing.

Bedi, S.M., C. Azzarri, B.H. Kotu, L. Kornher, and J. von Braun. 2022. "Scaling-Up Agricultural Technologies: Who Should Be Targeted?" *European Review of Agricultural Economics* 49 (4): 857–875.

Beintema, N., A. Nin-Pratt, and G.-J. Stads. 2020. *ASTI Global Update 2020.* Agricultural Science and Technology Indicators (ASTI) Program Note (September). Washington, DC: IFPRI.

Birner, R., K. Davis, J. Pender, et al. 2009. "From Best Practice to Best Fit: A Framework for Designing and Analyzing Pluralistic Agricultural Advisory Services Worldwide." *Journal of Agricultural Education and Extension* 15 (4): 341–55.

Boucher, S. R., M.R. Carter, J.E. Flatnes, T.J. Lybbert, J.G. Malacarne, P. Mareyna, and L.A. Paul. 2024. "Bundling Genetic and Financial Technologies for More Resilient and Productive Small-Scale Farmers in Africa." *The Economic Journal* 134 (662): 2321–2350.

Bouis, H., J. Foley, K. Lividini, et al. 2024. "Biofortification: Future Challenges for a Newly Emerging Technology to Improve Nutrition Security Sustainably." *Current Developments in Nutrition* 8 (12): 104478.

Brouwer, H., J. Woodhill, M. Hemmati, K. Verhoosel, and S. Van Vugt. 2019. *The MSP Guide: How to Design and Facilitate Multi-Stakeholder Partnerships.* Rugby, UK: Practical Action Publishing.

Byerlee, D., G. Alex, and R.G. Echeverrıa. 2002. "The Evolution of Public Research Systems in Developing Countries: Facing New Challenges." In *Agricultural Research Policy in an Era of Privatization*, eds. D. Byerlee and R.G. Echeverrıa, 19–34. Oxon, UK: CABI.

Byerlee, D., and J. Lynam. 2020. "The Development of the International Center Model for Agricultural Research: A Prehistory of the CGIAR." *World Development* 135: 105080.

Carden, F., N.M. Beintema, A. Admassie, L. Katera, T. Mboghoina, and C. Onyekwena. 2019. *Informing Policy with Agricultural R&D Evidence: An ASTI Pilot Project in Ethiopia, Nigeria, and Tanzania*, vol. 1860. Washington, DC: IFPRI.

CGIAR TAC (Technical Advisory Committee). 1985. *Report of the External Program Review of the International Food Policy Research Institute.* Rome: TAC Secretariat.

Chen, C., J. Gao, H. Cao, and W. Chen. 2024. "Unpacking the Agricultural Innovation and Diffusion for Modernizing the Smallholders in Rural China: From the Perspective of Agricultural Innovation System and Its Governance." *Journal of Rural Studies* 110: 103385.

Collinson, M.P., ed. 2000. *A History of Farming Systems Research.* Wallingford, UK: CABI.

Davies, J., Y. Maru, A. Hall, et al. 2018. "Understanding Innovation Platform Effectiveness through Experiences from West and Central Africa." *Agricultural Systems* 165: 321–334.

Davis, K., S. Babu, and C. Ragasa. 2020. *Agricultural Extension: Global Status and Performance in Selected Countries.* Washington, DC: IFPRI.

Davis, K., J. Ekboir, W. Mekasha, C.M.O. Ochieng, D.J. Spielman, and E. Zerfu. 2007. "Strengthening Agricultural Education and Training in Sub-Saharan Africa from an Innovation Systems Perspective: Case Studies of Ethiopia and Mozambique." IFPRI Discussion Paper 736. IFPRI, Washington, DC.

Devaux, A., M. Ordinola, and D. Horton, eds. 2011. *Innovation for Development: The Papa Andina Experience.* Lima, Peru: International Potato Center.

Donovan, J.A., P. Coaldrake, P. Rutsaert, et al. 2022. *Market Intelligence for Informing Crop-Breeding Decisions by CGIAR and NARES.* Market Intelligence Brief 1. Montpellier, France: CGIAR.

Duflo, E. 2004. "Scaling Up and Evaluation." In *Annual World Bank Conference on Development Economics,* eds. F. Bourguignon and B. Pleskovic, 341–369. Washington, DC: World Bank; New York: Oxford University Press.

Dzanku, F.M., P. Zambrano, U. Wood-Sichra, J.B. Falck-Zepeda, J.A. Chambers, H. Hanson, and P. Boadu. 2018. "Adoption of GM Crops in Ghana: Ex Ante Estimations for Insect-Resistant Cowpea and Nitrogen-Use Efficient Rice." IFPRI Discussion Paper 1775. IFPRI, Washington, DC.

Eicher, C.K. 1984. "International Technology Transfer and the African Farmer: Theory and Practice." Working Paper 3/84. Department of Land Management, Harare, Zimbabwe.

Falck-Zepeda, J.B., and M. Gouse. 2017. "Regulation of GMOs in Developing Countries: Why Socioeconomic Considerations Matter for Decision Making." In *Genetically Modified Organisms in Developing Countries: Risk Analysis and Governance,* eds. A.A. Adenle, E.J. Morris, and D.J. Murphy, 91–102. Cambridge, UK: Cambridge University Press.

Falck-Zepeda, J.B., J. Yorobe Jr., B.A. Husin, et al. 2012. "Estimates and Implications of the Costs of Compliance with Biosafety Regulations in Developing Countries." *GM Crops & Food* 3 (1): 52–59.

Fan, S., and P. Hazell. 2001. "Returns to Public Investments in the Less-Favored Areas of India and China." *American Journal of Agricultural Economics* 83 (5): 1217–1222.

Fan, S., D. Nyange, and N. Rao. 2005. "Public Investment and Poverty Reduction in Tanzania: Evidence from Household Survey Data." Development Strategy and Governance Division Discussion Paper 18. IFPRI, Washington, DC.

FAO (Food and Agriculture Organization of the United Nations). 2022. *Assessing Agricultural Innovation Systems for Action at Country Level: A Preliminary Framework*. Rome.

Forsythe, L., O. Olaosebikan, B. Teeken, et al. 2024. "A Case of Transdisciplinarity and Collaborative Decision Making: The Co-Construction of Gendered Food Product Profiles." *Journal of the Science of Food and Agriculture* 104 (8): 4485–4497.

Fuglie, K., M. Gautam, A. Goyal, and W.F. Maloney. 2019. *Harvesting Prosperity: Technology and Productivity Growth in Agriculture*. Washington, DC: World Bank.

Fuglie, K., and N. Rada. 2016. "Economies of Size in National Agricultural Research Systems." In *Agricultural Research in Africa: Investing in Future Harvests*, eds. J. Lynam, N. Beintema, J. Roseboom, and O. Badiane, 59–82. Washington, DC: IFPRI.

Fuglie, K., K. Wiebe, S. Prager, T.B. Sulser, N. Cenacchi, C. Bonilla-Cedrez, and D. Willenbocke. 2022. "Quantitative Analysis to Inform Priorities for International Agricultural Research." IFPRI Discussion Paper 2133. IFPRI, Washington, DC.

Gebreyes, M., K. Mekonnen, P. Thorne, et al. 2021. "Overcoming Constraints of Scaling: Critical and Empirical Perspectives on Agricultural Innovation Scaling." *PLoS ONE* 16 (5): e0251958.

GFRAS (Global Forum for Rural Advisory Services). 2012. "Building Knowledge Systems in Agriculture: Five Key Areas for Mobilising the Potential of Extension and Advisory Services." Position Paper. Lindau, Switzerland.

Gillespie, S., P. Menon, and A.L. Kennedy. 2015. "Scaling Up Impact on Nutrition: What Will It Take?" *Advances in Nutrition* 6 (4): 440–451.

Gouse, M., D. Sengupta, P. Zambrano, and J.F. Zepeda. 2016. "Genetically Modified Maize: Less Drudgery for Her, More Maize for Him? Evidence from Smallholder Maize Farmers in South Africa." *World Development* 83: 27–38.

Griliches. Z. 1964. "Research Expenditures, Education, and the Aggregate Agricultural Production Function." *American Economic Review* 54 (6): 961–974.

Hall, A. 2005. "Capacity Development for Agricultural Biotechnology in Developing Countries: An Innovation Systems View of What It Is and How to Develop It." *Journal of International Development* 17: 611–630.

Hartkamp, A.D., J.W. White, A. Rodríguez Aguilar, M. Bänziger, G. Srinivasan, G. Granados, and J. Crossa. 2000. *Maize Production Environments Revisited: A GIS-Based Approach*. El Batan, Mexico: CIMMYT.

Halperin, B., B. Ho, J.A. List, and I. Muir 2022. "Toward an Understanding of the Economics of Apologies: Evidence from a Large-Scale Natural Field Experiment." *Economic Journal* 132: 273–298.

Hassall, K.L., F. Baudron, C. MacLaren, et al. 2023. "Construction of a Generalised Farm Typology to Aid Selection, Targeting and Scaling of Onfarm Research." *Computers and Electronics in Agriculture* 212: 108074.

Hayami, Y., and V.W. Ruttan. 1971. "Induced Innovation in Agricultural Development." Discussion Paper No. 3. Center for Economic Research, Department of Economics, University of Minnesota, Minneapolis.

Hazell, P., and C. Ramaswamy. 1991. *Green Revolution Reconsidered: The Impact of the High Yielding Rice Varieties in South India*. Baltimore and London: John Hopkins University Press for IFPRI.

He, T., and C. Li. 2020. "Harness the Power of Genomic Selection and the Potential of Germplasm in Crop Breeding for Global Food Security in the Era with Rapid Climate Change." *The Crop Journal* 8 (5): 688–700.

Herrero, M., P.K. Thornton, D. Mason-D'Croz, et al. 2021. "Articulating the Effect of Food Systems Innovation on the Sustainable Development Goals." *Lancet Planetary Health* 5 (1): e50–e62.

Jayne, T.S., S. Zingore, A.I. Niang, C. Palm, S. Shah, and P. Sanchez. 2023. "Building Twenty-First Century Agricultural Research and Extension Capacity in Africa." *European Review of Agricultural Economics* 50 (5): 1824–1846.

Johnson, M., S. Benin, X. Diao, and L. You. 2011. "Prioritizing Regional Agricultural R&D Investments in Africa: Incorporating R&D Spillovers and Economywide Effects." ASTI Conference Working Paper 15. IFPRI, Washington, DC; Forum for Agricultural Research in Africa, Accra.

Klerkx, L., and D.G. Cremaschi. 2025. "Evolving Agrifood Innovation Systems." In *The Agricultural Technology and Innovation Outlook*. Rome: FAO, forthcoming.

Klerkx, L., B. Mierlo, and C. Leeuwis. 2012. "Evolution of Systems Approaches to Agricultural Innovation: Concepts, Analysis and Interventions." In *Farming Systems Research into the 21st Century: The New Dynamic*, eds. I. Darnhofer, D. Gibbon, and B. Dedieu, 457–483. Dordrecht, Netherlands: Springer.

Komarek, A.M., H. Kwon, B. Haile, C. Thierfelder, M.J. Mutenje, and C. Azzarri. 2019. "From Plot to Scale: Ex-Ante Assessment of Conservation Agriculture in Zambia." *Agricultural Systems* 173: 504–518.

Lam, D.P., B. Martín-López, A. Wiek, E.M. Bennett, N. Frantzeskaki, A.I. Horcea-Milcu, and D.J. Lang. 2020. "Scaling the Impact of Sustainability Initiatives: A Typology of Amplification Processes." *Urban Transformations* 2: 1–24.

Linn, J.F., ed. 2012. *Scaling Up in Agriculture, Rural Development, and Nutrition.* 2020 Focus Brief 19. Washington, DC: IFPRI.

List, J. 2024. "Optimally Generate Policy-Based Evidence Before Scaling." *Nature* 626: 491–499.

Lividini, K., J.L. Fiedler, F.F. De Moura, M. Moursi, and M. Zeller. 2018. "Biofortification: A Review of Ex-Ante Models." *Global Food Security* 17: 186–195.

Lynam, J. 2016. *Balancing International Public Goods and Accountability: Exploring the Impact of IFPRI's Policy Research on Science, Technology, and Innovation.* Independent Impact Assessment Report No. 43. Washington, DC: IFPRI.

Maryono, M., A.M. Killoes, R. Adhikari, and A.A. Aziz. 2024. "Agriculture Development through Multi-Stakeholder Partnerships in Developing Countries: A Systematic Literature Review." *Agricultural Systems* 213: 103792.

Meinzen-Dick, R.S., T. Falk, R. Sanil, et al. 2024. *Scaling Up Experiential Learning for Water Management.* Scaling Up Experiential Learning Tools Project Note 2. Washington, DC: IFPRI.

Mishra, K., R.A. Gallenstein, A.G. Sam, M.J. Miranda, P. Toledo, and E. Mulangu. 2023. "Does Bundling Credit with Index Insurance Boost Agricultural Technology Adoption? Evidence from Ghana." *Agricultural Economics* 54 (6): 778–792.

Naseem, A., D.J. Spielman, and S.W. Omamo. 2010. "Private-Sector Investment in R&D: A Review of Policy Options to Promote Its Growth in Developing-Country Agriculture." *Agribusiness* 26 (1): 143–173.

Nin-Pratt, A., and E. Magalhaes. 2018. "Revisiting Rates of Return to Agricultural R&D Investment." IFPRI Discussion Paper 1718. IFPRI, Washington, DC.

Nin-Pratt, A., G.-J. Stads, L. de los Santos, and G. Munoz. 2023. *Unlocking Innovation: Assessing the Role of Agricultural R&D in Latin America and the Caribbean.* Washington, DC: Inter-American Development Bank.

Occelli, M., J. Sellare, K.D. Sousa, et al. 2024. "Group-Based and Citizen Science On-Farm Variety Selection Approaches for Bean Growers in Central America." *Agricultural Economics* 55 (2): 270–295.

Omamo, S.W., X. Diao, S. Wood, et al. 2006. *Strategic Priorities for Agricultural Development in Eastern and Central Africa.* IFPRI Research Report 150. Washington, DC: IFPRI.

Pauw, K., and J. Thurlow. 2015. "Prioritizing Rural Investments in Africa: A Hybrid Evaluation Approach Applied to Uganda." *European Journal of Development Research* 27: 407–424.

Pender, J. 2005. "Econometric Methods for Measuring Natural Resource Management Impacts: Theoretical Issues and Illustrations from Uganda." In *Natural Resource Management in Agriculture: Methods for Assessing Economic and Environmental Impacts*, eds. B. Shiferaw, H.A. Freeman, and S.M. Swinton, 127–154. Wallingford, UK: CABI.

Pingali, P.L. 2022. "Green Revolution: Impacts, Limits, and the Path Ahead." *PNAS* 109 (31): 12302–12308.

Ragasa, C. 2016. "Organizational and Institutional Barriers to the Effectiveness of Public Expenditures: The Case of Agricultural Research Investments in Nigeria and Ghana." *European Journal of Development Research* 28 (4): 660–689.

Rice, B., B. Kramer, and C. Trachtman. 2024. *Barriers to Using Market Intelligence in Plant Breeding: Evidence from a Survey of Breeding Professionals*. Montpellier, France: CGIAR.

Rosegrant, M.W., J. Koo, N. Cenacchi, et al. 2014. *Food Security in a World of Natural Resource Scarcity: The Role of Agricultural Technologies*. Washington, DC: IFPRI.

Ruhinduka, R.D., J.B. Falck-Zepeda, U. Wood-Sichra, et al. 2020. "Ex Ante Economic Assessment of Impacts of GM Maize and Cassava on Producers and Consumers in Tanzania." IFPRI Discussion Paper 1911. IFPRI, Washington, DC.

Schut, M., J. Cadilhon, M. Misiko, and I. Dror. 2018. "Do Mature Innovation Platforms Make a Difference in Agricultural Research for Development? A Meta-Analysis of Case Studies." *Experimental Agriculture* 54 (1): 96–119.

Schut, M., J. Colomer, E. Proud, et al. 2024. "Innovation Portfolio Management for Responsible Food Systems Transformation in the Public Sector: Lessons, Results and Recommendations from CGIAR." *Agricultural Systems* 216: 103907.

Schut, M., C. Leeuwis, and G. Thiele. 2020. "Science of Scaling: Understanding and Guiding the Scaling of Innovation for Societal Outcomes." *Agricultural Systems* 184: 102908.

Sinclair, F., and R. Coe. 2019. "The Options by Context Approach: A Paradigm Shift in Agronomy." *Experimental Agriculture* 55 (Suppl 1): 1–13.

Spielman, D.J. 2005. "Innovation Systems Perspectives on Developing Countries." ISNAR Discussion Paper 2. IFPRI, Washington, DC.

Spielman, D.J., and R. Birner. 2008. "How Innovative Is Your Agriculture? Using Innovation Indicators and Benchmarks to Strengthen National Agricultural Innovation Systems." Agriculture and Rural Development Discussion Paper 41. World Bank, Washington, DC.

Spielman, D.J., and D. Kelemework. 2009. "Measuring Agricultural Innovation System Properties and Performance: Illustrations from Ethiopia and Vietnam." IFPRI Discussion Paper 851. IFPRI, Washington, DC.

Spielman, D.J., E. Lecoutere, S. Makhija, and B. Van Campenhout. 2021. "Information and Communications Technology (ICT) and Agricultural Extension in Developing Countries." *Annual Review of Resource Economics* 13: 177–201.

Spielman, D.J., C. Ragasa, and R. Rajalahti. 2012. "Designing Agricultural Research Linkages within an AIS Framework." In *Agricultural Innovation Systems: An Investment Sourcebook*, Module 4: Agricultural Research within an Agricultural Innovation System, 277–288. Washington, DC: World Bank.

Stevenson, J., B. Vanlauwe, K. Macours, et al. 2019. "Farmer Adoption of Plot- and Farm-Level Natural Resource Management Practices: Between Rhetoric and Reality." *Global Food Security* 20: 101–104.

Thornton, P., D. Mason-D'Croz, C. Kugler, R. Remans, H. Zornetzer, and M. Herrero. 2024. "Enabling Food System Innovation: Accelerators for Change." *Global Food Security* 40: 100738.

TAP (Tropical Agriculture Platform). 2016. *Common Framework on Capacity Development for Agricultural Innovation Systems: Conceptual Background*. Wallingford, UK: CABI.

Turner, J.A., A. Horita, S. Fielke, et al. 2020. "Revealing Power Dynamics and Staging Conflicts in Agricultural System Transitions: Case Studies of Innovation Platforms in New Zealand." *Journal of Rural Studies* 76: 152–162.

van Etten, J., K. de Sousa, J.E. Cairns, et al. 2023. "Data-Driven Approaches Can Harness Crop Diversity to Address Heterogeneous Needs for Breeding Products." *PNAS* 120 (14): e2205771120.

Walker, T.S., J. Alwang, and A. Alene. 2015. "Varietal Adoption, Outcomes and Impact." In *Crop Improvement, Adoption and Impact of Improved Varieties in Food Crops in Sub-Saharan Africa*, eds. T.S. Walker and J. Alwang, 388–405. Oxfordshire, UK: CABI.

Woltering, L., K. Fehlenberg, B. Gerard, J. Ubels, and L. Cooley. 2019. "Scaling: From 'Reaching Many' to Sustainable Systems Change at Scale: A Critical Shift in Mindset." *Agricultural Systems* 176: 102652.

World Bank. 2012. *Agricultural Innovation Systems: An Investment Sourcebook*. Washington, DC.

Zhang, W., G. Cao, X. Li, et al. 2016. "Closing Yield Gaps in China by Empowering Smallholder Farmers." *Nature* 537: 671–674.

Zhou, Y., and S. Babu, eds. 2015. *Knowledge Driven Development: Private Extension and Global Lessons*. Cambridge, MA: Academic Press.

Chapter 18

FINANCING

From Supporting Agricultural Production to Transforming Food Systems

Eugenio Díaz-Bonilla

Eugenio Díaz-Bonilla is a special advisor to the Inter-American Institute for Cooperation on Agriculture (IICA) and a visiting senior research fellow, Director General's Office, IFPRI.

Key messages

- Theory, research, and practice on the financing of agriculture and food production and consumption have evolved over the past 50 years, from a focus on primary production toward a broader conception of food systems financing.
- Under this broad conception, financial flows that support investment and development in agriculture and food systems include consumer spending, food systems operators' spending, international development funds, national budgets, and banks and capital markets.
- The recent shift toward food systems approaches has expanded the objectives of financing to include environmental and social objectives, and led to new institutional arrangements and instruments for financing.
- Those financing approaches include expansion and better use of international development funds (including the International Monetary Fund's Special Drawing Rights), blended public–private finance, thematic bonds, efforts to improve fiscal conditions in developing countries, and new roles for central, public, and multilateral banks.

A combination of approaches will be needed to finance food systems, and research and policy analysis must inform these choices:

- **Designing and implementing national programs.** Food systems transformation will require integrated national operational programs, including options for optimizing and repurposing public budgets, in general. Policy research can contribute to developing quantifiable objectives, estimated costs, budget analysis, and appropriate institutional and financing arrangements.

- **Implementing policy interventions for consumption and production.** National programs will require adequate macroeconomic policies, regulations, and incentives to shape spending by consumers and the operations of economic actors in food systems.
- **Using international development funds strategically.** These funds can help mobilize other financial flows, especially if they are well coordinated by adequate national programs.
- **Mobilizing funds from banking systems and capital markets.** To reorient lending toward agriculture, rural development, and food systems transformation goals, researchers can help design improved regulations, instruments, and operations for the banking system and to support the development of a robust pipeline for capital investments, including leveraging technologies by CGIAR Centers and national partners.

Policy debates in low- and middle-income countries (LMICs) have long focused on the financing of agricultural and food production and, for the poor and vulnerable, consumption. Over the past 50 years, both theory and practice have evolved around what to finance—and how to do it—driven by changes in the role of agriculture in supporting economic growth and rural development and in the perceptions of what constituted the most critical food and nutrition problems in LMICs. This evolution progressed from financing investments in primary agricultural production for increasing the food supply (mainly defined in terms of calories) toward a more complex understanding of agrifood demand and supply issues, along with a broader view of what financing means in the context of food systems.

This chapter provides a historical review, examines some key financial challenges in LMIC agrifood systems, and describes related policy research, highlighting many of IFPRI's contributions as the Institute completes 50 years. Looking to the future, the chapter discusses possible policy options, and related research topics, to mobilize the financing needed to achieve food systems transformation that will support economic growth and employment and lead to greater sustainability, poverty reduction, equity, food security, and nutrition.

This chapter employs a broad notion of "financing" that considers six flows of funds (following Díaz-Bonilla et al. 2021; Díaz-Bonilla 2021a)

(Box 18.1, Figure B18.1). These are consumption expenditures (flow 1 in the figure); financial flows of operators within food value-chain systems (flow 2); international development funds (including multilateral, bilateral, and philanthropic) (flow 3); national government budgets (flow 4); banking systems (flow 5); and capital markets (flow 6). These different flows originate with different actors and decision-makers, are channeled through specific operational or institutional mechanisms, may use different financial instruments, and have specific recipients whose behavior will affect outcomes. Thus, the public and private actions needed to reorient and scale up financing for food systems activities may vary by type of flow and require separate considerations for the within-flow components (that is, originators, institutions, and operational mechanisms, instruments, and recipients). Box 18.1 presents a more detailed description of the flows, including preliminary estimates of the global monetary values involved.

Evolving views on what to finance and how

Financing of primary agricultural production within import-substitution industrialization strategies

In the post–World War II era, standard development thinking expected the agriculture sector to support industrialization by transferring labor, supplying cheap food and raw materials, expanding exports to generate foreign currency, and providing savings to finance investments in the industrial sector and public infrastructure, an approach known as import-substitution industrialization (see Chapter 3) (Johnston and Mellor 1961). While the overall approach implied a tax on agriculture to assist industrial development, governments also implemented a mix of compensatory interventions to ensure a supply of cheap food and agricultural exports. With some funding from international development funds, many LMIC governments created agrarian reform entities and national agricultural research institutes and also supported public and parastatal enterprises operating in agricultural input and product markets, with varied impact on net financial flows to the agriculture sector (flows 3 and 4).

Another key policy intervention aimed to promote agricultural production operated through the banking system (flow 5), which historically has been the central focus of debates on financing. A typical program used loans from central banks ("rediscounts") and/or loans from international development institutions channeled to public agricultural banks. Those public banks, in turn, provided loans to farmers on concessional terms, including subsidized interest

BOX 18.1 Financial flows

The broad view of the financial transactions in food systems includes six flows, based on the financial flows used by social accounting matrices (SAMs) for whole economic systems (Pyatt and Round 1985). SAMs highlight the fact that there is no "free money"—that is, reallocating or scaling up funds for one activity implies reducing funds for other activities, with general equilibrium effects. Including consumer and other internal food system flows in our framework, rather than just the external flows often considered by financing discussions, acknowledges the enormous size and influence of these flows on food systems investment and development. This broad conception of financial flows is also consistent with the Paris Agreement's objective to align all financial flows with climate change mitigation, adaptation, and resilience (Article 2, paragraph 1c), which guides the estimates of climate funding conducted by the United Nations Framework Convention on Climate Change (UNFCCC 2021). The UN Food Systems Summit hub is working to estimate the value of some of these flows, starting with international development funds (IDF) and public expenditures.

FIGURE B18.1 Flow of funds for food systems

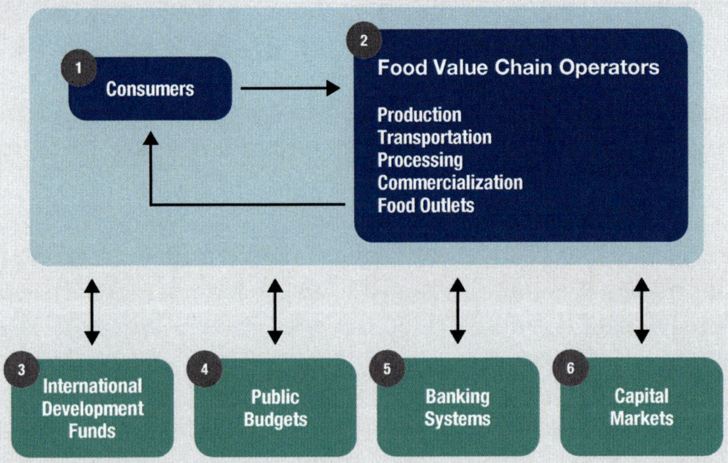

Source: Díaz-Bonilla (2021a) and Díaz-Bonilla et al. (2021).

FOOD EXPENDITURES by consumers (flow 1) are estimated at about U$10 to $14 trillion globally during recent years (see for instance, Díaz-Bonilla 2023a, 2023b). These expenditures are the sales of food systems operators across the value chain, who range from small-scale farmers and rural laborers to input suppliers, processors, transport operators, retailers, and food-service providers to large supermarkets and food outlets. Flow 1 is thus the main source of **operators' current expenditures, investments, and profits or savings** (flow 2). Flows 1 and 2 correspond to actors internal to food systems. This financing is supplemented by the four other flows operated by actors whose main activities are external to food systems.

INTERNATIONAL DEVELOPMENT FUNDS (IDF) (flow 3) include financial flows from multilateral financial and technical cooperation organizations, bilateral aid agencies, and philanthropic groups. IDF is the smallest of the flows, with average annual disbursements of about $12 billion (current value) for agriculture, forestry, and fishing (AFF) from 2015 to 2020. Estimates for all IDF-funded mitigation and adaptation activities in agriculture, forestry, land use, and natural resource management (AFOLU), a category that overlaps with AFF, averaged only about $9.1 billion for 2017–2018. Multilateral climate funds, such as the Global Environment Facility and the Green Climate Fund, provided just $3 billion annually for all sectors in 2017–2018, of which less than $0.8 billion was for AFOLU.

NATIONAL GOVERNMENT BUDGETS, including expenditures and revenues, are a large external flow (flow 4). For the 2015–2019 period, LMICs had average expenditures of more than $8 trillion (but excluding China, only $5 trillion), of which about $400 billion *each* was for AFF and social safety nets for the poor (or, excluding China, about $125 billion and $260 billion, respectively). There are no estimates for overall expenditures on food systems.

BANK LOANS (flow 5) in LMICs (excluding China) are estimated at only $87 billion on average for 2015–2019, of which $9.5 billion was for AFF (Díaz-Bonilla 2021a, 2023a, and 2023b). There are no estimates for food systems.

CAPITAL MARKETS (flow 6) include a variety of investors (such as pension funds, impact investors, and venture capitalists), institutional mechanisms, and financial instruments (including different equity, bond, project lending, and hybrid instruments). Capital markets differ from banking systems in that they act as an intermediary for savings and investments and offer risk management instruments, but do not provide check deposit and payment services. The largest share of capital market investments is made in high-income countries, and the amounts oriented toward agriculture and food systems transformation are small (see Díaz-Bonilla 2023a, 2023b).

rates, for specific agricultural products and on the condition that they apply Green Revolution technologies (improved seeds, fertilizers, and other inputs) (see Chapter 17).

These public credit programs were part of a general development approach relying on directed and subsidized financing and interest rate caps, which was criticized as "financial repression," with negative impacts on growth and a tendency to foster corruption (Adams et al. 1984; McKinnon 1973; World Bank 1986). It was argued that low repayment rates undermined the banking system and led to fiscally unsustainable subsidies to farmers and banks. At the same time, interest rate ceilings discouraged rural savings and the development of financial instruments such as savings deposits for rural families. By tying loans to specific inputs, investments, or crops, these programs limited farmers' flexibility in responding to market opportunities. They also appeared to primarily benefit large farmers and had negative macroeconomic consequences as well: when financed through money creation by the central bank, these programs could contribute to inflation, as happened in Brazil in the 1970s (World Bank 1986); and when the subsidies were paid from the public budget, they increased the fiscal deficit, consequently either driving up inflation or expanding public debt. In both cases, banks would then become more dependent on government support, creating a potentially vicious cycle of rising deficits and inflation.

Integrated rural development: Focus on poverty and geographic regions

The realization that the production-focused development approach was not reducing widespread rural poverty led to a shift toward investment programs focused on the accumulation of human and physical capital by the rural poor (Chenery et al. 1974).[1] In the late 1970s and 1980s, the World Bank along with most multilateral development banks (flow 3) promoted integrated rural development (IRD) programs, and many LMICs adopted this approach, reorienting government expenditures (flow 4) and banking operations (flow 5) to support these programs. The IRD programs aimed to transcend the narrow focus of supporting *agricultural production*, which tended to favor more modern and commercial farms, by shifting to *agricultural producers*—with emphasis on poor and small-scale farmers. The programs also expanded the focus

1 Redistributing *investments* was considered better than the alternatives of redistributing *incomes* to the poor for consumption (which was deemed not fiscally sustainable) or redistributing *assets* such as land (which faced strong political resistance) (see Chenery et al. 1974).

beyond the farm to rural areas where the poor live, recognizing the need for public investments in infrastructure and services such as health and education (Yudelman 1976).

Soon after IFPRI's creation, researchers began critically analyzing the IRD approach. For instance, it was argued that IRD programs could not reduce rural poverty because of their narrow focus on primary agricultural production (Ahmed 1977). At least two additional components were needed: support to non-agricultural rural enterprises to generate employment and expansion of public sector work programs for landless workers, who were even poorer than the small-scale farmers targeted by the IRD programs.

By the 1990s, the IRD programs were also criticized on other grounds, including the complex inter-agency arrangements needed to implement their various productive, infrastructure, and social components, which hampered their execution (World Bank 1988). These critiques are relevant to the current discussions about the implementation of broad food systems programs, which face even more complex problems of institutional coordination. A more general concern was that IRD programs were undertaken within the framework of import-substitution strategies, which were shown to constrain agricultural production despite compensatory public interventions to support farmers.

Structural adjustment and financial liberalization

Criticisms of import-substitution strategies had emerged early in the 1970s, emphasizing the negative impacts on growth and poverty of protecting capital-intensive industrialization (Balassa et al. 1971; Little et al. 1970). It was argued that those problems could not be resolved simply by improving agriculture sector policies or reallocating investments to the poor through IRD programs. Rather, a rethinking of development strategies was needed: LMICs should take advantage of opportunities in international trade and let markets operate more freely by eliminating major government interventions. IFPRI contributed to this debate with a series of analyses, some conducted jointly with the World Bank, arguing that protectionist trade and macroeconomic policies had tilted price incentives against the agriculture sector, impairing its growth (Bautista and Valdés 1993; Krueger et al. 1988).[2] For investments in the agriculture sector to be effective in reducing poverty, the damaging macroeconomic and trade policies had to be changed first.

2 One of the policy studies by the World Bank and IFPRI was provocatively titled *The Plundering of Agriculture in Developing Countries* (Schiff and Valdés 1994).

While these analyses drew attention to the limitations of import substitution, it took a series of global crises to force a change in development strategies. The oil price shocks of the 1970s led to high inflation that was countered with punishingly high interest rates in the United States and other major economies, leading to global recessions in 1980 and 1982, the collapse of commodity prices in the mid-1980s, and a debt crisis in many LMICs. In response, the World Bank and other multilateral development banks (flow 3) reduced financing for productive sectors, including agriculture,[3] in the 1980s and 1990s, and shifted their lending to structural adjustment programs (for general economic policies) and sector adjustment programs, including agriculture (see also Chapter 3).

The World Bank's agriculture sector adjustment programs were conditional on policy changes that included the elimination or scaling down of public sector agricultural agencies (flow 4), with the expectation that the private sector would step in. As part of those programs, many public agricultural banks (flow 5) were restructured or eliminated during the 1990s, primarily in Latin America and Africa, although some continued operating, particularly in Asia (Díaz-Bonilla 2015). In addition, lending from banking systems (flow 5) to agriculture dropped in the 1990s in Latin America and Africa, regions that had closed many public agricultural banks (Giehleret al. 2005). However, in Asia, which had opted to restructure those banks (with successful examples such as the Bank Rakyat Indonesia), lending to the agriculture sector was broadly maintained.[4]

At the same time, deteriorating public sector finances led to cuts in government expenditures (flow 4), including support for agriculture. The declines were most pronounced in Latin America and the Caribbean and in sub-Saharan Africa, where the ratio of agricultural expenditures to agricultural GDP in the 1990s dropped to 55–57 percent of the levels in the previous decade. Smaller declines occurred in other regions (Díaz-Bonilla 2015).

An evaluation of the new policies in Africa, conducted by IFPRI, found that the structural and sectoral adjustment programs led to the elimination of some inefficient and contradictory public sector interventions (Kherallah et al. 2002). But when public institutions that had provided credit, technical

3 The World Bank's agricultural lending, including for IRD, declined from about US$5 billion (in constant 2001 US dollars) and some 30 percent of total World Bank lending in the late 1970s and first half of the 1980s to about $1.5 billion and 7 percent of total lending by the early 2000s (Lipton and Paarlberg 1990). Similar declines occurred in other multilateral institutions and bilateral donors (a longer discussion is found in Díaz-Bonilla 2015).

4 A more detailed discussion of public banks in different regions is found in Díaz-Bonilla (2015).

assistance, and other key inputs to agricultural production were dismantled, private sector institutions that could provide similar services and inputs often failed to emerge, with negative impacts on agriculture.

Another line of research at IFPRI focused on the impacts on growth, productivity, and poverty reduction of different types of public expenditures. A detailed review of studies (Mogues et al. 2012) aimed to answer those questions. It showed that aggregate public spending for agriculture appeared to have relatively modest impacts on rural welfare, agricultural and economic growth, and poverty reduction. This was due to the mixture high- and low-payoff budget items—while some public expenditures including agricultural R&D and infrastructure showed substantial positive returns for agricultural productivity, growth, and poverty reduction, others did not. However, the ranking of the different expenditures varied by country, and some showed declining results over time. The policy recommendation drawn from this review was to analyze the impacts of the different components of public expenditures in each country context, and then reallocate funds to those with higher payoffs, which this research helped to identify (see also Fan 2008).

Other developments in financial flows during the 1980s and 1990s

While public expenditures and bank lending for food and agriculture were being reduced, other forms of financing were emerging—to support both small producers and enterprises and poor consumers and to support value chain development.

Microcredit. In 1983, the Grameen Bank, the world's first microcredit institution, was created in Bangladesh to offer small loans (part of flow 5), a new financial approach to fighting poverty. Microcredit institutions lend small amounts with few formal requirements, relying on community or reputational guarantees and allowing poor people to continue borrowing if they stay current on their payments. As this model was rapidly adopted in other LMICs, IFPRI established a multicountry research program on microcredit—Rural Finance Policies for Food Security for the Poor (1993–2001)—which introduced the idea of the triangle of microfinance (Figure 18.1) (Zeller and Meyer 2002). The triangle's three sides—financial sustainability of the microlenders, scope of coverage, and welfare impacts on the poor— illustrated the importance of maintaining balance and evaluating potential trade-offs across those three areas.

Along with other studies, this work showed that the poor need more than just loans for productive purposes; they also need broader financial services

FIGURE 18.1 The triangle of microfinance

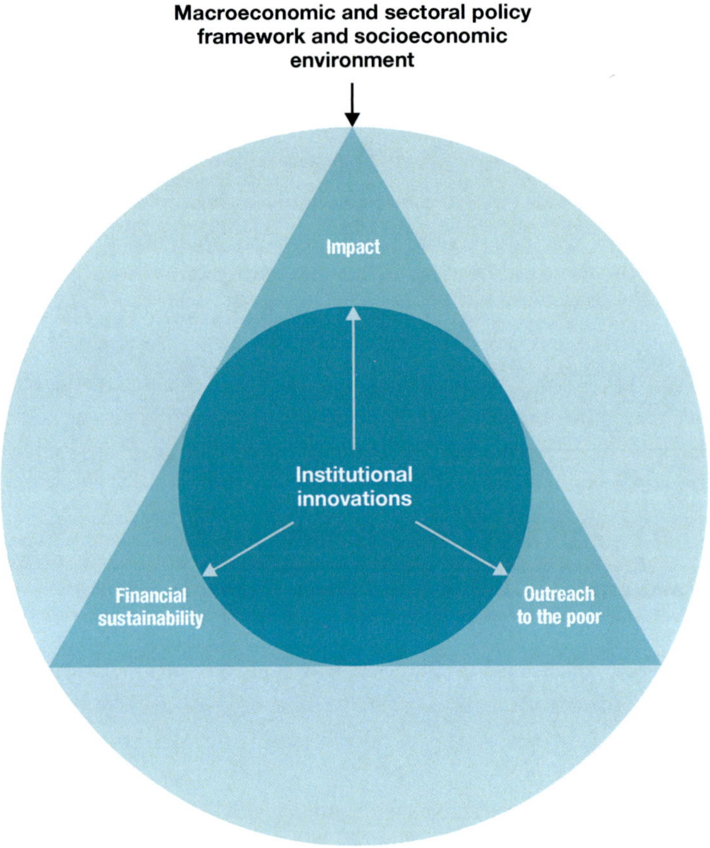

Source: Zellner and Meyer (2002).

to cope with the volatility of incomes and food consumption, protect human capital, and manage risk. It also highlighted the importance of adequate pricing of financial services, unlike the traditional approach built on (unsustainable) subsidized interest rates. Particularly for microcredit, it was argued that individual guarantees, rather than groups offering reciprocal guarantees, are often the most effective approach, and that, given the fungibility of credit, such guarantees need to be backed by the borrower's whole livelihood, not just the specific project being financed. This research also showed that although microcredit helped to improve some welfare indicators for the poor, it was not a silver bullet to address poverty, especially for the very poor, for whom allowable loans were too small to make a difference. Moreover, microcredit was not necessarily a viable solution for small-scale farmers if lending was linked to seasonal productive activities that could not accommodate the continuous repayment schedules of microcredit. An impact evaluation of IFPRI's research on microcredit found that it had a positive impact on the sector by demonstrating microcredit's effectiveness and promoting best practices (Alwang and Puhazhendhi 2002).

Microcredit was also part of a new paradigm of lending for agriculture that focused on creating a sustainable financial sector and financial institutions; treated borrowers and savers as clients rather than beneficiaries; developed financial products specifically for the rural population (not only loans but also deposits to help with payments and to manage risks through savings); and priced products and services to cover costs and risks, ensuring sustainability of the financial institutions.[5]

Cash transfers and safety nets. Cash transfer and safety net programs can support consumption and production expenditures within food systems (see Chapter 12). In the second half of the 1990s, Mexico and Brazil began providing cash transfers to poor households (flow 4 financing going mainly to flow 1), in a marked shift away from the costly universal food subsidies that were widely implemented following food price shocks in the 1970s.[6] Women are prioritized as recipients of cash transfers, as these programs aim to break the intergenerational transfer of poverty within families by building human capital in the young. The programs are often implemented with specific conditionalities, such as school attendance requirements and regular health checkups for children. Cash transfer programs appear to generate positive local growth-multiplier effects in poor communities through cash inflows that support local economic activity. IFPRI's research helped to inform the design of these programs, including their focus on women, and contributed to their short- and long-term sustainability (Behrman 2010).

Since these programs were first instituted in Latin America, they have been adopted in many developing regions as part of more sophisticated safety nets

---

5 Díaz-Bonilla (2015) includes a longer comparison of the "old" and "new" paradigms of agricultural finance through the banking system, based on the sources mentioned there.

6 IFPRI conducted a large study of those food subsidy programs in 13 LMICs in Africa, Asia, and Latin America from 1978 to 1986 (Pinstrup-Andersen, ed. 1988) (see Chapter 11). The case studies showed that many programs had important fiscal costs: for instance, it was estimated that in 1980–1981, Egyptian food subsidies were 20 percent of current expenditures and 7 percent of GDP. Universal food programs also had high fiscal costs in Bangladesh, China, Mexico, Peru, Sri Lanka, and Zambia, among others; the research also showed that well-targeted subsidy programs for the poor, such as in Colombia, were fiscally more sustainable.

designed to combat chronic poverty and hunger. These include nutritional programs for women and children, school lunches, food-for-work programs, and other interventions (all part of flow 4 and financing flow 1) (see Chapters 11, 12, and 23). These safety nets can also be scaled up and expanded when crises occur, such as natural disasters or the recent pandemic. For subsistence farmers, cash transfers coupled with support for productive and environmental activities may be a better way to address poverty and livelihood problems than either subsidized credit schemes or microcredit (that is, flow 4 replaces flow 5 to support the productive activities of small farmers in flow 2; see also Box 18.3).

Agricultural value chains. As primary food and agricultural products were increasingly incorporated into more complex food systems that included packaging, transportation, processing, marketing, and distribution, agricultural value chains became a focus of policy analysis and practice (see Chapter 7). Research began to explore supply-chain and value-chain lending as a means of financing small-scale farmers and small and medium enterprises (SMEs). In value chain–based schemes, international development funding and/or bank loans go to agrifood operators (traders, processors, supermarkets, and input and equipment suppliers) (from external flow(s) 3 and/or 5 to flow 2), which then on-lend to farmers (an intra-flow 2 operation) (de Brauw and Swinnen 2023). This innovation—along with contract farming, which serves as a guarantee allowing farmers to borrow from banks—has helped expand financing to agricultural and food production.

Foreign direct investments. During the 1990s and 2000s, economic liberalization and the growing demand for convenience foods and food consumption outside the home, which accompanied urbanization and women's greater participation in the labor force, led to increases in foreign direct investment (FDI) in food systems (part of flow 2), first in Latin America and later in other regions (see Chapter 7). For instance, annual inflows of FDI into food, beverage, and tobacco operations increased in Africa from US$14 billion (in constant 2015 US dollars) in the 1990s to about $50 billion in the 2010s, and in Latin America from $63 billion to $240 billion over the same period. Although there is no comparable quantitative data prior to 1990, the food industry experienced several waves of transformation, as the dismantling of public sector enterprises in the 1980s and 1990s, along with other economic and demographic changes, eventually led to a further expansion of private sector operators. Many of these have been multinational firms, such as agroindustry processors, supermarkets, and more recently, food outlets, including restaurants and fast-food chains (see Chapter 7) (Barrett et al. 2022; Reardon and Timmer 2012).

Together these developments underscored an obvious fact: most financing for agricultural production and food systems occurs as flows within value chain operations, and most of these flows depend on consumer expenditures (see Box 18.1).

Food systems as the focus of financing—the current approach

The concept of "food systems" has existed for some time, with various definitions, but the current conceptualization and its place in the policy agenda is largely the result of a series of developments that began with the United Nations' approval of the 2030 Agenda Sustainable Development Goals (SDGs) in 2015.[7] The 2019 United Nations conference identified agrifood systems as a priority area for achieving the SDGs, and in 2021, countries participating in the United Nations Food Systems Summit (UNFSS) presented "national pathways" for the transformation of their food systems, thus helping advance the 2030 Agenda.[8]

A parallel development with implications for agriculture and food systems was the 2015 Paris Agreement on climate change, which committed all signatory countries to prepare mitigation plans (Nationally Determined Contributions [NDCs]) and adaptation plans (National Adaptation Plans [NAPs]), with agriculture as an important component. Efforts have since been made to expand the conceptual focus of climate finance from agriculture to food systems, particularly following agreement on the Declaration on Sustainable Agriculture, Resilient Food Systems and Climate Action[9] at COP28 in 2023.

In sum, the combination of the 2030 Agenda and climate negotiations have broadened the focus of policy and research from financing agricultural and rural development, with an emphasis on staple products for calorie-rich diets and some cash crops for export, toward food systems transformation for healthy and diversified diets, environmentally sustainable production, and inclusive livelihoods along the whole value chain.

---

7 In the 1970s, and in reaction to the limited focus on economic growth, the "basic needs" approach highlighted the importance of health, education, and other objectives (Streeten and Burki 1978). This approach led to the UN Human Development Indicators (first published in 1990); then to the Millennium Development Goals, approved in 2000; and subsequently to the SDGs, established under the 2030 Agenda in 2015.

8 IFPRI participated in the debates about financing the transformation of food systems, producing a variety of analyses (Díaz-Bonilla et al. 2021; on SDG2, Díaz-Bonilla 2021a; and on climate change, Díaz-Bonilla and Echeverría 2022) and also contributed to the publication *Food Finance Architecture: Financing a Healthy, Equitable, and Sustainable Food System* (World Bank 2023).

9 www.cop28.com/en/food-and-agriculture

Several lines of work relevant to financing have been undertaken in connection with the UNFSS and climate negotiations. One effort aims to estimate the costs of food systems transformation,[10] which is a first step toward analyzing how to finance it. Several global estimates exist (FOLU 2019; Laborde and Torero 2023; Ruggeri Laderchi et al. 2024), but a 2023 evaluation of progress toward achieving the UNFSS goals revealed that many governments had not completed similar analyses at the country level.[11]

A second line of work involves estimating the financial flows currently going to food systems, as the basis for the subsequent analysis of the gaps between flows and estimated costs for food systems transformation. This work is being conducted by the UN Food Systems Hub[12] through a working group led by the International Fund for Agricultural Development and the World Bank, which has developed the Financial Flows to Food Systems framework (3FS) (IFAD and World Bank 2023). Based in part on IFPRI work (Díaz-Bonilla and Centurión 2024),[13] 3FS classifies international development funds and government budgets (flows 3 and 4) into five groups of interconnected expenditures for food systems that can be mapped to equivalent ministerial entities in many developing countries: agricultural development and value chains; infrastructure for food systems; nutrition and health; social assistance (including emergency food assistance); and climate change and natural resources. These groupings allow some cross-country comparability for flows 3 and 4. The work to complete those estimates at the country level has started in some pilot countries. The conceptual framework to estimate the monetary value of other flows (particularly the banking system and capital markets) is still being developed but is hampered by the lack of systematic data collection with a food systems focus.[14]

A third area of work relates to the need for specific programs for food systems transformation in individual countries, bringing together the national pathways outlined by the UNFSS with the NDCs and NAPs of the climate

10 This is different from the debate about the "true costs of food."

11 The evaluation found that among the countries surveyed, only 29 percent mentioned having an "Investment plan/implementation plan," and just 33 percent acknowledged "Costing of the pathway/implementation plan," according to the United Nations (UN 2023).

12 The UNFSS Hub was created by the UN Secretary General to coordinate the work of UN agencies (and engage with other international institutions) in support of specific national programs aimed at improving food systems. A specific working group (co-chaired by IFAD and the World Bank) focuses on financial issues (www.unfoodsystemshub.org/). The author has been on the advisory committee to that work and the Inter-American Institute for Cooperation in Agriculture (IICA) was supporting the work in countries in Latin America and the Caribbean.

13 The original IFPRI work was done in 2018, but it was published just recently.

14 FAOSTAT presents country data about the banking systems' portfolio of loans for agriculture, forestry, and fishing, but there is no similar data exercise for whole agrifood systems.

agreement. Currently, there are no broadly agreed-upon operational templates for how this could or should be done, although there are some initial efforts (such as the "roadmap" outlined in FAO 2023) that can be expanded and applied at the country level.

Other topics since the 2010s: Funding, instruments, and operations

This century's socioeconomic trends and major events, including the 2008–2009 global financial crisis and the 2020 COVID-19 pandemic, have sparked a variety of financial developments, including the use of new instruments and tools that can support food systems transformation.

Digital communications and financial services. The rapid growth of digital communications, including the internet and ever-more powerful mobile phones, could substantially reduce information and transaction costs across food systems and is generating many important changes, including in financing. As of 2022–2023, 67 percent of people globally were using the internet, and there were 111 mobile cellular subscriptions for every 100 people (World Bank database).[15] Even in low-income countries, about 40 percent of people use the internet, and about 60 percent have mobile cellular subscriptions.

IFPRI's research on the impact of digitalization on agriculture and food systems began in the 2010s. For instance, a set of policy briefs titled *Innovations in Rural and Agriculture Finance* covered, among other topics, mobile banking and payment systems in some LMICs (such as M-Pesa in Kenya) that reduced transaction costs for rural populations (Kloeppinger-Todd and Sharma 2010). Other work included the use of mobile phones for new agricultural insurance schemes (see Chapter 10) (Ceballos et al. 2020); the application of novel ways to assess risks in rural lending (Hernandez and Torero 2014); and the extension of digital interventions for operations in agrifood value chains beyond primary agriculture (Ambler et al. 2023). A recent evaluation of the use of digital tools for agrifood systems in Africa argued that although impacts at scale may not yet be evident these tools hold great promise for the future (Abate et al. 2023). These analyses also note that the success of digital solutions depends on overall investments in infrastructure, macroeconomic policies, regulations, and a business environment that supports the existence of truly profitable activities in food value chains. Without an enabling environment that facilitates the emergence of viable business opportunities, digital financial options that simply act

15 This means that there may be more than one subscription or one mobile phone per person, not that everyone has subscriptions or mobile phones.

as a channel of funds may not have major impacts. However, digital options that help manage risks and reduce transaction costs for the intermediation of funds can indeed support the transformation of food systems.

Conscientious investment instruments. In part related to the 2030 Agenda and the Paris Agreement, a global trend emerged toward investments with broader environmental and social objectives. This led to the development of new instruments (mainly for flows 5 and 6), including thematic bonds and "impact investment" funds focused on objectives related to the SDGs and, more generally, investment options and financial services with "environmental, social, and governance" purposes (though this trend is now in retreat). Even more recently, sustainability-linked loans and bonds have been developed to finance decarbonization transition plans. These feature variable interest rates tied to emissions-reduction goals or supply chain sustainability metrics.

International development funds. The Paris Agreement negotiations in 2015 generated a proliferation of climate-related international funds (flow 3), including the Global Environment Facility, the Green Climate Fund, the Special Climate Change Fund, the Least Developed Countries Fund, and the Adaptation Fund. These provide relatively small amounts of financial support and are characterized by complex governance structures, modalities, and objectives (Watson and Schalatek 2021) (see Box 18.1). Climate finance outcomes at COP29 in 2024 are summarized in Box 18.2.

In the case of multilateral development banks, there have been suggestions about how to increase their leverage ratio and lending—for instance, by giving more weight in risk evaluations to the built-in guarantees offered by member countries (such as "preferred creditor status" and "callable capital"); by expanding co-lending operations with private capital investors; and by making greater use of guarantees (in addition to direct lending), among other actions. If those changes are not enough to increase lending by multilateral development banks, the possibility of increases in their capital must also be considered.

In addition, the 2008–2009 global financial crisis and the 2020 COVID-19 pandemic led to two special allocations of the International Monetary Fund's Special Drawing Rights (SDRs)[16] to increase financial support to the world economy (the value of the new SDRs was about $280 billion in 2009 and $650 billion in 2021). There is an ongoing dialogue about how to use the SDRs more effectively by reallocating some percentage from countries that do not need them

16 The SDR is an international reserve asset, based on a basket of five currencies: the US dollar, the euro, the Chinese renminbi, the Japanese yen, and the British pound sterling. Countries holding SDRs can convert them into any of the five currencies and can keep that currency permanently at extremely low interest rates.

BOX 18.2 Climate financing

Held in Azerbaijan in 2024, the 29th Conference of the Parties has been called the "finance COP." The financial outcomes included the following.

First, a new target was set for climate finance for LMICs, known as the New Collective Quantified Goal on Climate Finance, which raised the yearly global goal from its previous level of US$100 billion to $300 billion by 2035 from international development funds, including multilateral development banks, bilateral funding, and philanthropy (flow 3). The negotiators also made a commitment to work together to mobilize funds from *all sources* to increase finance to LMICs to $1.3 trillion per year by 2035. This will require an analysis of the current levels and applications of all six flows.

Second, meaningful steps were taken to fully operationalize the Fund for Responding to Loss and Damage, with some initial funding allocated at COP29 (flow 3). Progress to advance this longstanding effort, which was first proposed in the 2007 Bali Action Plan, builds on the agreement at COP27 to create a new financial facility dedicated to remediating loss and damage following a push from a coalition of vulnerable countries at COP26.

Third, agreement was also reached on two building blocks needed for functioning carbon markets: (1) the operation of country-to-country trading, including how countries will authorize trading carbon credits, track these operations, and ensure environmental integrity through transparent technical reviews (Paris Agreement, Article 6.2), and (2) the standards for a centralized carbon market under the United Nations (Article 6.4). Implementation of Article 6 can help to scale up and reorient other financial flows, particularly from banking systems (flow 5) and capital markets (flow 6).

However, the withdrawal of the United States from the Paris Agreement will have a large impact on the implementation of these agreements.

to vulnerable LMICs that require financial support.[17] However, many of the alternatives under discussion to expand multilateral and bilateral international development funds are now uncertain, given the decisions of the current US government to withdraw from the Paris Agreement, suspend bilateral aid, and review its participation in multilateral development banks.

---

17 This debate needs to go beyond the reallocation of SDRs for the Poverty Reduction and Growth Trust and the Resilience and Sustainability Trust, and consider allocating SDRs to multilateral development banks to expand their lending capacity, or creating a guarantee fund at the IMF to allow LMICs to issue perpetual bonds to replace shorter-term debt and finance programs to achieve the SDGs and climate-change objectives (Díaz-Bonilla 2021a, 2021b; von Braun and Díaz-Bonilla 2021).

Mobilization of private sector funds. The Third International Conference on Financing for Development, held in Addis Ababa in 2015, highlighted the concept of "blended finance"—using international development funds (flow 3) or government funds (flow 4) to mobilize additional private sector investments (flows 2, 5, and 6). This approach is part of different forms of de-risking operations, such as absorbing part of the losses or providing technical assistance and support to the entities operating the investments to reduce transaction costs.[18]

Other ideas have focused on using incentive and regulatory frameworks to steer private sector investments toward development goals, including climate goals. One suggestion emerging from climate negotiations was to require public disclosure of climate-related financial risks by firms, banks, and other investors, with the expectation that greater transparency would increase the costs of detrimental investments and thereby reduce them (see the work of the Task Force on Climate-Related Financial Disclosures,[19] created by the G20's Financial Stability Board). Some central banks and regulators began reviewing their monetary and financial supervision functions in the context of climate change risks, which could also constrain operations with detrimental climate impacts.[20]

Expansion of fiscal space in LMICs. The negative shocks of the last two decades and the increase in public debt levels have led to different initiatives to improve fiscal conditions in LMICs. For instance, the Common Framework for Debt Treatment initiative[21] focuses on reducing the debt burden in low-income countries and expanding these countries' fiscal space. But the process is advancing slowly and should also be extended to middle-income countries. Other approaches consider the inclusion of clauses in public loans and bonds to allow debt suspension or stretching of payments when systemic shocks occur, such as climate disasters, pandemics, and even GDP fluctuations, and to allow broader use of debt-for-climate and debt-for-nature swaps (Georgieva et al. 2022). All these efforts will help to reduce fiscal pressures in LMICs and could facilitate the expansion of public financing of food systems transformation.

Central banks and public development banks. The 2008–2009 financial crisis and the COVID-19 pandemic revived the role of central banks as purveyors of credit and liquidity, similar to the role played by many developmental

18 "De-risking" a financial transaction may take other forms and not necessarily involve the use of blended finance.

19 www.fsb-tcfd.org/

20 See the work of the Network of Central Banks and Supervisors for Greening Financial Systems, launched in 2017: www.ngfs.net/en

21 https://clubdeparis.org/en/communications/page/common-framework

central banks in the 1950s through the 1970s. At the same time, there has been new interest in the positive role that public development banks can play when managed with incentives, performance metrics, and controls to avoid past problems. An example is the Finance in Common initiative of public development banks.[22]

Current and future challenges to financing food systems transformation

The historical review has shown the variety of topics considered under the general idea of "financing." Over the next 25 years, a combination of approaches will certainly be needed to finance food systems transformation. This section focuses on four topics and considers IFPRI's possible contribution through research in these important areas: (1) integrated national programs, (2) enabling macroeconomic and regulatory frameworks, (3) optimization of public budgets, and (4) greater mobilization of funds from international development institutions, banking systems, and capital markets.

Integrated national programs

Efforts to generate and direct hundreds of billions of dollars in funding for food systems transformation will be of little use without operational programs at the country level, which are lacking in many LMICs. Integrated investment programs must be designed with quantifiable objectives, instruments, and activities; an adequate institutional setting; estimated costs; and financing options. These country-specific programs should also make detailed analyses of the current financial flows to and within national food systems and their uses; of how those flows affect program objectives; and of how funds could be reallocated and/or scaled up to cover the estimated costs. As yet, the work on estimating current financial flows to food systems is just starting.

A significant constraint to advancing food systems transformation in LMICs is institutional weakness and fragmentation, which hamper the implementation of necessary multisector and multi-actor programs (see Chapter 15). The problems experienced by IRD programs show that successful implementation requires not only adjustments in the organizational charts and functions of institutions but also a fully articulated vision and program, linked to the budget and with specific local operational systems that deliver goods and services. Currently, there are no broadly accepted templates for designing and implementing these programs.

22 https://financeincommon.org/

IFPRI's contribution. IFPRI research can make an important contribution to the design of operational programs for food systems at the national level with clear objectives, policy instruments, estimated costs, and adequate institutional arrangements and financing. In particular, IFPRI's suite of global and national quantitative models can help to better analyze the different pathways and trade-offs across objectives and instruments, within the budget constraints imposed by national accounts. Finally, insights from IFPRI's work on political economy and institutions could also improve the design of the organizational and operational aspects of those programs, increasing the likelihood of their successful implementation.

Macroeconomic policies, regulations, and incentives

Such country-specific programs must be supported by macroeconomic policies that promote growth with low inflation, as well as regulatory and incentive policies that provide an adequate enabling environment for the decisions of all food systems actors, beginning with consumers (flow 1).[23] Governments can steer food consumption and production flows away from obesogenic products and toward healthier and more sustainable diets by influencing consumer decisions through prices, incomes, preferences, markets, and food environments (see Chapter 12). Tools include taxes and subsidies, income support with a nutrition focus for poor and vulnerable populations, and nutritional information and regulations. Governments already apply regulations and controls on producers' activities (flow 2) to promote health, nutrition, and food safety. Other policy interventions may be needed to address climate objectives (such as stopping deforestation and reducing food loss and waste) and social ones (such as protecting vulnerable communities from land displacement).

IFPRI's contribution. IFPRI has years of experience in analysis of macroeconomic and trade policies in relation to agriculture that can be extended to food systems. The Institute can also contribute to the analysis of better regulatory frameworks for consumption and production.

Optimization of public budgets

Integrated national programs require a careful review of public budget expenditures and revenues (flow 3) to determine the level and quality of utilization of fiscal funds available for food systems transformation. These analyses should look at promising proposals to increase or redirect financing, such as repurposing of agricultural support (see Chapters 16 and 17), and should also consider other

23 Of course, peace and good governance are overall requirements.

expenditures relevant to food systems, such as social protection (see Chapter 11), infrastructure, and productive support in general (see Chapter 17), as well as reconsidering fossil fuel subsidies, which in some countries exceed spending for agricultural support or safety nets for the poor. [24] Revenues will have to be increased in cases where reallocating, better targeting, and improving the design and structure of current expenditures do not free up sufficient funds.

IFPRI's contribution. As mentioned in the historical review, IFPRI has worked on the growth, productivity, and poverty-reduction impacts of different types of public expenditures, and more recently on the repurposing of agricultural support. These analyses can be extended to the different aspects of expenditures and revenues affecting food systems.

Scale-up and strategic use of international development funds

International development funds (flow 3) are relatively small, but they can mobilize other financial flows (particularly flows 5 and 6) through various forms of de-risking and by pioneering new instruments and investment structures (see Box 18.1). As long as the current political and economic environment limits options for increasing these funds, efforts should concentrate on optimizing their use. Strong country-based programs and institutions can help ensure that multilateral, bilateral, and philanthropic organizations better coordinate their own operations to avoid fragmentation and isolated initiatives.

IFPRI's contribution. The different applications of IFPRI's work discussed in this chapter (such as those focusing on integrated programs for food system transformation, macroeconomic and regulatory frameworks, and public budgets) could also improve how countries and international organizations program and use scarce international development funds.

Mobilization of funds from the banking system and capital markets

The banking system and capital markets (flows 5 and 6) are also key sources of financing, but information on existing flows to food systems is limited, including whether they align with the SDGs and climate change objectives. Reorienting bank and capital market funds toward desired food systems goals will require adequate macroeconomic, regulatory, and incentive frameworks, as mentioned before; but some specific interventions, discussed here, will be needed as well.

---

24 Díaz-Bonilla and Centurión (2022) compare social protection and fossil fuel subsidies in some countries in Central America.

Bank lending to the agriculture sector (flow 6) is affected not only by macroeconomic volatility but also by the dispersion and small scale of agrifood system operators (which increases transaction costs), the presence of covariant risks (weather, prices, pests, seasonality of production) that increase risks of default on loans (see Chapter 10), and investment regulations that are geared toward the urban sector (see Chapter 17). Historically, bank operations, and the issues they face, have been at the center of debates about agricultural financing. This broad subject is briefly summarized in Box 18.3, along with a more detailed discussion of some potential interventions, including the role of central banks for expanded lending to small-scale farmers and small and medium food system operators. However, for subsistence farmers and other poor operators in food value chains, loans—even with preferential terms—may not be the best instrument to support a just transition toward improved food systems. These populations would benefit more from cash transfers and other safety nets that include poverty, productivity, environmental, and nutritional components (see Chapter 11).

In the case of global capital markets (flow 6), IDFs could be used more strategically to leverage and mobilize private funds for food systems. For LMICs, the lack of a pipeline of investable options, with an attractive risk–reward profile, has been consistently identified as a key obstacle to scaling up the participation of capital market operators. That constraint could be alleviated by using IDFs or government funds to create and finance dedicated facilities for project preparation, incubation, and acceleration to develop a robust pipeline of investable opportunities (including individual projects, impact investment funds, and green bonds), particularly those involving small-scale farmers, women, youth, and micro, small, and medium enterprises (Díaz-Bonilla et al. 2018).

IFPRI's contribution. As was the case with microfinance, IFPRI can analyze policies, regulations, and instruments for the banking system to finance agricultural and rural development and the transformation of food systems. In addition to that research on more macro issues, the Institute can expand its current work on value chain finance insurance, and several other microeconomic topics.

Regarding capital markets, IFPRI could continue and expand its current efforts to articulate technological innovations from CGIAR and other sources with investable vehicles for the private sector, through project preparation and implementation teams and funds. An example is the Climate Smart Food System Fund, which aims to expand investments using CGIAR technology and financial support from KfW, the German Development Bank.

BOX 18.3 Banking systems

Fostering financial inclusion and expanding banking operations for the transformation of food systems will require a holistic analysis of banking systems (flow 5) and other financial services to inform effective policies. This analysis should consider (1) the source of funds, (2) types of institutions, (3) instruments on both the lending side and the deposit side, and (4) the enabling environment and other supporting policies (for greater detail on these four aspects, see Díaz-Bonilla 2015).

FUNDING. Bank funds may be sourced from deposits, central bank rediscounts, mandates to allocate a percentage of deposits to financing food systems or to buy dedicated government bonds, external loans, capital-market instruments, and the public budget. Bank deposits are usually short term, and therefore, credit for long-term investment may require dedicated funding from public fiscal or monetary sources or intermediation in capital markets.

Reviving the development role of central banks could provide additional funding. Within the framework of a consistent monetary program that controls inflation, central banks can offer specific lines of credit to finance loans targeted to small-scale farmers and small and medium enterprises (SMEs) in food value chains, including women and youth, and focused on improved technologies that address economic, social, and environmental objectives.

INSTITUTIONS. These include public development banks and a variety of private banks—from microfinance, community, and cooperative banks to large commercial banks. Well-managed development banks can be powerful instruments for addressing market failures that affect agricultural and rural financial markets and climate finance, as well as for crowding-in private sector funds from commercial banks and private investors by using blended finance and de-risking arrangements with their public capital.

INSTRUMENTS. Governments in LMICs tend to revert to subsidized interest rates to expand lending to farmers, although these can have negative impacts on growth and financial markets. To devise more adequate interventions, governments must address several factors that determine the effective interest rate charged to farmers, without interfering with market-determined interest rates. Other forms of financing also need to be supported, such as supply-chain or value-chain lending, which will require the adaptation of regulations and operational mechanisms. Other financial instruments are also crucial, such as developing appropriate options for rural populations and SMEs to manage payments and savings. These can include simplified checking and savings deposits, as savings are an important risk-mitigation tool for rural households.

ENABLING ENVIRONMENT AND SUPPORTING POLICIES FOR LENDING TO FARMERS AND SMES. Governments can play a key role in improving lending conditions for small-scale farmers and food system SMEs. For instance, lending regulations designed for the urban sector can be adjusted to account for seasonal cash flows in the agriculture sector. Governments can also offer transparent fiscal subsidies to financial institutions to cover the relatively high administrative and transaction costs of small loans, or to deploy trained extensionists to help small operators prepare bankable projects and complete necessary paperwork.

Other government investments that can facilitate lending to small-scale farmers include improvements in land legislation and public agencies that handle land titling (see Chapter 6), innovative insurance and guarantee schemes (see Chapter 10), and government support for critical public goods such as infrastructure, agricultural R&D, small irrigation schemes, drought-resistant seeds, improved sanitation and preventive health services, and weather forecasts (see Chapter 17), which can diminish risks and help to lower overall interest rates. Digital technologies can also improve financing for small operators by lowering transaction costs, expanding the geographic coverage of financial services, facilitating identification and monitoring of clients (thereby improving repayment rates), generating information about potential clients, and otherwise facilitating the links between supply and demand of different financial services.

Both the operations of the banking systems and capital markets could also benefit from IFPRI's work on digitalization that may help expand financing by lowering transaction costs, reaching isolated areas, gathering and analyzing information about potential clients, and improving the links between supply and demand of financial services in general.

In sum, IFPRI can build on its long history of policy and institutional research on financing for agrifood systems in LMICs—including its contributions on financial flows related to integrated rural development, macroeconomic and trade policies, value chains, microfinance, insurance, safety nets, and digitalization—to help optimize, reallocate, and scale up financial flows for the desired transformation of food systems. As during its first half-century, IFPRI will continue to contribute research and analysis of evidence-based policy options to help countries make investments that support the achievement of crucial development goals and sustainably improve lives and livelihoods.

References

Abate, G.T., K. Abay, J. Chamberlin, et al. 2023. "Digital Tools and Agricultural Market Transformation in Africa: Why Are They Not at Scale Yet, and What Will It Take to Get There?" *Food Policy* 116: 102439.

Adams, D.W., D.H. Graham, and J. von Pischke, eds. 1984. *Undermining Rural Development with Cheap Credit*. Boulder, CO: Westview Press.

Ahmed, R. 1977. "Agriculture in Integrated Rural Development: A Critique." *Food Policy* 2 (2): 140–147.

Alwang, J., and V. Puhazhendhi. 2002. "The Impact of the International Food Policy Research Institute's Research Program on Rural Finance Policies for Food Security for the Poor." Impact Assessment Discussion Paper 16. IFPRI, Washington, DC.

Ambler, K., J.R. Bloem, A. de Brauw, S. Islam, and J. Wagner. 2023. *Financial Access and Digital Services within Agri-Food Value Chains in Bangladesh*. IFPRI Project Note. Washington, DC: IFPRI.

Balassa, B. 1971. *The Structure of Protection in Developing Countries*. Baltimore: Johns Hopkins University Press.

Barrett, C.B., T. Reardon, J. Swinnen, and D. Zilberman. 2022. "Agri-Food Value Chain Revolutions in Low- and Middle-Income Countries." *Journal of Economic Literature* 60 (4): 1316–1377.

Bautista, R.M., and A. Valdés. 1993. "The Bias against Agriculture: Trade and Macroeconomic Policies in Developing Countries." San Francisco: International Center for Economic Growth; Washington, DC: IFPRI.

Behrman, J.R. 2010. "The International Food Policy Research Institute (IFPRI) and the Mexican PROGRESA Anti-Poverty and Human Resource Investment Conditional Cash." *World Development* 38 (10): 1473–1485.

Ceballos, F., B. Kramer, A. Mishra, M. Robles, and M.S. Toor. 2020. *Picture-Based Crop Insurance (PBI): Using Farmers' Smartphone Pictures to Reduce Basis Risk and Costs of Loss Verification*. Washington, DC: International Initiative for Impact Evaluation (3ie).

Chenery, H., M. Ahluwalia, C. Bell, J. Dulloy, and R. Jolly. 1974. *Redistribution with Growth*. Oxford, UK: Oxford University Press.

de Brauw, A., and J. Swinnen. 2023. "Building Inclusive Value Chains for Smallholders: The Role of Finance." In *Handbook of Microfinance, Financial Inclusion and Development*, eds. R.J. Cull and V. Hartarska, 181–193. Northhampton, MA: Edward Elgar Publishing for the World Bank.

Díaz-Bonilla, E. 2015. *Macroeconomics, Agriculture, and Food Security: A Guide to Policy Analysis in Developing Countries*. Washington, DC: IFPRI.

Díaz-Bonilla, E. 2021a. *Financing SGD2 and Ending Hunger*. Food Systems Summit Brief. Bonn, Germany: United Nations Forum on Sustainability Standards.

Díaz-Bonilla, E. 2021b. “Using the New IMF Special Drawing Rights for Larger Purposes: Guaranteeing ‘Pandemic Recovery Bonds’.” *IFPRI Blog*, October 22.

Díaz-Bonilla, E. 2023a. “Transformation of Food Systems: How Can It Be Financed?” *Frontiers of Agricultural Science and Engineering* 10 (1): 109–123.

Díaz-Bonilla, E. 2023b. “Financing the Transformation of Food Systems: A Flow of Funds Approach.” FSEC Working Paper. Food System Economics Commission (FCES), Oslo.

Díaz-Bonilla, E., and R.G. Echeverría. 2022. “Climate Finance: Funding Sustainable Food Systems Transformation.” In *2022 Global Food Policy Report: Climate Change and Food Systems*, 48–57. Washington, DC: IFPRI.

Díaz-Bonilla, E., and M. Centurión. 2022. “Improving Livelihoods and Reducing Outmigration from the Northern Triangle in Central America: The Potential Role of Cash Transfers in Expanded Social Safety Nets.” LAC Working Paper 27. IFPRI, Washington, DC.

Díaz-Bonilla, E., and M. Centurión. 2024. “Analisis del Gasto Publico para el Desarrollo Agropecuario, Rural y la Seguridad Alimentaria y Nutricional.” LAC Working Paper 31. IFPRI, Washington, DC.

Díaz-Bonilla, E., M. Loboguerrero, L. Verchot, E. Viglizzo, and A. Mirzabaev. 2018. *Financing “A Sustainable Food Future.”* T20 Argentina: Food Security and Sustainable Development Task Force Brief. Washington, DC: IFPRI.

Díaz-Bonilla, E., J. Swinnen, and R. Vos. 2021. “Financing the Transformation to Healthy, Sustainable, and Equitable Food Systems.” In *Global Food Policy Report 2021: Transforming Food Systems after COVID-19*, 20–23. Washington, DC: IFPRI.

Fan, S., ed. 2008. *Public Expenditures, Growth, and Poverty: Lessons from Developing Countries.* Baltimore: Johns Hopkins University Press for IFPRI.

FAO (Food and Agriculture Organization of the United Nations). 2023. *Achieving SDG 2 without Breaching the 1.5°C Threshold: A Global Roadmap*, Part 1. Rome.

FOLU (Food and Land Use Coalition). 2019. *Growing Better: Ten Critical Transitions to Transform Food and Land Use.* Global Consultation Report of the Food and Land Use Coalition. London.

Georgieva, K., M. Chamon, and V. Thakoor. 2022. “Swapping Debt for Climate or Nature Pledges Can Help Fund Resilience.” *IMF Blog*, December 14.

Giehler, T., H. Seibel, and S. Karduck. 2005. “Reforming Agricultural Development Banks.” Working Paper 2005,7. Development Research Center, University of Cologne, Germany.

Hernandez, M.A., and M. Torero. 2014. “Parametric versus Nonparametric Methods in Risk Scoring: An Application to Microcredit.” *Empirical Economics* 46 (3): 1057–1079.

Johnston, B., and J. Mellor. 1961. “The Role of Agriculture in Economic Development.” *American Economic Review* 51 (4): 566–593.

IFAD (International Fund for Agricultural Development) and World Bank. 2023. *3FS Tracking Financial Flows to Food Systems: Factsheet.* Rome: IFAD.

Kherallah, M., C. Delgado, E. Gabre-Madhin, N. Minot, and M. Johnson. 2002. *Reforming Agricultural Markets in Africa.* Baltimore: Johns Hopkins University Press for IFPRI.

Kloeppinger-Todd, R., and M. Sharma. 2010. *Innovations in Rural and Agriculture Finance.* 2020 Vision Focus 18. Washington, DC: IFPRI.

Krueger, A., M. Schiff, and A. Valdés. 1988. "Agricultural Incentives in Developing Countries: Measuring the Effects of Sectorial and Economywide Policies." *World Bank Economic Review* 2 (3): 255–271.

Laborde, D., and M. Torero. 2023. "Modeling Actions for Transforming Agrifood Systems." In *Science and Innovations for Food Systems Transformation*, eds. J. von Braun, K. Afsana, L.O. Fresco, and M.H.A. Hassan. Cham, Switzerland: Springer.

Little, I., T. Scitovsky, and M. Scott. 1970. *Industry and Trade in Some Developing Countries.* Paris: Organisation for Economic Co-operation and Development.

McKinnon, R. 1973. *Money and Capital in Economic Development.* Washington, DC: The Brookings Institution.

Mogues, T., B. Yu, S. Fan, and L. McBride. 2012. "The Impacts of Public Investment in and for Agriculture: Synthesis of the Existing Evidence." IFPRI Discussion Paper 1217. IFPRI, Washington, DC.

Pinstrup-Andersen, P., ed. 1988. *Food Subsidies in Developing Countries: Costs, Benefits, and Policy Options.* Baltimore: Johns Hopkins University Press for IFPRI.

Pyatt, G., and J.I. Round, eds. 1985. *Social Accounting Matrices: A Basis for Planning.* Washington, DC: World Bank.

Reardon, T., and P. Timmer. 2012. "The Economics of the Food System Revolution." *Annual Review of Resource Economics* 4 (1): 225–264.

Ruggeri Laderchi, C., H. Lotze-Campen, F. DeClerck, et al. 2024. *The Economics of the Food System Transformation.* Global Policy Report. Food System Economics Commission, Berlin.

Schiff, M., and A. Valdés. 1994. *The Plundering of Agriculture in Developing Countries.* World Bank Report 013. Latin America and the Caribbean Region Department. Washington, DC: World Bank.

Streeten, P., and S.J. Burki. 1978. "Basic Needs: Some Issues." *World Development* 6 (3): 411–421.

UN (United Nations). 2023. "Secretary-General's Call to Action for Accelerated Food Systems Transformation." UN Food Systems Summit + 2 Stocktaking Moment, July 26.

UNFCCC (United Nations Framework Convention on Climate Change). 2021. *UNFCCC Standing Committee on Finance: Fourth (2020) Biennial Assessment and Overview of Climate Finance Flows.* Bonn, Germany.

von Braun, J., and E. Díaz-Bonilla. 2021. "Perpetual Bonds Can Help States Fight Hunger" [letter]. *Financial Times*, June 6.

Watson, C., and L. Schalatek. 2021. *The Global Climate Finance Architecture.* Climate Finance Fundamentals No. 2. Washington, DC: Heinrich Böll Stiftung; London: Overseas Development Institute.

World Bank. 1986. *World Development Report 1986.* New York: Oxford University Press.

World Bank. 1988. *World Bank Experience with Rural Development: 1965–1986.* Washington, DC.

World Bank. 2023. *Food Finance Architecture: Financing a Healthy, Equitable, and Sustainable Food System.* Washington, DC.

Yudelman, M. 1977. "The Role of Agriculture in Integrated Rural Development Projects: The Experience of the World Bank." In *Decision-Making and Agriculture, Papers, and Reports, 16th International Conference of Agricultural Economists*, eds. T. Dams and K.E. Hunt, 444–464. Oxford: Agricultural Economics Institute.

Zeller, M., and R.L. Meyer, eds. 2002. *The Triangle of Microfinance: Financial Sustainability, Outreach, and Impact.* Washington, DC: IFPRI.

SECTION VI

Regional Developments and Priorities

Chapter 19

AFRICA

John Ulimwengu, Steven Were Omamo, Ousmane Badiane, and Samuel Benin

John Ulimwengu is a senior research fellow, Development Strategies and Governance (DSG) Unit, IFPRI; **Steven Were Omamo** is director, DSG, and director for Africa, IFPRI; **Ousmane Badiane** is executive chairperson, AKADEMIYA2063; and **Samuel Benin** is a senior research fellow, DSG, IFPRI.

Africa's food systems have undergone significant transformations over the past four decades, with notable improvements in agricultural productivity and food security, but persistent challenges remain (Table 19.1). Across the continent, agricultural productivity, measured as value added per worker in agriculture, forestry, and fishing, increased from US$1,657 in the mid-1980s to $3,315 in recent years, likely driven by technology adoption, investment, and policy reforms that have improved efficiency in the sector. However, Africa's reliance on imported foods has increased, with the share of cereals that are imported (the dependency ratio) up from 39 percent in the 1985–2000 period to 46.6 percent for 2016–2023. While the overall per capita food supply has become more stable, this progress remains fragile in the face of climate change, economic shocks, and political instability.

Poverty and undernutrition remain pressing concerns. Although at the continental level, poverty has fallen from 49.3 percent to 31.3 percent of the population, regional disparities mean that rates of poverty and undernutrition are substantially higher in some places (Table 19.1). In comparison, South Asia also faces notable poverty and undernutrition challenges but has made more rapid progress than Africa; between the 1990s and early 2020s, the extreme poverty rate in South Asia fell from nearly 47 percent to around 12 percent (World Bank 2023). Within Africa, Central Africa still struggles with high poverty rates and slow reduction in child stunting. Eastern Africa has made steady improvements in food production, yet food insecurity and undernutrition remain high in rural areas. In Southern Africa, the situation is relatively better, with higher agricultural productivity and lower stunting and wasting rates. However, recurrent climate shocks, rapid urbanization, and changing diets, which have increased demand for cereals like wheat and rice that are not widely produced locally, mean that Southern Africa is highly dependent on cereal imports even as it remains vulnerable to climate-induced shocks to food production. Northern Africa also benefits from high agricultural productivity but has the highest dependency on cereal imports, making it particularly vulnerable to global market fluctuations. Western Africa has seen moderate progress in reducing poverty and food insecurity but continues to face volatility in food availability because of fluctuations in agricultural yields and political instability.

Child malnutrition remains a major concern across Africa. Stunting among children under age five has declined from 38.7 percent to 30.2 percent, but rates remain unacceptably high, particularly in Central Africa (34.9 percent) and parts of Eastern Africa. Wasting has also decreased, from 9.0 percent to 6.7 percent, reflecting some progress in addressing acute malnutrition, but regional disparities persist.

Over the past decades, IFPRI as well as other international organizations have offered research-based solutions to Africa's development challenges. IFPRI's contributions to policy research for agriculture and food systems on the continent have evolved from an initial focus on food security, agricultural productivity, and rural development to include the effective design and implementation of safety net programs (see Chapter 11), agro-parks (see Chapter 16), and national development strategies, as well as linkages between agriculture and broader development outcomes, including sustainability, equity, and health (see Chapter 3) (Diao et al 2007; Hazell et al. 2010).

Policy evolution in response to agrifood systems challenges

Africa's agrifood policy landscape has evolved in response to complex challenges, including food insecurity, climate change, and socioeconomic disparities (Figure 19.1). Three major phases define these shifts in policy frameworks: the post-independence era, structural adjustment programs, and the Comprehensive Africa Agriculture Development Programme (CAADP) framework.

In the post-independence era of the 1960s and 1970s, African countries prioritized industrialization as the key driver of economic development, at the expense of developing the agriculture sector (see Chapters 3 and 18). To support industrial growth, governments implemented import-substitution policies that heavily taxed agricultural exports and controlled agricultural markets through state marketing boards. These strategies generated revenue for

TABLE 19.1 Continental and regional food systems variables, 1985–2023

	Period	Agrifood systems greenhouse gas emissions per person (kT CO^2eq)	Agriculture, forestry, and fishing value added per worker (constant 2015 US$)	Cereal import dependency ratio (%)	Food supply variability per capita (kcal/person/day)	Poverty headcount ratio (%)	Stunting in children under 5 years (%)	Wasting in children under 5 years (%)
Africa	1985–2000	3775.6	1657.2	39.0	47.5	49.3	38.7	9.0
	2001–2015	2958.0	2214.7	42.7	44.6	38.7	34.4	8.0
	2016–2023	2583.1	3315.6	46.6	37.4	31.3	30.2	6.7
Central Africa	1985–2000	6945.2	1447.7	51.1	48.7	43.6	39.8	8.4
	2001–2015	4849.2	1420.8	51.5	48.4	40.6	34.9	8.1
	2016–2023	4339.1	1730.6	47.4	40.3	34.8	30.5	8.2
Eastern Africa	1985–2000	2282.4	1200.2	28.3	43.5	61.1	46.2	8.6
	2001–2015	1921.0	1481.1	33.8	44.5	47.7	40.3	6.9
	2016–2023	1815.6	3184.3	37.5	41.0	44.2	36.9	4.4
Northern Africa	1985–2000	1048.4	3380.0	59.2	44.0	5.8	25.8	4.9
	2001–2015	1319.3	5944.4	56.7	37.9	4.0	23.4	8.0
	2016–2023	1472.4	8507.1	63.0	32.7	1.4	11.9	2.4
Southern Africa	1985–2000	10256.2	2773.2	52.2	52.8	50.5	33.5	8.2
	2001–2015	8298.8	3280.7	60.6	47.4	28.5	30.9	4.2
	2016–2023	6631.1	3550.6	67.2	43.9	34.3	27.8	2.8
Western Africa	1985–2000	2260.7	1159.8	33.0	49.3	59.5	36.6	12.1
	2001–2015	1812.1	1497.8	36.1	44.2	43.0	32.1	9.5
	2016–2023	1545.6	2234.0	40.9	31.9	23.6	27.6	8.5

Source: The Food Systems Dashboard. https://www.foodsystemsdashboard.org

Note: Methodological notes on how the indicators were computed are provided in the Dashboard.

FIGURE 19.1 Africa food policy timeline

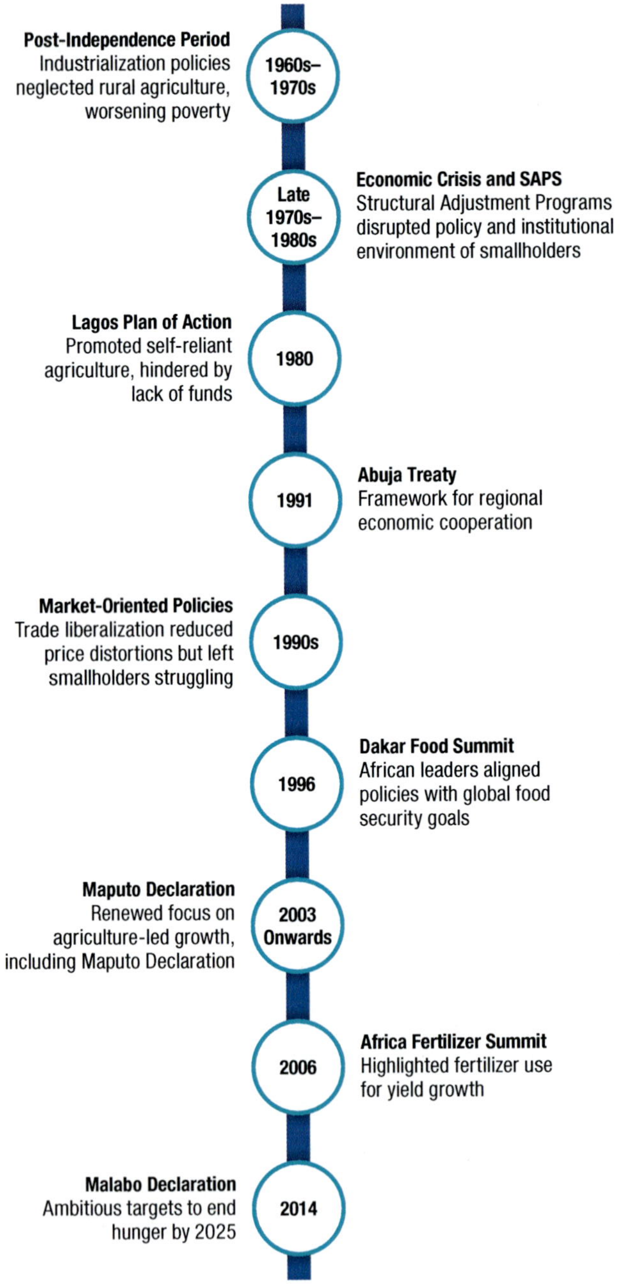

Source: Authors' compilation from various sources.

industrial projects by promoting the growth of export-oriented cash cropping within relatively small geographic areas but failed to boost overall agricultural productivity, thus deepening rural poverty and increasing socioeconomic inequalities (Badiane and Makombe 2014). These policies weakened the main source of livelihoods and growth for most of the population. Agricultural productivity stagnated, rural incomes declined, and food production suffered. As a result, poverty and food insecurity worsened, especially in rural areas, and overall economic growth slowed due to underinvestment in the agriculture sector, which was critical for broad-based development. The competitiveness of African agriculture was further eroded by exchange rate distortions and trade restrictions, leaving the sector under-resourced and vulnerable to external shocks (Krueger et al. 1988).

The economic crises of the 1970s and 1980s triggered the introduction of structural adjustment programs (SAPs), driven by the World Bank, International Monetary Fund, and major bilateral donors. Economic and sectoral reforms under these programs sought to liberalize markets, reduce subsidies, and enhance private sector participation in a shift away from import substitution and government intervention in agricultural markets (see Chapters 3 and 18).

Although SAPs did improve macroeconomic stability and laid the groundwork for long-term recovery (Badiane et al. 2020), they had mixed results across Africa in the short and medium term. While export-oriented farmers benefited from improved prices and exchange rate devaluations, the removal of agricultural subsidies and public investments disproportionately harmed smallholders (IMF 1987; World Bank 2004). Moreover, weak markets, physical and institutional infrastructure, and capacity gaps hindered the private sector's ability to fill the void left by state withdrawal. In Ghana, currency devaluation under SAPs boosted earnings for cocoa farmers by making exports more competitive, but producer price increases did not keep pace with devaluation, reducing real incomes for farmers (Green 1998). In Zambia, the removal of fertilizer and maize subsidies severely impacted small-scale farmers, leading to declining yields and worsening food insecurity (MacKenzie 1993). Similarly, in Tanzania, market liberalization led to the collapse of cooperative unions that had previously supported smallholders, leaving them vulnerable to price fluctuations and exploitation by private traders (Taube 1993). Despite these challenges, SAPs contributed to macroeconomic stability in some countries, such as Uganda, where liberalization attracted investment and spurred growth in cash crop production (Green 1998). IFPRI's report on the impacts of these policies in sub-Saharan Africa, *The Road Half Travelled* (Kherallah et al. 2002), highlighted the mixed outcomes of liberalization and advocated for more balanced reforms that would support vulnerable populations while fostering economic efficiency.

The Comprehensive Africa Agriculture Development Programme (CAADP), launched by the African Union in 2003, brought a new focus on agriculture, positioning the sector as a key driver of development and sustainability and highlighted the need for evidence-based policymaking. Two main targets for the CAADP member states were set by the 2003 Maputo Declaration—achieving 6 percent annual agricultural growth and allocation of 10 percent of national budgets to agriculture—with the aim of halving poverty by 2015. The Maputo Declaration was directly aligned with the Millennium Development Goals (MDGs), particularly MDG 1, which likewise aimed to halve extreme poverty and hunger by 2015. By prioritizing agriculture, a sector central to livelihoods and food security in Africa, the Declaration sought to make meaningful progress toward that goal. However, most African countries will fail to meet the CAADP targets by 2025. Few have consistently allocated 10 percent of national budgets to agriculture, and agricultural growth has often fallen short of the 6 percent target. Weak implementation capacity, limited political commitment, underinvestment, and broader governance and macroeconomic challenges contributed to their poor performance.

The 2014 Malabo Declaration added more ambitious goals, including tripling intra-African trade in agricultural goods and services, building resilience to climate and other shocks, and eliminating hunger. Major accompanying policies included regional trade facilitation through the African Continental Free Trade Area (AfCFTA), promotion of climate-smart agriculture practices, and enhanced accountability through the mechanisms of the Biennial Review and Malabo Scorecard. While Benin, Burkina Faso, Cabo Verde, Comoros, Ethiopia, Malawi, Niger, and Zimbabwe have achieved some of the targets, especially those related to policymaking processes, most African governments have not met their budget commitments, and no country will achieve all the goals of this Declaration by 2025. According to the 4th CAADP Biennial Review Report, while no country remained on track across all commitment areas throughout the four review cycles, several countries, including Rwanda, Morocco, and Uganda, demonstrated substantial progress in their agricultural growth rates and policy coordination efforts over the course of CAADP (AUC 2024). Some nations have shown incremental progress in areas such as evidence-based policymaking, institutional reforms, and agricultural productivity. The report highlights that 13 countries were actively implementing recommendations from previous Biennial Review cycles, indicating a growing culture of accountability and learning.

At the Extraordinary Summit of the African Union held in Kampala in January 2025, African leaders prepared a new 10-year CAADP Strategy and Action Plan (2026–2035) that lays out another ambitious framework for achieving sustainable, inclusive, and resilient agrifood systems. The Kampala Declaration emphasizes stronger governance, institutional capacity development, and improved accountability mechanisms through enhanced monitoring, evaluation, and reporting systems. Especially critical is a focus on bridging persistent implementation gaps that impeded progress during the Maputo and Malabo Rounds, with effective implementation through National Agrifood Systems Investment Plans prioritized in the Declaration, Strategy, and Action Plan.

IFPRI's contributions to policy research

IFPRI, along with other CGIAR Centers, has contributed to Africa's agrifood policy landscape through research on policy reforms and innovation and through capacity development. National research institutions, development agencies, and private sector stakeholders have also played vital roles in assessing, implementing, and scaling up innovations in agrifood systems across the continent. IFPRI's early research has been instrumental in shaping understanding of market dynamics, pricing policies, input subsidies, and the effects of trade liberalization conducted around 2010. A notable example is IFPRI's analysis of domestic agricultural marketing and pricing policies. By examining the impacts of various government interventions—including marketing boards, buffer stocks, and pricing mechanisms—researchers highlighted the trade-offs between stabilization and efficiency and provided evidence-based recommendations for policy reforms to balance efficiency with inclusivity (Christiaensen 2017). IFPRI's work on social protection and safety nets has also informed policies. For example, research on Ethiopia's Productive Safety Net Programme provided evidence-based insights to refine program mechanisms for targeting, improve program efficiency, and increase impacts on poverty reduction and agricultural productivity (see Chapter 11) (Gilligan et al. 2008; Hoddinott et al. 2012).

The IFPRI 2020 Vision Initiative, launched in the 1990s, helped cement the Institute's influence in the region. Through a series of forward-looking conferences and research programs across Africa that addressed food security, nutrition, health, rural poverty, and sustainable agriculture, the Initiative shaped national and regional agendas. This work facilitated dialogue among policymakers, researchers, and stakeholders, fostering a shared commitment to evidence-based solutions (von Braun et al. 2005).

IFPRI's regional and country offices in Africa have played a foundational role in localizing policy research and facilitating country-led food systems transformation through direct engagement with policymakers, stakeholders, and regional institutions. By providing evidence-based research tailored to local contexts, IFPRI has strengthened national capacities for informed policy decisions, increasing the likelihood that governments will adopt evidence-based policies (IFPRI 2024). Following the launch of CAADP, IFPRI and other CGIAR Centers, in partnership with the African Union and several of Africa's regional economic communities, established the Regional Strategic Analysis and Knowledge Support System (ReSAKSS).[1] ReSAKSS supports African policymakers in evidence-based planning, implementation, and mutual review of progress toward CAADP goals. In 2020, the ReSAKSS system and other related programs, along with many staff working on them, were transferred to AKADEMIYA2063, a new African institution based in Kigali, Rwanda, thus expanding locally driven agrifood policy analysis and capacity strengthening activities (Makombe et al. 2024). IFPRI continues to provide technical support to CAADP, including data and analysis for the program's Biennial Review reports and the development of the New Strategy and Action Plan (2026–2035).

Transforming Africa's agrifood systems requires sustained collective action. Research breakthroughs—such as CGIAR and national center efforts to develop disease-resistant, high-yield rice; drought-tolerant maize; and biofortified crops rich in essential micronutrients—are vital. National research institutions, private agribusinesses, and international donors play crucial roles in scaling these innovations. In addition, official development assistance has supported capacity-development programs, extension services, and market infrastructure, ensuring that improved technologies and practices reach farmers and increase agricultural productivity. By identifying binding constraints and high-potential opportunities to overcome these constraints, as well as highlighting critical spatial and temporal trade-offs, policy research can play a critical role in ensuring that these innovations are effectively scaled up and adopted by farmers (see Chapter 17).

1 CGIAR Centers involved include the International Livestock Research Institute and the International Water Management Institute, as well as IFPRI. Regional economic communities involved include the Common Market for Eastern and Southern Africa (COMESA), Economic Community for West African States (ECOWAS), and Southern Africa Development Community (SACD).

Anticipated challenges and opportunities through 2050

Africa's agrifood systems will continue to face profound challenges over the next 25 years, including further demographic shifts and urbanization, inequality, and climate change. But there are also promising opportunities to be found in technological advancements and developing value chains, markets, and trade.

Africa's population is projected to nearly double to 2.5 billion by 2050, of whom more than 60 percent will be under age 25. This will make it the youngest continent, significantly increasing food demand and straining agrifood systems (AUC 2024). Urbanization will continue to transform dietary patterns, increasing demand for processed and convenience foods and disrupting traditional production and marketing practices. At the same time, climate change, rising temperatures, and resource degradation will increasingly threaten agricultural productivity unless adaptive measures are scaled up. These include climate-smart agriculture practices such as drought-tolerant crop varieties, improved water management (for example, drip irrigation and rainwater harvesting), agroforestry, conservation agriculture, index-based weather insurance, and the use of climate information services to guide planting and harvesting decisions (see Chapters 4, 5, and 10). Strengthening extension services and investing in research and innovation tailored to local agroecological conditions are also essential to support farmers in adapting to changing climatic conditions (see Chapters 8 and 9). The social and political pressures and disruptions generated by these trends are likely to persist (Reardon et al. 2021; FAO 2019).

Smallholder farmers, women, and youth face systemic inequities that limit their productivity and participation in agrifood value chains (see Chapter 6). These groups often lack access to financing, information and new technology, and markets. Addressing these inequities will require targeted investments in infrastructure, innovation, and above all, inclusive policies to bridge rural–urban disparities (IFPRI 2007).

Current trends also offer promising opportunities. Urbanization and economic growth can support agro-industrialization and growth in intra-Africa trade. The AfCFTA, which entered into effect in 2021, is expected to enhance intra-African trade, support development of regional value chains, and expand market access. By 2050, a well-integrated regional market could improve access to affordable and diverse diets as well as fostering sustainable agrifood systems (Reardon et al. 2021; AUC 2024) by reducing food waste, improving diet diversity, strengthening climate resilience, and reducing carbon footprints (AfDB 2017). However, agrifood systems transformation must balance growth and efficiency with food security, sustainability, and inclusivity, especially for small-scale farmers and traders.

Technological innovation can play a central role in overcoming productivity bottlenecks in Africa's agrifood systems. Innovations in digital tools, biotechnology, and precision farming can enhance efficiency and sustainability. For example, mobile platforms like M-Pesa (Kenya) and Esoko (Ghana) already improve smallholders' access to markets, credit, and weather data. Climate-smart practices, such as drought-resistant crops and integrated pest management, can mitigate climate risks (FAO 2019) (see Chapters 4 and 5). Additionally, adoption of improved plant and animal genetic materials, such as stress-tolerant cassava in Nigeria or heat-resistant cattle breeds in the Sahel that are tailored to withstand climate stress, degraded soils, and evolving pest and pathogen pressures, will be crucial for sustainable agricultural intensification and for building resilient agrifood systems (Jain et al. 2023) (see also Chapter 9). Restoring soil health and mitigating desertification are critical to ensuring long-term agricultural viability in arid and semi-arid regions. This includes the Sahel, one of the world's most climate-vulnerable regions, where Africa's Great Green Wall initiative aims to ensure long-term agricultural viability (Partey et al. 2018).

Policy and research needs: A roadmap for the next 25 years

The new CAADP Strategy and Action Plan (2026–2035) provides a framework for achieving sustainable, inclusive, and resilient agrifood systems in Africa. It outlines key policy and research priorities, with an emphasis on actionable strategies and collaborative opportunities for the coming decade.

Actionable policy and research priorities

Achieving CAADP's vision will require a systemic approach that integrates policy research for production, processing, trade, and consumption. Targeted research on key components of Africa's agrifood systems can address pressing food and nutrition challenges. This includes developing and scaling climate-smart innovations not only in crops—such as drought-tolerant maize, biofortified cassava, and high-yield, disease-resistant rice—but also in livestock and fisheries, including improved animal health systems, heat-tolerant cattle breeds, and sustainable aquaculture practices. Research by CGIAR Centers, national agricultural institutions, and other partners plays a vital role in both technological innovation and policy analysis. IFPRI's work, in particular, can guide the design of policies and programs that acknowledge the potential trade-offs among climate-smart agriculture goals: reducing emissions and environmental degradation, increasing productivity, and strengthening resilience.

Capacity development also remains a critical priority. IFPRI's portfolio of research initiatives in Africa features significant capacity development activities at individual, organizational, and institutional levels, which will continue in coming years. At the continental level, IFPRI played a vital role in conceiving, funding, and championing the Collaborative Masters Programme in Agricultural and Applied Economics, which was launched in 2005. With strong leadership from the African Economic Research Consortium (AERC), this program has produced successive cohorts of African researchers and policymakers who have shaped food policy on the continent. Continued investments in human, organizational, and institutional capacity, including the use of digital tools, advanced analytics, and sustainable practices, can empower stakeholders to implement innovative food systems strategies and policies effectively.

Inclusive policies promoting engagement of women and youth in food value chains are vital to Africa's agrifood transformation. Targeted interventions to integrate these groups can foster equitable growth (see Chapter 14). For smallholder farmers, appropriate financing mechanisms, such as blended finance and de-risking instruments, are needed to catalyze investments in agribusiness (see Chapters 17 and 18). Especially critical will be laws and regulations promoting inclusion, public–private collaborations, and farmer and trader collective action in commercializing agrifood value chains.

Strengthening governance and accountability frameworks is equally important. Effective policies, investments, and interventions in Africa's agrifood transformation depend on transparent, well-functioning institutions (see Chapter 15). Without good governance, even well-designed inclusion-promoting policies and financial mechanisms may fail due to corruption, misallocation of resources, weak enforcement, and/or lack of monitoring. The CAADP Biennial Review Process and mutual accountability mechanisms exemplify Africa's embrace of evidence-based policymaking. Its advanced monitoring and evaluation tools can enhance transparency, align stakeholders, and track progress against CAADP's goals.

Collaborative research opportunities and IFPRI's role

Collaboration across and among stakeholders is at the heart of CAADP's approach. Partnerships with entities including the African Union Commission, national governments, and development partners can align resources and priorities across organizations. The rise of national and regional research institutions highlights Africa's capacity to develop locally tailored solutions. Institutions such as AKADEMIYA2063 and AERC and regional initiatives such as the African Network of Agricultural Policy Research Institutes are well equipped to provide localized research capacity and generate context-specific insights.

IFPRI's research priorities align closely with CAADP's strategic objectives and broader African priorities:

- **Climate resilience:** IFPRI's work on climate-smart technologies and sustainable practices addresses the dual challenge of productivity and environmental protection, supporting CAADP's goals for climate-resilient agrifood systems.
- **Nutrition security:** IFPRI's research on biofortification (HarvestPlus), nutrient-rich crops, and nutrition-sensitive policies aligns with CAADP's emphasis on reducing malnutrition and improving public health.
- **Inclusive markets and trade:** IFPRI's work on market efficiency and trade systems is reflected in CAADP's support for the AfCFTA agreement. Policies to reduce trade barriers and enhance regional value chains can improve market access and affordability.
- **Rural transformation:** IFPRI's focus on fostering rural economic development through capacity development, infrastructure, and inclusive growth strategies is congruent with CAADP's commitment to integrating smallholder farmers into value chains.
- **Governance and accountability:** IFPRI's focus on institutional development and evidence-based frameworks helps strengthen governance systems and supports CAADP's mutual accountability processes by providing a platform for transparent monitoring and policy dialogue.

As Africa looks ahead to 2050, research must expand into underexplored areas. These include circular economy approaches, which can reduce food waste and optimize resource use (Abe-Inge et al. 2024); and water-energy-food nexus approaches, which can support integrated management of essential resources—water for irrigation, renewable energy for farming, and sustainable food production—helping to mitigate climate impacts and enhance food security (see Chapter 5) (Mperejekumana 2024). The integration of innovative technologies, including artificial intelligence (AI), digital platforms, and precision agriculture, is increasingly transforming African agriculture. For example, AI-powered tools are now being used for real-time crop health assessments and early detection of livestock diseases in parts of Kenya and Nigeria, offering promising solutions to productivity and climate-related challenges. However, these advances also present challenges, such as access to data, gaps in infrastructure, and capacity constraints. Collaborative research by CGIAR Centers is instrumental in overcoming these barriers. A notable example is the International Livestock Research Institute's pioneering work in gene editing to boost livestock productivity and resistance to diseases such as East Coast fever (ILRI 2024). Strengthening partnerships with regional initiatives such as AfCFTA and the Common Africa Agro-Parks can help ensure that research translates into actionable strategies that foster resilient and inclusive food systems.

The path forward for Africa's agrifood policy

The forward-looking perspective of the CAADP 2026–2035 Strategy and Action Plan reflects Africa's determination to harness agriculture as a driver of economic growth, food and nutrition security, and climate resilience. It underscores the necessity of systemic, inclusive, and evidence-based approaches to agrifood transformation. Investments in climate-resilient technologies, inclusive value chains, and robust governance frameworks are essential to address the demographic, sustainability, and climate challenges projected for 2050. The integration of emerging technologies with evidence-based policies offers a pathway to enhance productivity and sustainability. Collaborative efforts remain central to achieving these goals.

IFPRI's ongoing research is already contributing significantly to these strategic areas. For example, its work on climate-resilient food systems in Ethiopia (Murgatroyd et al. 2025), gender-inclusive extension in Kenya (Timu et al. 2024), and seed systems governance in Nigeria (Takeshima et al. 2025) informs both national policies and regional platforms with rigorous, context-sensitive evidence. In the area of nutrition, IFPRI's biofortification monitoring across sub-Saharan Africa (Friesen et al. 2024) and its empowerment metrics tailored to African contexts (Lentz 2025) are shaping how policymakers integrate food security with health and gender objectives.

Looking ahead, IFPRI is well positioned to help African Union member states and regional economic communities translate the CAADP 2026–2035

Strategy into actionable, country-specific pathways. Through foresight modeling, integrated food system diagnostics, and inclusive policy design, IFPRI's future research can help reconcile trade-offs across CAADP's priorities—such as balancing productivity with environmental sustainability or expanding market access while safeguarding vulnerable populations. Moreover, IFPRI's engagement in platforms such as the CAADP Biennial Review and the AfCFTA provides opportunities to further institutionalize the use of data and evidence in agricultural planning and accountability processes. By continuing to support African stakeholders in navigating complex policy choices with robust, locally grounded analysis, IFPRI's research and engagement can help shape a more resilient, equitable, and food-secure future for the continent.

By learning from past experiences, fostering innovation, and promoting accountability, Africa can achieve its vision of a sustainable, resilient, and food-secure region. The next 25 years present a pivotal opportunity to reimagine Africa's agrifood systems as engines of inclusive growth and resilience. Research institutions including IFPRI, international partners, and African institutions will play a vital role in supporting this transformation through research, capacity development, and policy analysis.

References

Abe-Inge, V. R. Aidoo, E.M. Kwofie, and J.M. Ulimwengu. 2024. "A Nutrition-Sensitive Circular Bioeconomy for Food Systems Transformation in Africa." In *Advancing the Climate and Bioeconomy Agenda in Africa for Resilient and Sustainable Agrifood Systems: ReSAKSS 2024 Annual Trends and Outlook Report*, eds. G. Tadesse, K. Glatzel, and M. Savadogo, 98–115. Kigali: AKADEMIYA2063; Washington, DC: IFPRI.

AfDB (African Development Bank). 2017. "Intra-African Trade Is Key to Sustainable Development: African Economic Outlook." May 23.

AUC (African Union Commission). 2024. *Fourth CAADP Biennial Review Report on the Implementation of the Malabo Declaration*. Addis Ababa: African Union.

Badiane, O., J. Collins, and J.M. Ulimwengu. 2020. "The Past, Present, and Future of Agriculture Policy in Africa." In *Sustaining Africa's Agrifood System Transformation: ReSAKSS 2020 Annual Trends and Outlook Report*, eds. D. Resnick, X. Diao, and G. Tadesse. Washington, DC: IFPRI; Kigali: AKADEMIYA2063.

Badiane, O., and T. Makombe. 2014. "The Theory and Practice of Agriculture, Growth, and Development in Africa." UNU-WIDER Working Paper 2014/061. UNU-WIDER, Helsinki.

Christiaensen, L. 2017. "Agriculture in Africa: Telling Myths from Facts: A Synthesis." *Food Policy* 67: 1–11.

Diao, X., P. Hazell, D. Resnick, and J. Thurlow. 2007. *The Role of Agriculture in Development: Implications for Sub-Saharan Africa*. Research Report 153. Washington, DC: IFPRI.

FAO (Food and Agriculture Organization of the United Nations). 2019. *Climate-Smart Agriculture: Policies, Practices, and Financing for Food Security, Adaptation, and Mitigation*. Rome.

Friesen, V.M., B. Mudyahoto, A.M. Nyangaresi, and M.N.N. Mbuya. 2024. "Monitoring Biofortification Program Performance and Potential for Impact: Indicators, Methods, and Learnings from the Commercialization of Biofortified Crops Program in Six Countries across Africa and Asia." *Current Developments in Nutrition* 8 (12): 104498.

Gilligan, D., J. Hoddinott, and A. Taffesse. 2008. "The Impact of Ethiopia's Productive Safety Net Programme and Its Linkages." *Journal of Development Studies* 45: 1684–1706.

Green, R.H. 1998. "A Cloth Untrue: The Evolution of Structural Adjustment in Sub-Saharan Africa." *Journal of International Affairs* 52 (1): 207–230.

Hazell, P., C. Poulton, S. Wiggins, and A. Dorward. 2010. "The Future of Small Farms: Trajectories and Policy Priorities." *World Development* 38 (10): 1349–1361.

Hoddinott, J., G. Berhane, D. Gilligan, N. Kumar, and A. Taffesse. 2012. "The Impact of Ethiopia's Productive Safety Net Programme and Related Transfers on Agricultural Productivity." *Journal of African Economies* 21: 761–786.

IFPRI (International Food Policy Research Institute). 2007. *The Role of Agriculture in Development: Implications for Sub-Saharan Africa*. Washington, DC.

IFPRI. 2024. *IFPRI's Country Programs: Facilitating Country-Led Food Systems Transformation*. IFPRI Brochure. Washington, DC.

ILRI (International Livestock Research Institute). 2024. "Revolutionary Field Test of Gene-Edited Cattle against East Coast Fever." August 9.

IMF (International Monetary Fund). 1987. "Agricultural Structural Policies." In *Growth-Oriented Adjustment Programs*, eds. V. Corbo, M. Goldstein, and M. Khan. Washington, DC.

Jain, M., C.B. Barrett, D. Solomon, and K. Ghezzi-Kopel. 2023. "Surveying the Evidence on Sustainable Intensification Strategies for Smallholder Agricultural Systems." *Annual Review of Environment and Resources* 48: 347–369.

Kherallah, M., C.L. Delgado, L. Christopher, E.Z. Gabre-Madhin, N. Minot, and M.E. Johnson. 2000. *The Road Half Travelled: Agricultural Market Reform in Sub-Saharan Africa*. Food Policy Report 10. Washington, DC: IFPRI.

Lentz, E., N. Jensen, W. Lepariyo, S. Narayanan, and E. Bageant. 2025. "Adapting the Women's Empowerment in Nutrition Index: Lessons from Kenya." *World Development* 188: 106887.

Mackenzie, F. 1993. "Exploring the Connections: Structural Adjustment, Gender, and the Environment." *Geoforum* 24 (1): 13–34.

Makombe, T., L. Butake, and S. Benin. 2024. "AKADEMIYA2063: A Successful Case of Localization." *IFPRI Blog*, January 30.

Mperejekumana, P., L. Shen, and S. Zhong. 2024. "Exploring the Potential of Decentralized Renewable Energy Conversion Systems on Water, Energy, and Food Security in Africa." *Energy Conversion and Management* 315: 118757.

Murgatroyd, A., T.S. Thomas, J. Koo, K.M. Strzepek, and J. Hall. 2025. "Building Ethiopia's Food Security Resilience to Climate and Hydrological Change." *Environmental Research: Food Systems* 2 (1): 015008.

Partey, S.T., R.B. Zougmoré, M. Ouédraogo, and B.M. Campbell. 2018. "Developing Climate-Smart Agriculture to Face Climate Variability in West Africa: Challenges and Lessons Learnt." *Journal of Cleaner Production* 187: 285-295.

Reardon, T., D. Tschirley, O. Badiane, et al. 2021. "Rapid Urbanization and Food System Transformation in Africa." *Global Food Security* 31: 100585.

Resnick, D., D. Diao, and G. Tadesse. "The Past, Present and Future of Agriculture Policy in Africa." In *Sustaining Africa's Agrifood System Transformation: ReSAKSS 2020 Annual Trends and Outlook Report*, eds. D. Resnick, X. Diao, and G. Tadesse, 9–25. Washington, DC: IFPRI; Kigali: AKADEMIYA2063.

Takeshima, H., C. Ragasa, T. Bamiwuye, et al. 2025. "Seed Certification, Certified Seeds Use and Yield Outcomes in Nigeria: Insights from Nationally-Representative Farm Panel Data and Seed Company Location Data." *Agricultural Systems* 224 (March): 104268.

Taube, G. 1993. "Social Dimensions of Adjustment (SDA): Conceptual Issues and Empirical Evidence from Tanzania and Other African Countries." *Africa Spectrum* 28 (3): 233–257.

Timu, A.G., D. Manoti, A. Shee, and L. You. 2024. "Impacts of Gender-Inclusive Extension Approaches on Farmer Understanding and Willingness to Pay for Bundled Financial Services." *Current Research in Environmental Sustainability* 8 (2024): 100268.

von Braun, J. 2005. *The IFPRI 2020 Vision for Food, Agriculture, and the Environment.* Washington, DC: IFPRI, 2005.

World Bank. 2004. *A Critical Review of the Literature on Structural Adjustment and the Environment.* Washington, DC.

World Bank. 2023. *South Asia Economic Focus, April 2023: Toward Faster, Cleaner Growth.* Washington, DC.

Chapter 20

CENTRAL ASIA

Kamiljon Akramov, Isabel Lambrecht, Suresh Babu, and Sharanya Rajiv

Kamiljon Akramov is a senior research fellow, Development Strategies and Governance Unit (DSG), IFPRI; **Isabel Lambrecht** is a research fellow, DSG, IFPRI-Dushanbe; **Suresh Babu** is a senior research fellow and head of capacity strengthening, DSG, IFPRI; and **Sharanya Rajiv** is a research analyst, DSG, IFPRI.

Since the countries of Central Asia gained independence from the Soviet Union in 1991, the region's food systems have undergone significant transformations shaped by political and economic reforms, institutional shifts, globalization, climate change, and labor migration. At the beginning of their transition to market-based economies, Central Asian countries experienced significant economic decline, poverty, and food insecurity. The region faced immediate policy, governance, institutional, and capacity challenges and largely lacked the macroeconomic setting and enabling environment necessary for a successful transition (Rozelle and Swinnen 2004). As a result, agricultural production and the broader local economies were severely disrupted by the change.

Economic growth in the region has since rebounded and passed the pre-transition levels due to market-oriented reforms, which, coupled with agricultural transformation and some diversification, have led to substantial reductions in poverty, food insecurity, and undernutrition. Food policy research has played a crucial role in the region's evolution, with international organizations, universities, and research institutes developing evidence to inform policy reforms and promote sustainable agricultural development and food security throughout Central Asia. The collaborative research conducted by international research centers supported by CGIAR and other international organizations and universities helped to initiate a food and agriculture policy dialogue and coordination among the Central Asian countries to achieve regional food and nutrition security. In particular, IFPRI helped to carry out a review of the policy reform process and agrifood sector development by convening external technical groups as well as local and international researchers working in the region to identify promising options for agrifood policy reforms (Babu

BOX 20.1 IFPRI in food policy dialogue in Central Asia

In 1997, the World Bank invited IFPRI to provide policy analysis training to a group of analysts from the Central Asia region. This led to the development of a working relationship between IFPRI researchers and the agricultural ministries of the five Central Asian countries to help them with policy analysis, work that was supported through World Bank–funded agrarian reform projects. At the same time, CGIAR, with the International Center for Agricultural Research in Dry Areas (ICARDA) taking the lead, established technical collaboration with the national agricultural research systems (NARS) in these countries. It also established a regional program management unit in Tashkent, Uzbekistan, which provided logistical support to CGIAR's research and capacity development activities. IFPRI, together with ICARDA, the International Water Management Institute, the International Maize and Wheat Improvement Center, the International Potato Center, and the Alliance of Bioversity International and CIAT, participated in this consortium to support the policy aspects of the agricultural transition process while also working independently with country-level collaborators in the region. Collaborative research conducted during the program's first decade established IFPRI as a reliable partner for national agrifood system policymakers in the region. Later, in the 2010s, two CGIAR Research Programs led by IFPRI—Policies, Institutions, and Markets (PIM) and Agriculture for Nutrition and Health (A4NH)—continued this work, providing research and capacity strengthening support to Central Asian governments, research centers, and individual researchers.

and Djalalov 2006; Babu and Pinstrup-Anderson 2000; Babu and Tashmatov 2000). IFPRI also helped to create a network of researchers and experts who can analyze the challenges and policy issues pertinent to food and nutrition security in the national and regional contexts (see Box 20.1).

Despite significant progress in improving food and nutrition security since the initial transition, serious challenges remain, including vulnerability to climate change impacts, limited availability and management of water resources, limited local employment opportunities, heavy reliance on international remittances, unstable commodity markets, undiversified trade flows, poor diets, and the high cost of healthy diets. Addressing these issues will require further research and policy focus on sustainable agriculture, rural entrepreneurship, regional cooperation, technological innovations in food production and marketing, and greater diversification toward high-value agrifood chains.

Food systems challenges in the early years of independence

As Central Asian countries began their transition in the early 1990s, agriculture dominated the region's economy, contributing about one-third of gross domestic product (GDP) and providing more than 50 percent of employment. Thus, a major policy issue for the transition was the impact of market-oriented agrarian reforms on rural poverty and food security.

During the first decade of independence, aggregate output and productivity declined substantially, with agrifood production falling precipitously as the sector was especially hard hit by the shift away from Soviet-era central planning (Pomfret 2010, 2016). Across the region, gross agricultural output dropped by nearly 40 percent between 1990 and 1998. Improving agricultural performance was essential to boost household incomes, especially in poor rural areas reliant on agriculture, but required changes to producer incentives, a shift to a profit orientation, and individual involvement (Lerman et al. 2004; Lerman and Sedik 2018). To make this transition, governments in the region had to tackle several immediate challenges: distorted prices and incentives, farm reorganization and land reform, irrigation and water management, and the need for market institutions (Rozelle and Swinnen 2004).

Liberalization of input and output prices

Under the planned economy of the Soviet era, central planners attempted to stimulate food production by lowering input prices and raising output prices. Collective and state farms received critical inputs—such as seeds, machinery, and fertilizers—at subsidized rates, and could sell their outputs to state procurement agencies at above-market prices. The elimination of these subsidies and premiums following independence significantly affected the relative prices of inputs and outputs, leading to severe impacts on agrifood production and productivity in the region, as in other post-Soviet countries (Rozelle and Swinnen 2004). Kazakhstan and Kyrgyzstan implemented bold price reforms that increased incentives for profit-oriented individual farms, while Tajikistan, Turkmenistan, and Uzbekistan delayed implementing reforms (Akramov 2011; Anderson and Pomfret 2003; Anderson and Swinnen 2009). Despite differences in the speed and coverage of price reforms, most input and output prices in Central Asian agriculture are now liberalized and determined by market forces, with some minor exceptions.

Land rights, tenure reforms, and productivity impacts

The institutional reforms and liberalization of input and output prices led to the collapse of large collective farms, a drop in agricultural production, and loss of employment for those formerly working in state and collective farms and state-owned enterprises. Households saw their incomes drop while they also faced limited access to food, which led to a surge in poverty, food insecurity, and malnutrition (Babu and Tashmatov 2000). Compared with large collective farms that employed hired labor and were run by outside managers, individual or family farms were easier to organize and operate. However, the move toward smaller, individual farms brought significant problems in accessing agricultural extension services and input and output markets, as well as in managing critical infrastructure, such as on-farm and common pool irrigation networks. Central Asia's agriculture sector officially recognizes two major types of individual farms: homestead plots and peasant farms, differentiated mainly by their commercial orientation, size, and legal status. Homestead plots are generally smaller and more subsistence-oriented than peasant farms, although there is some overlap between the two groups. The third officially recognized farm type is the agricultural enterprise, or corporate farm, a category that includes state, cooperative (collective), and large private farms.

While all Central Asian governments recognized the need for the individualization of agricultural production, the pace and extent of land tenure reform differed significantly across the region. Kyrgyzstan and Kazakhstan began land reforms in the mid-1990s, eventually introducing private land ownership. Uzbekistan initiated limited land reforms in the 1990s, but by offering long-term leases to farmers, the government was able to control the structure and size of farms and ensure their continued dependence on the state. Tajikistan initiated reforms in 1992, but notable progress began only after the state granted land use rights in 1998. Turkmenistan also initiated land reforms; in 1996, the government began leasing land to farmers but maintained control over land, inputs, and output quotas.[1] Today, all agricultural land effectively remains state-owned in Tajikistan, Turkmenistan, and Uzbekistan, with farmers holding only long-term use rights (Lerman and Sedik 2018).

Land tenure reforms led to significant shifts in agricultural land distribution among different farm types. The total area of agricultural land allocated to agricultural enterprises declined significantly throughout the region, except in Kazakhstan. The share of arable land in individual use (homestead plots and peasant farms) increased from 5 percent in 1991 to nearly 80 percent in 2010, with a corresponding increase in the share of agricultural production generated by these farms (Lerman and Sedik 2018). However, land reforms remain

1 For more on these reforms, see EBRD (2002), Akramov and Omuraliev (2009), Djanibekov et al. (2012), Akramov and Shreedhar (2012), and Ni et al. (2004).

incomplete in some Central Asian countries. While small homestead farms enjoy complete freedom to farm, larger peasant farms still face some restrictions on crop choice and marketing decisions (Lerman and Sedik 2018).

Partial productivity analyses for Central Asia show considerable complexity in the relationship between land and labor productivity as well as substantial differences across countries (Figure 20.1). Although both land and labor productivity dropped at the beginning of the transition across all countries, the subsequent rise in individual cultivation of land coincided with an increase in land and labor productivity, marking a turning point for Central Asia's agriculture sector. Research shows that individual farms have higher levels of productivity and efficiency than large enterprise farms, underscoring the importance of individual incentives for farm efficiency (Lerman et al. 2004; Lerman and Sedik 2018). Research comparing the early and comprehensive land reforms in Kyrgyzstan and Kazakhstan with the slower and incomplete reforms in Uzbekistan, Tajikistan, and Turkmenistan found that early reformers experienced more significant improvements in agricultural productivity and rural incomes (Babu and Tashmatov 2000). Kyrgyzstan saw an early rebound (1995) in labor and land productivity, while productivity in Kazakhstan, Turkmenistan, and Uzbekistan rebounded later (1997). Tajikistan, which experienced a civil war during the transition period, did not see a rebound until 1999 and did not reach pre-transition productivity levels until 2007 (Lerman and Sedik 2018).

The productivity growth patterns illustrated in Figure 20.1 show that output per hectare is higher and has grown faster after the turning point in land-constrained countries (Tajikistan and Uzbekistan), where small homestead farms produce a substantial share of crop and livestock production, than in land-abundant countries (Kazakhstan and Turkmenistan), where large farms still dominate agriculture. Conversely, output per worker is higher in land-abundant countries than in land-constrained countries, although labor productivity growth trends appear similar. Another interesting observation is that land and labor productivity growth in Kyrgyzstan slowed compared to other countries. One explanation is that farms in Kyrgyzstan are too small to mobilize the investment and agricultural machinery needed to support more productive farming practices (Guadagni and Fileccia 2009; Mogilevskii et al. 2017).

While some policy reforms remain incomplete, recently, growing agricultural land acquisitions by large domestic and foreign investors are creating new policy challenges (Hofman 2016, 2021; Visser and Spoor 2011). These investments may lead to land consolidation and higher land and labor productivity by reducing transaction costs and benefiting from economies of scale.

FIGURE 20.1 Land and labor productivity in Central Asian agriculture, 1980–2022

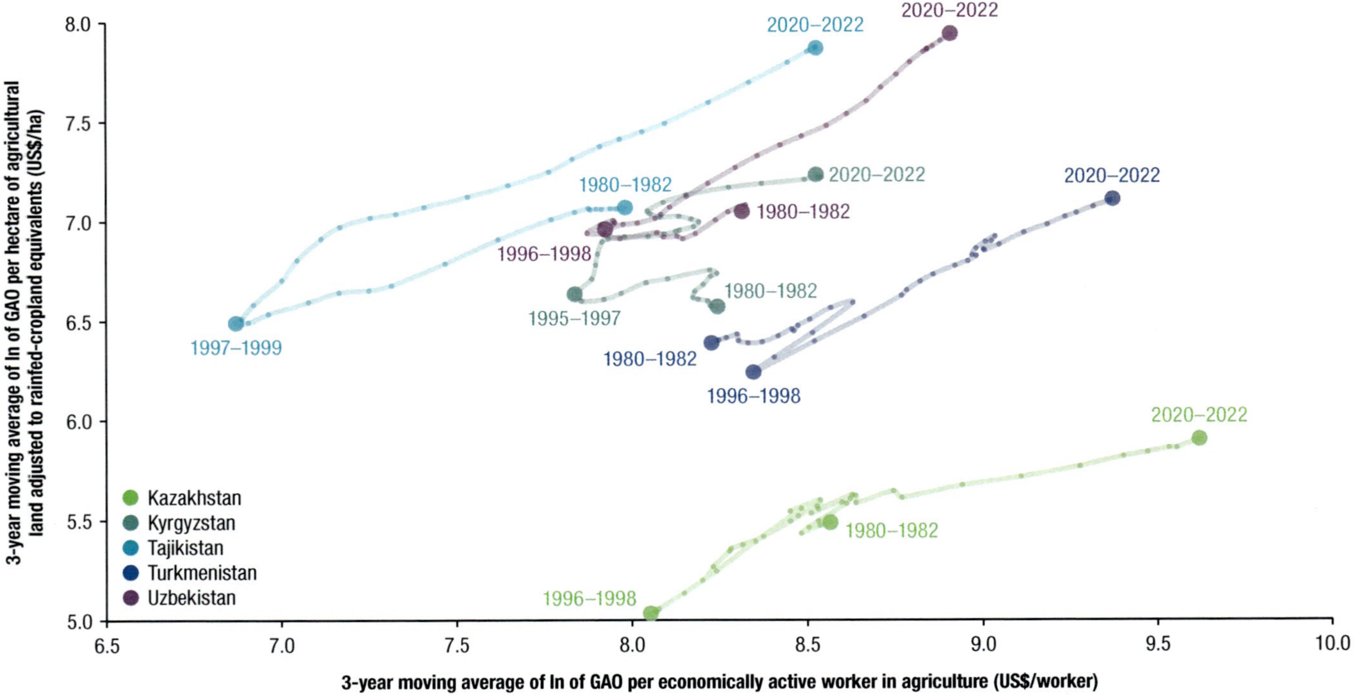

Source: Authors' depiction using data from USDA-ERS, accessed 2024.
Note: GAO = gross agricultural output; ln = natural log.

However, land consolidation also risks a negative impact on employment, as larger farms may require less labor. If the benefits of increased productivity exacerbate income inequality and do not reach the most vulnerable households, this consolidation will not help to address rural poverty.

Water management

Irrigation plays a crucial role in Central Asia's agricultural systems due to the region's generally dry climate (Babu and Djalalov 2006). Kyrgyzstan, Tajikistan, Turkmenistan, and Uzbekistan irrigate more than 80 percent of arable land. Northern Kazakhstan is the exception; reliant on rainfed wheat production, the country irrigates only 7 percent of arable land. The individualization of farming has required reconsideration of irrigation water management policies. Extensive irrigation systems were built throughout the region as part of the significant Soviet-era investments in agricultural infrastructure (Anderson and Swinnen 2009). These systems were designed for large-scale collective farms and consequently are poorly aligned with the water management needs of today's smaller farms. The governments understood these needs and recognized that centralized management of the irrigation infrastructure was not possible in the new individualized agricultural systems. They initiated the decentralization of water management responsibilities to the local communities and promoted collective action among farmers (Babu and Akramov 2022). Water users' associations, found in various forms worldwide, began emerging in Central Asia in the late 1990s and early 2000s to provide micro-level management of water resources and collect irrigation service fees from farmers. However, these associations have struggled to generate adequate revenue and capital for the maintenance and operation of irrigation infrastructure, placing a significant financial burden on government budgets.[2] Managing local irrigation networks and water allocation among farmers thus remains a major challenge.

Institution building

Amid the more immediate need for reforms to price policies, land tenure, and farm organization, Central Asian countries were slow to build institutions to foster competition, define and enforce property rights and contracts, ensure access to credit, and provide information to farmers and other food system actors. These institutional limitations had predictably negative effects on long-term investment and short-term production decisions (EBRD 2002;

2 Relevant studies include Akramov and Omuraliev (2009), Akramov and Shreedhar (2012), Djanibekov (2008), Babu and Akramov (2022), and Frenken (2013).

Rozelle and Swinnen 2004; Spoor 1999). Ambitious efforts to reorganize and privatize farms were often constrained by the lack of functioning agricultural input, output, and financial markets, and real privatization was stalled in the early years of the transition period. Since then, governments have created legal frameworks and promoted market institutions to nurture competition, protect property rights, and enforce contracts. Similarly, options for rural credit, severely limited in the early stages of independence, have expanded in recent years (Akramov et al. 2023).

During the early years of the transition, comparative analyses provided policy recommendations for improving agricultural productivity and food security, such as reforms to incentives, subsidy reductions, and establishment of market institutions (Anderson and Swinnen 2009; Rozelle and Swinnen 2004). These studies highlighted examples of reforms that reduced distortions to farm incentives, such as the relatively quick liberalization implemented in Kyrgyzstan and the reforms introduced to reduce state control of the cotton sector in Uzbekistan, which have both led to improvements in investment, productivity, and much-needed diversification of agricultural production. The research on Kyrgyzstan and Tajikistan helped to identify focus areas for development partners' support programs. These studies emphasized the need for an integrated approach to improving food and nutrition security that considers macro, sectoral, and household-level components (Akramov and Omuraliev 2009; Akramov and Shreedhar 2012). Efforts have also been made to build local capacity for policy analysis and implementation, including training local researchers and policymakers to develop and execute effective food policies (Babu and Tashmatov 2000).

Evolution of food systems challenges

Since the late 1990s, renewed economic growth, combined with improvements in labor and land productivity, has led to a more than twofold increase in gross agricultural output (Akramov et al. 2023). As agriculture and other sectors rebounded, research attention shifted toward addressing the detrimental effects of climate change and its impact on land degradation and water scarcity, improving the efficiency of agricultural production, and achieving better nutrition. Demographic trends have come under greater scrutiny as well, as labor migration and remittances have become an important coping strategy for poor rural households (Atamanov and Van Dan Berg 2012). Moreover, while poverty and food insecurity have decreased significantly, they remain major concerns throughout the region.

Nutrition and poverty

Despite being agrifood producers, Central Asian countries rely on food imports to meet at least a part of their needs, leveraging regional trade based on comparative advantages in the production of various crops. For example, Kazakhstan exports wheat to other Central Asian countries and imports fruits and vegetables from Kyrgyzstan and Uzbekistan. Rising global food prices and supply chain disruptions in recent years have increased the risk of food insecurity and malnutrition, especially for poor people, who obtain most of their calories from staples (especially wheat). In 2022, 13.1 percent of children under five years old in Tajikistan were stunted, and 9.3 percent of Tajikistan's population faced undernourishment. At the same time, micronutrient deficiencies, overweight and obesity, and diet-related noncommunicable diseases are increasingly common throughout Central Asia, with more than half of all adults overweight or obese (WHO 2025).

Targeted interventions, including social safety nets, nutrition programs, and initiatives to improve rural livelihoods, have been highlighted as key for supporting vulnerable populations (EBRD 2002). Spending on social safety nets in Central Asian countries is relatively high by international comparison, but its effectiveness is low due to limited coverage and adequacy. For example, existing safety nets in Tajikistan do not enable households to invest and enhance resilience or provide sufficient benefits to improve economic access to nutritious diets. Also, targeting of social safety nets remains problematic in the region (Dugarova 2019; Kawabata et al. 2020; World Bank 2022).

Climate change and natural resources

Climate change is already affecting Central Asia's food systems, with an increase in the frequency and intensity of extreme events, notably heatwaves, droughts, and dust storms, in recent decades. Research by IFPRI and partners helped to estimate the costs of these impacts and develop policy recommendations to address their adverse effects on food systems. Land degradation reduces agricultural incomes by lowering crop yields and livestock productivity and increasing production costs. Collaborative research between IFPRI and the University of Bonn estimated the annual cost of land degradation in Central Asia at about US$5.85 billion between 2001 and 2009, equivalent to 3 percent of GDP in Kazakhstan and Uzbekistan, 4 percent in Turkmenistan, and 10–11 percent in Kyrgyzstan and Tajikistan (Mirzabaev et al. 2016).

Central Asia's limited water resources are being further reduced by climate change. Changes in critical seasonal runoff (meltwater) and precipitation, related in part to earlier and greater glacier melt and snowmelt and increased evapotranspiration, aggravate regional water shortages and limit the availability of irrigation water in the summer. Making food systems resilient to the changing climate will require water-saving irrigation technologies, such as laser leveling, drip and sprinkler irrigation, drought and heat-tolerant or resistant crops, and climate-resilient crop varieties. Research conducted by the International Water Management Institute in the Karshi region of Uzbekistan shows that improved irrigation practices, such as drip irrigation, can reduce water use by approximately 30 percent compared to traditional irrigation practices (Djumaboev, Manthrithilake et al. 2019; Djumaboev, Yuldashev et al. 2019).[3] These technologies were projected to improve crop yields in pump-irrigated areas, as well as reduce energy use and carbon emissions, with estimates suggesting that optimized irrigation could save a half trillion liters of water and 259 gigawatt hours of electricity. These practices could offer a promising path forward for other Central Asian countries, where pump irrigation now covers about 2.84 million hectares, or about 18 percent of the irrigated area.

Climate change and associated extreme weather events affect agricultural production and productivity in Central Asia, leading to declines and volatility in crop yields (Bobojonov and Aw-Hassan 2014; Sommer et al. 2013). Crop diversification is a feasible and cost-effective means to improve the resilience of the region's food systems, and diversification can help farmers use land and water more efficiently and increase their net returns per hectare. Evidence from the region also suggests that agricultural diversification can improve dietary diversity, nutritional outcomes, and intrahousehold equality for women and children (Takeshima et al. 2020, 2022, 2024). Central Asian governments are now pursuing diversification strategies that emphasize high-value agriculture, such as horticulture and oilseed production. In partnership with the Asian Development Bank and the CAREC Institute, IFPRI research on Central Asia's agrifood systems has provided policy-relevant information that can help design interventions for diversification and revitalization of the region's agriculture sector (CAREC et al. 2019).

Migration and remittances

Central Asia's food systems are also being shaped by migration and other demographic trends. A shortage of domestic employment opportunities and historically low wages have led to significant labor migration from Kyrgyzstan,

3 The evidence-based policy recommendations of the CGIAR Research Program on Water, Land, and Ecosystems made a significant contribution to policy change and development and adoption of the state program on water-saving technologies. For more details, see www.cgiar.org/annual-report/performance-report-2020/ water-technologies-shift-energy-policies-in-uzbekistan/

Tajikistan, and Uzbekistan, mainly to Russia. This trend has increased over the last 15 to 20 years, with remittances increasingly contributing to economic growth, poverty reduction, and food security in these countries. In mid-2020, an estimated 6.6 million individuals migrated from the region's five countries, accounting for more than 10 percent of the region's total population.[4] Evidence suggests that remittances increase household incomes, improve food security, and strengthen child nutritional outcomes, and that they outweigh the costs of migration (Azzarri and Zezza 2011; Ghimire et al. 2023). However, labor migration reduces the agricultural income of migrant-sending households, and the benefits from migration decline with farm size. Women in migrant households are more likely to be involved in decisions about productive activities on the household farm, control income, own assets, and achieve workload balance compared to women in non-migrant households (Slavchevska et al. 2021). In addition, women in migrant households are more involved in labor-intensive activities as a result of the loss of male labor (Atamanov and Van Dan Berg 2012; Bloem and Lambrecht 2023; Ghimire et al. 2023), which has important implications for designing gender-inclusive agricultural and nutrition programs and policies in the region.

Future food system challenges and directions for food policy research

By 2050, Central Asia's food systems will likely face multiple interconnected challenges driven by climate change, remaining issues in land use, the market environment and farmers' incentive structure, demographic shifts, socioeconomic trends, and geopolitical factors. Policy research can provide evidence to help address these challenges.

Invest in climate change mitigation and adaptation

Central Asia is highly vulnerable to climate change and increasingly faces climate-driven water scarcity, land degradation, and extreme weather events. Rising temperatures are already significantly above the global mean (Yao et al. 2021), and projected impacts suggest that temperatures will rapidly increase even further. A combination of changes in precipitation and increasing glacier melt will likely aggravate regional water shortages. Currently, Turkmenistan and Uzbekistan are severely water scarce, with withdrawal levels for renewable water resources exceeding 100 percent (Akramov et al. 2022). Other

---

4 Data from the Migration Data Portal, accessed March 7, 2025. www.migrationdataportal.org.

Central Asian countries will also experience a steady decline in the per capita availability of renewable water resources due to degradation of land and water resources. This scenario poses significant risks to agrifood systems, particularly crop yields, given the region's arid climate and reliance on snow-fed rivers. These threats urgently require mitigation and adaptation strategies.

Food policy research can help address these challenges by examining governance and financial arrangements for the implementation of sector-specific strategies at the regional, national, and local levels, identifying innovations for managing climate change–related production and economic risks, and investigating drivers and barriers to the adoption of innovative climate-smart agricultural and water-saving technologies.

Address constraints to land access and land consolidation

Although land tenure reforms have significantly improved access to land, increased employment, and reduced poverty, Central Asia still faces significant issues related to land fragmentation as well as growing concerns about involuntary consolidation. Larger farms benefit from higher labor productivity (and, therefore, income), while land fragmentation increases transaction costs, for instance, due to the need for multiple small farmers to agree on land use. Small farms may also have less access to credit, as lenders may be cautious about providing loans to small, fragmented farms and may not adopt modern technologies and machinery because of the difficulty of coordinating across multiple small peasant farms.

Future research should examine various policy options, including strengthening the regulatory frameworks for leasing and selling land to ensure efficiently functioning land markets, identifying innovative land tenure policies that can secure farmers' property rights, and creating economic incentives for market-based (voluntary) land consolidation. Research can help policymakers better understand the complex trade-offs between promoting economic efficiency and ensuring equitable distribution, as land consolidation policies aimed at maximizing productivity could exacerbate inequalities and poverty. It can also assist in addressing land tenure and structural transformation challenges associated with land consolidation, which can significantly impact rural livelihoods, food security, and environmental sustainability.

Identify policies for better nutrition and inclusivity

As the region's population grows, rising food demand and changing diets will place greater pressure on food systems. Central Asia's population rose to more than 82 million in 2024, an increase of more than 60 percent since 1990, and it is expected to exceed 112 million by 2050.[5] The share of working-age people is projected to increase even faster, a trend that will likely exacerbate socioeconomic challenges in the region, as many young people and women struggle to obtain employment and earn adequate income even now (Akramov et al. 2020). The high cost of healthy diets, inadequate nutrition knowledge, and unhealthy consumption habits have led to rising trends in overweight, obesity, and other nutrition-related health issues in recent years (Akramov et al. 2024). Addressing these obstacles to healthy diets and creating employment opportunities for youth and women will be essential to strengthening the inclusiveness of food systems, including improving incomes, equity, and nutrition for poor people. Of note is that rising incomes are generally associated with higher dietary diversity and increasing demand for livestock products, fruits, vegetables, and other nutritious foods (Krivonos and Kuhn 2019). Policy research can support these goals by examining and identifying feasible policy, institutional, technological, and behavioral innovations to meet the growing demand for food and help transform food systems in the region for better nutrition and diet quality.

Increase resilience to market volatility

Central Asian countries are vulnerable to volatility in international commodity markets, primarily due to their heavy reliance on exports of just a few commodities, including oil, gas, gold, and cotton, and imports of key food staples such as wheat. This reliance leads to large fluctuations in government revenue as commodity prices rise and fall, impacting economic stability, investment decisions, and social welfare. For example, Kazakhstan, a major oil exporter, is heavily affected by global oil prices. Price fluctuations for cotton, a key commodity in Tajikistan and Uzbekistan, significantly impact the livelihoods of farmers and textile industry workers. Kyrgyzstan, Tajikistan, and Uzbekistan import a substantial share of wheat, and thus are strongly affected by changes in wheat prices. In addition, close economic and trade connections with Russia, a country highly dependent on oil and gas exports, increase Central Asia's vulnerability to market volatility. Similarly, geopolitical and trade tensions linked to Russia negatively affect the region as trade and remittance flows decline (Akramov et al. 2019).

Central Asian countries can reduce their reliance on commodity exports by investing in other sectors, such as manufacturing and agrifood processing.

---

5 World Population Review, accessed February 5, 2025. https://worldpopulationreview.com/continents/central-asia.

They can also foster independence from the Russian economy by diversifying their international trade partners. In recent years, Central Asian countries have begun exporting agricultural products to China. For example, according to the Ministry of Agriculture of Kazakhstan, the country's grain exports to China have reached 1.4 million tons and are expected to reach 2 million tons in the near future. The Ministry of Agriculture of Uzbekistan reports that the value of its edible fruit and vegetable exports to China topped US$150 million in 2023. Rising demand for food products in China and elsewhere in Asia may provide an opportunity for Central Asian countries to further diversify their agricultural production and increase export earnings, especially from oilseeds, horticulture, and livestock (Burunciuc et al. 2018).

Food policy research can help the Central Asian governments prioritize improved connectivity to better integrate into global value chains and overcome geographic disadvantages. Research can also evaluate the region's remaining trade barriers and recommend reforms for trade facilitation, diagnose obstacles or constraints in individual value chains, and explore the merits of alternative solutions that use data-driven approaches.

Promote food system efficiency and growth

The outcomes of agricultural reforms and reductions in market distortions have varied across Central Asian countries. In addition, evidence suggests that adoption of innovations is slow in the region's agrifood systems due to unfavorable policy, regulatory, and market environments. Future food policy research can help identify remaining bottlenecks and provide insights into what catalyzes agrifood system development. Research should also explore the public foundations necessary for private sector agricultural growth and how government budgets can support public goods investments—especially in infrastructure and institutions—as well as in improving the quality and efficiency of agrifood systems, including through better product traceability. Research can also help to identify the barriers to adoption of innovative practices and technologies by conducting enabling environment diagnostics and dynamic causal evaluations.

Generate data and analysis

Central Asian countries lack reliable data and analysis on important aspects of food systems, including climate change impacts; land, labor, and water productivity; costs and benefits of climate-smart agriculture and water-saving technologies; and nutrition and dietary practices at the national, community, and household levels. For example, a nationally representative agricultural household survey has never been conducted in Turkmenistan and Uzbekistan.[6] In Tajikistan, the most recent publicly available nationally representative survey data that included detailed information on household agricultural activities were collected in 2007 and 2009.[7] This lack of information poses a significant challenge to understanding how the agrifood system functions in these countries and developing appropriate evidence-based policy measures. IFPRI and other research institutions can help address these needs by generating evidence-based policy recommendations, supporting countries in their collection of data, strengthening countries' analytical and applied policy research capacity, and examining political economy and policy coordination problems related to food systems.

Emphasize capacity building

Given the transboundary nature of many challenges in the region, such as labor migration, climate change, and water management, it is crucial to advance a regional approach to capacity building. Collaborative research and capacity-building activities that bring together stakeholders from across the region have been important components of IFPRI's research and outreach in Central Asia, beginning with IFPRI's early work convening stakeholders to identify research priorities for advancing food policy reforms (Babu and Tashmatov 2000). Building on promising examples such as this can enhance the relevance and impact of policy research. Capacity building and collaborative research not only improve research quality but also foster strong policy research linkages across countries in the region, paving the way for more coordinated and effective responses to shared challenges in the future.

6 The National Statistical Committee of the Republic of Uzbekistan regularly conducts household budget surveys. However, these survey data are not publicly available.

7 The Leibniz Institute for East- and Southeast European Studies (Germany), in cooperation with the SHARQ Research Center based in Dushanbe, Tajikistan, conducted a household survey using the sample of the 2007 and 2009 surveys. However, this survey focused on migration and remittances, not agricultural activities (https://lambda.ios-regensburg.de/tajikistan-household-panel-survey-2011). In 2017, the Food and Agriculture Organization of the United Nations and the Agency on Statistics under the President of the Republic of Tajikistan conducted a nationally representative rural household survey. However, these survey data are not publicly available.

References

Akramov, K.T. 2011. "International Food Prices, Agricultural Transformation, and Food Security in Central Asia." *Development in Practice* 21 (4–5): 741–754.

Akramov, K., K. Chen, A. Park, J. Ilyasov, and S. Eltazarov. 2023. "Agricultural Development and Cooperation in Central Asia." In *Sino-Asia Agricultural Trade and Development Cooperation: Progress, Challenges and Outlook*, eds. K. Chen and R. Mao, 113–160. Montreal, Quebec: Royal Collins Publishing Group.

Akramov, K., K. Djumaboev, and R. Romashkin. 2022. "Regional Developments: Promoting Climate-Smart Practices and Crop Diversification in Central Asia." In *Global Food Policy Report 2022: Climate Change and Food Systems*, 124–127. Washington, DC: IFPRI.

Akramov, K., J. Ilyasov, E. Tsvetnov, and A. Park. 2019. "Regional Developments: Growing Connectivity Including Transportation, Investment, and Trade in Central Asia." In *Global Food Policy Report 2019*, 86–117. Washington, DC: IFPRI.

Akramov, K., I. Lambrecht, and S. Pechtl. 2024. "Regional Developments: Food Systems for Healthy Diets and Nutrition in Central Asia." In *Global Food Policy Report 2024: Food Systems for Health Diets and Nutrition*, 96–102. Washington, DC: IFPRI.

Akramov, K., and N. Omuraliev. 2009. "Institutional Change, Rural Services, and Agricultural Performance in Kyrgyzstan." IFPRI Discussion Paper 904. IFPRI, Washington, DC.

Akramov, K., R. Romashkin, A. Park, and J. Ilyasov. 2020. "Central Asia." In *Global Food Policy Report 2020*, 74–76. Washington, DC: IFPRI.

Akramov, K., and G. Shreedhar. 2012. "Economic Development, External Shocks, and Food Security in Tajikistan." IFPRI Discussion Paper 1163. IFPRI, Washington, DC.

Anderson, K., and R. Pomfret. 2003. *Consequences of Creating a Market Economy*. Cheltenham, UK: Edward Elgar Publishing.

Anderson, K., and J. Swinnen. 2009. "Distortions to Agricultural Incentives in Eastern Europe and Central Asia." Agricultural Distortions Working Paper 63. World Bank, Washington, DC.

Atamanov, A., and M. van dan Berg. 2012. "Heterogeneous Effects of International Migration and Remittances on Crop Income: Evidence from the Kyrgyz Republic." *World Development* 40 (3): 620–630.

Azzarri, C., and A. Zezza. 2011. "International Migration and Nutritional Outcomes in Tajikistan." *Food Policy* 36 (1): 64–70.

Babu, S.C., and K. Akramov. 2022. "Agrarian Reforms and Food Policy Process in Transition Economies: Lessons from Tajikistan Agriculture." *Central Asian Journal of Water Research* 8 (1): 27–48.

Babu, S.C., and S. Djalalov. 2006. *Policy Reforms and Agriculture Development in Central Asia*. New York: Springer.

Babu, S.C., and P. Pinstrup-Andersen. 2000. "Achieving Food Security in Central Asia: Current Challenges and Policy Research Needs." *Food Policy* 25 (6): 629–635.

Babu, S.C., and A. Tashmatov, eds. 2000. *Food Policy Reforms in Central Asia: Setting the Research Priorities*. Central Asia Policy Brief No. 1. Washington, DC: IFPRI.

Bloem, J.R., and I. Lambrecht. 2023. "Income Aspirations, Migration, and Investments On and Off the Farm: Evidence from Rural Tajikistan." IFPRI Discussion Paper 2209. IFPRI, Washington, DC.

Bobojonov, I., and A. Aw-Hassan. 2014. "Impacts of Climate Change on Farm Income Security in Central Asia: An Integrated Modeling Approach." *Agriculture, Ecosystems & Environment* 188: 245–255.

Burunciuc, L., W. Fengler, W. Schloemer, and D. Taglioni. 2018. "How Uzbekistan Is Transforming into an Open Economy." Commentary. Brookings, December 20.

CAREC Institute, IFPRI, and Asian Development Bank. 2019. *Agriculture Development in Central Asia Regional Economic Cooperation Program Member Countries: Review of Trends, Challenges, and Opportunities*. Manila.

Djanibekov, N. 2008. "A Model-Based Analysis of Land and Water Use Reforms in Khorezm: Effects on Different Types of Agricultural Producers." In *Continuity and Change: Land and Water Use Reforms in Rural Uzbekistan*, eds. P. Wehrheim, A. Schoeller-Schletter, and C. Martins, 43–62. Halle, Germany: Leibniz Institute of Agricultural Development in Central and Eastern Europe.

Djanibekov, N., K. van Assche, I. Bobojonov, and J.P.A. Lamers. 2012. "Farm Restructuring and Land Consolidation in Uzbekistan: New Farms with Old Barriers." *Europe-Asia Studies* 64 (6): 1101–1126.

Djumaboev, K., H. Manthrithilake, J. Ayars, T. Yuldashev, B. Akramov, R. Karshiev, and D. Eshmuratov. 2019. "Growing Cotton in Karshi Steppe, Uzbekistan: Water Productivity Differences with Three Different Methods of Irrigation." In *Proceedings of 9th International Micro Irrigation Conference, Aurangabad, India*, 391–397. Bangalore: Ivy League.

Djumaboev, K., T. Yuldashev, B. Holmatov, and Z. Gafurov. 2019. "Assessing Water Use, Energy Use, and Carbon Emissions in Lift Irrigated Areas: A Case Study from Karshi Steppe in Uzbekistan." *Irrigation and Drainage* 68 (3): 409–419.

Dugarova, E. 2019. "Poverty Reduction through Social Protection and Labor Policies in the Former Soviet Union." In *Social Policy, Poverty, and Inequality in Central and Eastern Europe and the Former Soviet Union*, eds. S. An, T. Chubarova, B. Deacon, and P. Stubbs, 123–145. New York: Columbia University Press.

EBRD (European Bank for Reconstruction and Development). 2002. *Transition Report 2002: Agriculture and Rural Transition*. London.

Frenken, K. 2013. *Irrigation in Central Asia in Figures: AQUASTAT Survey 2012*. Water Reports No. 39. Rome: Food and Agriculture Organization of the United Nations.

Ghimire, T., P. Harou, and S. Balasubramanya. 2023. "Migration, Gender Labor Division and Food Insecurity in Tajikistan." *Food Policy* 116: 102438.

Guadagni, M., and T. Fileccia. 2009. *The Kyrgyz Republic: Farm Mechanization and Agricultural Productivity*. FAO Investment Centre Country Highlights Paper. Rome: Food and Agriculture Organization of the United Nations.

Hofman, I. 2016. "Politics or Profits along the 'Silk Road': What Drives Chinese Farms in Tajikistan and Helps Them Thrive?" *Eurasian Geography and Economics* 57 (3): 457–481.

Hofman, I. 2021. *Chinese Cotton Diplomacy in Tajikistan: Greasing the Ties by Reviving the Cotton Economy*. China Law and Development Research Brief 8/2021. Oxford, UK: Oxford University.

Kawabata, M., A. Berardo, P. Mattei, and S. de Pee. 2020. "Food Security and Nutrition Challenges in Tajikistan: Opportunities for a Systems Approach." *Food Policy* 96: 101872.

Krivonos, E., and L. Kuhn. 2019. "Trade and Dietary Diversity in Eastern Europe and Central Asia." *Food Policy* 88: 101767.

Lerman, Z., C. Csaki, and G. Feder. 2004. *Agriculture in Transition: Land Policies and Evolving Farm Structures in Post-Soviet Countries*. Lanham, MD: Lexington Books.

Lerman, Z., and D. Sedik. 2018. "Transition to Smallholder Agriculture in Central Asia." *Journal of Agrarian Change* 18 (4): 904–912.

Mirzabaev, A., J. Goedecke, O. Dubovyk, U. Djanibekov, B.L. Quang, and A. Aw-Hassan. 2016. "Economics of Land Degradation in Central Asia." In *Economics of Land Degradation and Improvement: A Global Assessment for Sustainable Development*, eds. E. Nkonya, A. Mirzabaev, and J. von Braun, 261–290. London: Springer Nature.

Mogilevskii, R., N. Abdrazakova, A. Bolotbekova, S. Chalbasova, S. Dzhumaeva, and K. Tilekeyev. 2017. "The Outcomes of 25 Years of Agricultural Reforms in Kyrgyzstan." IAMO Discussion Paper No. 162. Institute of Agricultural Development in Transition Economies, Halle, Germany.

Ni, L., K. Akramov, and S. Fan. 2024. "Land Tenure Change and Agricultural Production and Productivity in Uzbekistan." In *New Uzbekistan: The Third Renaissance*, eds. B. Mirkasimov and R. Pomfret, 135–163. London and New York: Routledge.

Pomfret, R. 2010. "Central Asia after Two Decades of Independence." Working Paper No. 2010/53. UNU-WIDER, Helsinki.

Pomfret, R. 2016. "Modernizing Agriculture in Central Asia." *Global Journal of Emerging Market Economies* 82 (2): 104–125.

Rozelle, S., and J. Swinnen. 2004. “Success and Failure of Reform: Insights from the Transition of Agriculture.” *Journal of Economic Literature* 42 (2): 404–456.

Slavchevska, V., C. Doss, M. Hillesland, and E. Mane. 2021. “The Impacts of Rural Outmigration on Women’s Empowerment Evidence from Nepal, Senegal, and Tajikistan.” IFPRI Discussion Paper 2099. IFPRI, Washington, DC.

Sommer, R., M. Glazerina, T. Yuldashev, et al. 2013. “Impact of Climate Change on Wheat Productivity in Central Asia.” *Agriculture, Ecosystems & Environment* 178: 78–99.

Spoor, M. 1999. “Agrarian Transition in Former Soviet Central Asia: A Comparative Study of Kazakhstan, Kyrgyzstan and Uzbekistan.” ISS Working Paper 298. Institute for Social Studies, The Hague, Netherlands.

Takeshima, H., K. Akramov, J. Ilyasov, Y. Liu, and T. Ergasheva. 2022. “Agriculture-Nutrition Linkages, Cooking-Time, Intrahousehold Equality among Women and Children: Evidence from Tajikistan.” *European Journal of Development Research* 34: 940–977.

Takeshima, H., K. Akramov, A. Park, J. Ilyasov, Y. Liu, and T. Ergasheva. 2020. “Agriculture–Nutrition Linkages with Heterogeneous, Unobserved Returns and Costs: Insights from Tajikistan.” *Agricultural Economics* 51 (4): 553–565.

Takeshima, H., I. Lambrecht, K. Akramov, and T. Ergasheva. 2024. “Nutrition-Sensitive Agriculture Diversification and Dietary Diversity Panel Data Evidence from Tajikistan.” IFPRI Discussion Paper 2249. IFPRI, Washington, DC.

USDA-ERS (US Department of Agriculture, Economic Research Service). International Agricultural Productivity data. Accessed October 2024. www.ers.usda.gov/data-products/international-agricultural-productivity

Visser, O., and M. Spoor. 2011. “Land Grabbing in Post-Soviet Eurasia: The World’s Largest Agricultural Land Reserves at Stake.” *Journal of Peasant Studies* 38 (2): 299–323.

World Bank. 2022. *Social Protection for Recovery: Europe and Central Asia Economic Update.* Fall 2022. Washington, DC.

Yao, J., Y. Chen, J. Chen, et al. 2021. “Intensification of Extreme Precipitation in Arid Central Asia.” *Journal of Hydrology* 598: 125760.

Chapter 21

SOUTH ASIA

Shahidur Rashid, S. Mahendra Dev, P.K. Joshi, and Purnima Menon

Shahidur Rashid is director for South Asia, Development Strategies and Governance Unit, IFPRI-India; **S. Mahendra Dev** is a distinguished professor, ICFAI School of Social Sciences, Hyderabad, and Chairman of the Institute for Development Studies, Andhra Pradesh, India; **P.K. Joshi** is vice president, National Academy of Agricultural Sciences, and former director for South Asia, IFPRI; and **Purnima Menon** is senior director for Food and Nutrition Policy, IFPRI.

Food systems in South Asia have evolved tremendously over the past 50 years, beginning with the technological innovations of the Green Revolution and then applying lessons learned from that period to address present-day challenges to agriculture, nutrition, and health. The last half-century has been marked by progress in establishing agricultural growth linkages, policymaking and investments in agriculture, and institutional innovations that left an enduring legacy in the region. Policy research has played a critical role in shaping national policies on food security, rural development, and nutrition over this period. Looking ahead to 2050, research on climate change, digitalization, and diets and nutrition will be needed to support South Asia in continuing to build sustainable and resilient food systems that nourish more than 2 billion people while delivering equitable and sustainable outcomes for the region.

Evolutions in policy and research over the past 50 years

As one of the most notable events in the history of agriculture over the past century, the Green Revolution arguably had the greatest impact in South Asia. The new technologies, and the associated policies and institutions, were central to the region's success in increasing productivity, reducing poverty, and transforming its chronic food deficit into a food surplus in less than two decades (Hazell 2009), although many have since raised concerns over the environmental issues created by these technologies and the uneven distribution of their benefits.

The Green Revolution first began to take hold in South Asia in the 1970s, when the region was heavily dependent on food aid imports and facing chronic food shortages related to the extreme poverty, food insecurity, and livelihood vulnerability prevalent under colonial rule. Prior to this period, none of the critical components of the Green Revolution—fertilizer, irrigation, and modern seeds—existed in the region, nor did any of its countries have the policies, institutions, or infrastructure needed to promote these new technologies (Rashid et al. 2008).[1] Trade and exchange policies were restrictive, infrastructure was weak, and institutions for risk mitigation, such as credit and insurance, and incentives for farmers to adopt new technologies, such as subsidies and pricing policies, were nonexistent.

During this time, international development organizations including CGIAR and IFPRI made major contributions through analysis of policies to promote the Green Revolution and agricultural growth in the region. The resulting evidence base from South Asia constituted largely new work in the agriculture and development policy field. Studies by IFPRI and others covered the diffusion and impacts of modern inputs, role of infrastructure, social protection and related institutions, market liberalization, and agriculture-nutrition linkages, among other topics (Figure 21.1).[2]

Of this body of work, research in three main areas has proven especially influential: agricultural growth linkages, returns on public investment in agriculture, and innovation in social protection and nutrition programs.

Agricultural growth linkages

Early efforts to promote agricultural growth in South Asia were influenced by the theory of agriculture-led development, a then-novel approach that posited agriculture as the primary driver of structural transformation in developing economies (Johnston and Mellor 1961) and marked a major shift from prior development thinking (see Chapters 3 and 17).[3] The trajectory of countries such as Japan had supported the theory, but South Asia became a testing ground to study it from the very start of the region's growth process. Research

1 For instance, when India embarked on promoting the Green Revolution, the country was using less than 1 million metric tons of nitrogen-phosphorus-potassium fertilizer in 1965/66, a rate that jumped to more than 5 million tons in 1980, about 18 million tons in 2000, and roughly 33 million tons in 2020 (FAOSTAT).

2 This large body of literature includes A. Ahmed (1992), A. Ahmed and Sampath (1992), R. Ahmed (1979b, 1985) Desai (1991), Dorosh and Valdes (1990), Edirisinghe (1988), Goletti et al. (1991, 2000), Meinzen-Dick (1996), Pinckney et al. (1988), Rogers (1988), and Stone (1979).

3 For more on this topic, see Balassa (1978), Bhagwati and Srinivasan (1978), Prebisch (1951, 1964), Singer (1986), and Singer and Alizadeh (1988).

FIGURE 21.1 IFPRI research on key policy issues in South Asia, number of publications (1975–2000)

Source: Authors' calculations based on data from CGSpace repository.
Note: Only peer-reviewed research publications are included in the figure.

on the linkages between agricultural growth and other sectors demonstrated that agricultural growth not only increased farm income but also generated higher nonfarm income, which in turn led to both higher consumption and faster reductions in poverty (Hazell and Haggblade 1990, 1991a). Agricultural growth linkages remain foundational to the region's development even today, with almost all countries in the region emphasizing agricultural growth linkages in their development strategies.

Investments in agriculture

Research focused on infrastructure, institutions, and policy constraints to agricultural growth made lasting contributions to the national policies and strategies of South Asia's governments. In Bangladesh, for example, research on the effects of a dual pricing system on foodgrain supply, distribution, and consumption in the late 1970s (Ahmed 1979a) and a subsequent study on food policy served as the catalyst for instituting the country's Food Planning and Monitoring Unit, which continues to act as the coordinating body for food security programs involving 16 different ministries. A study on the impacts of rural infrastructure in Bangladesh, which found associations between infrastructure and higher household income and income from agriculture, fisheries, and livestock (Ahmed and Hossain 1990), led to the development of the predecessor to the Local Government of Rural Development, an important unit within the country's Ministry of Planning.

Novel research work on returns to public spending in agriculture demonstrated that during the 1970s and 1980s, input subsidies in fertilizer, irrigation, and credit, along with investment in agricultural research, contributed significantly to enhancing agricultural productivity and growth. However, during the 1990s and early 2000s, subsidies crowded out other investments in agriculture, and their contributions to agricultural productivity and growth decelerated. Later research found that agriculture-related public investment in roads and agricultural R&D had much higher returns for both poverty reduction and agricultural growth than did investments in power or irrigation.[4] Demand for this type of research and analysis remains strong in South Asia, and the methods used by these studies have since been replicated in other countries around the world.

Institutional innovations in social protection and nutrition

Beginning in the 1970s, the region's approach to social protection shifted from a singular focus on averting famine to the broader aims of protecting the poor and addressing food insecurity. Research by IFPRI and others focused on restructuring the region's food rationing programs, exploring policy options for dealing with shocks, and examining the efficiency of social protection programs for reducing subsidies, targeting approaches, and linking these programs with overall development objectives (see Chapter 11).[5] These studies consistently found that traditional public distribution and rationing programs were expensive and plagued with corruption, with little reaching the intended beneficiaries. In Bangladesh, IFPRI's research led to the dismantling of the rationing system and the introduction of Food For Education, the first program of its kind (IFPRI 2013). All of the region's countries ultimately shifted from using universal subsidies to targeted food subsidies.

Over the last several decades, many countries in South Asia have drawn on policy research to expand their social protection programs to include nutrition, gender, and agriculture (see Chapters 11, 12, and 14). Policy research by IFPRI and partners has supported decision-making on the design, delivery, and prioritization of social protection programs. For instance, an impact evaluation of the

4 This literature includes Bathla et al. (2018, 2020), Fan and Hazell (2001), Fan et al. (2008), and Joshi et al. (2015).

5 Research on subsidies included A. Ahmed (1992), R. Ahmed (1988), Edirisinghe (1988), Kumar (1979), Rogers (1988), and Sahn (1988); on targeting, A. Ahmed (2000) and Babu (2002); and on linking to development objectives, A. Ahmed and Del Ninno (2002) and George (1979).

operational effectiveness of different transfer modalities in Bangladesh found that combining cash transfers with behavior change communication (BCC) led to significant improvements in child nutrition, along with various spillover benefits for nutrition knowledge and practices (A. Ahmed, Hoddinott et al. 2024; A. Ahmed et al. 2025; Roy et al. 2019). These results led Bangladesh's government to include nutrition BCC in a safety net program benefiting more than a million rural women. Many countries in the region have also introduced programs to strengthen the scale, quality, and impact of school meals, such as PM-POSHAN (formerly the Mid-Day Meal Scheme) in India and the School Health Promotion Program in Bangladesh. IFPRI's research suggests that India's school meals program has likely contributed to intergenerational reductions in child stunting (Chakrabarti et al. 2021); the implications for policy are yet to be determined.

Overall, nutrition has been rising on the policy agenda—and across sectors—in the region for the last two decades. IFPRI's contributions have kept pace with these changes, helping to support policymakers in the region with research and data. For example, Partnerships and Opportunities to Strengthen and Harmonize Actions for Nutrition (POSHAN), a partnership-based evidence-to-policy initiative that undertook research and policy analyses on maternal and child nutrition (POSHAN 2025), supported India's nutrition community in identifying strategic policy directions (Frongillo and Escobar-Alegria 2021) and monitoring the country's nutrition progress (NITI Aayog and IFPRI 2022); similar work is now underway across the region (IFPRI 2025). Linkages between agriculture and nutrition have been at the heart of the evidence-building efforts as well. IFPRI's early work on tackling the agriculture-nutrition disconnect in India (Kadiyala et al. 2012, 2014) set the stage for a variety of agriculture-nutrition research and engagement efforts, including through women's livelihoods program platforms. In Bangladesh, the Agriculture, Nutrition, and Gender Linkages (ANGeL) project, a partnership of the US government, Helen Keller International, and IFPRI, identified actions and investments that could leverage agricultural development for improved nutrition and improve pathways to women's empowerment, particularly within agriculture, paving the way for broader efforts to design, implement, and scale countrywide interventions to improve nutrition and women's empowerment. In India, studies focused primarily on women's livelihood group programs, and shed light on the impact of efforts leveraging these platforms to improve household welfare and diets.[6]

6 For more on ANGeL, see A. Ahmed et al. (2023), A. Ahmed, Coleman et al. (2024), and R. Ahmed et al. (2023); on the project in Bihar, see Raghunathan et al. (2023); for more on the India study with PRADAN, see, Kumar et al. (2024), Raghunathan et al. (2022), and Scott et al. (2022).

BOX 21.1 School-based meal programs

School-based meal programs are a powerful tool that can lead to positive impacts for both nutrition and education in South Asia. India is home to the region's largest school meal program, PM-POSHAN (formerly the Mid-Day Meal Scheme), and other countries in the region have also implemented diverse types of school meal programs. Early research in Bangladesh demonstrated the beneficial impact of nutritious biscuits in shaping educational outcomes for girls (Ahmed and Arends-Kuenning 2006). More recently, evidence has pointed to potential intergenerational benefits for the children of school meal recipients in India: using nationally representative data from 1993 to 2016, researchers found that the school meals expansion supported intergenerational improvements in child linear growth, providing a further impetus to this agenda (Chakrabarti et al. 2021). Policy engagement and efforts to improve the nutrition of school-age children have also identified new opportunities for school-based efforts to improve healthy diets for adolescents and young adults, an important agenda for the future.

Emerging trends: Challenges and opportunities

Over the last several decades, environmental degradation and climate change, digitalization in agriculture, and changes in diets and nutrition have emerged as trends that present both challenges to and opportunities for improving food systems in South Asia. These trends interact with the persistent gender and social equity issues in the region, including low empowerment and participation of women in the labor force, youth employment challenges, and rapidly increasing urbanization.

Environment and climate change

The extensive body of research that emerged from the Green Revolution documented not only its beneficial impacts but also negative consequences, such as resource degradation and equity implications, that remain relevant today (Hazell 2009; Hazell et al. 1991b; Hossain 1988). In the 1970s, policy attention was focused on the need to increase food production and feed the growing number of hungry people in the region. Even as it became clear that these policies had caused negative environmental impacts, it proved difficult to reform them (Rashid et al. 2007, 2008). For example, although both irrigation and fertilizers played a crucial role in increasing agricultural productivity, irrigation subsidies are a major cause of the region's alarming rate of groundwater depletion—South Asia extracts 25 percent of global groundwater, and India is the world's largest user of groundwater for irrigation—and fertilizer subsidies have contributed to soil degradation and reduced productivity of farmlands. The financial cost of these policies has also become unsustainably high: India's input subsidies for fertilizer, irrigation, and electricity rose from US$25 billion in 2011 to $48 billion in 2022–2023. Fertilizer subsidies account for 45.6 percent of all subsidies in India, and power subsidies account for another 30 percent (Sengupta 2024). Research suggests that countries should adopt strategies to reallocate these resources to tackle emerging policy challenges, such as climate adaptation and mitigation (see Chapter 18).

In recent decades, South Asia has become a hotspot of climate change, with the region facing significant climate-induced risks that are intensified by existing vulnerabilities. It has already experienced a rise in temperatures, accelerated rates of glacial melting and ground water depletion, and unpredictable patterns in rainfall and heat, leaving a large portion of the population exposed to climate hazards (IFPRI 2022). These trends have profound implications for the future of food production, health and nutrition, and the overall sustainability of food systems in the region.

AI and digitalization in agriculture

As optimism grows around the potential of AI and digital technologies to help transform food systems, South Asian countries have adopted a range of policies and programs to incorporate digital technologies into agriculture and food systems. Bangladesh, India, and Pakistan have adopted national governance plans and/or policies on digital and AI technologies, as well as forming country and corporate partnerships to integrate AI into agriculture. Programs adopted by South Asian countries under these policies have already shown success: in India, more than 17 million farmers and 250,000 traders have registered with the Electronic National Agriculture Market (e-NAM), a virtual platform to facilitate online trading of agricultural commodities, ensure transparent pricing, and enable farmers to receive better prices (MoA&FW 2025). In 2024, e-NAM's total trade volume reached approximately $7.7 million.

Dietary transition and agriculture-nutrition linkages

Despite impressive economic growth, rapid reductions in poverty and stunting,[7] and the introduction of many nutrition-focused public policies, large population sizes mean that South Asia experiences some of the highest burden

7 For documentation, see Avula et al. (2022), Bhutta et al. (2020), Headey et al. (2017), and Nisbett et al. (2017, 2023).

TABLE 21.1 Prevalence of malnutrition in South Asia, relative to least-developed countries, low- and middle-income countries, and the world, 2022

Indicators	South Asia	LDCs	LMICs	World
Unable to afford a healthy diet	74	72	68	42.1
Child stunting (< 5 years of age)	33	29	26	21.8
Child wasting (<5 years of age)	14	10	7	7.0
Anemia in women (ages 15 to 49)	53	45	38	30.0
Overweight/obesity (<5 years of age)	4	3	5	6.0

Source: FAO et al. (2024).
Note: LDC=least-developed countries; LMICs=low- and middle-income countries.

of malnutrition in the world (Table 21.1). More than 53 percent of the region's population, or 1.07 billion people, cannot afford a healthy diet; about half of women ages 15 to 49 are affected by anemia; and more than 30 percent of children under the age of five are stunted (FAO et al. 2024). Along with these high rates of undernutrition, the prevalence of overweight, obesity, and non-communicable diseases among adults is already high, while that among children is on the rise (FAO et al. 2024).

The underlying causes of the persistence of poor nutrition in the region are complex and extend beyond income and information. These new challenges have only recently received attention from the international development community, both in South Asia and the rest of the world, and further policy attention is greatly needed. For example, most countries in South Asia have not yet developed any policies or regulations for the growing market for ultra-processed foods, a serious oversight given the rapid growth of this sector of the food industry.[8] Consumption of unhealthy foods have become more prevalent in South Asia as incomes have risen and the share of staple foods in diets has declined. The ultra-processed food market in the region has more than tripled in just 10 years—rising from $19 billion in 2015 to more than $70 billion in 2025—and is likely to accelerate further in coming years (Figure 21.2) (Euromonitor International 2024). From a food and nutrition security standpoint, this poses major challenges for health outcomes, along with employment and economic well-being. More broadly, improving health and nutrition outcomes remains a major policy challenge, which, given the sheer number of

8 Some policies have been adopted—such as a ban on ultra-processed food sales within 50 meters of school premises and the launch of the Eat Right Campaign in India, a ban on ultra-processed food in schools in Punjab and Sindh in Pakistan, and a national policy on ultra-processed food in schools in Sri Lanka—but there is little to no evidence on their impacts.

FIGURE 21.2 Growth in ultra-processed food market in South Asia, 2015–2025

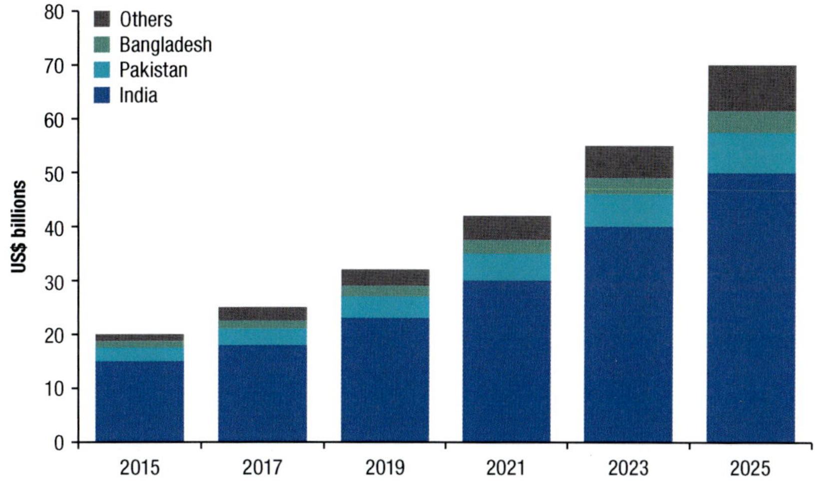

Source: Authors' construction based on reports from Euromonitor International (2024) and the Food Safety and Standards Authority of India (various years).

Note: "Others" includes Bhutan, Maldives, Nepal, and Sri Lanka.

undernourished children and women, is important not only for the region's countries but also for achieving the Sustainable Development Goals on health and nutrition.

Future food systems research

Looking ahead to 2050, and building on the challenge areas, we note that these also present policy research opportunities for South Asia: climate change, AI and digital efforts in agriculture, and diets and nutrition without losing focus on addressing underlying poverty, gender and social inequities.

Climate change. In the next 25 years, robust policies and innovations will be needed to tackle climate adaptation and mitigation in the region's food systems. Existing research from IFPRI and partners points toward necessary policy actions, as well as pathways for financing programs that could help accelerate a climate-positive food systems transformation (IFPRI 2022). Research partnerships will need to be expanded beyond ministries of agriculture and national think tanks to include ministries of environment, commerce, and planning, as well as other national institutions—such as NITI Aayog in India and planning commissions in other countries—that coordinate across multiple ministries.[9] At the country level, evidence is needed on the effectiveness of various pilot programs to address climate threats, as well as on adaptive approaches that combine traditional social protection and livelihood interventions with mechanisms to rapidly address disaster- and climate-related shocks (World Bank 2021). This is especially crucial in a region where gender and other social disparities persist despite overall economic progress.

AI and digitalization in agriculture. Policy research is needed to ensure that AI and digital technologies include everyone in South Asia, even the region's poor. Barriers to achieving this goal include bottlenecks in institutions and infrastructure, combined with market failures within food systems such as limited digital infrastructure, concentration of market power, and issues with data governance and privacy. Inability to tackle these challenges can lead to the "digital divide," a negative outcome that was evident in the region's education sector during the COVID-19 pandemic. While students attending private school or those from richer households could attend virtual classes, students in public schools had no such ability. Innovative programs include the Drone Didi program in India, which aims to provide drones to 15,000 women's self-help groups between 2024 and 2026. After receiving necessary training, women pilots will provide rental services to farmers. The underlying idea is to achieve the twin objectives of improving income and enhancing women's empowerment. Other programs have either been rolled out or are in development to ensure the inclusive and sustainable adoption of AI in agriculture. Generating evidence on the operational effectiveness of the program will be critical—both to improve program design and to generate evidence on broader policy actions that will be critical to fostering inclusion, including through targeted support to smallholder farmers, youth, women, and other vulnerable groups, and through skills development (OECD 2022; World Bank 2019, 2021).

Dietary transition and agriculture-nutrition linkages. Future research on diets and nutrition must cut across the entire food system—from production to consumption—and beyond. Evidence-based insights are needed on how major agricultural subsidies can be reshaped to encourage the production of nutrient-rich foods, while avoiding negative impacts on the availability of

9 In the case of the livestock sector, for example, R&D, extension, and other routine operations fall under the livestock (or livestock and fisheries) ministry; food safety falls under the ministry of health; policies related to greenhouse gas reductions are implemented by the environment/climate change ministry; broader policy or program designs are carried out by the planning ministry; and budgets for the programs are approved by the finance ministry.

South Asia's predominant staple foods. Between the farm and the consumer, research will be essential to strengthen value chains and market opportunities for nutritious and often perishable commodities such as fruits, vegetables, animal-source foods, and other nutrient-rich foods. These foods are currently limited in diets due to limited affordability and accessibility. It will be imperative to identify behavior change innovations, including policy levers such as food labels, taxes, and nudges using digital technology, to redirect consumer preferences and choices toward healthy diets. In a changing social environment, identifying policies and interventions, including for men's engagement, to reshape gender dynamics across the food system will be crucial (Kumar et al. 2024). Finally, efforts to identify the most efficient and effective strategies to address the affordability of healthy diets must remain a top priority for South Asia, since millions of people continue to face economic precarity and poverty in the region. A key challenge for South Asia is whether it can avoid the nutrition trajectories of regions that face major public health challenges from overweight and obesity.

A new strategy for South Asia. Over the past 50 years, research by IFPRI and others in South Asia has contributed toward a better understanding of the linkages between agriculture, nutrition, and health (see Chapter 2). The Institute supported advancement of these efforts through large-scale outreach and new initiatives, notably the 2020 Vision Initiative conference on agriculture, nutrition, and health; numerous nutrition initiatives; and the launching of HarvestPlus in the region. However, most of this research and outreach was carried out in partnership with agriculture ministries or national agricultural research systems.

Given the new realities, IFPRI and other research partners will need to consider new ways of doing business in South Asia. Instead of solely relying on donor funding, organizations should explore opportunities to broaden partnerships and mobilize resources from within the region for future research. In addition to working more with non-agricultural ministries within the regions, IFPRI and others must also strengthen partnerships with regional bodies such as ASEAN, BIMSTEC, and SAARC, and foster triangular South-South cooperation. The challenges of food systems transformation extend far beyond agricultural ministries, and research organizations such as IFPRI need to adjust accordingly to enhance their relevance and funding opportunities for critical research. There is growing demand for IFPRI's research from governments and the private sector, and donors will likely find it strategic to invest in regional body and South-South partnerships that address challenges in South Asia's food systems.

As we look to the future, new partnerships and ways of conducting policy research will be essential to generate enduring impacts for policymaking and academic scholarship and will pave the way for new thinking on sustainable food systems transformation. South Asia's successes have tremendous insights to offer the world, and IFPRI's efforts in the region will also be oriented toward strengthening efforts to translate evidence from the regional to the global level.

References

Ahmed, A. 1992. "Operational Performance of the Rural Rationing Program in Bangladesh." Food Policy in Bangladesh Working Paper 5. IFPRI, Washington, DC.

Ahmed, A. 2000. "Trends in Consumption, Nutrition, and Poverty." In *Out of the Shadow of Famine: Evolving Food Markets and Food Policy in Bangladesh*, eds. R. Ahmed, S. Haggblade, and T.E. Chowdhury, 101–120. Baltimore: Johns Hopkins University Press for IFPRI.

Ahmed, A., and M. Arends-Kuenning. 2006. "Do Crowded Classrooms Crowd Out Learning? Evidence from the Food for Education Program in Bangladesh." *World Development* 34 (4): 665–684.

Ahmed, A., F. Coleman, J. Ghostlaw, et al. 2024. "Increasing Production Diversity and Diet Quality: Evidence from Bangladesh." *American Journal of Agricultural Economics* 106 (3): 1089–1110.

Ahmed, A., F. Coleman, J. Hoddinott, et al. 2023. "Comparing Delivery Channels to Promote Nutrition-Sensitive Agriculture: A Cluster-Randomized Controlled Trial in Bangladesh." *Food Policy* 111: 102484.

Ahmed, A., and C. Del Ninno. 2002. "The Food for Education Program in Bangladesh: An Evaluation of Its Impact on Educational Attainment and Food Security." FCND Discussion Paper 138. IFPRI, Washington, DC.

Ahmed, A., M. Hidrobo, J. Hoddinott, B. Kolt, S. Roy, and S. Tauseef. 2025. "Sustainable Poverty Reduction through Social Assistance: Modality, Context, and Complementary Programming in Bangladesh." *American Economic Journal: Applied Economics* 17 (2): 102–126.

Ahmed, A., J. Hoddinott, S. Roy, and E. Sraboni. 2024. "Transfers, Nutrition Programming, and Economic Well-Being: Experimental Evidence from Bangladesh." *World Development* 173: 106414.

Ahmed, A., and R.K. Sampath. 1992. "Effects of Irrigation-Induced Technological Change in Bangladesh Rice Production." *American Journal of Agricultural Economics* 74 (1): 144–157.

Ahmed, R. 1979a. *Foodgrain Supply, Distribution, and Consumption Policies within a Dual Pricing Mechanism: A Case Study of Bangladesh.* Research Report 8. Washington, DC: IFPRI.

Ahmed, R. 1979b. "Structure, Dynamics, and Related Policy Issues of Fertilizer Subsidy in Bangladesh." In *Fertilizer Pricing Policy in Bangladesh*, ed. B. Stone, 117–180. Occasional Paper 9. Washington, DC: IFPRI.

Ahmed, R. 1985. In *Agricultural Change and Rural Poverty: Variations on a Theme by Dharm Narain*, eds. J.W. Mellor and G.M. Desai, 124–127. Baltimore: Johns Hopkins University Press.

Ahmed, R. 1988. "Structure, Costs, and Benefits of Food Subsidies in Bangladesh." In *Food Subsidies in Developing Countries: Costs, Benefits, and Policy Options*, ed. P. Pinstrup-Andersen, 219–228. Baltimore: Johns Hopkins University Press for IFPRI.

Ahmed, R., J. Ghostlaw, and E. Jones. 2023. "Making a Difference: The Agriculture, Nutrition, and Gender Linkages (ANGeL) Project." *IFPRI Blog*, June 26.

Ahmed, R., and M. Hossain. 1990. *Developmental Impact of Rural Infrastructure in Bangladesh.* Research Report 83. Washington, DC: IFPRI.

Avula, R., P.H. Nguyen, L.M. Tran, et al. 2022. "Reducing Childhood Stunting in India: Insights from Four Subnational Success Cases." *Food Security* 14 (4): 1085–1097.

Babu, S.C. 2002. "Social Safety Nets for Poverty Reduction in South Asia: Global Experiences." *Sri Lankan Journal of Agricultural Economics* 5: 1–9.

Balassa, B. 1978. "Exports and Economic Growth: Further Evidence." *Journal of Development Economics* 5 (2): 181–189.

Bathla, S., P.K. Joshi, and A. Kumar. 2020. "Contextualising Public Intervention in Agriculture and Welfare Effects." In *Agricultural Growth and Rural Poverty Reduction in India: Targeting Investments and Input Subsidies*, eds. S. Bathla, P.K. Joshi, and A. Kumar, 19–28. Berlin: Springer.

Bathla, S., A. Kumar, and P.K. Joshi. 2018. "Regional Income Inequalities and Public Investments in Rural India." *Agricultural Economics Research Review* 32 (1): 13–27.

Bhagwati, J.N., and T.N. Srinivasan. 1978. *Trade Policy and Development*, vol. 90. Washington, DC: World Bank.

Bhutta, Z.A., N. Akseer, E.C. Keats, et al. 2020. "How Countries Can Reduce Child Stunting at Scale: Lessons from Exemplar Countries." *The American Journal of Clinical Nutrition* 112: 894S–904S.

Chakrabarti, S., S.P. Scott, H. Alderman, P. Menon, and D.O. Gilligan. 2021. "Intergenerational Nutrition Benefits of India's National School Feeding Program." *Nature Communications* 12 (1): 4248.

Desai, G.M. 1991. "Issues and Themes in Growth of Fertilizer Use in India: An Agenda for Further Research and Future Policies." *Journal of the Indian Society of Agricultural Statistics* 43 (1): 81–102.

Dorosh, P.A., and A. Valdés. 1990. *Effects of Exchange Rate and Trade Policies on Agriculture in Pakistan*. Research Report 84. Washington, DC: IFPRI.

Edirisinghe, N. 1988. "Food Subsidy Changes in Sri Lanka: The Short-Run Effect on the Poor." In *Food Subsidies in Developing Countries: Costs, Benefits, and Policy Options*, ed. P. Pinstrup-Andersen, 253–266. Baltimore: Published for IFPRI by Johns Hopkins University Press.

Euromonitor International. 2024. Various industry reports on processed food, snack food, and foodservice markets in South Asia. www. euromonitor.com

Fan, S., and P.B.R. Hazell. 2001. "Returns to Public Investment in the Less-Favored Areas of India and China." *American Journal of Agricultural Economics* 83 (5): 1217–1222.

Fan, S., B. Yu, and A. Saurkar. 2008. "Public Spending in Developing Countries: Trends, Determination, and Impact." In *Public Expenditures, Growth, and Poverty*, ed. S. Fan, 20–55. Baltimore: Johns Hopkins University Press.

FAO (Food and Agriculture Organization of the United Nations), IFAD (International Fund for Agricultural Development), UNICEF, WFP (World Food Programme), and WHO (World Health Organization). 2024. *The State of Food Security and Nutrition in the World 2024: Financing to End Hunger, Food Insecurity and Malnutrition in All Its Forms*. Rome: FAO.

Frongillo, E.A., and J.L. Escobar-Alegria. 2021. "Advancing Use of Nutrition Knowledge to Improve Practice by Policy and Program Communities in India during a Political Transition." *Current Developments in Nutrition* 5 (10): nzab120.

George, P.S. 1979. *Public Distribution of Foodgrains in Kerala: Income Distribution Implications and Effectiveness*. Research Report 7. Washington, DC: IFPRI.

Goletti, F., R. Ahmed, and N. Chowdhury. 1991. "Optimal Stock for the Public Food Grain Distribution System in Bangladesh." Food Policy in Bangladesh Working Paper 4. IFPRI, Washington, DC.

Goletti, F., A.S.M. Jahangir, and S. Zohir. 2000. "Prospects for Rice Exports in Bangladesh." In *Out of the Shadow of Famine: Evolving Food Markets and Food Policy in Bangladesh*, eds. R. Ahmed, S. Haggblade, and T.-e.-E. Chowdhury, 261–277. Baltimore: Johns Hopkins University Press.

Hazell, P.B.R. 2009. "The Asian Green Revolution." IFPRI Discussion Paper 911. IFPRI, Washington, DC.

Hazell, P.B.R., and S. Haggblade. 1991a. "Rural-Urban Growth Linkages in India." *Indian Journal of Agricultural Economics* 46 (4): 515–529.

Hazell, P.B.R., C. Ramasamy, and P.K. Aiyasamy. 1991b. *The Green Revolution Reconsidered: The Impact of High-Yielding Rice Varieties in South India*. Washington, DC: IFPRI.

Hazell, P.B.R., and S. Haggblade. 1990. *Rural-Urban Growth Linkages in India*, vol. 430. Washington, DC: World Bank Publications.

Headey, D.D., and H.H. Alderman. 2019. "The Relative Caloric Prices of Healthy and Unhealthy Foods Differ Systematically across Income Levels and Continents." *Journal of Nutrition* 149 (11): 2020–2033.

Headey, D., J. Hoddinott, and S. Park. 2017. "Accounting for Nutritional Changes in Six Success Stories: A Regression-Decomposition Approach." *Global Food Security* 13: 12–20.

Hossain, M. 1988. *Nature and Impact of the Green Revolution in Bangladesh*. Research Report 67. Washington, DC: IFPRI.

IFPRI. 2013. *IFPRI in Bangladesh: Highlights of Research and Capacity Strengthening*. Washington, DC.

IFPRI. 2022. *2022 Global Food Policy Report: Climate Change and Food Systems*. Washington, DC.

IFPRI. 2025. "South Asia Nutrition Knowledge Initiative (SANI)." www.ifpri.org/project/south-asia-nutrition-knowledge-initiative-sani/

Johnston, B.F., and J.W. Mellor. 1961. "The Role of Agriculture in Economic Development." *The American Economic Review* 51 (4): 566–593.

Joshi, P.K., P. Kumar, and S. Parappurathu. 2015. "Public Investment in Agricultural Research and Extension in India." *European Journal of Development Research* 27 (3): 438–451.

Kadiyala, S., J. Harris, D. Headey, S. Yosef, and S. Gillespie. 2014. "Agriculture and Nutrition in India: Mapping Evidence to Pathways." *Annals of the New York Academy of Sciences* 1331 (1): 43–56.

Kadiyala, S., P.K. Joshi, S.M. Dev, T.N. Kumar, and V. Vyas. 2012. "A Nutrition Secure India: Role of Agriculture." *Economic and Political Weekly* 47 (8): 21–25.

Kumar, N., A. Banerjee, and A.R. Quisumbing. 2024. "Task or Time? Comparing Methods for Measuring the Gender Distribution of Work." IFPRI Discussion Paper 2277. IFPRI, Washington, DC.

Kumar, N., K. Raghunathan, A. Quisumbing, et al. 2024. "Women Improving Nutrition through Self-Help Groups in India: Does Nutrition Information Help?" *Food Policy* 128: 102716.

Kumar, S.K. 1979. *Impact of Subsidized Rice on Food Consumption and Nutrition in Kerala*. Research Report 5. Washington, DC: IFPRI.

Meinzen-Dick, R.S. 1996. *Groundwater Markets in Pakistan: Participation and Productivity*. Research Report 105. Washington, DC: IFPRI.

MoA&FW (Ministry of Agriculture & Farmers' Welfare). 2025. *Strengthening India's Agricultural Backbone: Key Achievements and Government Initiatives*, New Delhi: Press Information Bureau.

Nisbett, N., J. Harris, D. Headey, et al. 2023. "Stories of Change in Nutrition: Lessons from a New Generation of Studies from Africa, Asia and Europe." *Food Security* 15 (1): 133–149.

Nisbett, N., M. van den Bold, S. Gillespie, et al. 2017. "Community-Level Perceptions of Drivers of Change in Nutrition: Evidence from South Asia and Sub-Saharan Africa." *Global Food Security* 13: 74–82.

NITI Aayog and IFPRI. 2022. *Preserving Progress on Nutrition in India: POSHAN Anhiyaan in Pandemic Times*. New Delhi: NITI Aayog.

OECD (Organisation for Economic Co-operation and Development). 2022. *The Digitalisation of Agriculture: A Literature Review and Emerging Policy Issues*. OECD Food, Agriculture and Fisheries Papers 176. Paris.

Pinckney, T.C., N. Hamid, M. Klaus, and A. Valdés. 1988. *Wheat Storage Policy in Pakistan: Implications of Price Policy and Production Instability*. Washington, DC: IFPRI.

POSHAN. 2025. "Publications and Resources." Accessed April 2025. https://poshan.ifpri.info/publications-and-resources/

Prebisch, R. 1951. *Theoretical and Practical Problems of Economic Growth*. United Nations Economic Commission for Latin America and the Caribbean (ECLAC). New York: United Nations.

Prebisch, R. 1964. *Towards a New Trade Policy for Development*. New York: United Nations Conference on Trade and Development (UNCTAD).

Ragunathan, K., N.Kumar, S. Gupta, T. Chauhan, A.K. Kathuria, and P. Menon. 2023. "Learning Together: Experimental Evidence on the Impact of Group-Based Nutrition Interventions in Rural Bihar." *World Development* 168: 106267.

Raghunathan, K., N. Kumar, S. Gupta, et al. 2022. "Scale and Sustainability: The Impact of a Women's Self-Help Group Program on Household Economic Well-Being in India." *Journal of Development Studies* 59 (4): 490–515.

Rashid, S., R. Cummings Jr., and A. Gulati. 2007. "Grain Marketing Parastatals in Asia: Results from Six Case Studies." *World Development* 35 (11): 1872–1888.

Rashid, S., A. Gulati, and S.M. Dev. 2008. "Parastatals and Food Policies: The Indian Case." In *From Parastatals to Private Trade: Lessons from Asian Agriculture*, 51–76. Baltimore: Johns Hopkins University Press; New Delhi: Oxford University Press.

Rogers, B.L. 1988. "Pakistan's Ration System: Distribution of Costs and Benefits." In *Food Subsidies in Developing Countries: Costs, Benefits, and Policy Options*, ed. P. Pinstrup-Andersen, 242–252. Baltimore: Johns Hopkins University Press.

Roy, S., M. Hidrobo, J. Hoddinott, and A. Ahmed. 2019. "Transfers, Behavior Change Communication, and Intimate Partner Violence: Postprogram Evidence from Rural Bangladesh." *Review of Economics and Statistics* 101 (5): 865–877.

Sahn, D.E. 1988. "The Effect of Price and Income Changes on Food-Energy Intake in Sri Lanka." *Economic Development and Cultural Change* 36 (2): 315–340.

Scott, S., S. Gupta, P. Menon, et al. 2022. "A Quasi-Experimental Evaluation of a Nutrition Behavior Change Intervention Delivered Through Women's Self-Help Groups in Rural India: Impacts on Maternal and Young Child Diets, Anthropometry, and Intermediate Outcomes." *Current Developments in Nutrition* 6 (6): nzac079.

Sengupta, R. 2024. "Agricultural Subsidies in India," Dec. 2. International Institute for Sustainable Development. Policy Analysis.

Singer, H.W. 1986. "Raúl Prebisch and His Advocacy of Import Substitution." *Development & South-South Cooperation* 2 (1): 1–5.

Singer, H.W., and P. Alizadeh. 1988. "Import Substitution Revisited in a Darkening External Environment." In *Policies for Development: Essays in Honour of Gamani Corea*, 60–86. London: Palgrave Macmillan UK.

Stone, B. 1979. "Fertilizer Development and Price in International Perspective." In *Fertilizer Pricing Policy in Bangladesh*, ed. B. Stone, 17–116. Occasional Paper 9. Washington, DC: IFPRI.

World Bank. 2019. *Future of Food: Harnessing Digital Technologies to Improve Food System Outcomes*. Washington, DC.

World Bank. 2021. *The Converging Technology Revolution and Human Capital: Potential and Implications for South Asia*. South Asia Development Forum. Washington, DC.

Chapter 22

EAST AND SOUTHEAST ASIA

Kevin Z. Chen, Yunyi Zhou, Ke Kou, and Shenggen Fan

Kevin Z. Chen is a senior research fellow, Development Strategies and Governance Unit (DSG), IFPRI, Beijing, and Qiushi chair professor, Zhejiang University, Hangzhou; **Yunyi Zhou** is an assistant research fellow, School of Economics, Sichuan University, Chengdu; **Ke Kou** is a research assistant, DSG, IFPRI, Beijing; and **Shenggen Fan** is a member, CGIAR Integrated Partnership Board, and chair professor, China Agricultural University, Beijing.

Over the past half century, the food systems of East and Southeast Asia[1] (ESA) have been shaped by rapid economic growth and structural transformation, intensifying climate- and weather-induced extreme events, and the region's unique agricultural landscape and value chain dynamics. Rapid urbanization and the concurrent dietary change (that is, increased consumption of processed and animal-source foods, with robust demand for rice) has exacerbated challenges within the nexus of climate change, natural capital, and sustainable food systems. By aligning research outputs, financial programs, and regional cooperation initiatives with national policy frameworks in the region, organizations such as the Asian Development Bank (ADB), the Association of Southeast Asian Nations (ASEAN), the Food and Agriculture Organization of the United Nations (FAO), the International Fund for Agricultural Development (IFAD), IFPRI, and CGIAR have facilitated improvements in food security, nutrition, and livelihoods in ESA countries, though important challenges remain. Looking ahead, food systems transformation in this region will require a multisectoral approach that includes cross-cutting foresight and integrated approaches that combine disruptive technologies, participatory governance, and scalable solutions supported by sustainable financing mechanisms. Successful experiences in the region, such as poverty reduction and low-carbon agricultural practices, can provide models for policies and actions in other regions.

1 East Asian countries are China, Japan, Democratic People's Republic of Korea, Republic of Korea, and Mongolia. Southeast Asian countries are Brunei, Cambodia, Indonesia, Lao PDR, Malaysia, Myanmar, the Philippines, Singapore, Thailand, Timor-Leste, and Viet Nam.

Challenges over the past five decades

Since the 1960s, ESA has experienced unprecedented food systems transformation and improvements in nutrition, supported by rapid economic growth in East Asia. China's economic reforms and market opening since the late 1970s further boosted the region's economic growth and structural transformation; for example, the region's importance as a source of final demand and as a hub for regional production networks has grown substantially (UNESCAP 2019). Southeast Asia, where nutrition status was among the worst worldwide until the early 1990s, is now among the world's fastest-growing areas (although with varied levels of development across different countries) (Choiruzzad 2024). As a result, ASEAN member states reduced the region's extreme poverty from 47 percent in 1990 to 15 percent in 2015 (ASEAN 2025), China helped 800 million people move out of extreme poverty over the past 40 years (World Bank 2022), and the Global Hunger Index score for ESA improved from 28.5 in 1990 to 8.3 in 2024 (von Grebmer et al. 2015; WHH et al. 2024).

Nevertheless, food systems transformation and rural development in ESA have faced significant climate, demographic, and economic challenges. With the majority of the region's population and economic activities concentrated in coastal areas and river basins, regional food systems have been particularly vulnerable to droughts, floods, heat waves, typhoons, and rising sea levels (Chen and Zhan 2022). These intensifying climate impacts, coupled with global economic uncertainties over the past 50 years—including trade frictions, geopolitical conflicts, and pandemics—have degraded natural capital. This pressure has been compounded by the region's demographic change: in Southeast Asia, the population has nearly doubled since 1980, and in the developing ESA countries, the share of population aged 65 and above tripled to 12.1 percent, while the rural population fell by half since 1960 (World Bank 2025a). Together, these changes have driven competitive consumption of natural resources across sectors and geographic areas (Papilo et al. 2022), impeding the realization of greater inclusiveness, resilience, and sustainability (Lee et al. 2024; Zhou and Chen 2023).

On the production side, ESA has long been crucial to global food security, contributing 20–30 percent of global cereal production and 50–80 percent of global aquaculture output since 1960. Recent data show the region produces more than half of the world's rice, 80 percent of aquatic foods, and significant quantities of sugar crops, meat products, and eggs (World Bank 2025a). Despite this large role, food production in the region relies heavily on small-scale household farming and on heavy use of synthetic nitrogen fertilizer to sustain yields amid climate shocks and constraints on other resources. Limited diversity of production and large postharvest losses have been difficult to

FIGURE 22.1 Changing dietary composition in East and Southeast Asia, 1975 and 2022

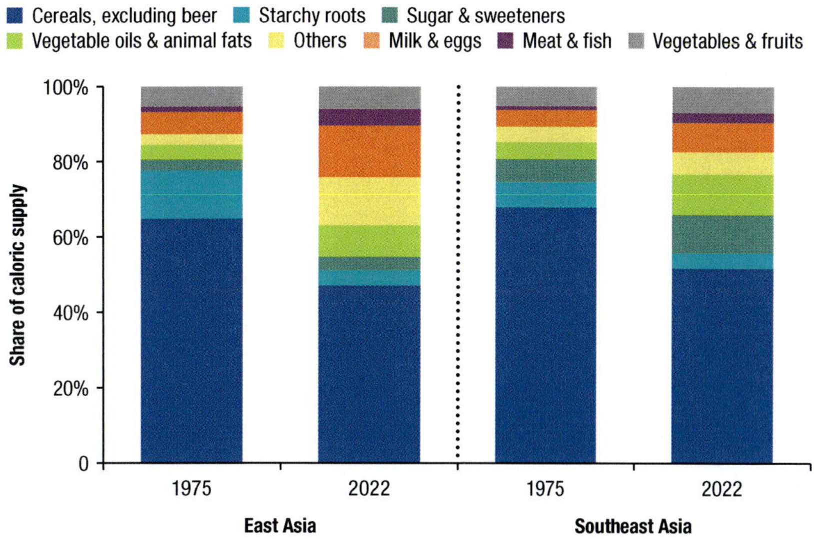

Source: Data from FAO (2024).

address (Bennett et al. 2022; Sundram 2023). Efforts toward regional integration have been hindered by inconsistent adherence to standards for good agricultural practices, which constrains farmers' access to lucrative markets and limits broader food systems integration (GIZ 2024).

On the demand side, urbanization and income growth during the region's structural transformation have driven significant dietary changes, with a growing shift toward animal-source foods (Figure 22.1). Simultaneously, industrialization of domestic food sectors and powerful marketing by transnational food and beverage companies have fueled demand for ultra-processed foods and beverages (Baker and Friel 2016), often at the cost of human health and biodiversity along value chains (Kesse-Guyot et al. 2023; Leite et al. 2022). Despite significant strides in reducing poverty and hunger, residents in the region (especially low-income groups) have remained vulnerable, facing multiple burdens of malnutrition (Figure 22.2) and unable to afford healthy diets (FAO 2024). Furthermore, given the reliance of agriculture on weather conditions and chemical inputs, small-scale rural producers and vulnerable food consumers remain exposed to value chain vulnerabilities, including price inflation for fertilizers (nitrogen and phosphates) and food commodities, which heightens concerns about systemic crises in the region's food systems (Jones and Nti 2022).

FIGURE 22.2 Undernourishment and obesity in East and Southeast Asia

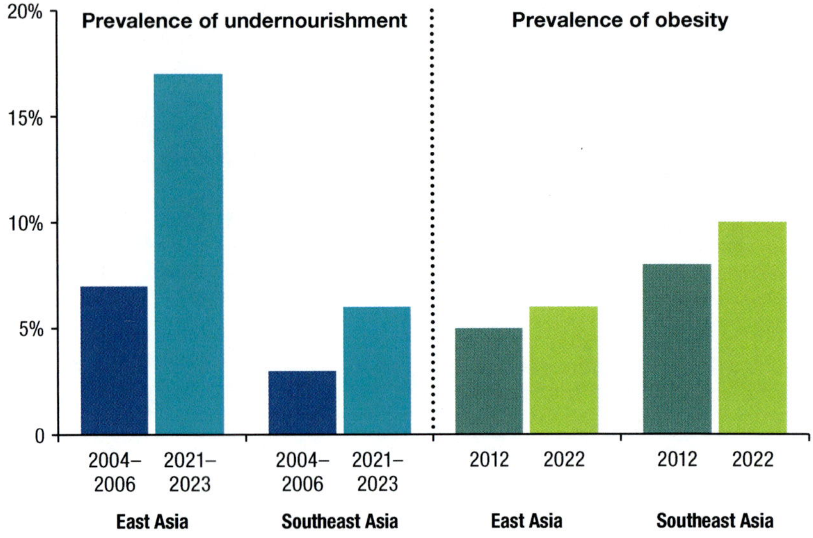

Source: Data from FAO et al. (2024).
Note: The estimates for the years 2004–2006 and 2021–2023 are calculated based on three-year averages.

Financing and investment have been recognized as critical levers for addressing these long-term challenges and building the resilience of food systems, yet significant funding gaps persist. For instance, in China and Viet Nam, more than 60 percent of national agricultural budgets between 2010 and 2020 were allocated to infrastructure (especially for irrigation) and production subsidies (World Bank 2022). Because these subsidies are recognized as often encouraging unsustainable farming practices, redirecting these funds to capacity building and the application, dissemination, and scaling of agricultural technologies and innovations would likely better contribute to sustainable and resilient food systems. However, budget constraints—including declining foreign direct investment for agriculture (Figure 22.3) and rising government debt in many low- and middle-income countries in the region—as well as structural challenges, such as the urban bias of many policies and investments, have limited the scope of food systems financing (Chen et al. 2023) (see also Chapter 18). Given the limitations of public financing, engagement of the private sector and civil society in developing climate-resilient infrastructure and related initiatives has become increasingly critical (Chiengkul 2025) (see Chapters 4 and 5).

FIGURE 22.3 Agriculture, forestry, and fishing FDI flows to ASEAN, 2013–2022

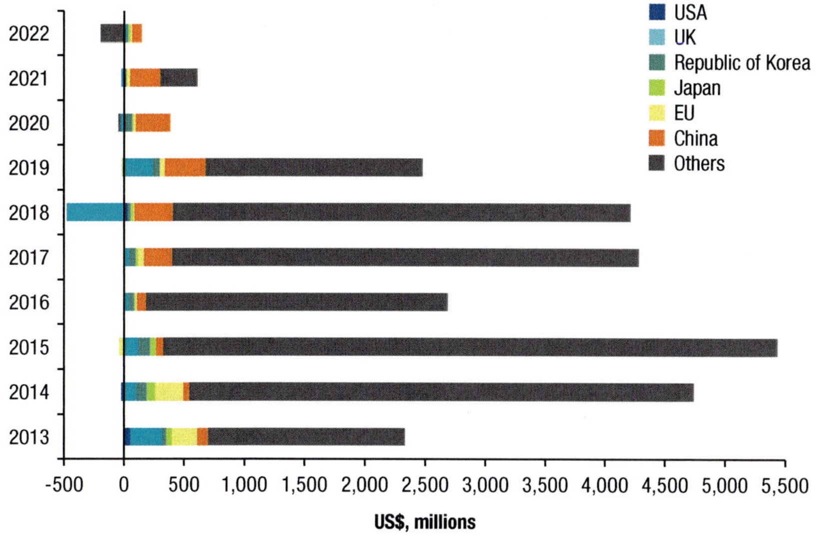

Source: Data from ASEAN (2023).
Note: FDI = foreign direct investment. Negative FDI usually indicates reverse investment or disinvestment.

Evolution of research and policy responses

As empirical evidence and foresight projections regarding ESA have accumulated over the past five decades, integrated rural development[2] and sustainable food systems transformation have emerged as two key approaches to addressing these daunting challenges (see Chapter 3).

Development and research organizations

Over this period, IFPRI and other research and development organizations made an important shift from supporting one-way knowledge transfer to approaches that combine participatory methods and technological innovation to facilitate food systems synergies (Habermann et al. 2023). These efforts address risks and resilience-building strategies for food systems transformation and urban and rural development, including increased emphasis on identifying and adapting to local contextual factors that shape outcomes. The ADB has

2 Integrated rural development is a comprehensive strategy that enhances rural livelihoods by combining agricultural productivity, infrastructure development, education, healthcare, economic diversification, and other interconnected factors (see Chapter 3).

recognized that a holistic vision is necessary to navigate the nexus of climate change, natural capital, and sustainable food systems (ADB 2021), and as a multilateral development finance institution, the ADB plays a crucial role in funding food systems transformation and the low-carbon development initiatives of Asia's low- and middle-income countries. The World Bank also helps fund food systems transformation through the Food Systems 2030 Umbrella Multi-Donor Trust Fund (FS2030), established in 2020. This initiative complements other global efforts through financial and technical support for sustainable agriculture reforms across 30 countries. In collaboration with the FAO, the initiative developed a toolkit for repurposing agricultural policies, while IFAD contributed by tracking financial flows through its Financial Flows into Food Systems Model (3FS). Repurposing projects financed by FS2030 are now being implemented in Indonesia, driving policy reforms for more resilient and sustainable food systems. At the same time, countries including Cambodia, Lao PDR, and Viet Nam have joined the One Health Global Community of Practice, which is promoting healthy and sustainable diets (World Bank 2025b).

Policy evolution

In parallel with the evolution of research on food systems, policies have evolved from a traditionally narrow focus on increasing agricultural production to a systems focus on multiple, integrated objectives: from reducing hunger ("eating enough") to enhancing dietary quality ("eating well") to advancing nutrition and health ("eating nutritious and healthy diets") and ultimately to incorporating ecological sustainability ("eating green"). Among Asian countries, China exemplifies this shift through its transformation from traditional farmland-based agricultural policies to a broader strategy that leverages diverse land resources while balancing food security, nutritional needs, and ecological modernization.

Beginning in the late 1960s, the Green Revolution in ESA significantly enhanced wheat and rice productivity through the introduction of high-yielding crop varieties, intensive fertilizer and pesticide use, improved irrigation, and other technological advances (Swastika et al. 2024). The significant increase in agricultural productivity raised farmers' incomes and reduced famine across the region (Pingali 2012; Pingali and Abraham 2022). From its establishment in 1975, IFPRI aligned with these efforts, focusing on improving agricultural productivity among smallholder farmers by creating more supportive enabling environments. These included policies promoting land and water rights, pro-poor technologies, and access to appropriate input and output support, which further amplified the gains from Green Revolution technologies across Asia (IFPRI 2015a).

Economic reforms in the 1980s spurred further growth in the agriculture sector. In China, the Household Responsibility System incentivized productivity and sparked remarkable growth in the agriculture sector. In Viet Nam, the transition away from collective farming enabled rapid productivity gains, diversification of agricultural production, and growth in agricultural exports (Lin 1987; Pingali and Abraham 2022; Pingali and Xuan 1992). During this period, IFPRI facilitated Viet Nam's agricultural transformation through its Rice Market Monitoring and Policy Options Study, launched in 1995 with ADB support (IFPRI 2015b). IFPRI research influenced key decisions to relax rice export quotas, remove internal trade restrictions, and involve the private sector in exporting. Research from other institutes, including the Harvard Institute for International Development and the Centre for International Economies in Australia, likewise contributed to Viet Nam's transition from a command economy to a market economy. These reforms ultimately strengthened Viet Nam's agriculture sector, benefiting farmers and supporting the country's broader economic growth (Ryan 1999). As hunger became a less pressing concern for many ESA nations in this century, food policies shifted toward enhancing quality. For example, Thailand's From Farm to Table initiative in 2004 established a comprehensive food safety monitoring and control system, and China's revised Food Safety Law in 2015 reinforced its focus on food quality.

Addressing poverty and livelihoods

In recent decades, low- and middle-income countries in the region have gradually incorporated human capital development into poverty reduction policies that work through food systems transformation and integrated rural development initiatives. For example, China introduced the Targeted Poverty Alleviation policy, which contributed to meeting the goal of eradicating absolute poverty by 2020 and shrank the country's urban–rural income gap. In addition to its focus on increasing incomes, China's strategy has paid particular attention to improving the health and nutritional status of impoverished rural groups (NCR 2021). Viet Nam transitioned to commercial agriculture between 1986 and 2000, and then began pursuing sustainability and resilience through its National Targeted Program on New Rural Development. This program has improved rural infrastructure, increased household incomes, and reduced regional socioeconomic disparities. Similarly, the Philippines' Pantawid Pamilyang Pilipino Program has tackled poverty through conditional cash transfers, aiming to reduce inequality by prioritizing children's health and education.

Climate change and sustainability

As intensified climate- and weather-induced extreme events become common, ESA governments have generally supported policy efforts to adapt to and mitigate adverse climate impacts on food security, agricultural livelihoods, and other food-system-related components of well-being. In the past decade, many have submitted pledges under the United Nations Framework Convention on Climate Change and announced low-carbon development strategies, guidelines, and action plans that include the agriculture sector (Zhou et al. 2023). For instance, Cambodia, Lao PDR, Mongolia, Myanmar, the Philippines, and Viet Nam have set agriculture-specific greenhouse gas emissions targets. Viet Nam's National Climate Change Adaptation Plan identifies sustainable rice production and reduced agrochemical use as key goals to achieve by 2030. China achieved its goal of zero growth in the use of chemical fertilizers and pesticides by 2020 and has been examining the potential for reducing emissions through rural revitalization to meet its pledge to reach peak carbon emissions by 2030 and achieve carbon neutrality by 2060. CGIAR's MITIGATE+ Research Initiative (2022–2024) explored innovative low-carbon development potential rooted in local conditions and has identified China and Viet Nam as pilot nations for this work. This global initiative aims to ensure that a low-emissions food systems approach will continue to be part of the global political agenda based on rigorous science, good governance, and the principles of gender and social equity.

Regional integration

Regional economic integration has been another cornerstone of food and agricultural policies. ASEAN's agricultural cooperation, which began in 1968, has expanded beyond production and supply concerns to encompass food security, safety, halal food standards, crop and livestock development, agricultural training, cooperatives, and joint initiatives for agricultural and forest products. Frameworks including the Integrated Food Security Framework and the Vision and Strategic Plan for ASEAN Cooperation in Food, Agriculture, and Forestry (2016–2025) articulate a vision for a competitive, inclusive, and sustainable regional food system.

The establishment of the ASEAN Free Trade Area in 1993 and the inclusion of agricultural products in 1995 marked significant milestones. ASEAN's deeper cooperation through free trade agreements and the Regional Comprehensive Economic Partnership (RCEP) has enhanced regional agricultural trade and market integration. Initiatives including the ASEAN Plus Three Cooperation Strategy and the Lancang–Mekong Cooperation mechanism have further strengthened collaboration. The RCEP is projected to lower agricultural tariffs among member states, with China's tariffs on RCEP agricultural imports expected to drop from 14 percent to below 4 percent. This reduction, coupled with nontariff concessions, enhances trade prospects and promotes agricultural investment (see Chapter 16). RCEP also facilitates cross-border e-commerce and technical cooperation in areas such as pest control and animal health, helping to create pathways for sustainable agricultural growth (ASEAN-China Center 2023).

IFPRI's work helps to facilitate integration across ESA. For instance, the Agricultural Transformation and Market Integration in the ASEAN Region project (2016–2022)[3] informed improvements in policies, food safety, and investment in agrifood industries, reinforcing market integration and regional investments. Building on this, IFPRI has been a part of the ASEAN-CGIAR Innovate for Food Regional Program (2023–2027),[4] with a focus on strengthening agrifood trade and value chains across the ASEAN member states to ensure the region remains competitive and resilient within the global agrifood system.

Toward 2050

ESA's food systems are at a critical juncture in the face of climate- and weather-related risks, demographic shifts, and structural transformation, but there are also promising opportunities. In Southeast Asia, rapid population growth is expected to increase the subregion's rice demand by 18 percent by 2040 (Yuan et al. 2022) and overall food demand by 40 percent by 2050 (Goyal et al. 2023). In the context of increasing demand, constraints on cropland expansion, and current yield trends, Cambodia, Myanmar, the Philippines, and Thailand are likely to see considerable gaps between average farmer rice yields and yield potential, and Indonesia and the Philippines are likely to increase their dependence on rice imports by 2040 to meet demand (Yuan et al. 2022). Given that vulnerable smallholders dominate agricultural production and vast low-income populations are put at risk by high food prices, the disproportionate impacts of climate change could increase the average poverty headcount in the region's rural areas. Concurrently, the nexus of climate

3 www.ifpri.org/project/agricultural-transformation-and-market-integration-asean-region-responding-food-security-and/

4 www.cgiar.org/news-events/news/asean-and-cgiar-launch-joint-program-on-accelerating-innovation-in-agri-food-systems/

change, natural capital, and sustainable food systems in the region could be affected by the aging of the region's population, with the share of people over 59 expected to rise from 9.8 percent in 2017 to 20.3 percent by 2050 (WHO 2025).

The region's changing diets and lifestyles will have diverse impacts, including potentially increasing the coexistence of undernutrition in rural areas with rising obesity and diet-related noncommunicable diseases in urban centers, both of which need to be addressed (see Chapter 12) (Pingali and Abraham 2022). For example, childhood overweight and obesity have become major public health concerns in China, affecting 6.8 percent (overweight) and 3.6 percent (obese) of children under 6 as well as 11.1 percent (overweight) and 7.9 percent (obese) of those aged 6–17 between 2015 and 2019, with these rates expected to rise in the coming years (Pan et al. 2021; Pan and Fang 2024). Recent national initiatives in China, including Healthy China 2030 and the implementation plan for the Year of Weight Management Campaign (2024–2026), explicitly target obesity control, diet improvement, and increased physical activity.

Food systems in the region will also be reshaped by emerging opportunities, such as technological innovations that offer new pathways for food systems efficiency, risk mitigation, and market integration (Mojica-Sevilla 2023; Sarmiento 2023). Examples include use of rural big data platforms and machinery automation for smart agriculture systems, and modern biotechnology to respond to diseases and climate change. At the same time, the demand for healthier and lower-emissions foods from the region's expanding urban middle-income populations will also create new opportunities within food systems (Yang et al. 2018). Projections for China suggest that a combination of improving agricultural technologies, reducing food loss and waste, and shifting dietary patterns could reduce greenhouse gas emissions from the country's food system by 47 percent by 2060 compared with 2020 (IFPRI et al. 2021), helping to reconcile food systems development efforts with sustainability. Future analysis must delve into understanding trade-offs and potential synergies between integrated rural development and food systems transformation, such as balancing the combination of improving agricultural efficiency and adopting sustainable diets or balancing goals of emissions reduction from agriculture, rural revitalization, and ecosystem conservation. For example, small-scale paddy rice farming is a significant contributor to greenhouse gas emissions from food systems in the ESA region. Traditional rice cultivation practices, namely continuous flooding irrigation, create anaerobic conditions conducive to growth of microbes that produce methane emissions. In contrast, circular agriculture that integrates rice cultivation with aquaculture, such as rice-fish systems, offers an opportunity for synergistic benefits (Kimura et al. 2022).

Aquaculture and palm oil production likewise present trade-offs. These sectors have played transformative roles in ESA's food systems as drivers of rural development, employment, and trade but have also raised environmental concerns and tensions. Several developing ESA countries have led the rapid expansion of farmed fish and aquatic products, with aquaculture production projected to grow by 21.1 percent in China, 26.2 percent in Indonesia, 22.3 percent in the Philippines, 15.6 percent in Thailand, and 13.1 percent in Viet Nam between 2020 and 2030 (FAO 2022). However, sustainability concerns (such as poor governance causing environmental degradation, habitat destruction, and overfishing) necessitate research into feed alternatives, improved disease management, and enhanced market access to reduce pressure on the region's environmental resources (FAO 2022). Similarly, palm oil production, which is dominated by Indonesia and Malaysia, has fueled rural development and growth in the global edible oil supply but has been linked to deforestation, biodiversity loss, and social conflicts (Pacheco et al. 2018).

Addressing these challenges through the emerging opportunities will require substantial policy innovations and regional cooperation, as well as responsive investments and financing. The ADB and other key development institutions in the region have shifted their strategic priorities toward integrated rural development. The ADB's investment in agriculture, natural capital, and rural development quadrupled between 2009 and 2020, reaching approximately US$6 billion (ADB 2021), and for food systems transformation in the Asia and Pacific region, it has now committed $40 billion by 2030. Initially dominated by irrigation investments, this portfolio now encompasses broader food systems priorities for the future, including (1) rural market infrastructure; (2) agro-industry, marketing, and trade; and (3) agricultural policy, institutional, and capacity development. To scale up these efforts, regional financial institutions, national governments, and private investors engage through substantial collaborative platforms, including South–South and trilateral cooperation, regional economic cooperation frameworks, and local partnerships. Capacity building throughout government, agricultural extension systems, and financial systems will be essential to streamline future food systems financing and funding.

While development organizations have provided valuable insights and funding in the region, policy implementation remains siloed and without coordinated structures or networks in many cases. Examples include interjurisdictional watershed conservation and food standards for trade integration that consider nutrition and health. Stronger transnational cooperation facilitated by international and regional organizations, along with collaboration among governments, private sector entities, and civil society, is essential for integrating action frameworks and identifying scalable solutions for greater synergies. Strengthening regional collaboration through organizations such as ASEAN is another area for future policy and research, as a means to facilitate integrated action frameworks and scaled-up solutions to shared challenges. Promising areas for collaboration include developing nutrition- and health-oriented food standards and controlling agricultural pests and diseases.

Future policies of ESA nations and policy research by IFPRI and other development organizations for the region's food systems transformation and integrated rural development must embrace two key perspectives. First, although widely recognized in research and increasingly considered in policies, the concept of agrifood systems remains fairly narrow and is skewed toward food security in the ESA political landscape. A systems perspective is necessary to capture interactions among food and nutrition security, health, and environmental sustainability within the context of economic, demographic, and dietary changes in the region. Second, a participatory approach that supports the agency of marginalized smallholders and local consumers can help policymakers design interventions responsive to local dietary cultures, production conditions, and diverse sectoral interests, ensuring greater equity for all (Nguyen et al. 2024). Together, these systematic, regional, and locally informed perspectives can help to navigate food-systems-related policy and research to support regional integration and local opportunities in ESA, aimed toward reaching the Sustainable Development Goals by 2050.

References

ADB (Asian Development Bank). 2021. *Financing Sustainable and Resilient Food Systems in Asia and the Pacific.* Manila.

ASEAN (Association of Southeast Asian Nations). 2023. *ASEAN Statistical Yearbook 2023.* Jakarta: ASEAN Secretariat.

ASEAN. 2025. "Rural Development and Poverty Eradication." https://asean.org/our-communities/asean-socio-cultural-community/rural-development-and-poverty-eradiction/

ASEAN–China Center. 2023. "China–ASEAN Agricultural Cooperation Has Broad Prospects: Record of China-ASEAN Agricultural Cooperation Symposium." *China Report.*

Baker, P., and S. Friel. 2016. "Food Systems Transformations, Ultra-Processed Food Markets and the Nutrition Transition in Asia." *Globalization and Health* 12: 80.

Bennett, A., S. Dubey, W.T.K. Lee, B. Damen, and C. Bucatariu. 2022. *FAO Regional Strategy on Food Loss and Waste Reduction in Asia and the Pacific.* Bangkok: Food and Agriculture Organization of the United Nations.

Chen, K.Z., and Y. Zhan. 2022. "East and Southeast Asia." In *2022 Global Food Policy Report: Climate Change and Food Systems,* 133–136. Washington, DC: IFPRI.

Chen, K.Z., Y. Zhou, and M. Rui. 2023. "East and Southeast Asia." In *2023 Global Food Policy Report: Rethinking Food Crisis Responses,* 108–111. Washington, DC: IFPRI.

Chiengkul, P. 2025. "Southeast Asia's Frustration with the State of Climate Finance." *Fulcrum,* January 21.

Choiruzzad, S.A.B. 2024. "Status of Food Security in East and Southeast Asia." In *The Water, Energy, and Food Security Nexus in Asia and the Pacific: East and Southeast Asia,* eds. E. Lee, B. Böer, L. Surendra, J.A. Chun, and M. Taniguchi, 83–112. Cham: Springer Nature Switzerland.

FAO (Food and Agriculture Organization of the United Nations). 2022. *The State of World Fisheries and Aquaculture 2022: Towards Blue Transformation.* Rome.

FAO. 2024. FAOSTAT. July update. www.fao.org/faostat/

FAO, IFAD (International Fund for Agricultural Development), UNICEF, WFP (World Food Programme), and WHO (World Health Organization). 2024. *The State of Food Security and Nutrition in the World 2024: Financing to End Hunger, Food Insecurity and Malnutrition in All Its Forms.* Rome: FAO.

GIZ (German Society for International Cooperation). 2024. Standards in the South-East Asian Food Trade. Project description. https://www.giz.de/en/worldwide/28747.html

Goyal, A., E. Lock, D. Moorthy, and R. Perera. 2023. "Saving Southeast Asia's Crops: Four Key Steps Toward Food Security." *Our Insights,* July 13. McKinsey & Company.

Habermann, B., R. Nehring, W. Zhang, et al. 2023. "A Conceptual Framework of Living Labs for People for Sustainable Food Systems." IFPRI Discussion Paper 2227. IFPRI, Washington, DC.

IFPRI. 2015a. *Investing in Sustainable Agricultural Production*. Making a Difference through Food Policy Research Brief. Washington, DC.

IFPRI. 2015b. *Unlocking Markets and Trade for Rural Development*. Making a Difference through Food Policy Research Brief. Washington, DC.

IFPRI; Academy of Global Food Economics and Policy, China Agricultural University; China Academy for Rural Development, Zhejiang University; Centre for International Food and Agricultural Economics, Nanjing Agricultural University; Institute of Agricultural Economics and Development, Chinese Academy of Agricultural Science. 2021. *2021 China and Global Food Policy Report: Rethinking Agrifood Systems for the Post-COVID World*. Be Washington, DC: IFPRI.

Jones, K., and F. Nti. 2022. *Impacts and Repercussions of Price Increases on the Global Fertilizer Market*. Washington, DC: U.S. Department of Agriculture.

Kesse-Guyot, E., B. Allès, J. Brunin, et al. 2023. "Environmental Impacts Along the Value Chain from the Consumption of Ultra-Processed Foods." *Nature Sustainability* 6 (2): 192–202.

Kimura, S., K. Chen, and B. Gong. 2022. *Circular Agriculture for Sustainable and Low-Carbon Development in the People's Republic of China*. Manila: Asian Development Bank.

Lee, E., B. Böer, L. Surendra, J.A. Chun, and M. Taniguchi. 2024. *The Water, Energy, and Food Security Nexus in Asia and the Pacific*. Cham, Switzerland: Springer Nature.

Leite, F.H.M., N. Khandpur, G.C. Andrade, K. Anastasiou, P. Baker, M. Lawrence, and C.A. Monteiro. 2022. "Ultra-Processed Foods Should Be Central to Global Food Systems Dialogue and Action on Biodiversity." *BMJ Global Health* 7 (3): e008269.

Lin, J.Y. 1987. "The Household Responsibility System Reform in China: A Peasant's Institutional Choice." *American Journal of Agricultural Economics* 69 (2): 410–415.

Mojica-Sevilla, F. 2023. *Philippines: Agricultural Biotechnology Annual*. Manila: Foreign Agricultural Service, United States Department of Agriculture.

NCR (New China Research). 2021. *Chinese Poverty Alleviation Studies: A Political Economy Perspective*. Beijing: New China Research of Xinhua News Agency.

Nguyen, P.H., K.Z. Chen, and J. Harris. 2024. "East and Southeast Asia." In *2024 Global Food Policy Report*, 108–114. Washington, DC: IFPRI.

Pacheco, P., P. Levang, G. Schoneveld, and A. Dermawan. 2018. "The Palm Oil Governance Complex: Progress, Problems and Gaps." In *Achieving Sustainable Cultivation of Oil Palm*, ed. A. Rival. Burleigh Dodds Series in Agricultural Science, 27. Cambridge, UK: Burleigh Dodds Science Publishing.

Pan, X.-F., and Z.-Z Fang. 2024. "Obesity in China: What We Know and What We Can Do." *Lancet Public Health* 9 (12): e1000–e1001.

Pan, X.-F., L. Wang, and A. Pan. 2021. "Epidemiology and Determinants of Obesity in China." *Lancet Diabetes & Endocrinology* 9 (6): 373–392.

Papilo, P., M. Marimin, E. Hambali, et al. 2022. "Palm Oil-Based Bioenergy Sustainability and Policy in Indonesia and Malaysia: A Systematic Review and Future Agendas." *Heliyon* 8 (10): e10919.

Pingali, P.L. 2012. "Green Revolution: Impacts, Limits, and the Path Ahead." *PNAS* 109 (31): 12302–12308.

Pingali, P.L., and M. Abraham. 2022. "Food Systems Transformation in Asia: A Brief Economic History." *Agricultural Economics* 53 (6): 895–910.

Pingali, P.L., and V.-T. Xuan. 1992. "Vietnam: Decollectivization and Rice Productivity Growth." *Economic Development and Cultural Change* 40 (4): 697–718.

Ryan, J.G. 1999. "Assessing the Impact of Rice Policy Changes in Viet Nam and the Contribution of Policy Research." Impact Assessment Discussion Paper 8. IFPRI, Washington, DC.

Sarmiento, P. 2023. "Innovative Tech with AI Ease Farmers' Woes across SE Asia." *China Daily*, Sept. 21.

Sundram, P. 2023. "Food Security in ASEAN: Progress, Challenges and Future." *Frontiers in Sustainable Food Systems* 7: 1260619.

Swastika, D.K.S., A. Priyanti, A.M. Hasibuan, et al. 2024. "Pursuing Circular Economics through the Integrated Crop-Livestock Systems: An Integrative Review on Practices, Strategies and Challenges Post Green Revolution in Indonesia." *Journal of Agriculture and Food Research* 18: 101269.

UNESCAP (UN Economic and Social Commission for Asia and the Pacific). 2019. *China's Economic Transformation: Impacts on Asia and the Pacific*. Bangkok.

von Grebmer, K., J. Bernstein, A. de Waal, N. Prasai, S. Yin, and Y. Yisehac. 2015. *2015 Global Hunger Index: Armed Conflict and the Challenge of Hunger*. Washington, DC: IFPRI, Welthungerhilfe, and Concern Worldwide.

WHH (Welthungerhilfe), Concern Worldwide, and Institute for International Law of Peace and Armed Conflict (IFHV). 2024. *2024 Global Hunger Index: How Gender Justice Can Advance Climate Resilience and Zero Hunger*. Bonn/Berlin: WHH; Dublin: Concern Worldwide; Bochum, Germany: IFHV.

WHO (World Health Organization). 2025. "Ageing and Health in the South-East Asia Region." www.who.int/southeastasia/health-topics/ageing

World Bank. 2022. *Reforms for Recovery. World Bank East Asia and Pacific Economic Update (October)*. Washington, DC: World Bank.

World Bank. 2025a. World Bank Open Data. https://data.worldbank.org

World Bank. 2025b. *Food Systems Are Being Transformed*. Food Systems 2030 Annual Report. Washington, DC.

Yang, J., J.G. Siri, J.V. Remais, et al. 2018. "The Tsinghua–Lancet Commission on Healthy Cities in China: Unlocking the Power of Cities for a Healthy China." *Lancet* 391 (10135): 2140–2184.

Yuan, S., A.M. Stuart, A.G. Laborte, et al. 2022. "Southeast Asia Must Narrow Down the Yield Gap to Continue to Be a Major Rice Bowl." *Nature Food* 3 (3): 217–226.

Zhou, Y., and K.Z. Chen. 2023. "Food Price Inflation in East and Southeast Asia: Situation, Driving Forces, and the Outlook." *Studies in Agricultural Economics* 125: 77–86.

Zhou, Y., Z. Chen, and K.Z. Chen. 2023. "Building Climate-Resilient Food Systems in East and Southeast Asia: Vulnerabilities, Responses, and the Outlook." *Frontiers of Agricultural Science and Engineering* 10 (1): 16–30.

Chapter 23

LATIN AMERICA AND THE CARIBBEAN

Valeria Piñeiro, Brian McNamara, Joaquín Arias Segura, and Eugenio Díaz-Bonilla

Valeria Piñeiro is regional representative for Latin America and the Caribbean, Markets, Trade, and Institutions Unit (MTI), IFPRI; **Brian McNamara** is a program manager, MTI, IFPRI; **Joaquín Arias Segura** is a senior analyst in agrifood development, Inter-American Institute for Cooperation on Agriculture (IICA); and **Eugenio Díaz-Bonilla** is a special advisor to the Inter-American Institute for Cooperation on Agriculture (IICA) and a visiting senior research fellow, Director General's Office, IFPRI.

Over the past five decades, Latin America and the Caribbean (LAC) has experienced significant changes in its food systems, while also facing both long-standing and new challenges, including food insecurity, poverty, malnutrition, inequality, and problems related to environmental sustainability and agricultural productivity. As the world's largest net food-exporting region and one endowed with rich natural resources, LAC plays a key role in global food security and nutrition, with potential to contribute even more in the future. The region also plays a critical role in stabilizing the global climate and in conserving biodiversity. To continue making these contributions to global food security and environmental sustainability, however, the region must overcome the hurdles created by economic and political instability, climate change, and deep-rooted structural inequalities spanning social, economic, gender, and health dimensions.

Food policy research has played a crucial role in addressing these challenges, providing insights that have guided strategies and policies to shape the region's agrifood sector. This chapter draws on research by IFPRI and regional partners to outline the evolving food systems landscape in LAC and present evidence-based policy options and research priorities for the years ahead.

Evolution of food systems challenges

Food security, diets, and nutrition

The recent evolution of food systems challenges in the LAC region reflects a complex interplay of socioeconomic, environmental, and demographic factors. In the 1970s and 1980s, the region grappled with poverty-driven food insecurity, exacerbated by economic crises, political instability, and unequal land distribution. Rapid urbanization during this period strained rural food production systems and, in some countries, increased reliance on imported food. More than 82 percent of the LAC population now lives in urban areas, up from less than half in 1960 (Busso et al. 2023). While agricultural policies in the 1970s and early 1980s focused on increasing agricultural productivity, the policy focus shifted in the 1980s and 1990s toward addressing undernutrition (diets lacking essential nutrients and calorie deficits in some countries), particularly among children and rural populations.

The food security landscape has changed dramatically since the early 2000s. From 2000 to 2013, as gross domestic product (GDP) per capita rose steadily, undernourishment rates consistently fell. However, since then, an economic slowdown in the region, the COVID-19 pandemic, and geopolitical crises have led to increased rates of undernourishment. Despite some economic recovery in recent years, undernourishment remains elevated compared to its 2013 levels, suggesting that economic growth alone may not be sufficient to address hunger and food security (Figure 23.1).

Over the same period, rising incomes and dietary shifts contributed to an alarming increase in obesity and related noncommunicable diseases. Stalled progress in reducing undernutrition, combined with the rapid increase in overweight and obesity, has created a "double burden" of malnutrition (see Chapter 12). Deficiencies in zinc, iron, and vitamin B12 among children and women of childbearing age remain a persistent problem, as are high rates of chronic diseases such as anemia, although reliable data for the region have not been collected for nearly a decade (Soto-Méndez and Boy 2024). Figure 23.2 illustrates these nutrition challenges, showing rates of obesity and child stunting across different subregions of LAC. The double burden and micronutrient deficiencies are most acute in Mexico and Central American countries, where rates of both obesity and child stunting are above the regional average. In contrast, South America struggles with high rates of obesity, but the rate of child stunting is low, indicating that undernutrition is a less pressing concern.

FIGURE 23.1 GDP and prevalence of undernourishment in LAC, 2000–2024

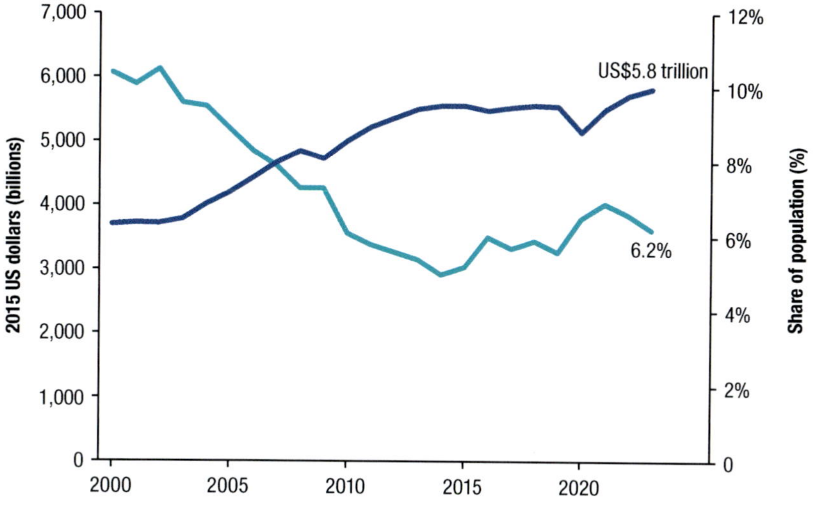

Source: Based on FAOSTAT data. www.fao.org/faostat/

Note: Undernourishment is an indicator of hunger and food insecurity developed by the Food and Agriculture Organization of the United Nations (FAO) and based on the Food Insecurity Experience Scale (FIES).

FIGURE 23.2 Prevalence of adult obesity and child stunting, 2000–2022

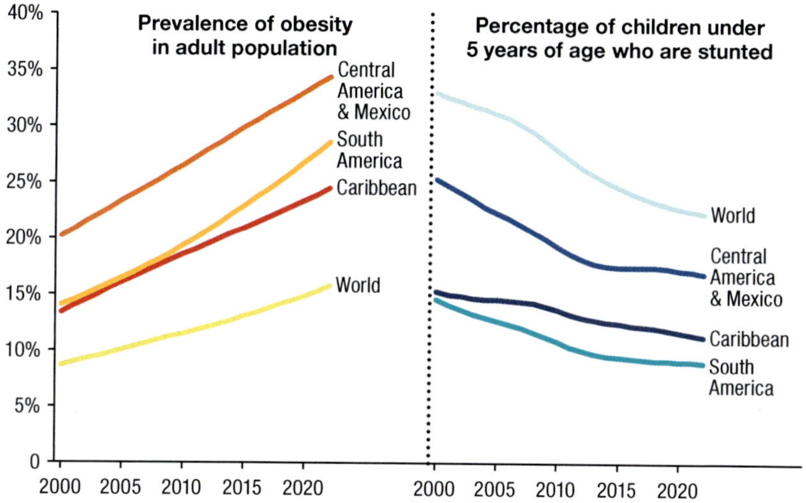

Source: Based on FAOSTAT data. www.fao.org/faostat/

Healthy diets are unaffordable for many people in the region (Figure 23.3). Despite the relatively moderate cost of healthy diets (under $3.00 PPP per person per day in some countries), persistent poverty and income disparities have put better diets out of reach for 182 million people in the region (FAO et al. 2025). Recent years, particularly 2021 and 2022, have seen rising healthy diet costs, reflecting post-pandemic economic strains and geopolitical developments.

LAC's agriculture sector

LAC's agriculture sector is a major global supplier that provides both food and employment for the region's population. Food exports from the region increased rapidly between 2000 and 2020, rising from US$35 billion to $156 billion (Piñeiro et al. 2023). Today, LAC supplies 50 percent of corn, 70 percent of soybean meal and oil, 60 percent of soybeans, and 36 percent of bovine and poultry meat exports globally, along with other critical food products. About 43 percent of global agricultural exports are projected to come from LAC by 2032, reflecting its pivotal role in international food security (Piñeiro et al. 2023). These exports stabilize global markets against shocks and support food security in other regions, as occurred at the beginning of the Russia–Ukraine war (Glauber et al. 2023).

Production varies significantly across the LAC region. Figure 23.4 shows production divided by subregion for cereals, fruits, and vegetables. More than 75 percent of cereal production is in the Southern Cone,[1] led by Brazil and Argentina. This subregion is characterized by highly efficient large-scale agriculture, producing about 24 metric tons of cereals per hectare, that drives exports from the region. The Southern Cone also leads fruit and vegetable production, though other subregions play a large role. The Andean region[2] contributes about 20 percent of regional fruit production, while Mexico contributes 15 percent, and Central American countries such as Guatemala are leading exporters of bananas. Mexico has become one of the region's largest producers of vegetables, reaching some 16 million tons per year, about 9 percent of the total. Andean countries, with their cooler growing areas at higher altitudes, are also well suited to vegetable farming and contribute about 9 percent of production.

LAC's agriculture sector is also an important source of livelihoods, accounting for more than half of all employment in Ecuador, Guatemala, and Nicaragua and more than 40 percent in Honduras and Peru (FAO 2025). However, inequality in agricultural land distribution is greater in LAC

1 The Southern Cone countries are Argentina, Brazil, Chile, Paraguay, and Uruguay.

2 The Andean countries included here are Bolivia, Chile, Colombia, Ecuador, Peru, and Venezuela.

FIGURE 23.3 High cost of healthy diets

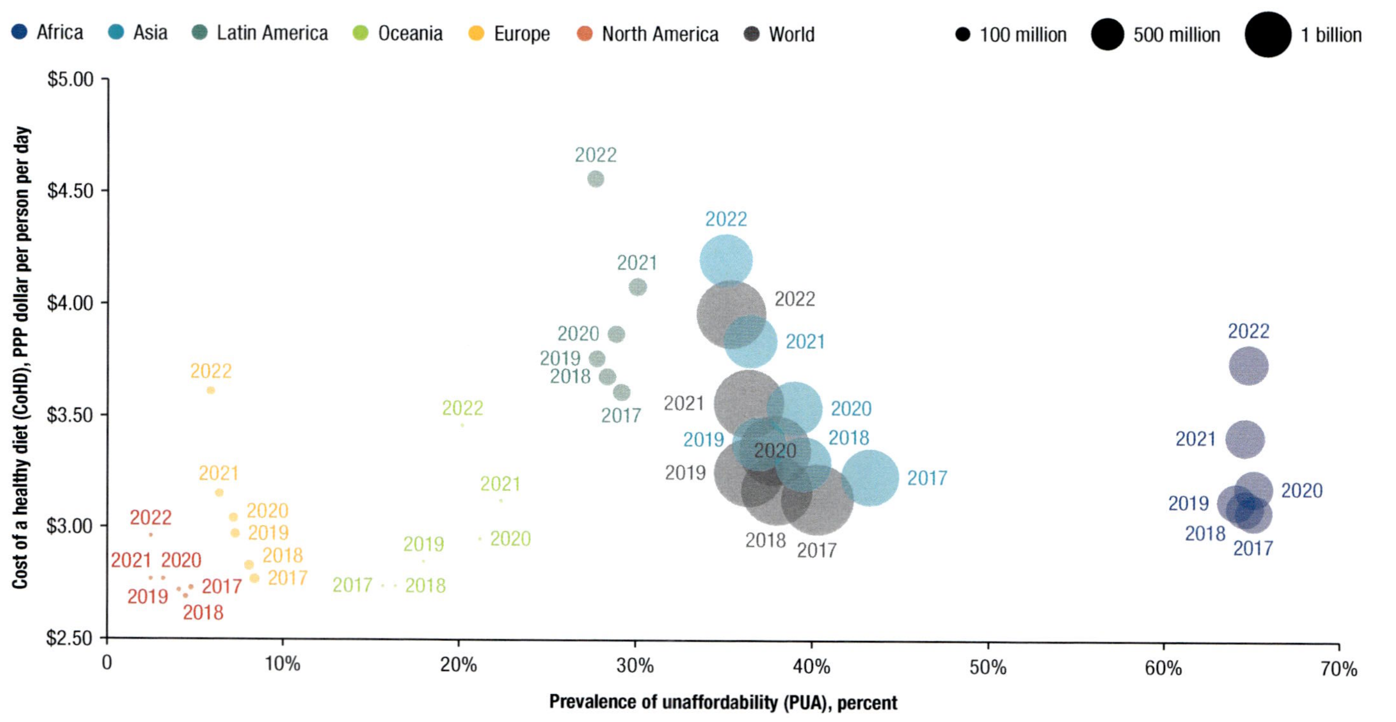

Source: Based on FAOSTAT data. www.fao.org/faostat/

FIGURE 23.4 Production of crops across subregions

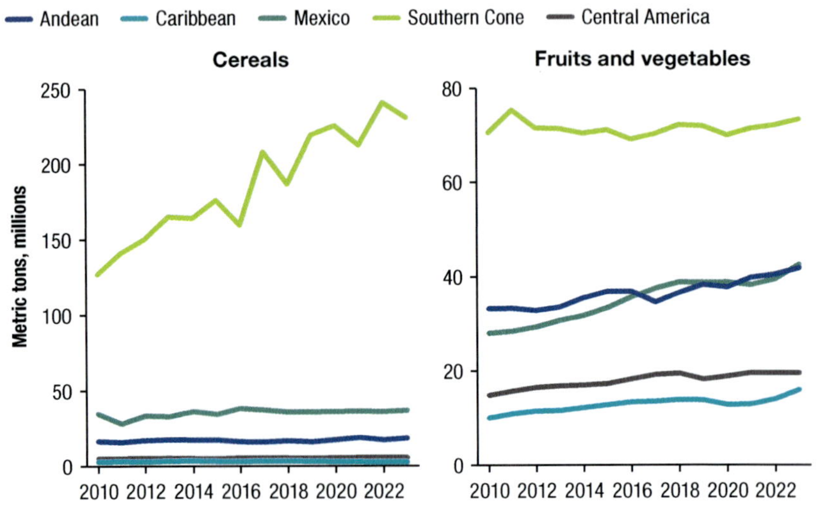

Source: Based on FAOSTAT data. www.fao.org/faostat/

than any other region, with a predominance of small farms. In Guatemala, for example, 64 percent of farms are smaller than 1 hectare, while only 6.2 percent are larger than 5 hectares (Britos et al. 2022). This pattern of very small farms, common across Central America, Mexico, and much of the Andean region, limits labor productivity and the earning potential of smallholder farmers (Gáfaro et al. 2023). Smallholders also encounter barriers to accessing markets, finance, technology, and other resources (Berdegué and Fuentealba 2011).

Gender inequities are also stark. Recent analysis from Colombia showed that women in the agriculture sector earned 43 percent less than men on average. This disparity was significantly larger for higher skilled agricultural occupations and in rural areas, where women earn nearly 70 percent less (Rueda-Gallardo et al. 2025). Food and agricultural policy therefore has major implications for many households, and equity is a central concern.

Environmental services and sustainability

LAC is also the world's largest provider of ecosystem services, with significant shares of global renewable water reserves, forest carbon stocks, and plant and animal species (Pan et al. 2024). These provide key global and regional ecosystem services, such as regulating regional climates by maintaining hydrological cycles and supporting biodiversity hotspots that are crucial for ecological balance and agricultural resilience.

However, as food production in the region has increased, environmental sustainability has become a greater concern. Increased production in LAC over the last few decades has been associated with relatively low increases in greenhouse gas emissions from agriculture, forestry, and other land use change (AFOLU) compared to other regions, but deforestation, soil degradation, and water scarcity have emerged as major threats to sustainability (Piñeiro et al. 2023). Expansion of cropland and pastures has driven deforestation, and intensive farming practices have contributed to soil erosion and nutrient depletion in key agricultural areas (FAO 2024). In addition, climate change impacts—including rising temperatures, water scarcity, extreme weather events, and reduced agricultural yields—will exacerbate these threats. For example, coastal regions, which are home to large populations, will have to contend with the dual pressures of sea-level rise and saltwater intrusion, which will affect both food production and food systems livelihoods.

These challenges highlight the need for comprehensive rural development strategies that account for the environmental dimension of food systems. While the LAC region must continue to increase food production to meet the needs of a growing global population, it also needs to address deforestation, soil degradation, and water management for the sustainability of its food systems and the health of the planet (Piñeiro et al. 2023).

Innovations in food policy from LAC

The LAC region has shown leadership in pioneering innovative policies, backed by research, to address food systems challenges. In nutrition, for example, Guatemala's PROCOMIDA food-assisted maternal and child health and nutrition program reduced child stunting and led to changes in household diet decisions (Jensen et al. 2016; Olney et al. 2018). IFPRI researchers evaluated the program, providing evidence-based adjustments that enhanced its impact. Other policies have also promoted healthier diets. Chile introduced a comprehensive policy requiring front-of-package health labeling and marketing restrictions for unhealthy foods, while also removing junk food from schools. This policy was shown to contribute to a decline in consumption of sugary drinks and processed snacks among children, contributing to improved dietary patterns and public health outcomes (Fretes et al. 2025). Mexico overcame significant political hurdles to introduce a similar strategy in 2019 that required front-of-package labeling for food and beverages with high levels of calories, sugar, saturated fat, trans fat, and sodium, as well as those containing noncaloric sweeteners, in an attempt to promote healthy food choices among consumers (White and Barquera 2020).

In response to environmental challenges, LAC countries have adopted policies to promote sustainable intensification. For example, the Low-Carbon Agriculture Plan and the National Program for Strengthening Family Farming in Brazil use credit instruments, extension services, and social support to promote no-till planting, integrated crop-livestock-forest systems, reforestation, and other more sustainable land management strategies (Harfuch et al. 2024). The program had been implemented on more than 54 million hectares of agricultural land by 2024 (MAPA 2023). These and other examples can provide valuable ideas and lessons for countries and regions around the world facing similar challenges.

Future food systems challenges

The LAC region's food systems will face a range of complex challenges as 2050 approaches. Global demand for LAC's agricultural output and environmental services will continue to grow. Continuing urbanization within LAC is expected to intensify the demand for food, particularly processed and convenience foods, which could exacerbate dietary shifts and heighten the risk of health crises related to obesity and noncommunicable diseases (Fretes et al. 2024). At the same time, climate change, biodiversity loss, and environmental degradation, combined with unsustainable agricultural practices and deforestation, threaten long-term food security, both in the region and globally. Persistent inequalities in access to land, resources, and markets will likely continue to constrain the productivity of smallholder farmers, necessitating targeted investments in rural infrastructure, education, and technology to improve agricultural production and rural livelihoods. Ensuring access to technologies for all producers will be crucial to increasing opportunities and preventing the amplification of existing inequalities.

To address these challenges, food policy research must become more forward-looking, inclusive, and adaptable. Several key research priorities are outlined here.

Climate-smart and sustainable agriculture. Strengthening climate-smart and sustainable agriculture should be a priority, with research focusing on identifying climate-resilient crop varieties, promoting soil health, and scaling up agroecological approaches. This focus would leverage the region's abundant natural resources and role as a major global exporter, while building on recent successes in sustainable intensification. Research can identify and inform policies that incentivize farmers to adopt sustainable practices, which is crucial for long-term food security (see Chapters 5, 9, and 17) (Piñeiro et al. 2020).

Equitable access. Ensuring equitable access to resources, markets, and technologies for smallholders and women will be critical (IFPRI 2020). As a region with strong historic agricultural productivity growth but also high levels of inequality in income and asset ownership, policies to promote equitable growth can increase incomes, resilience, and affordability of healthy diets for the most vulnerable. Research can identify strategies to close opportunity gaps for smallholders and promote gender equity, thereby enhancing the overall resilience of food systems (see Chapters 6, 11, and 14).

Technology and innovation. Leveraging technology and innovation holds potential to revolutionize food systems through advances in digital tools, precision agriculture, and biotechnology. LAC has high rates of digital literacy and access to digital tools, along with many examples of programs that promote the adoption of technologies and practices (Shanahan and Bahia 2024). These strengths can be leveraged to expand access to and adoption of recent food systems innovations and develop lessons for other regions. For example, a review of technologies in the LAC livestock sector identified numerous existing innovations in silvopastoral practices,[3] and strategies for improved feeding and better data tracking, among others, which have been shown to produce multiple economic, social, and environmental benefits (Elverdin et al. 2025). Enabling policies should be implemented to scale up such innovations and maximize their benefits. Food policy research must ensure these technologies are accessible to smallholder farmers and contribute to inclusive development (see Chapters 11, 14, and 17).

Diets and nutrition. Addressing the double burden of malnutrition requires a focus on food environments, education, and consumer behavior. To confront these challenges, LAC has adopted a range of innovative approaches, including taxes on unhealthy foods, nutrition labeling, and regulations on the marketing of unhealthy foods to children, which should be further studied and implemented in other countries. Nutrition-sensitive social protection programs are another policy tool with potential to improve access to healthy diets and elicit dietary behavior change. Research can build on efforts in many LAC countries to identify policies that promote affordable, nutritious, and sustainably produced foods while addressing the drivers of unhealthy diets (see Chapter 12).

---

3 Silvopastoral practices are land management systems that integrate trees, forage crops, and livestock on the same land to enhance biodiversity, productivity, and sustainability.

Regional cooperation. Regional cooperation will be crucial for addressing global challenges such as climate change and food trade. Food policy research should support platforms that foster knowledge-sharing, harmonize policies, and improve trade efficiency across LAC (see Chapter 16). Institutions such as the Inter-American Institute for Cooperation on Agriculture have partnered with IFPRI to advance this agenda by launching a network of agricultural trade negotiators to align priorities and foster cooperation across the region (IICA 2023).

Financing. Resources must also be allocated effectively across sectors and actors to have the largest impact. While there is limited research on the financial requirements to transform food systems in LAC, one estimate suggests 0.52 percent of the region's GDP would need to be allocated annually to end hunger and malnutrition (FAO et al. 2024). More generally, the same authors observe that many plans have been proposed to achieve economic, social, and environmental goals in LAC food systems, but they often lack clear financing strategies. IFPRI researchers have noted several analytical challenges, including the need to take a holistic view of all financial flows, both internal and external to food systems (Díaz-Bonilla 2021). This comprehensive analysis is needed to make financing more accessible to small and medium producers to support their adoption of sustainable intensification technologies (see Chapter 18).

Foresight and impact evaluation. Research can inform policymakers and other actors about the most effective interventions for responding to economic shocks, climate risks, and crises. IFPRI, under the CGIAR Research Initiative on Foresight, has partnered with Brazil's Embrapa, the Grain Exchange, and the Institute for International Agricultural Negotiations to conduct frequent model-based analysis, build capacities among regional researchers, and provide policy guidance to MERCOSUR, the regional trade bloc (Gianatiempo and McNamara 2024). Economic and climate modeling is critical for anticipating future challenges and taking preemptive actions (see Chapters 3 and 4). Impact evaluations and literature syntheses can complement these efforts by identifying effective strategies for responding to anticipated challenges and improving resilience.

In sum, LAC has made significant strides in addressing food systems challenges over the past 50 years, with food policy research playing a central role in driving evidence-based solutions. Although the region faces a complex set of challenges as it looks toward 2050, including climate change, urbanization, inequality, and technological disruptions, the next 25 years present an opportunity to build on past successes, address persistent challenges, and ensure a more food-secure and sustainable future for the region.

IFPRI has already contributed significantly to all these areas through evidence-based solutions, analysis of drivers and trends, and development of innovations. Together with other research organizations and regional and national partners, IFPRI will continue to play a pivotal role in generating the knowledge and solutions needed to transform food systems in the LAC region.

References

Berdegué, J.A., and R. Fuentealba. 2011. "Latin America: The State of Smallholders in Agriculture." Paper presented at the *IFAD Conference on New Directions for Smallholder Agriculture*, Rome, January 24–25.

Britos, B., M.A. Hernandez, M. Robles, and D.R. Trupkin. 2022. "Land Market Distortions and Aggregate Agricultural Productivity: Evidence from Guatemala." *Journal of Development Economics* 155: 102787.

Busso, M., P.E. Carrillo, and J.P. Chauvin. 2023. *Rethinking Urban Migration: Policy Options for Cities in Latin America and the Caribbean*. Washington, DC: Inter-American Development Bank.

Díaz-Bonilla, E. 2021. "The Role of Institutional, Technological and Financial Innovation in the Transformation of Food Systems in Latin America and the Caribbean, in the Context of the 2030 Agenda and the COVID-19 Pandemic." In *Food Systems in Latin America and the Caribbean: Challenges in a Post-Pandemic World*, eds. J. Graziano da Silva, M. Jales, R. Papallo, et al., 71–94. Panama City: FAO and CIDES.

Elverdin, P., H. Miranda, F. Bert, et al. 2025. "Assessing Interventions to Facilitate the Adoption of Technologies and Practices That Promote Sustainability in the Bovine Sector at the Farm Level across Latin America and the Caribbean." LAC Working Paper. IFPRI, Washington, DC. Forthcoming.

FAO (Food and Agriculture Organization of the United Nations). 2024. *Sustainable Soil Management: Opportunities and Recommendations for Decision Makers*. Policy Guidance Note. Santiago, Chile.

FAO. 2025. "Hunger and Food Insecurity." Accessed April 8. www.fao.org/hunger/en

FAO, ECLAC (United Nations Economic Commission for Latin America and the Caribbean), IICA (Inter-American Institute for Cooperation on Agriculture), and WFP (World Food Programme). 2024. *Financing for Food Security and Nutrition in Latin America and the Caribbean*. Santiago, Chile: FAO.

FAO, IFAD (International Fund for Agricultural Development), PAHO (Pan American Health Organization), UNICEF, and WFP. 2025. *Latin America and the Caribbean Regional Overview of Food Security and Nutrition 2024: Building Resilience to Climate Variability and Extremes for Food Security and Nutrition*. Santiago, Chile: FAO.

Fretes, G., Q. Marshall, and J.L. Leroy. 2024. "Food Environments: Improving Their Healthfulness." In *Global Food Policy Report 2024: Food Systems for Healthy Diets and Nutrition*, 46–52. Washington, DC: IFPRI.

Fretes, G., N. Wilson, C. Corvala, C. Economos, and S. Chas. 2025. "Front-of-Pack Labels and Young Consumers: An Experimental Investigation of Nutrition and Sustainability Claims in Chile." *Food Quality and Preference* 127: 105432.

Gáfaro, M., A.M. Ibáñez, D. Sánchez-Ordoñez, and M.C. Ortiz. 2023. "Farm Size and Income Distribution of Latin American Agriculture: New Perspectives on an Old Issue." IDB Working Paper No. 01510. Inter-American Development Bank, Washington, DC.

Gianatiempo, J.P., and B. McNamara. 2024. "Mercosur Outlook Launch: Charting a Course for Sustainable Agricultural Growth amid Uncertainty." *IFPRI Blog*, October 16.

Glauber, J.W., D. Laborde Debucquet, V. Piñeiro, and A. Tejeda. 2023. "Can Agricultural Exports from Southern Cone Countries Make Up for Global Supply Disruptions Arising from the Russia-Ukraine War?" In *The Russia-Ukraine Conflict and Global Food Security*, eds. J. Glauber and D. Laborde, 181–186. Washington, DC: IFPRI.

Harfuch, L., R.C.A. Lima, and G. Dantas Lobo. 2024. "Challenges to Fostering Low-Carbon Agriculture through Public Policies and Support in Brazil." *IISD Trade and Sustainability Review: Trade in Food and Agriculture*, December 2. Winnipeg: International Institute for Sustainable Development.

IFPRI. 2020. *2020 Global Food Policy Report: Building Inclusive Food Systems*. Washington, DC.

IICA (Inter-American Institute for Cooperation on Agriculture). 2023. "En el Marco de la OMC, el IFPRI y el IICA Lanzan Red Latinoamericana de Negociadores Comerciales del Sector Agroalimentario." Press release, July 27. Turrialba, Costa Rica.

Jensen, M.L., E.A. Frongillo, J.L. Leroy, and C.E. Blake. 2016. "Participating in a Food-Assisted Maternal and Child Nutrition and Health Program in Rural Guatemala Alters Household Dietary Choices." *Journal of Nutrition* 146 (8): 1593–1600.

MAPA (Ministry of Agriculture and Livestock, Brazil). 2023. *ABC Plan: Ten Years of Success and a Brand New Sustainable Agriculture Production*. Brasilia.

Olney, D.K., J. Leroy, L. Bliznashka, and M.T. Ruel. 2018. "PROCOMIDA, a Food-Assisted Maternal and Child Health and Nutrition Program, Reduces Child Stunting in Guatemala: A Cluster-Randomized Controlled Intervention Trial." *Journal of Nutrition* 148 (9): 1493–1505.

Pan, Y., R.A. Birdsey, O.L. Phillips, et al. 2024. "The Enduring World Forest Carbon Sink." *Nature* 631 (8021): 563–569.

Piñeiro, V., J. Arias, J. Dürr, et al. 2020. "A Scoping Review on Incentives for Adoption of Sustainable Agricultural Practices and Their Outcomes." *Nature Sustainability* 3: 809–820.

Piñeiro, V., M. Piñeiro, E. Bianchi, et al. 2023. "From Farm to Table: Agrifood Systems and Trade Challenges in the Southern Cone." LAC Working Paper 30. IFPRI, Washington, DC.

Rueda-Gallardo, J., D. Vergara Mateus, V. Piñeiro, and L. Escalante. 2025. "Understanding the Gender Gap in the Colombian Agricultural Sector." LAC Working Paper 34. IFPRI, Washington, DC. Forthcoming.

Shanahan, M., and K. Bahia. 2024. *The State of Mobile Internet Connectivity 2024*. London: GSMA Intelligence.

Soto-Méndez, M.J., and E. Boy. 2024. "Introduction and Executive Summary for the Supplement Micronutrients in Latin America: Current State and Research." *Food and Nutrition Bulletin* 45 (Suppl 2): S4–S6.

White, M., and S. Barquera. 2020. "Mexico Adopts Food Warning Labels, Why Now?" *Health Systems & Reform* 6 (1): e1752063.

Chapter 24

MIDDLE EAST AND NORTH AFRICA

Sikandra Kurdi, Ganna Hassan, Nina Jovanovic, and Linda Steinhuebel-Rasheed

Sikandra Kurdi is country program leader for Egypt and a research fellow, Development Strategies and Governance Unit (DSG), IFPRI-Egypt; **Ganna Hassan** is research assistant, DSG, IFPRI-Cairo; **Nina Jovanovic** is an associate research fellow, DSG, IFPRI-Egypt; and **Linda Steinhuebel-Rasheed** is a research fellow, DSG, IFPRI-Egypt.

Food and agricultural policy in the Middle East and North Africa (MENA) region has oscillated over the past 50 years between prioritizing efficiency and economic growth, on the one hand, and ensuring national security through food self-sufficiency and broad-based provision of staple commodities, on the other. This chapter summarizes the historical trends in policies and outcomes as the region moved from heavy state interference to a period of liberalization. We look first at agricultural production and related policies and then at consumers and food and nutrition policy, and conclude with a look at emerging policy issues and research priorities.

Agricultural policy

1970s–1980s: Quest for self-sufficiency

The MENA region as a whole is scarce in arable land and water resources. Most MENA countries exceed ecologically sustainable rates of freshwater use and all countries except Iraq are below the absolute water scarcity threshold for renewable water resources (de Waal et al. 2023). As a result, countries in the region depend heavily on food imports, though the degree of import dependence has varied over time, driven by changing policies and external trends.

For analysis of this period, we follow a categorization from an early IFPRI report (Khalid 1984) that groups MENA countries as oil exporters, labor exporters, and those largely reliant on their own production. Figure 24.1 shows that import dependence in wheat has been high since the 1970s for oil-exporting countries, which benefited from foreign exchange earnings during the oil boom of the 1960s and 1970s, and also for a group of neighboring labor-exporting countries that were indirectly affected as the oil-exporting countries drew in labor from around the region. In Egypt, Jordan, Lebanon, and Yemen, remittances earned by labor migrants to the Gulf states and Libya replaced agricultural production as the main source of income in rural areas. Higher incomes in turn shifted consumption patterns, with imported wheat preferred over traditional local crops of barley, sorghum, and millet. At the same time, higher incomes in both oil-exporting and labor-exporting countries drove demand for meat and animal feed, increasing total grain consumption (Khalid 1984).

The increasing reliance on imports across the region led to government worries about potential vulnerability to external shocks. Thus, agricultural policies were designed to prioritize domestic production of strategic crops such as staple grains. Arab socialism, the leading ideology in the most populous Arab countries in the 1970s and 1980s, inspired agricultural policies that emphasized centralized planning. In Egypt, for example, farmers were given specific crop planting requirements and delivery quotas, with prices fixed separately for farmers and for consumers to promote production while keeping food prices low (Cassing et al. 2009). Subsidies were introduced for fertilizers and other inputs as well as tariffs to protect domestic livestock production.

FIGURE 24.1 Average wheat import dependency ratio by country group, 1970–2023

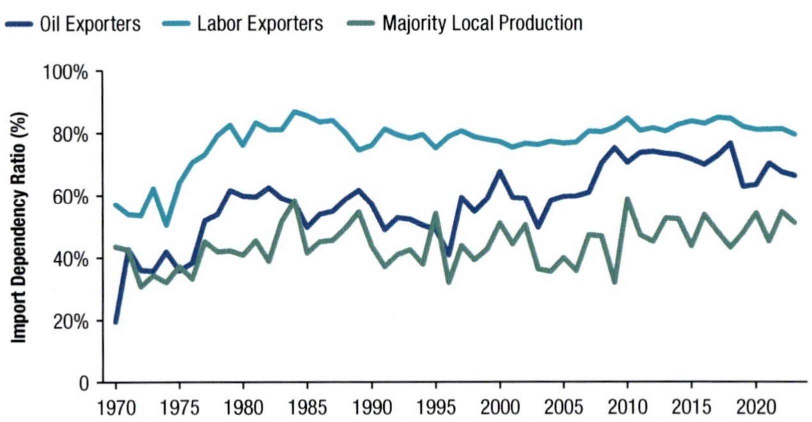

Source: FAOSTAT 2025.

Note: Oil exporters are Algeria, Iraq, Libya, Oman, and Saudi Arabia; Labor exporters are Egypt, Jordan, Lebanon, and Yemen. Majority local production countries are Morocco, Sudan, Syrian Arab Republic, and Tunisia.

In spite of this prioritization of agriculture, agricultural productivity growth remained low. The emphasis on centralized planning led to inefficient use of existing resources (Adams 1993), while constraints on arable land and water and a rapidly growing population made self-sufficiency goals difficult to achieve. Self-sufficiency in cereals dropped from 83 percent in 1970 to only 60 percent in 1980 (von Braun and de Haen 1983).

1990s–2000s: Limited liberalization

Global geopolitical shifts and declining oil prices in the late 1980s and 1990s created fiscal pressures and impetus for liberalization of agricultural policies. Liberalization led to a more efficient allocation of resources and higher overall production, though countries remained far from achieving self-sufficiency in key commodities.

In Egypt, for example, two rounds of agriculture and trade policy reforms included reductions of input subsidies and removal of area restrictions and price controls on most major crops (Ender and Holtzman 2003). But studies found that these reforms had only limited impact on production of key commodities: while the government's low procurement prices under centralized planning had penalized wheat producers, the agriculture sector as a whole had received net protection from trade restrictions; liberalization of trade thus diluted the impacts of agricultural reform (Hazell et al 1995).

FIGURE 24.2 Average wheat yield by country group, 1970–2023

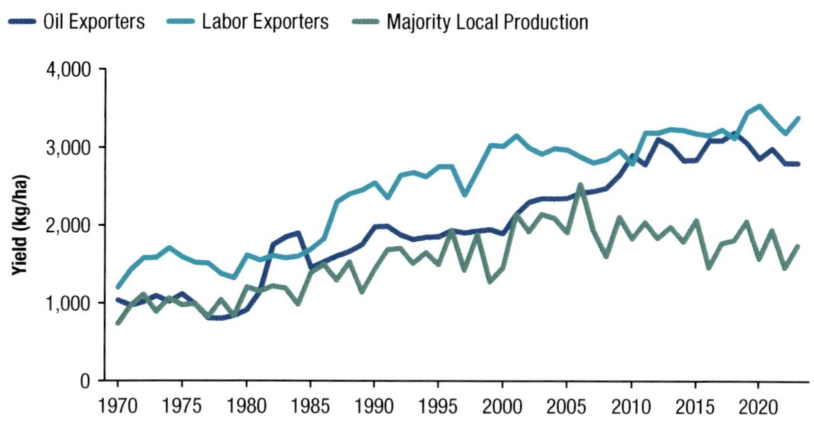

Source: FAOSTAT 2025.

Note: Oil exports are Algeria, Iraq, Libya, Oman, and Saudi Arabia; Labor exporters are Egypt, Jordan, Lebanon, and Yemen. Majority local production countries are Morocco, Sudan, Syrian Arab Republic, and Tunisia.

For wheat specifically, the liberalization of prices for suppliers was beneficial for production and self-sufficiency increased from 21 percent in 1986 to 47 percent in 1996, as the area planted doubled and yields increased (Kherallah et al. 2000). Surveys of wheat farmers showed their support for the reform process overall but also provided evidence of the high costs of self-sufficiency in wheat. Because only a minority of Egyptian wheat farmers are net sellers, the supply elasticity of wheat is low (that is, there is little production increase as prices rise) and further gains in self-sufficiency were unlikely to be achieved through government policy (Kherallah et al. 2000).

Figure 24.2 shows that the increase in wheat yields achieved across the region in the late 1980s and 1990s, especially in labor-exporting countries, leveled off around 2000. Correspondingly, as seen in Figure 24.1, import dependence declined slightly in the 1980s and early 1990s but far from enough to reverse the previous trend. Liberalization was also tempered by security concerns, as most countries maintained some direct price controls to ensure low prices for consumers (Minot et al. 2010; Nin-Pratt 2018).

In contrast to liberalization in pricing and input markets, the state continued to interfere heavily in water allocation, providing high subsidies for water used in agriculture and addressing water management as a problem of increasing supply rather than improving allocation efficiency (Hazell et al. 1995). Economists have repeatedly pointed to theory and modeling results showing the potential gains from more efficient water allocation. For example, modeling the impacts of water price reform in Morocco showed significant potential productivity gains from more efficient sectoral allocation of water resources (Löfgren et al. 1997). However, such recommendations have consistently run into political economy challenges. In contrast, central governments find it easier to invest in large water infrastructure, which provides a visible and politically valuable achievement (de Waal 2023).

2010s–2020s: Slow growth, vulnerability, and the Arab Spring

The limitations of the liberalization policies of the 1990s and early 2000s became clear in the 2011 Arab Spring uprisings. While long-term economic stagnation was a major driver of the uprisings, high food prices were also among the specific grievances voiced (Malik 2013). Economic growth rates had been low in the 2000s, and while there were gains in agricultural productivity, there was little accompanying reduction in poverty (Breisinger et al. 2012). In highly import-dependent, non-oil-exporting countries such as Yemen, poverty had increased sharply prior to the uprisings as a result of the 2007–2008 world food price crisis, while satisfaction with living standards had decreased in Egypt and Libya (Breisinger et al. 2011). The failures of agricultural policy were also directly implicated in the unrest in Syria, where a major drought from 2006 to 2010 followed a period of policies that prioritized groundwater extraction for cotton cultivation, shifting the country from being a net exporter of wheat to a net importer and displacing thousands of farmers (Souffiantini 2020).

In 2020, the Russia–Ukraine war and disruptions in exports from the Black Sea region caused another sharp increase in global wheat, fertilizer, and oil prices, with particularly severe impacts in MENA countries reliant on imports from the region. Like the previous shock, the 2020 crisis was most severe for import-dependent, non-oil exporters, in particular Egypt, Sudan, and Yemen, since oil and food prices increased together. While immediate policy responses focused on how to cope with high consumer prices, MENA countries have also been discussing renewed attempts to increase domestic agricultural production (Abay et al. 2022).

In sum, while agricultural liberalization in the MENA region resulted in improvements in input efficiency and modest gains in self-sufficiency, its overall impact was constrained by persistent institutional challenges and resource constraints, particularly water scarcity.

Food subsidies, social protection, and nutrition

1970s–2000s: High spending on food subsidies

To ensure access to staple foods in the MENA region, safety net systems have historically centered around food subsidies (Sdralevich et al. 2013). Although the subsidies are largely untargeted and inefficient, the system protects the poorest households from global price fluctuations and guarantees stable access to subsidized staple foods that make up a meaningful share of their consumption (Pinstrup-Anderson and Alderman 1988).

Egypt's ration card system is the largest food subsidy program in the region, covering more than three-quarters of the Egyptian population. Costs have been high, peaking at 10 percent of the state budget in the 1970s, and for decades, the system has been criticized for poor targeting even as the political economy incentives for subsidies have been recognized (Ahmed and Bouis 2002; Alderman 1987). Although attempts at direct reform in this period were met with strong political opposition, several rounds of cuts in the ration card rolls and reductions in the quantities of subsidized food did lower program costs (Abdallah and Al-Shawarby 2017). Regionally, in spite of limited reforms pushed during the liberalization period, subsidy spending was above average levels in the 2000s, ranging from 8 to 14 percent of gross domestic product (GDP) in MENA countries (Breisinger et al. 2012).

Beyond the fiscal burdens of national subsidy systems (exemplified by Egypt's ration card system), these subsidies have been criticized for promoting overconsumption of staple foods. Due to cultural norms, dietary patterns, and access to subsidized foods, MENA accounts for most of the countries worldwide with the highest shares of female overweight and obesity. An IFPRI study showed that Egypt's food subsidies made it easier for recipients to afford a low-diversity diet, thus contributing to malnutrition among children and mothers in beneficiary households (Ecker et al. 2016). Likewise in Egypt and Jordan, government spending on subsidies was correlated with larger increases in bodyweight (Abay et al. 2020).

2010s–2020s: Emergence of cash transfer programs

Since the Arab Spring, there has been a surge of new cash-based social protection systems in MENA, following the example of Latin America's conditional cash transfer programs. These cash-based programs, which provide poor households with regular direct payments that they are free to use as they prefer, have been shown to be cost effective and provide significant benefits for household consumption, in particular allowing for increased consumption of more nutritious food items (Breisinger et al. 2018; Kurdi 2021; Salti et al. 2022; Schwab 2020)(see Chapter 11). Some prominent cash-based programs are Egypt's Takaful, Jordan's program of the same name Takaful, Morocco's Tayssir, and Tunisia's National Program for Needy Families (Auktor and Loewe 2021; Rodriquez and Wai-Poi 2024; Thyen and Karadag 2021). This programming was further expanded as part of the COVID-19 response, in which MENA countries spent an average of 1.1 percent of GDP on social protection (Almenfi et al. 2020). Improvements in data systems and the introduction of proxy means tests have also allowed for more accurate targeting of poor households compared to the older subsidy systems (Breisinger et al. 2024).

Even in fragile and conflict-affected contexts in the MENA region, such as Sudan, Yemen, and refugee communities in Lebanon, a shift from in-kind food distribution to cash has been promoted as an approach that is more nutrition sensitive, less distortionary, and less costly to implement (see Chapters 11 and 13). Just prior to the ongoing civil war, Sudan launched a cash-based program called the Thamarat Family Support program (Breisinger et al. 2023), and in Yemen, the Cash for Nutrition program has been rolled out nationally and continues to function. Cash-based programming provided by humanitarian actors is also expanding in the region to meet needs in fragile settings, and these efforts are developing linkages with national social protection systems (Smith 2020).

At the same time, the political economy of food subsidies continues to be a major factor shaping policies. In the context of high inflation, households continue to prefer in-kind transfers over equivalently valued cash (Abay et al. 2023). In Egypt, for example, after discussing a plan to completely replace subsidies with cash in 2024, the government has so far refrained from making any policy changes.

Emerging research areas and looking forward

Looking toward the future, rigorous research to inform policy is particularly important in the following areas in the MENA region:

Food security in fragile settings. Ongoing conflicts and protracted fragility in Gaza, Iraq, Lebanon, Libya, Sudan, Syria, and Yemen have created significant hotspots for global food insecurity and hunger. Better evidence is needed on the impact of both agricultural and social protection interventions that can feasibly support food security in fragile and conflict-affected settings to inform international responses at the humanitarian-peace-development nexus (see Chapter 13) (Al Daccache et al. 2024; Hirvonen and Leight 2024; Lind 2022).

Climate-resilient agricultural systems. Projected increases in temperature and climatic variability in the MENA region will require adaptation in management practices. Anticipatory action is an emerging paradigm for climate crisis preparedness built around early warning and triggered funding mechanisms for mitigation, but research is needed to guide both public and private insurance investments in this area (de Brauw and Bloom 2024) (see Chapter 10). Farmers can also take direct actions to adopt new climate-smart practices. Research focused on understanding farmers' digital access, knowledge gaps, risk preferences, and other factors that determine behavior can facilitate interventions to promote broader adoption of climate adaptation measures (see Chapters 4, 5, 9, and 17) (Tabe-Ojong et al. 2024).

Resource governance. The water-energy-food-environment nexus emphasizes the need for agricultural innovations that perform positively across all four dimensions of sustainability (Bhaduri et al. 2015; Elsayed 2023). For example, solar-powered irrigation is linked to improved agricultural efficiency and productivity gains (Closas and Rap 2017; Ziadat and Sultan 2011). However, recent rapid expansion also poses big risks, particularly the accelerated depletion of nonrenewable groundwater resources (Balasubramanya et al. 2024). Remote-sensing and machine-learning tools can help to map the scope of solar irrigation adoption, while research on cooperative behavior, water pricing models that incorporate political economy constraints, and functioning of existing resource governance institutions can inform institutional development aimed at sustainable water governance (see Chapter 5).

Promotion of healthy diets. Given the increasing burden of obesity and noncommunicable diseases, understanding how to motivate healthier food choices is also a key area of emerging research. Across the region, initiatives such as education, labeling, and price interventions have been explored, but rigorous evidence on effectiveness of these efforts in shaping consumer dietary choices is limited (Al-Jawaldeh et al 2022). Nutrition-sensitive school feeding, which is less developed in the MENA region than other places but has strong community support, could also contribute to this agenda (Bliznashka et al 2024) (see Chapter 12).

Social protection for economic inclusion. While the expansion of cash transfer programs in MENA is an efficient way to protect consumption of the poorest, sustained poverty reduction is a greater challenge. Policy guidance and evaluations of social protection approaches that emphasize economic inclusion are needed (see Chapter 11). In the MENA region, the persistently low labor force participation rate for women is a particular concern for poverty reduction, so economic inclusion programming must also take into account social and legal barriers that exist to women's participation in formal sector work (see Chapter 14) (El-Enbaby et al. 2024; Assaad et al. 2020; Ridao-Cano 2023).

References

Abay, K., H. Ibrahim, C. Breisinger, and N. Bayasgalanbat. 2020. "Food Policies and Their Implications on Overweight and Obesity Trends in Selected Countries in the Near East and North Africa Region." MENA Regional Program Working Paper 30. IFPRI, Cairo.

Abay, K., L. Abdelfattah, M. Elkaramany, D. Elsabbagh, and S. Kurdi. 2023. "Nutrition-Sensitive Food Distribution amidst Inflationary Shock: Evidence from a Randomized Intervention in Egypt." IFPRI Discussion Paper 2218. IFPRI, Washington, DC.

Abay, K.A., C. Breisinger, J.W. Glauber, S. Kurdi, D. Laborde Debucquet, and K. Siddig. 2022. "The Russia–Ukraine Crisis: Implications for Global and Regional Food Security and Potential Policy Responses." MENA Working Paper 39. IFPRI, Cairo.

Al-Jawaldeh, A., M. Taktouk, S. Naalbandian, et al. 2022. "Sugar Reduction Initiatives in the Eastern Mediterranean Region: A Systematic Review." *Nutrients* 15 (1): 55.

Adams, R. 1993. "Bureaucrats and Peasants: The Dynamics of Local-Level Agricultural Bureaucracy in Rural Egypt." In *The Political Economy of Food and Nutrition Policies, ed.* P. Pinstrup-Andersen, 101–115. Baltimore: Johns Hopkins University Press for IFPRI.

Abdallah, M., and S. Al-Shawarby. 2017. "The Tamween Subsidy System in Egypt: Evolution and Recent Implementation Reforms." In *The 1.5 Billion People Question: Food, Vouchers, or Cash Transfers?*, eds. H. Alderman, U. Gentilini, and R. Yemtsov, 107–150. Washington, DC: World Bank Group.

Ahmed, A.U., and H.E. Bouis. 2002. "Weighing What's Practical: Proxy Means Tests for Targeting Food Subsidies in Egypt." *Food Policy* 27 (5–6): 519–540.

Al Daccache, M., B. Abi Zeid, L. Hojeij, G. Baliki, T. Brück, and H. Ghattas. 2024. "Systematic Review on the Impacts of Agricultural Interventions on Food Security and Nutrition in Complex Humanitarian Emergency Settings." *BMC Nutrition* 10 (1): 60.

Alderman, H. 1987. "Food Subsidies and State Policies in Egypt." In *Food, States, and Peasants: Analyses of the Agrarian Question in the Middle East*, ed. A. Richards, 183–200. New York and Abingdon, UK: Routledge.

Almenfi, M., U. Gentilini, I. Orton, and P. Dale. 2020. *Social Protection and Jobs Responses to COVID-19: A Real-Time Review of Country Measures*. Washington, DC: World Bank.

Assaad, R., R. Hendy, M. Lassassi, and S. Yassin. 2020. "Explaining the MENA Paradox: Rising Educational Attainment, Yet Stagnant Female Labor Force Participation." *Demographic Research* 43 (28): 817–850.

Auktor, G.V., and M. Loewe. 2021. "Subsidy Reforms in the Middle East and North Africa: Strategic Options and their Consequences for the Social Contract." Discussion Paper No. 12/2021. German Development Institute, Bonn.

Balasubramanya, S., D. Garrick, N. Brozović, et al. 2024. "Risks from Solar-Powered Groundwater Irrigation." *Science* 383 (6680): 256–258.

Bhaduri, A., C. Ringler, I. Dombrowski, R. Mohtar, and W. Scheumann. 2015. "Sustainability in the Water–Energy–Food Nexus." *Water International* 40 (5-6): 723–732.

Bliznashka, L., D. Elsabbagh, S. Kurdi, O. Ecker, and A. Gelli. 2024. "School Feeding for Improving Child Nutrition in Conflict-Affected Settings: Feasibility and Cost Efficiency of Alternative Models in Yemen." *Annals of the New York Academy of Sciences* 1540 (1): 251–264.

Breisinger, C., O. Ecker, P. Al-Riffai, and B. Yu. 2012. *Beyond the Arab Awakening: Policies and Investments for Poverty Reduction and Food Security*. Food Policy Report. Washington, DC: IFPRI.

Breisinger, C., X. Diao, M.H. Collion, and P. Rondot. 2011. "Impacts of the Triple Global Crisis on Growth and Poverty: The Case of Yemen." *Development Policy Review* 29 (2):155–184.

Breisinger, C., D. Gilligan, N. Karachiwalla, S. Kurdi, H. El-Enbaby, A. Jilani, and G. Thai. 2018. "Impact Evaluation Study for Egypt's Takaful and Karama Cash Transfer Program: Part 1: Quantitative Report." MENA Regional Program Working Paper 14. IFPRI, Washington, DC, and Cairo.

Breisinger, C., Y. Kassim, S. Kurdi, J. Randriamamonjy, and J. Thurlow. 2024. "From Food Subsidies to Cash Transfers: Assessing Economy-Wide Benefits and Trade-Offs in Egypt." *Journal of African Economies* 33 (2): 109–129.

Cassing, J., S. Nassar, G. Siam, and H. Mousa. 2009. "Arab Republic of Egypt." In *Distortions to Agricultural Incentives in Asia*, eds. K. Anderson and W.A. Masters, 71–98. Washington, DC: World Bank.

Closas, A., and E. Rap. 2017. "Solar-Based Groundwater Pumping for Irrigation: Sustainability, Policies, and Limitations." *Energy Policy* 104: 33–37.

de Brauw, A. and J. Bloem. 2024. *Anticipatory versus Post-Shock Transfers: Conceptual Analysis.* Fragility, Conflict, and Migration Initiative Technical Report. Montpellier, France: CGIAR.

de Waal, D., S. Khemani, A. Barone, and E. Borgomeo. 2023. *The Economics of Water Scarcity in the Middle East and North Africa: Institutional Solutions.* Washington, DC: World Bank.

Ecker, O., P. Al-Riffai, C. Breisinger, and R. El-Batrawy. 2016. *Nutrition and Economic Development: Exploring Egypt's Exceptionalism and the Role of Food Subsidies.* Washington, DC: IFPRI.

Elsayed, H. 2023. "Managing Water, Energy, and Food Nexus in Egypt." *Alternative Policy Solutions Commentaries*, January 17.

Ender, G., and J.S. Holtzman, eds. 2003. *Does Agricultural Policy Reform Work? The Impact on Egypt's Agriculture 1996–2002.* Rockville, MD and Washington, DC: Abt Associates Inc. and USAID.

El-Enbaby, H., D. Gilligan, N. Karachiwalla, Y. Kassim, and S. Kurdi. 2024. "Cash Transfers, Gender Norms, and Women's Control over Decision-Making in Egypt." *Economic Development and Cultural Change*, forthcoming.

FAO (Food and Agriculture Organization of the United Nations). 2025. FAOSTAT. www.fao.org/faostat/

Hazell, P.B., I. Soliman, G. Siam, and N.D. Perez. 1995. "Impact of the Structural Adjustment Program on Agricultural Production and Resource Use in Egypt." EPTD Discussion Paper 10. IFPRI, Washington, DC.

Hirvonen, K., and J. Leight. 2024. "The Effects of Cash and Cash Plus Interventions on Livelihoods Outcomes: A Systematic Review and Meta-Analysis." Powerpoint presentation given at IFPRI, Washington, DC, May 21.

Khalid, N. 1984. *Evolving Food Gaps in the Middle East/North Africa: Prospects and Policy Implications*. Research Report 47. Washington, DC: IFPRI.

Kherallah, M., H. Löfgren, P. Gruhn, and M.M. Reeder. 2000. *Wheat Policy Reform in Egypt: Adjustment of Local Markets and Options for Future Reforms*. Research Report 115. Washington, DC: IFPRI.

Kurdi, S. 2021. "The Nutritional Benefits of Cash Transfers in Humanitarian Crises: Evidence from Yemen." *World Development* 148: 105664.

Lind, J., R. Sabates-Wheeler, and C. Szyp. 2022. "Cash and Livelihoods in Contexts of Conflict and Fragility: Implications for Social Assistance Programming." BASIC Research Working Paper 8. Institute of Development Studies, Brighton, UK.

Löfgren, H., R. Doukkali, H. Serghini, and S. Robinson. 1997. "Rural Development in Morocco: Alternative Scenarios to the Year 2000." TMD Discussion Paper 17. IFPRI, Washington, DC.

Malik, A., and B. Awadallah. 2013. "The Economics of the Arab Spring." *World Development* 45: 296–313.

Minot, N., M.A. Chemingui, M. Thomas, R. Dewina, and D. Orden. 2010. *Trade Liberalization and Poverty in the Middle East and North Africa*. Washington, DC: IFPRI.

Nin-Pratt, A., H. El-Enbaby, J.L. Figueroa, H. Eldidi, and C. Breisinger. 2018. *Agriculture and Economic Transformation in the Middle East and North Africa*. Food Policy Report. Washington, DC: IFPRI.

Pinstrup-Anderson, P., and H. Alderman. 1988. "The Effectiveness of Consumer-Oriented Food Subsidies in Reaching Rationing and Income Transfer Goals." In *Food Subsidies in Developing Countries: Costs, Benefits, and Policy Options*, ed. P. Pinstrup-Andersen, 21–35. Baltimore: Johns Hopkins University Press for IFPRI.

Ridao-Cano, C., D. Moosa, M. Pallares-Miralles, and J. Pinxten. 2023. *Built to Include: Reimagining Social Protection in MENA*. Washington, DC: World Bank.

Rodriguez, L., and M. Wai-Poi. 2024. "Fiscal Policy, Poverty and Inequality in Jordan." Policy Research Working Paper 10732. World Bank, Washington, DC.

Salti, N., J. Chaaban, W. Moussa, A. Irani, R. Al Mokdad, Z. Jamaluddine, and H. Ghattas. 2022. "The Impact of Cash Transfers on Syrian Refugees in Lebanon: Evidence from a Multidimensional Regression Discontinuity Design." *Journal of Development Economics* 155: 102803.

Schwab, B. 2020. "In the Form of Bread? A Randomized Comparison of Cash and Food Transfers in Yemen." *American Journal of Agricultural Economics* 102 (1): 91–113.

Sdralevich, C., R. Sab, Y. Zouhar, and G. Albertin. 2013. *Subsidy Report in the Middle East and North Africa: Recent Progress and Challenges Ahead*. Washington, DC: International Monetary Fund.

Soffiantini, G. 2020. "Food Insecurity and Political Instability during the Arab Spring." *Global Food Security* 26: 100400.

Smith, G. 2020. *Linking CVA and Social Protection in the MENA Region: Cribsheet*. Oxfordshire, UK: The CALP Network.

Tabe-Ojong Jr., M.P., Y. Salama, K.A. Abay, et al. 2024. "Harnessing Digital Innovations for Climate Action and Market Access: Opportunities and Constraints in the CWANA Region." *Global Food Security* 41: 100763.

Thyen, K., and R. Karadag. 2021. "Between Affordable Welfare and Affordable Food: Internationalized Food Subsidy Reforms in Egypt and Tunisia." *Social Policy Administration* 55 (6): 1050–1065.

von Braun, J., and H. de Haen. 1983. *The Effects of Food Price and Subsidy Policies on Egyptian Agriculture*. Research Report 42. Washington, DC: IFPRI.

Ziadat, F.M., and K.A. Sultan. 2011. "Combining Current Land Use and Farmers' Knowledge to Design Land-Use Requirements and Improve Land Suitability Evaluation." *Renewable Agriculture and Food Systems* 26 (4): 287–296.

Made in United States
Orlando, FL
04 November 2026

90138309R00347